A HANDBOOK *of* PLANT BIOSYNTHETIC PATHWAYS

A HANDBOOK
of
PLANT BIOSYNTHETIC PATHWAYS

Dr Debabrata Chaudhuri MSc, PhD
Dr Dipak Kumar Kar MSc, PhD
Dr Soma Halder MSc, MTech, PhD

New Central Book Agency (P) Ltd
LONDON
HYDERABAD ERNAKULAM BHUBANESWAR
DELHI KOLKATA PUNE GUWAHATI

NCBA

REGD OFFICE
8/1 Chintamoni Das Lane, Kolkata 700 009, India
email: ncbapvtltd@eth.net

OVERSEAS
NCBA (UK) Ltd, 149 Park Avenue North, Northampton, NN3 2HY, UK
email: ncbauk@yahoo.com

EXPORT
NCBA Exports Pvt. Ltd, 212 Shahpur Jat, New Delhi 110 049
email: ncbaexp@ncbapvtltd.com

BRANCHES

208B Peacock Lane, Shahpur Jat
New Delhi 110 049
email: ncbadel@ncbapvtltd.com

House No. 3-1-315, 1st Floor
Nimboliadda, Kachiguda, **Hyderabad** 500 027
email: ncbahydb@ncbapvtltd.com

Shop Nos. 3 & 4, Vinayak Towers
681/B Budhwar Peth, Appa Balwant Chowk
Pune 411 002
email: ncbapune@ncbapvtltd.com

GSS Shopping Complex, 1st Floor
Opposite College Post Office, Convent Road
Ernakulam 682 035
email: ncbaernk@ncbapvtltd.com

Shop No. 15, Bharati Towers, Ground Floor
Forest Park, **Bhubaneswar** 751 001
email: ncbabub@ncbapvtltd.com

Radhamayee Complex (Second Floor)
Opp. S.B.I., Panbazar, Jaswanta Road
Guwahati 781 001

A HANDBOOK OF PLANT BIOSYNTHETIC PATHWAYS • D Chaudhuri, D K Kar, S Halder

First Published: 2008
Reprinted: January 2013

PUBLISHER AND TYPESETTER
New Central Book Agency (P) Ltd
8/1 Chintamoni Das Lane, Kolkata 700 009

PRINTER
New Central Book Agency (P) Ltd
Web-Offset Division, Dhulagarh, Sankrail, Howrah

COVER DESIGNER
Suman Dutta

TECHNICAL EDITOR
Dipan Roy

PROJECT TEAM
Suman Dutta, Prabhat Jas, Prasenjit Halder

ISBN: 978-81-7381-587-4

Price: ₹ 385.00

CONTENTS

PREFACE

The book covers biochemistry in general, plant biochemistry in particular, an area in the centre of plant science today. Most of the biosynthetic pathways of different plant metabolism are being covered. The book is centred on plant biosynthetic pathways highlighting the reactants, products, enzymes involved and their regulations. Chemistry of different biomolecules has been placed as an introductory chapter. Both anabolic (photosynthesis) and catabolic (respiration) pathways of carbohydrate metabolism, a primary metabolism of the living world, has been depicted first. Pathways of fatty acids and lipid metabolism follow next. Information pathways, the centre of attraction to biochemists, include nucleic acid metabolism followed by metabolism of amino acids and protein. Plant synthesises a vast array of secondary metabolites, useful for human existence, is also under the perview of this book. Biosynthesis of phytohormones, which regulate the growth and development of plant, has been placed at the end. A useful list of standard biochemical abbreviations has been added at last.

We express our sincere thanks to Mr Amitabha Sen and Dr (Mrs) Mita Sen of New Central Book Agency for their spontaneous acceptance and ardent enthusiasm in publishing this book. It has been a pleasure for us to work with Mr Soumen Paul, Mr Siladitya Sinha Roy and other members of the publishing house, specially Mr Suman Dutta, Mr Pradip Chowdhury, and Mr Prabhat Jas, all of whom took great care to produce the book.

Suggestions from readers will be highly appreciated.

Rathajatra
4 July 2008
Asutosh College
Kolkata

DEBABRATA CHAUDHURI
DIPAK KUMAR KAR
SOMA HALDER

CHAPTER

1

BIOMOLECULES

Biomolecules are organic compounds of large molecular weight (macromolecules) present as an essential component of living organisms. Biomolecules contain covalently bonded carbon backbones to which different kinds of **functional groups** (–COOH, $–NH_2$, –OH, etc.) are attached and determine the chemical properties of the molecules. Important biomolecules are **Nucleic acids, Proteins, Polysaccharides, Lipids**, etc., which are composed of small monomeric subunits.

Figures: 1.1 and 1.2

CARBOHYDRATES

Carbohydrates occur in nature as monosaccharides, oligosaccharides and polysaccharides. **Monosaccharides** are polyhydroxy aldose (glucose) or ketose (fructose) having empirical formula $(CH_2O)_n$. They may be triose (Glyceraldehyde, Dihydroxyacetone), tetrose (Erythrose and Erythrulose), pentose (Ribose and Ribulose), hexose (Glucose, Mannose, Galactose and Fructose), heptose (Sedoheptulose). Monosaccharides have at least one asymmetric carbon atom and exist in **Stereoisomeric forms**—D and L forms. Sugars with 4 or more carbon atoms may cyclised into closed ring 5-membered **Furanose** (Fructose) or 6-membered **Pyranose** (Glucose). Both the rings occur in two **Anomeric forms** (α and β) and are interconverted due to mutarotation. Many derivatives of the simple sugars (amino-sugars and their acetylated derivatives, etc.) are found in living cells.

Oligosaccharides are composed of a few monosaccharide units. Disaccharides are composed of two monosaccharides joined by glycosidic bond (Maltose, Lactose, Sucrose, Trehalose). Many oligosaccharides exist as glycoconjugates in which carbohydrate moiety may be covalently bound to a protein or lipid (Glycoprotein, Lipopolysaccharides).

Polysaccharides are homopolymers (Starch) or heteropolymers (Peptidoglycan) of monosaccharide residues in glycosidic linkage arranged in linear (Cellulose) or branched (Amylopectin) chains of varying length.

Figures: 1.3–1.16

FATTY ACIDS AND LIPIDS

Lipids are chemically diverse group of compounds. The simplest lipids (**Fat** and **Oil**) are triacylglycerides containing three fatty acid molecules (same or different types) esterified to the three hydroxyl groups of glycerol. Fatty acids are hydrocarbon derivatives and may be saturated or unsaturated. Important membrane lipids are **Phospholipids, Glycolipids.** In **Glycerophospholipids** the two hydroxyl groups (–OH) of glycerol are esterified by two fatty acid molecules and the third one by an alcohol through phosphodiester bond. **Sphingolipids** contain sphingosine, a long chain aliphatic amino-alcohol, but no glycerol. **Sterols** (Cholesterol) are lipids having steroid nucleus with alkyl side chains. Ubiquinine and plastoquinone are **isoprenoid derivatives** function as electron carrier.

Figures: 1.17–1.22

AMINO ACIDS AND PROTEINS

Twenty different **amino acids** are found as residues in proteins and contain an α-carboxyl group, an α-amino group and a distinctive R-group substituted on the α-carbon atom. On the basis of polarity and charge of R-group, amino acids are classified into nonpolar aliphatic, nonpolar aromatic, polar uncharged, polar negatively charged and polar positively charged.

Amino acids are covalently joined through peptide bonds to form **peptides** and **proteins** of long polypeptide chains. In contrast to simple proteins made up of only amino acids, conjugated proteins in addition contain metal ion (activator) or organic cofactor (coenzyme/prosthetic group). Proteins have four levels of structures. **Primary structure** refers to amino acid sequence, location of disulphide bonds and the entire peptide group with a rigid planar configuration. **Secondary structure** is the spatial relationship of adjacent amino acids in localised stretches. Two major types of secondary structures were predicted—**α-helix** and **β-conformation**. In α-helix the backbone of the polypeptide chain is extending into a helical structure with rigid peptide bonds while in β-conformation it extends into a zigzag structure which may be arranged side by side to form **β-pleated sheet. Tertiary structure** is the three dimentional conformation of the entire polypeptide chain can be analysed by studying the motifs and domains. **Quaternary** structure involves the spatial relationship of multiple associated polypeptide chains, protomers.

Figures: 1.23–1.31

NUCLEOTIDES AND NUCLEIC ACIDS

Nucleic acids are polymers of **nucleotides**, joined together by phosphodiester linkages. A nucleotide consists of a nitrogenous base (**Purines**—Adenine, Guanine ; or **Pyrimidines**—Thymine, Uracil, Cytosine), a pentose sugar and Phosphate group. Nucleic acids are of two types—**DNA** and **RNA**. Nucleotides of DNA differ from RNA—(i) Uracil is absent in DNA and thymine is absent in RNA. (ii) 2′-deoxyribose is present in DNA instead of ribose as in RNA.

DNA is a double helical structure of two antiparallel polynucleotide chains. Complementary base pairs, A = T and G ≡ C, are formed by hydrogen bonding within the helix, and the hydrophilic sugar-phosphate backbones are located on the outside. The base pairs are stacked perpendicular to the long axis, 3·4 Å apart and with 10 base pairs in each complete turn of the double helix.

RNA is single stranded polynucleotide chain often folded into hairpins, double-stranded regions and complex loops. Different types of RNA are mRNA, tRNA, rRNA.

Nucleotides carry chemical energy in cells, e.g., **ATP** and large free-energy change is associated with ATP hydrolysis. Nucleotides are components of many enzyme **co-factors** (NAD, NADP, FAD, Co-A). Nucleotides are involved in enzyme modification reactions—phosphorylation, adenylation, uridylylation, ribosylation. Nucleotides are also regulatory molecules or second messengers (cAMP, cGMP, ppGPP).

Figures: 1.32–1.45

METABOLISM OF BIOMOLECULES

Three types of nonlinear metabolic pathways exists—(i) Converging catabolic, (ii) Diverging anabolic and (iii) Cyclic. Biomolecules through **Catabolic pathways** deliver chemical energy, i.e., ATP, NADH, $NADPH_2$, $FADH_2$ which are used in **Anabolic pathways** to convert small precursor molecules into biomolecules.

Figures: 1.46 and 1.47

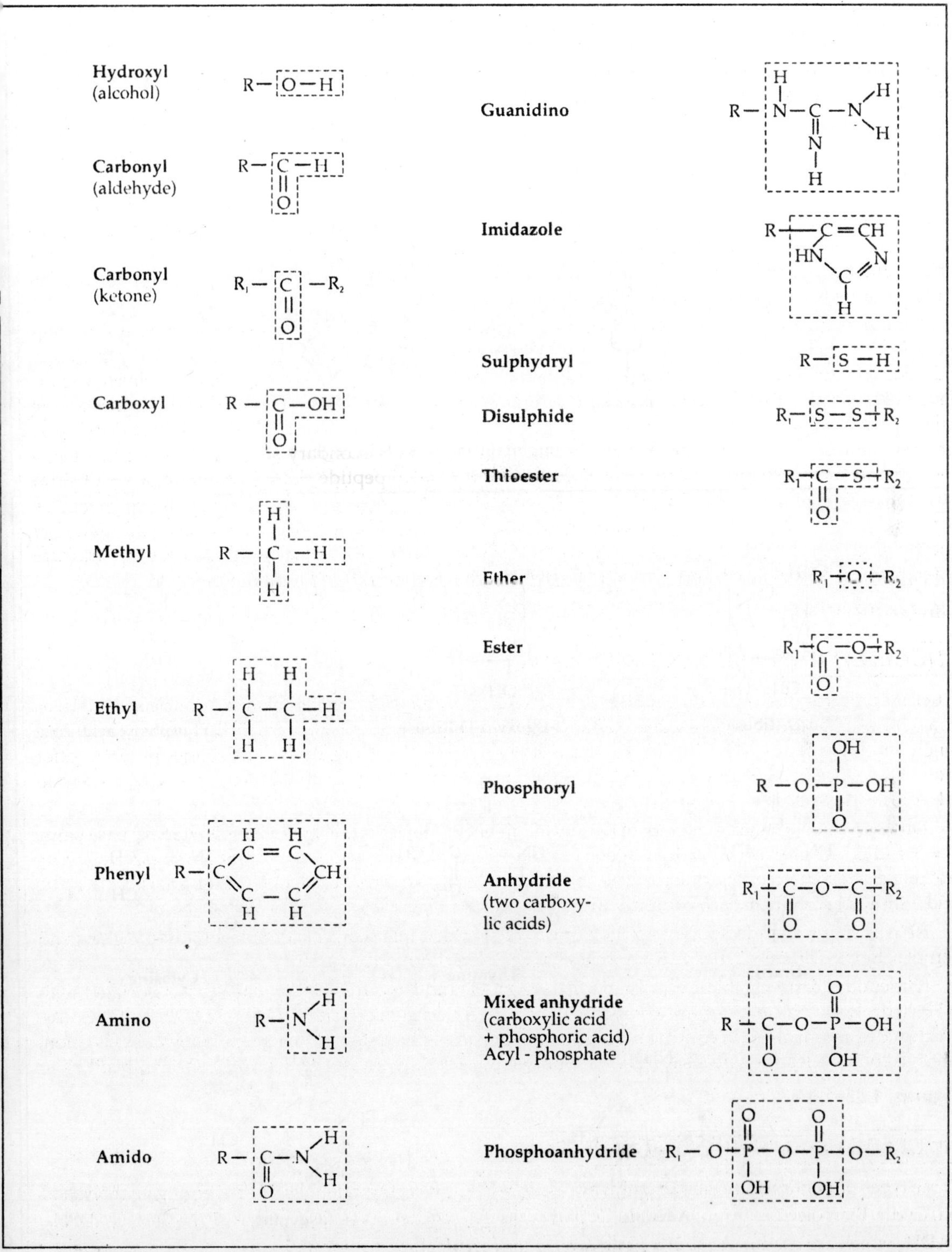

Fig. 1.1: Some common functional groups of biomolecules: R or R_1 and R_2 are substituents of each molecule which may be hydrogen or typical carbon-containing moiety.

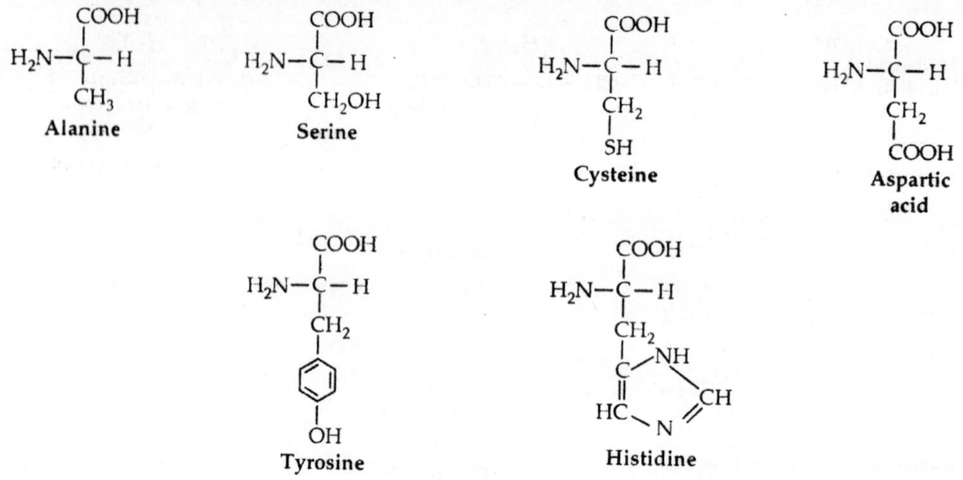

(a) Some of the important amino acids in proteins

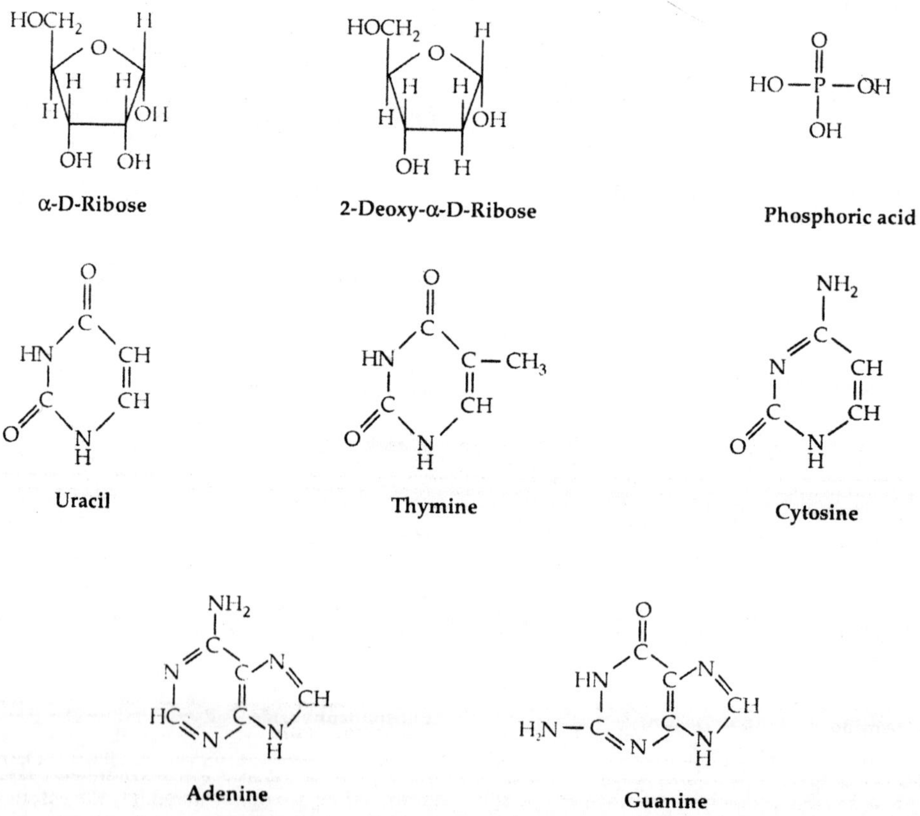

(b) The components of nucleic acids

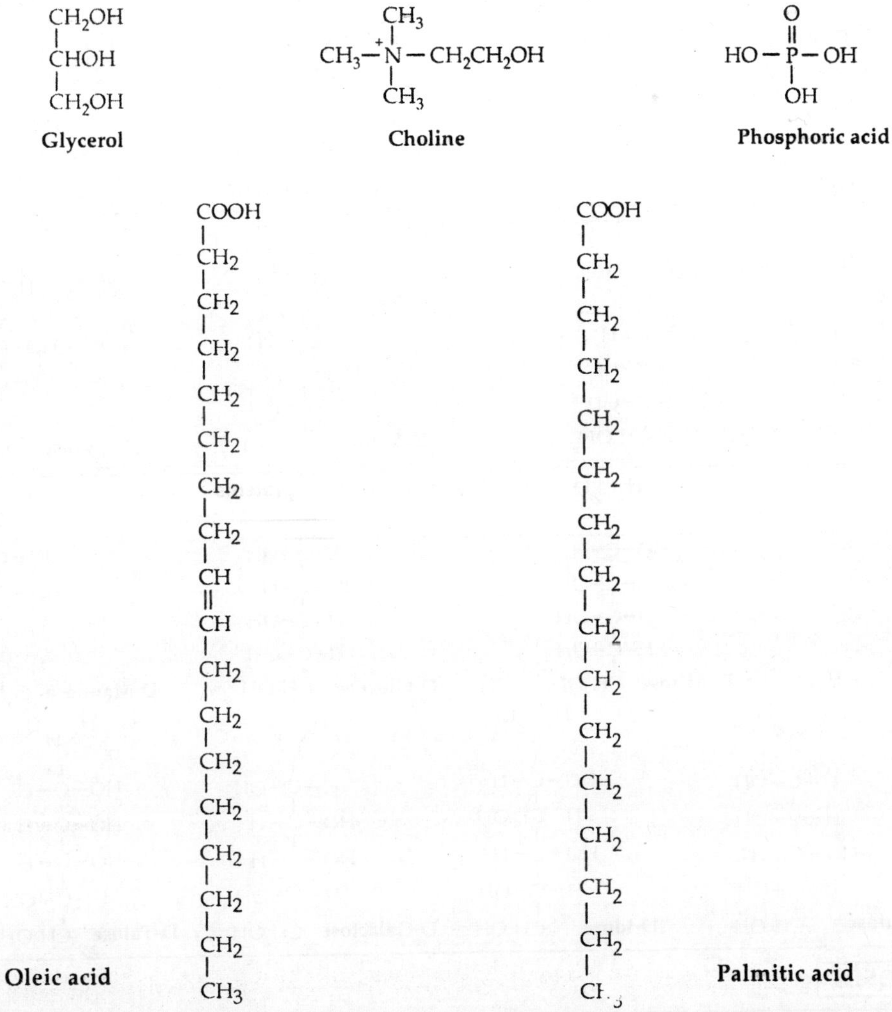

(c) Some components of lipid

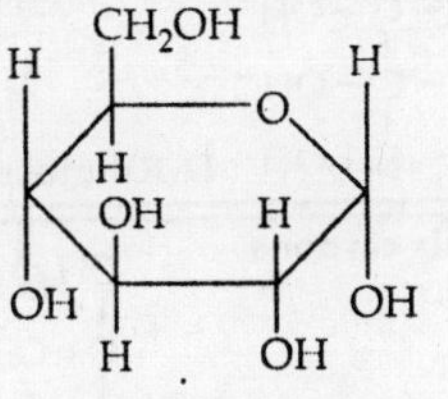

α-D-glucose

(d) The parent sugar molecule

Fig. 1.2: The organic molecules from which most cellular materials are constructed: (a) Six essential amino acids among 20 amino acids from which all proteins are built. **(b)** Two five-carbon sugars, phosphoric acid, and all the five nitrogenous bases from which all the nucleic acids are built. **(c)** Phosphoric acid, glycerol, choline, unsaturated fatty acid (oleic) or saturated fatty acid (palmitic)—these are five different components for many membrane lipids. **(d)** D-Glucose—is the parent sugar molecule from which most carbohydrates are derived.

(a) D-ALDOSES

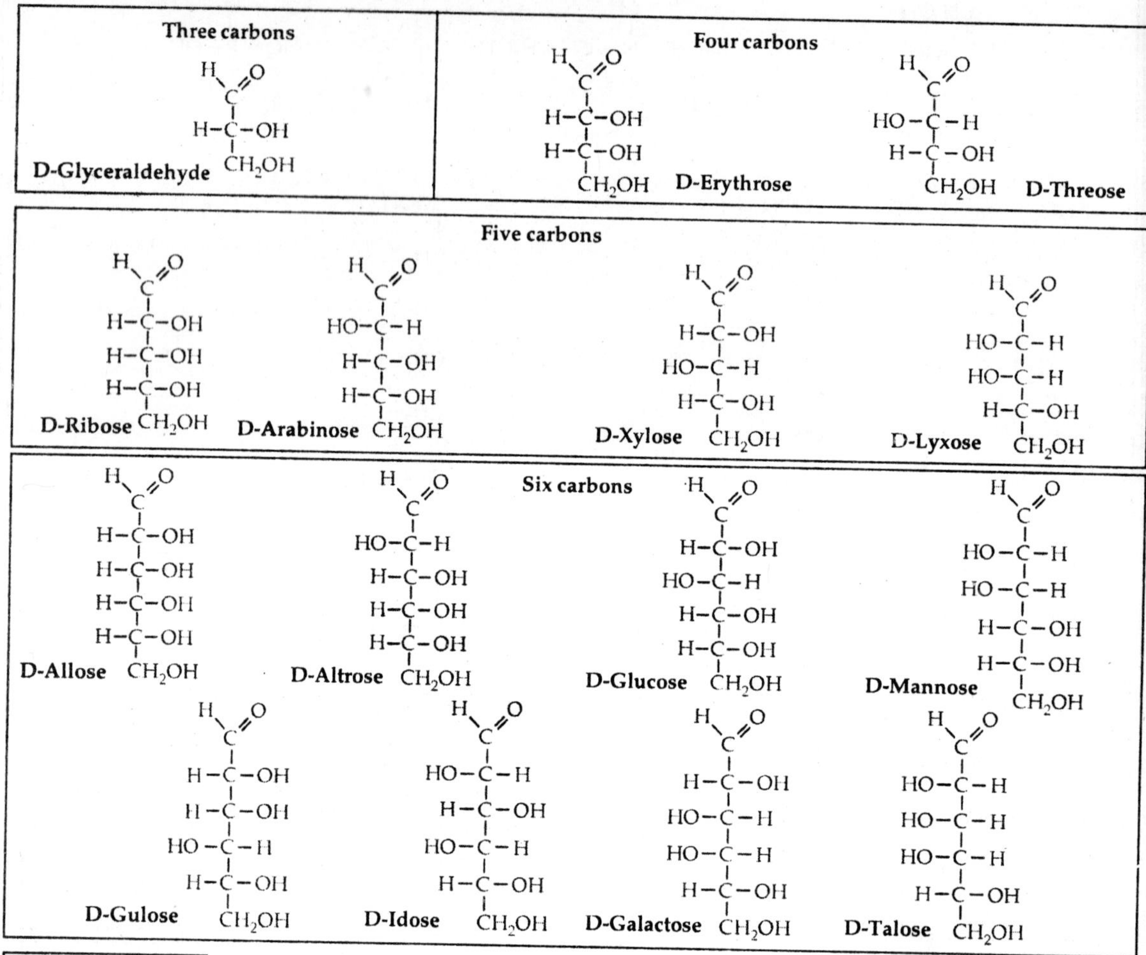

(b) D-KETOSES

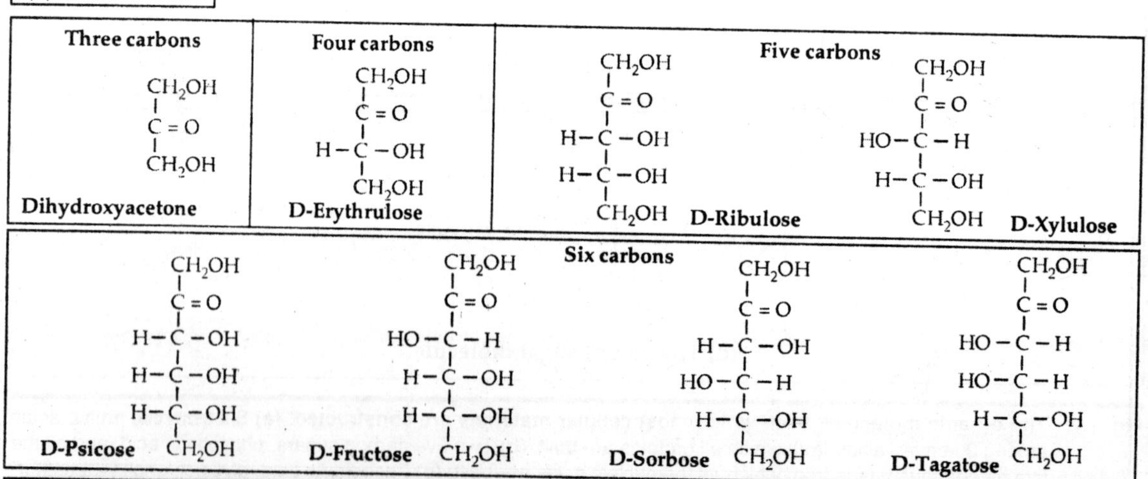

Fig. 1.3: The series of different Aldose and Ketose sugars : sugars having three to six carbon atoms showing the placement of hydroxyl and hydrogen atom.

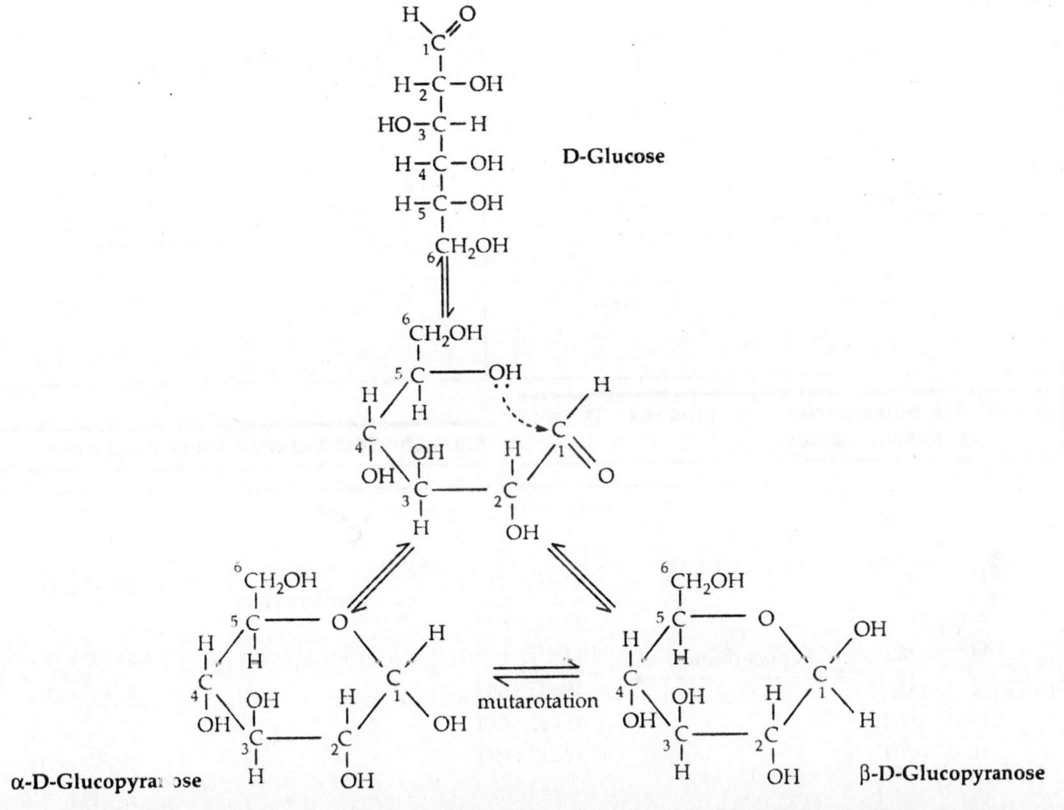

Fig. 1.4a: Mutarotation of two cyclic forms of D-Glucose : Reaction between the aldehyde group at C-1 and the hydroxyl group at C-5 forms a hemiacetal linkage, producing either of two stereoisomers, the α and β anomers, which differ only in the stereochemistry around the hemiacetal carbon. The interconversion of α and β anomers is called mutarotation.

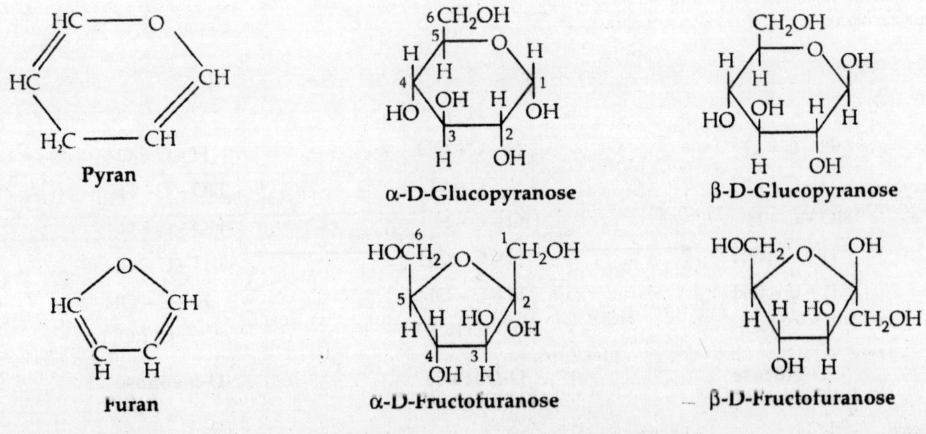

Fig. 1.4b: Pyranoses and Furanoses : The pyranose forms of D-glucose and the furanose forms of D-fructose are shown; structures may be compared with pyran and furan ring shown at left hand side.

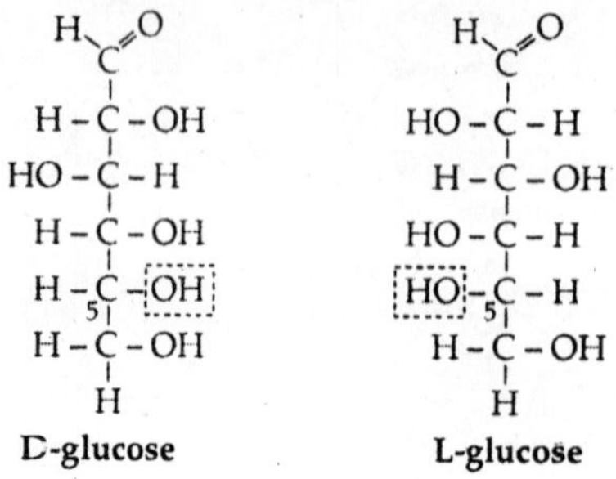

Fig. 1.5a: Stereoisomerism in glucose: D- and L-isomers of glucose.

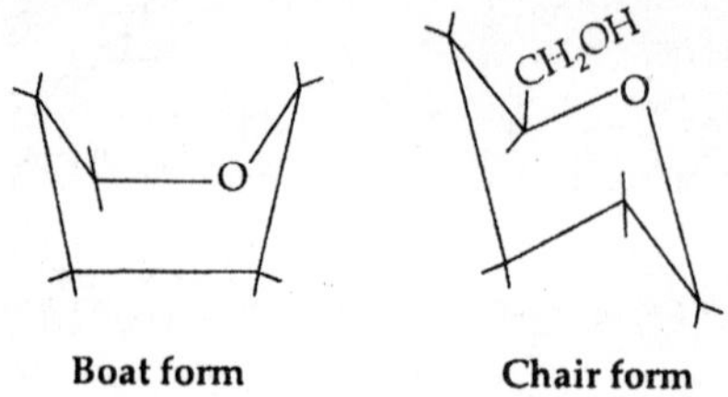

Fig. 1.5b: Boat and chair forms of pyranose.

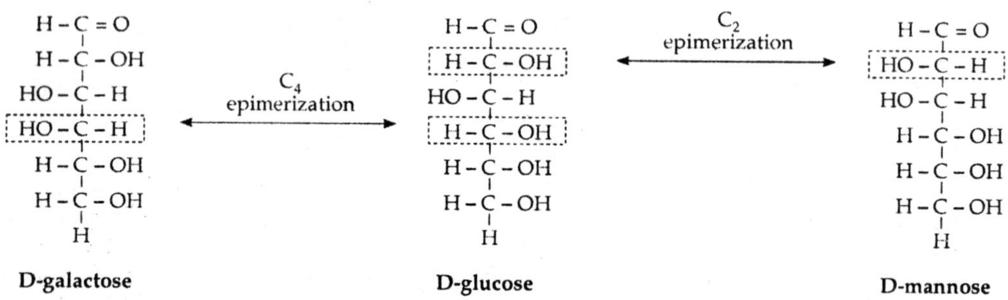

Fig. 1.5c: Epimerization of hexoses.

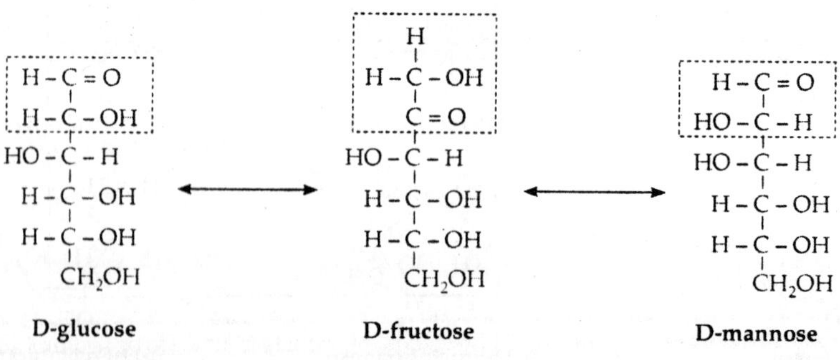

Fig. 1.5d: Tautomerization of hexoses.

(a) GLUCOSE FAMILY

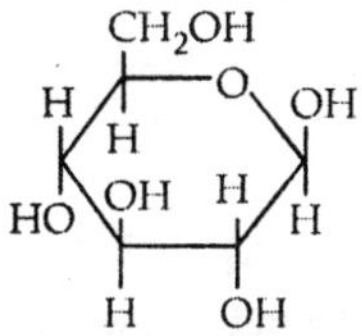
β-D-Glucose

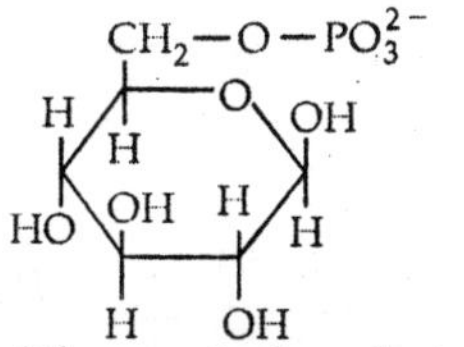
β-D-Glucose 6-phosphate

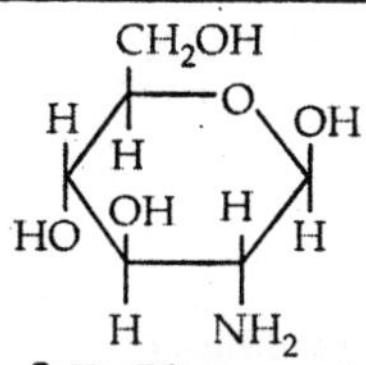
β-D-Glucosamine

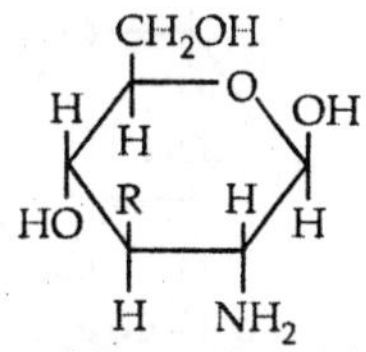
Muramic acid

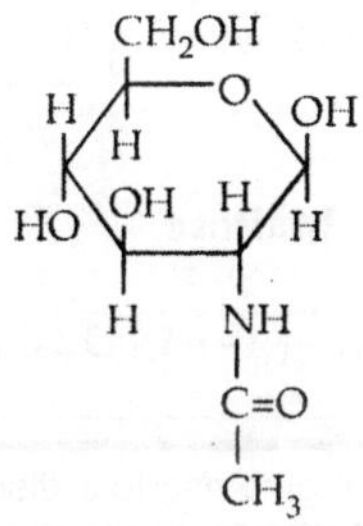
N-Acetyl-β-D-glucosamine

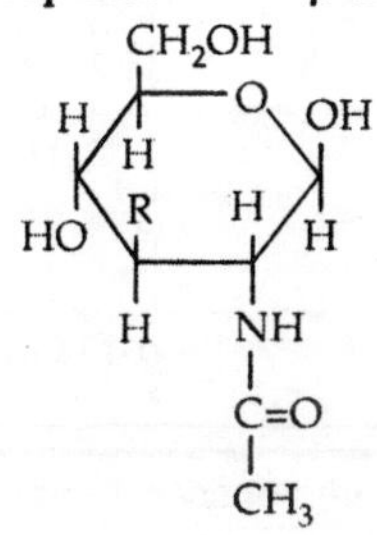
N-Acetylmuramic acid

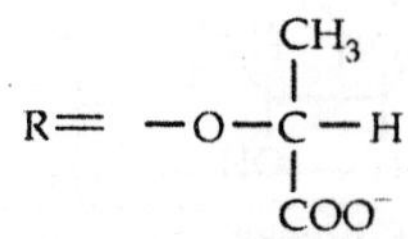

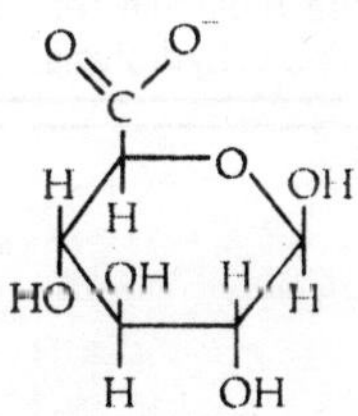
β-D-Glucuronate

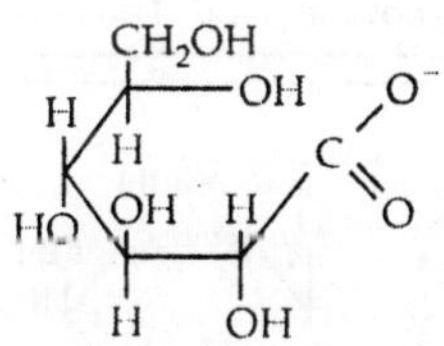
D-Gluconate

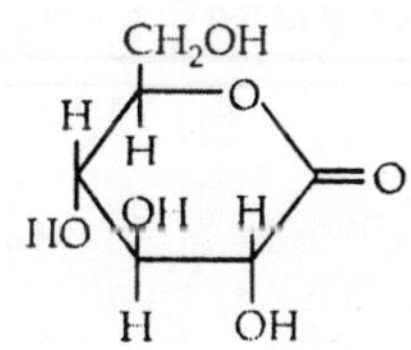
D-Glucono-δ-lactone

(b) AMINO SUGARS

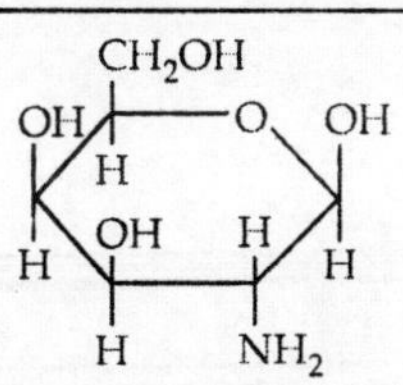
β-D-Galactosamine

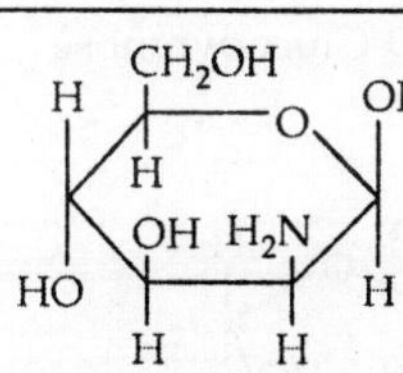
β-D-Mannosamine

(c) DEOXY SUGARS

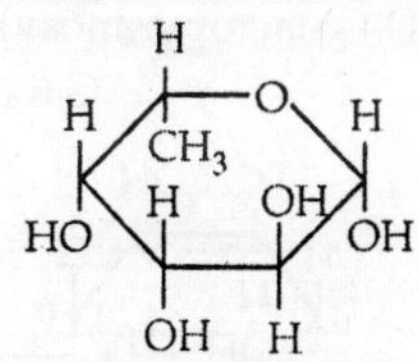
β-L-Fucose

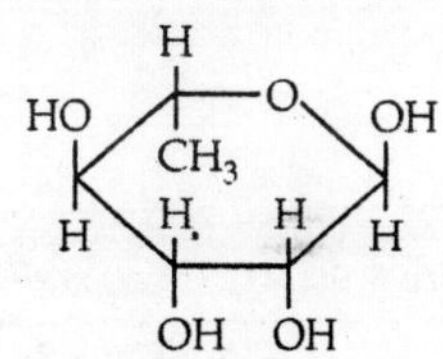
α-L-Rhamnose

(d) ACIDIC SUGARS

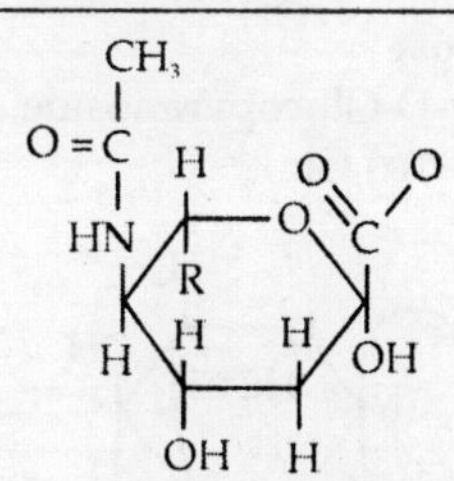
N-Acetylneuraminic acid
(Sialic acid)

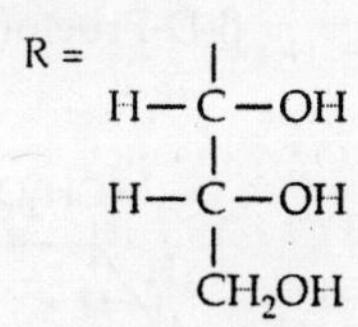

Fig. 1.6: Some hexose derivatives : In **Glucose-6-phosphate**, phosphate is attached by hydroxyl substitution ; Carboxylation of C_6 carbon results into **Glucuronate** formation ; **D-Glucono-δ-lactone** results from formation of an ester linkage between the C-1 carboxylate group and the C-5 hydroxyl group of D-Gluconate ; **Amino sugars** are formed by replacement of –OH group by $-NH_2$ group in parent hexose ; hydroxyl group substitution by –H results in **Deoxy sugar**, which are present in L-isomeric forms in nature. The **Acidic sugars** contain a carboxylate group which confers a negative charge at neutral pH.

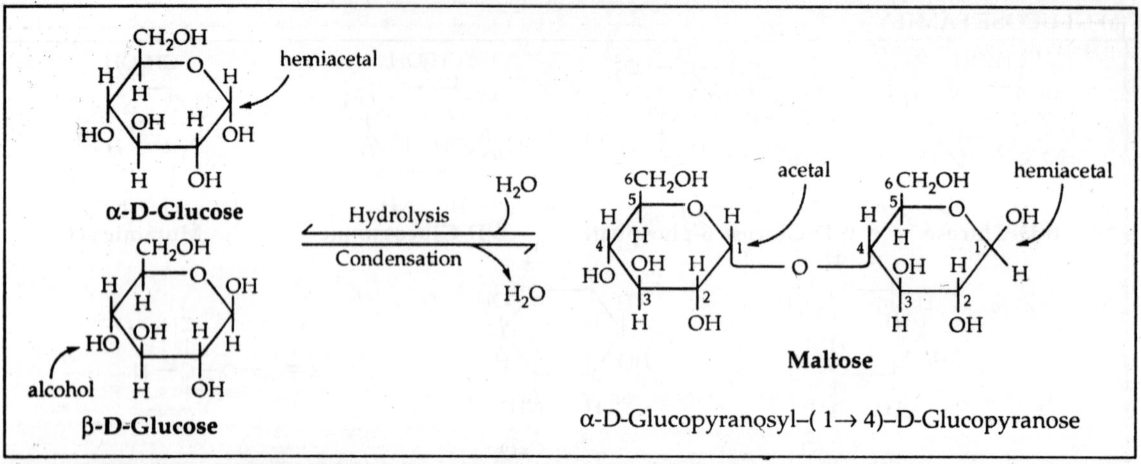

Fig. 1.7: Formation of Maltose : Two molecules of monosaccharides (D-Glucose) give rise to a disaccharide (Maltose) when an –OH (alcohol) group of one glucose molecule condenses with the hemiacetal of the other glucose molecule, with the elimination of H_2O and formation of O-glycosidic bond. The maltose molecule retains the hemiacetal at the C-1, which is not involved in glycosidic bond. The reversal of this reaction is hydrolysis.

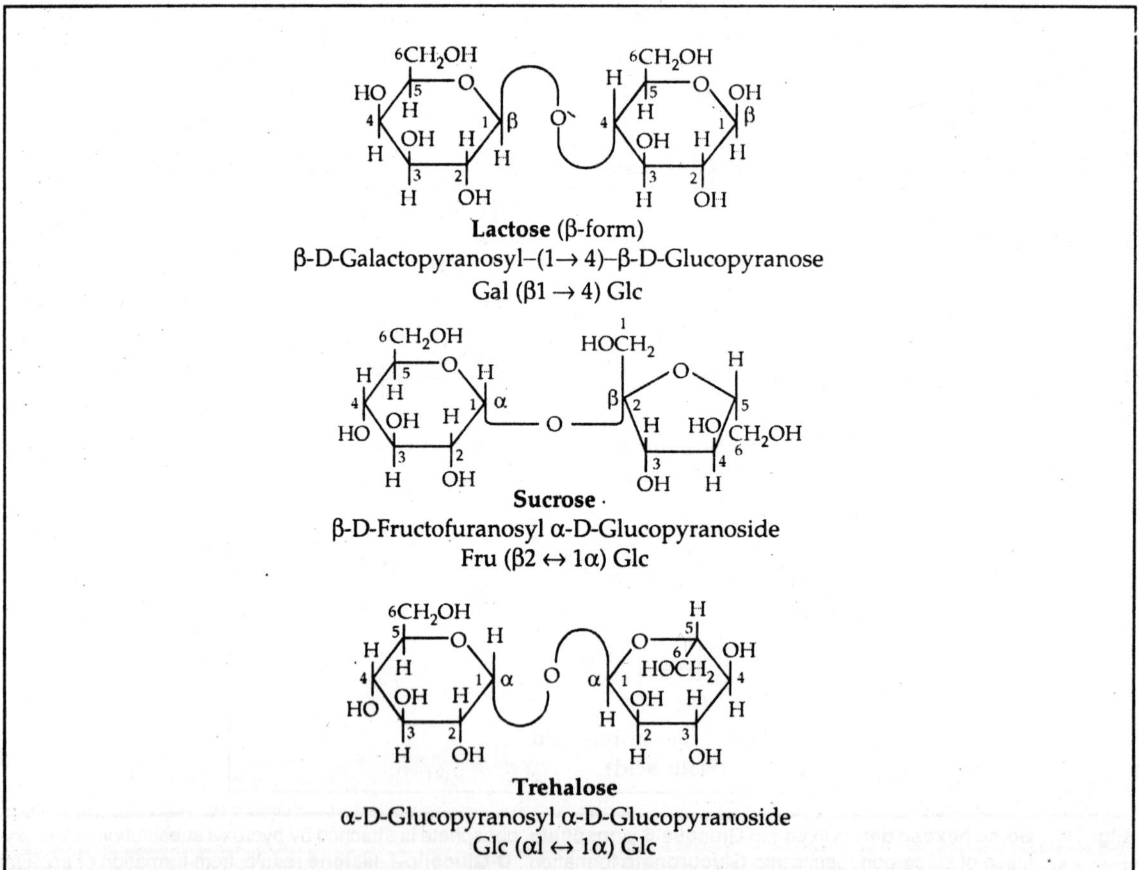

Fig. 1.8: Some common disaccharides : Lactose, Sucrose, Trehalose [Common name, systematic name and abbreviation are given for each].

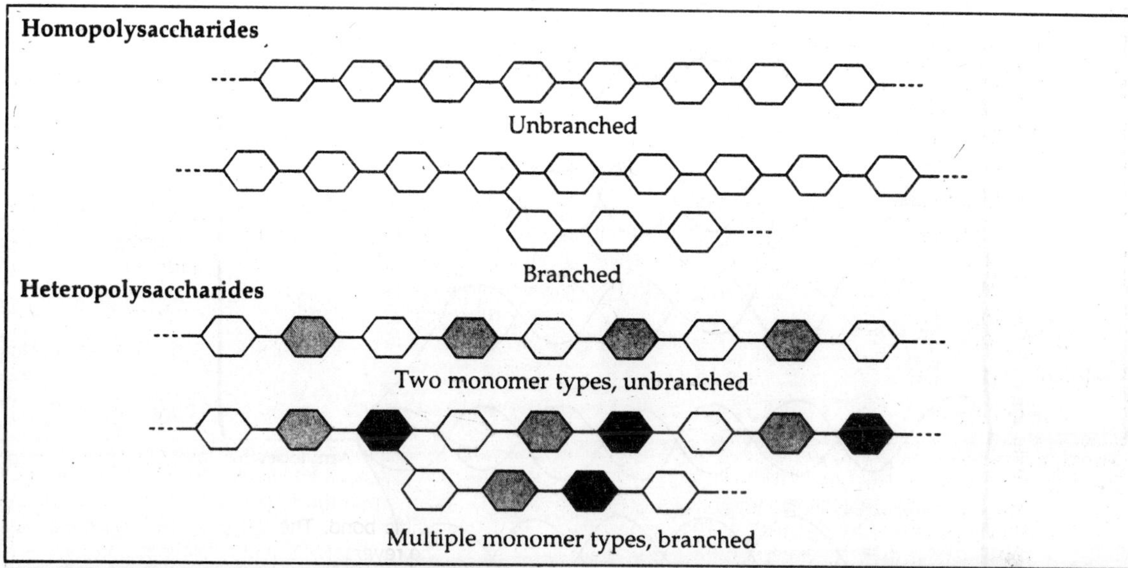

Fig. 1.9: Polysaccharides : These may be composed of one, two or several different monosaccharides, in straight chain or branched chains of varying length.

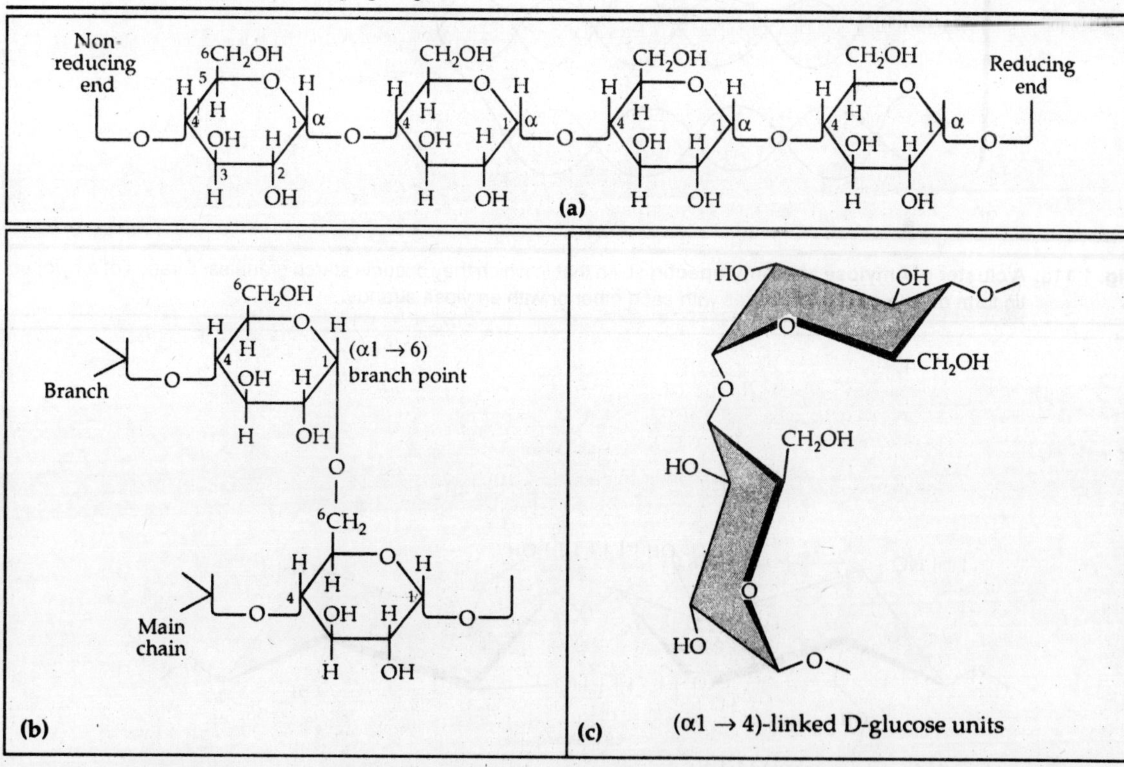

Fig. 1.10: Amylose and Amylopectin, the polysaccharides of starch : (a) A short segment of amylose, a linear polymer of D-glucose residues in $(\alpha 1 \rightarrow 4)$ linkage. A single chain can contain several thousand glucose residues. Amylopectin has stretches of similarly linked residues between branch points. **(b)** An $(\alpha 1 \rightarrow 6)$ branch point of amylopectin. **(c)** The structure of starch (amylose), in the most stable conformation of adjacent rigid chairs, the polysaccharide chain is curved, rather than linear as in cellulose.

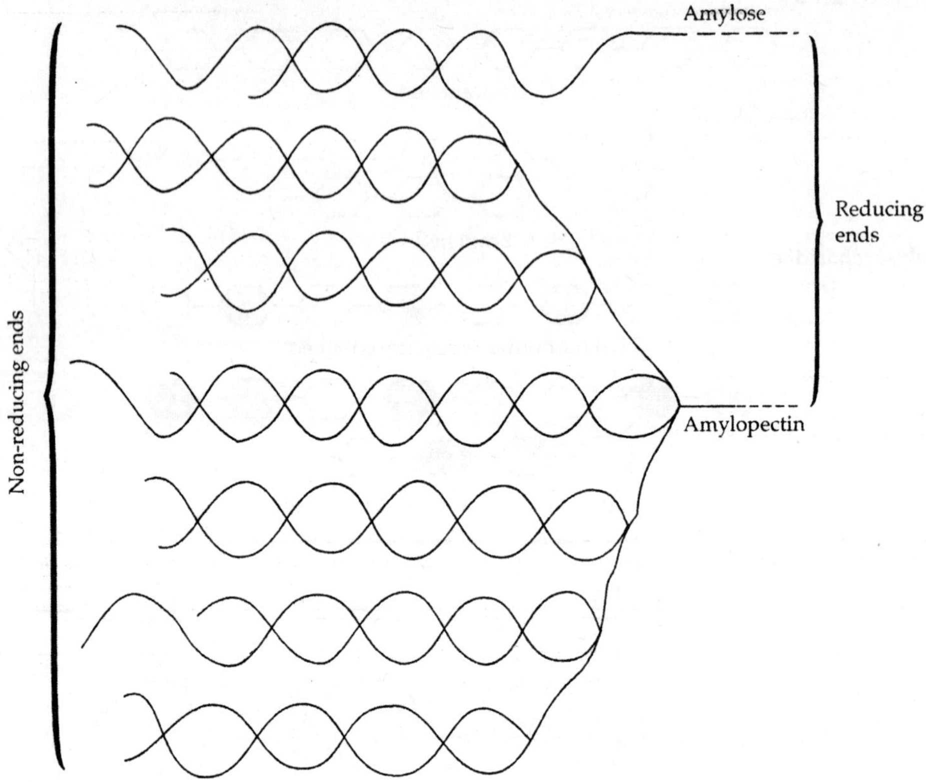

Fig. 1.11a: A cluster of Amylose and Amylopectin : Like that in which they occur in starch granules. Strands of amylopectin form double-helical structures with each other or with amylose strands.

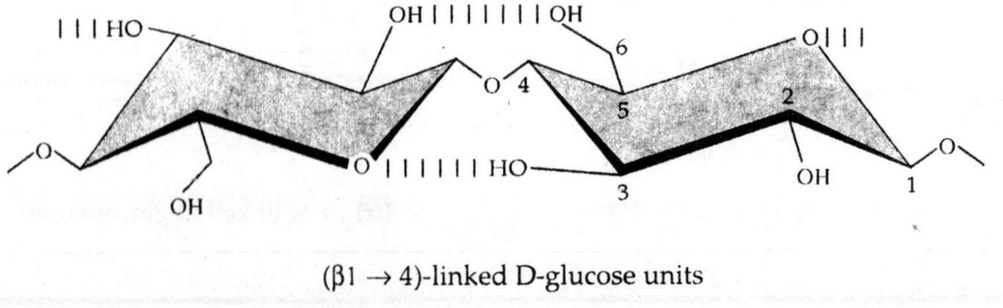

Fig. 1.11b: The structure of cellulose : Two units of a cellulose chain ; the D-glucose residues are in (β1 → 4) linkage. The rigid chair structures can rotate relative to one another.

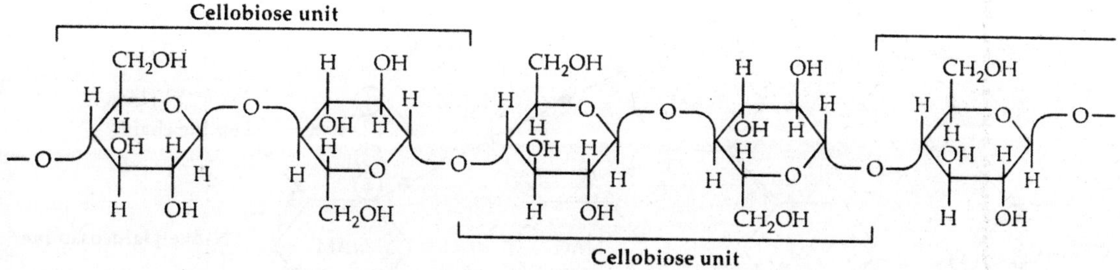

Fig. 1.11c: Structure of cellulose with repeating cellobiose unit having β-1 → 4 glycosidic bond between β-D-glucose units.

Polysaccharides	Linkage
(i) Glucans (Glucose)	
Starch	α-1 → 4
Cellulose	β-1 → 4
Chitin	β-1 → 4
Glycogen	α-1 → 4, α-1 → 6
(ii) Galactans (Galactose)	
Agar, Pectin, Galactan	—
(iii) Mannans (Mannose)	α-1 → 6
(iv) Xylan (Xylose)	β-1 → 4
(v) Fructan (fructose)	β-1 → 2

Fig. 1.12: Different linkage in polysaccharides.

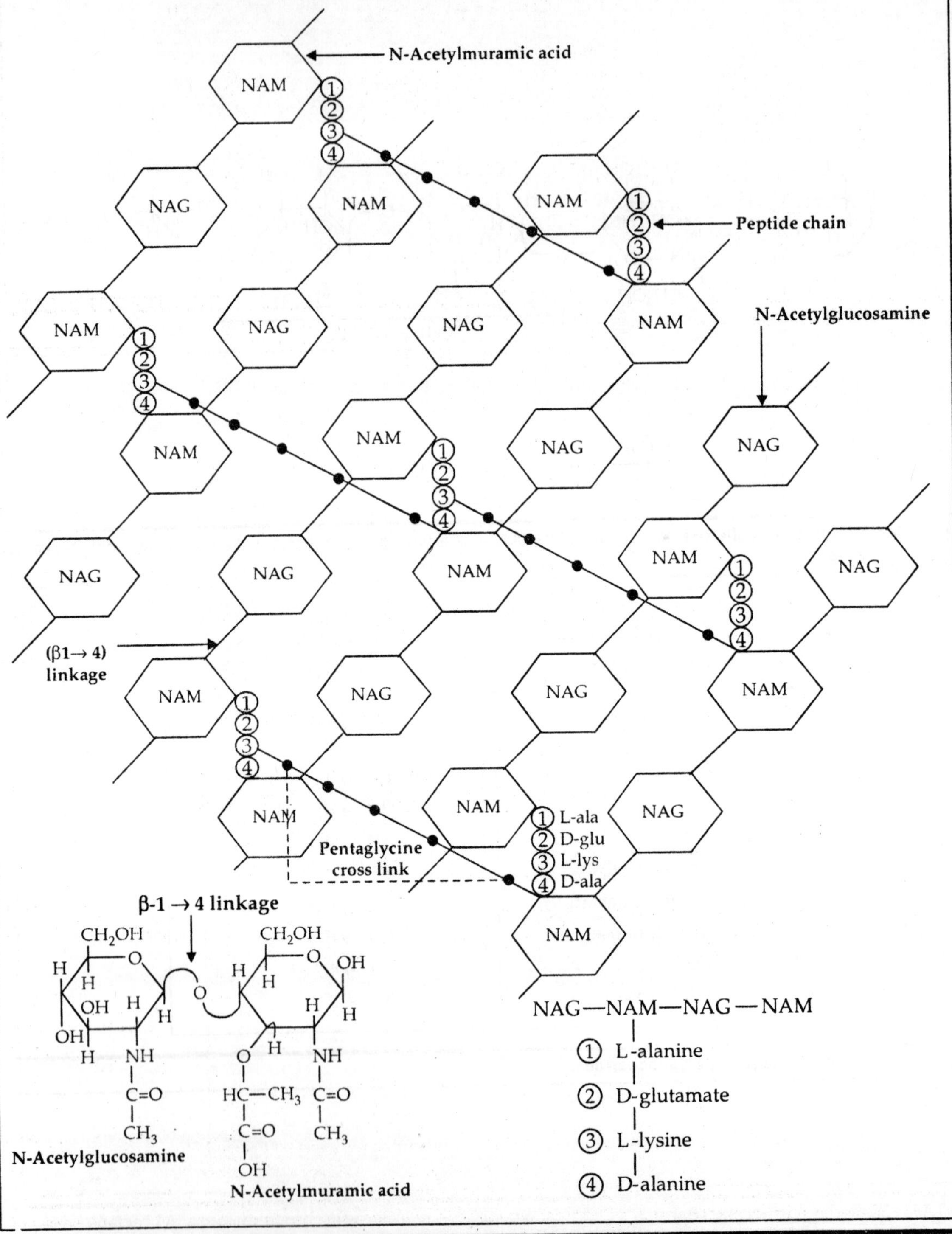

Fig. 1.13: Peptidoglycan: This is the molecular structure of peptidoglycan layer of cell wall of Bacteria. Peptides co-valently link N-Acetylmuramic acid residues in neighbouring polysaccharide chains. Gram +ve bacteria have pentaglycine chain in cross-link, Gram –ve bacteria lack the pentaglycine cross-link.

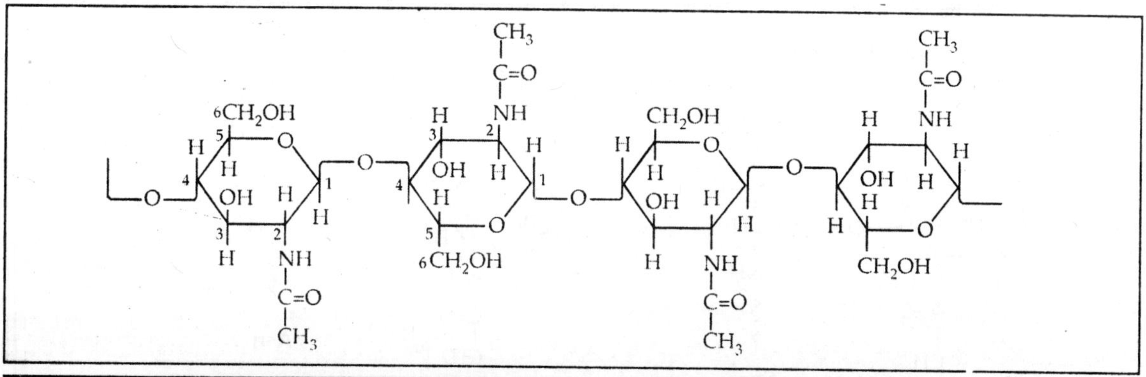

Fig. 1.14: A short segment of chitin : A homopolymer of N-acetyl-D-glucosamine units in (β1 → 4) linkage.

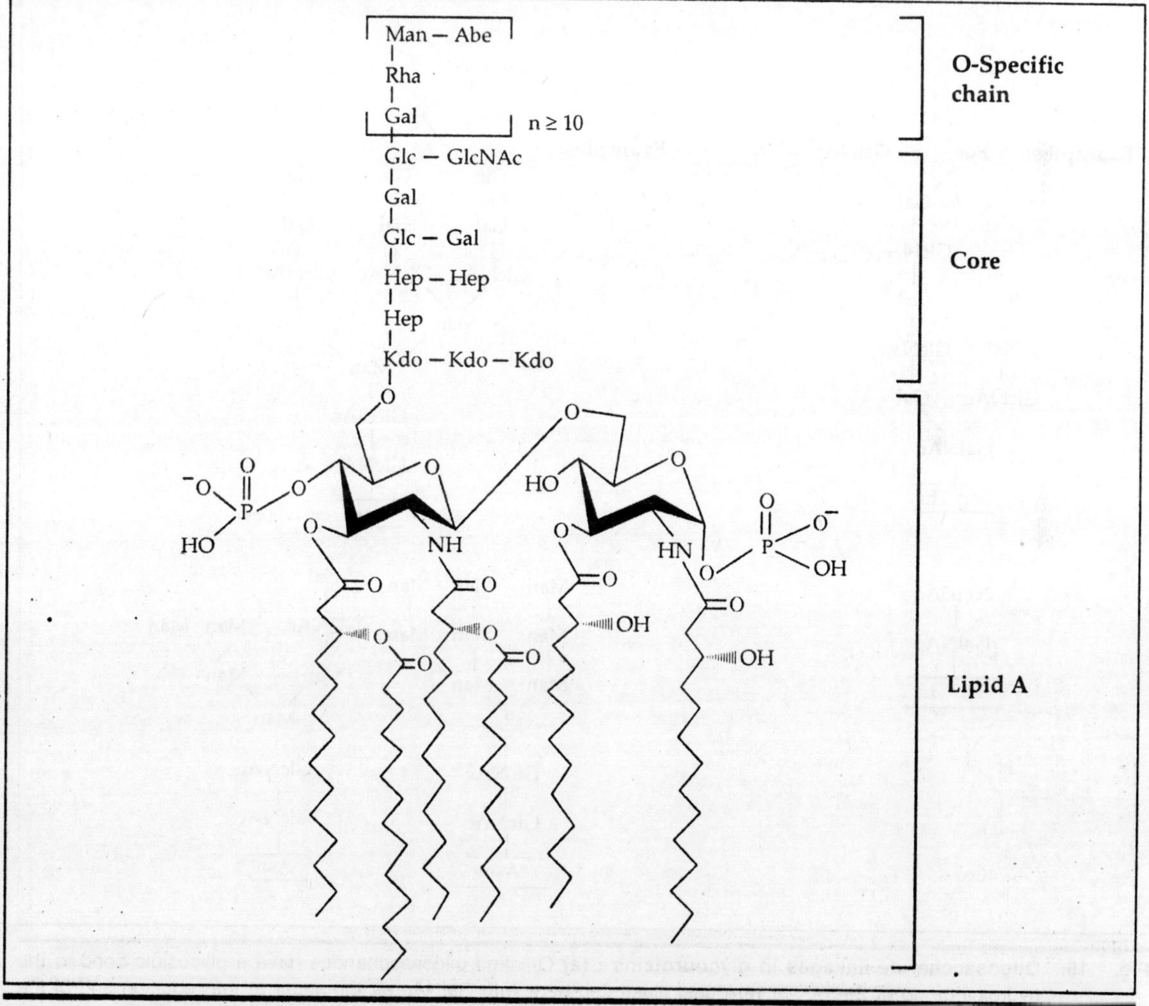

Fig. 1.15: Bacterial lipopolysaccharides : Chemical organisation of lipopolysaccharide molecule. All bacteria have in common a **lipid A** portion of six fatty acids, a **core** oligosaccharide and an **'O-specific' chain**, which is the principal determinant of serotype. [Kdo : ketodeoxyoctulosonic acid (2-keto-3-deoxy-D-manno-octonic acid, Hep : L-glycero-D-mannoheptose, Glc : Glucose, Gal : Galactose, GlcNAc : N-Acetylglucosamine, Rha : Rhamnose, Man : Mannose, Abe : Abequose (3,6 dideoxyhexose).

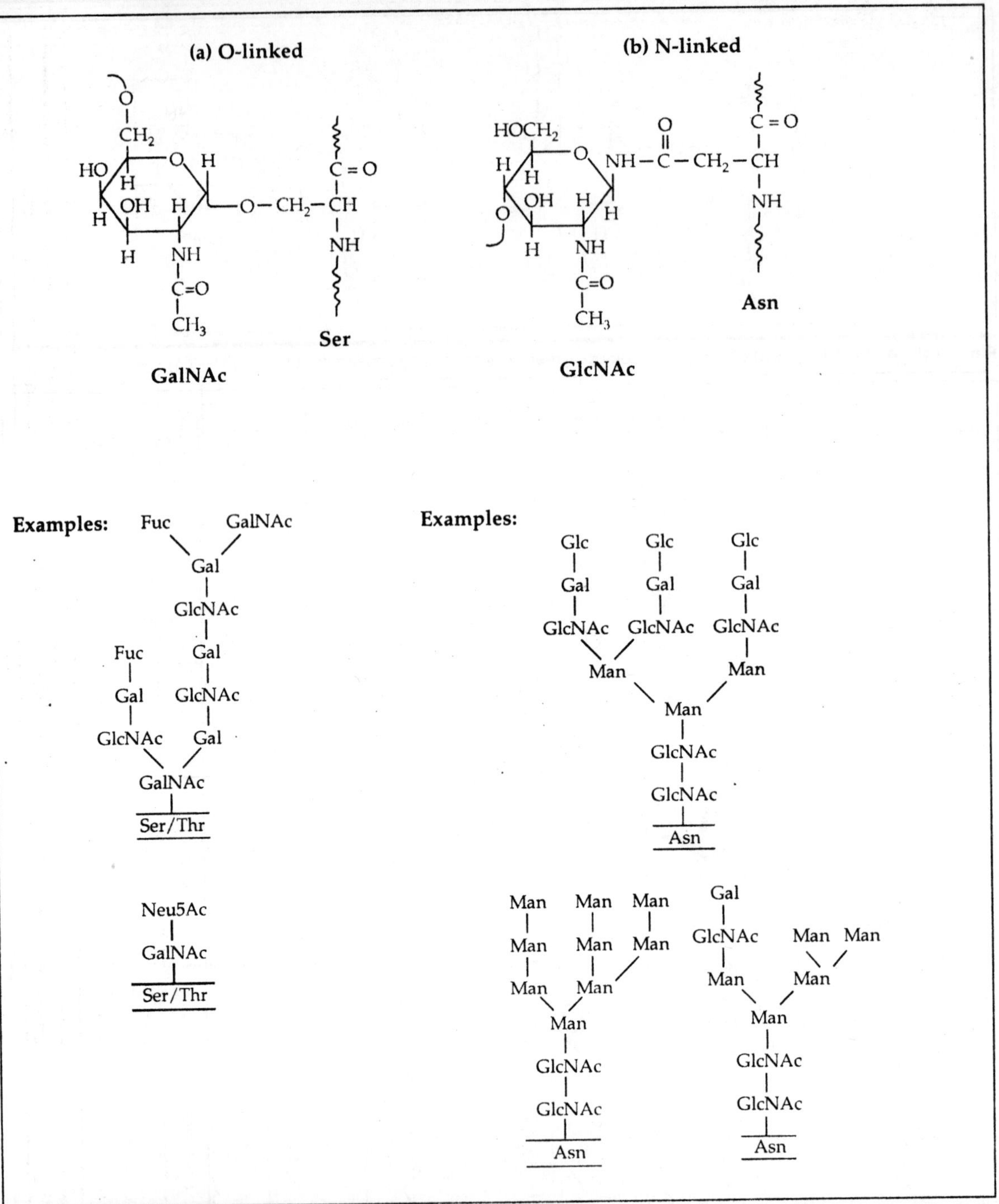

Fig. 1.16: Oligosaccharide linkages in glycoproteins : (a) O-linked oligosaccharides have a glycosidic bond to the hydroxyl group of Ser or Thr residues, illustrated here with GalNAc as the sugar at the reducing end of the oligosaccharide. One simple chain and another complex chain are shown. **(b)** N-linked oligosaccharides have an N-glycosyl bond to the amide nitrogen of an Asn residue, illustrated here with GlcNAc as the terminal sugar. Three common types of oligosaccharide chains that are N-linked in glycoproteins are shown. [Glc : D-glucose, GlcNAc : N-acetyl-D-glucosamine, Man : D-mannose, Gal : D-galactose, Neu5Ac : N-acetyl-neuraminic acid, Fuc : fucose].

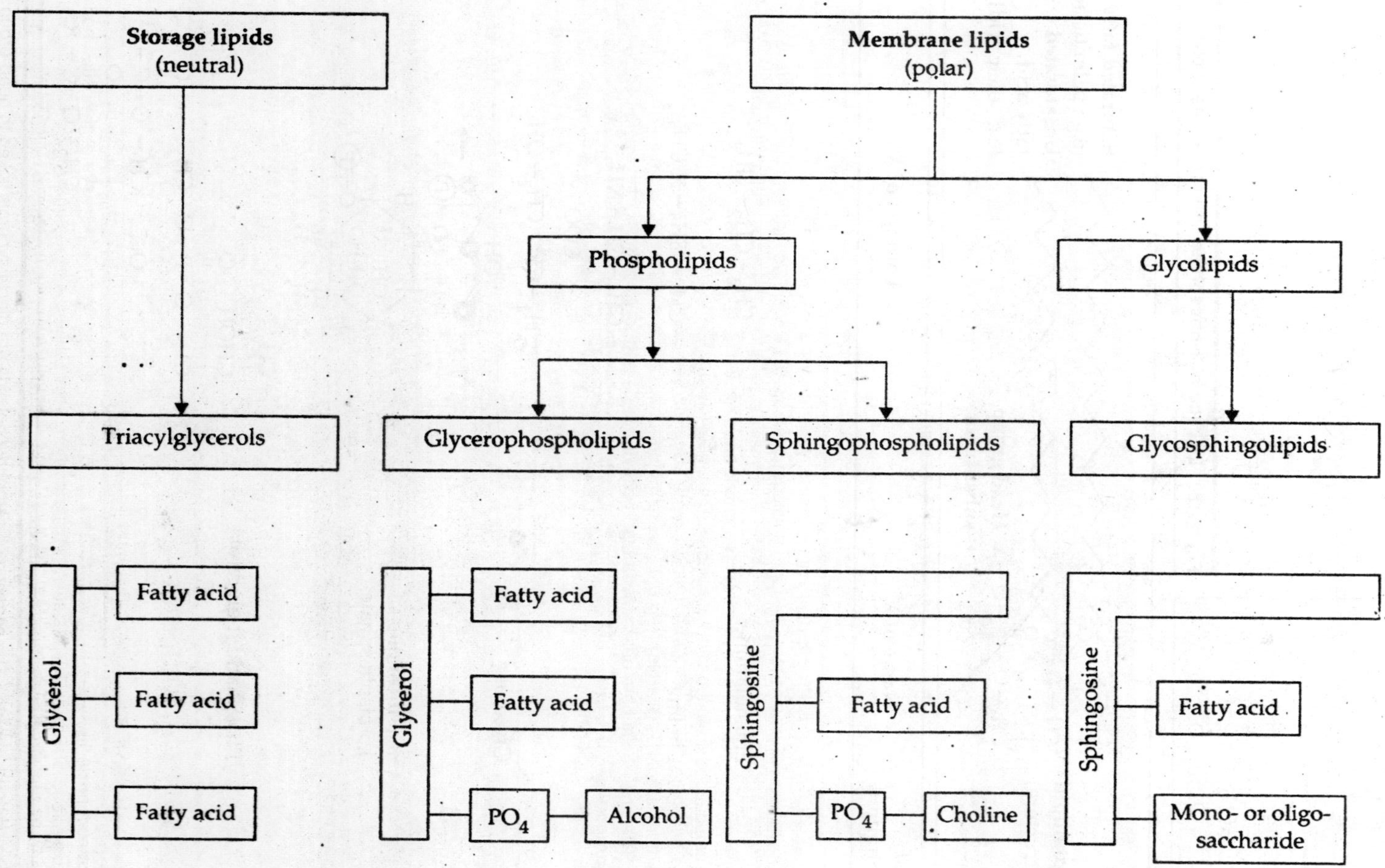

Fig. 1.17: The principal classes of storage and membrane lipids : All the lipids shown here have either glycerol or sphingosine as the backbone.

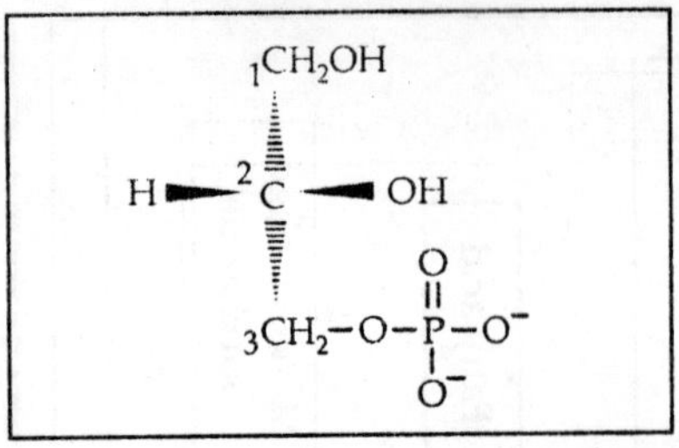

Fig. 1.18: L-Glycerol 3-phosphate : The backbone of phospholipids.

Glycerophospholipid (general structure)

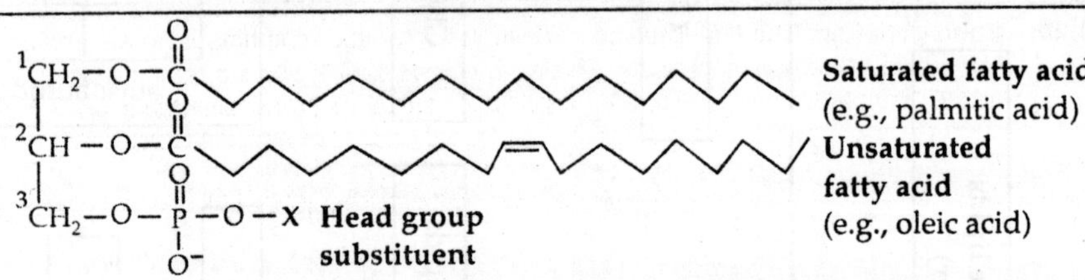

Name of Glycerophospholipid	Name of X	Formula of X
Phosphatidic acid	—	$-H$
Phosphatidyl-ethanolamine	Ethanolamine	$-CH_2-CH_2-\overset{+}{N}H_3$
Phosphatidyl-choline	Choline	$-CH_2-CH_2-\overset{+}{N}(CH_3)_3$
Phosphatidyl-serine	Serine	$-CH_2-CH(COO^-)-\overset{+}{N}H_3$
Phosphatidyl-glycerol	Glycerol	$-CH_2-CH(OH)-CH_2-OH$
Phosphatidyl-inositol 4,5-bisphosphate	*myo*-Inositol 4,5-bisphosphate	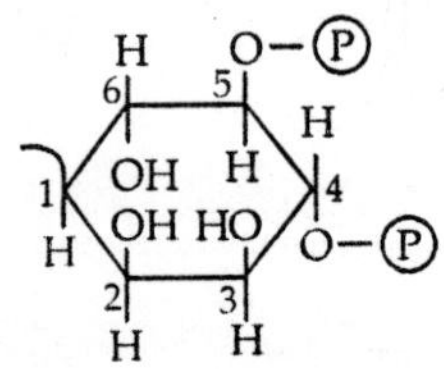
Cardiolipin	Phosphatidyl-glycerol	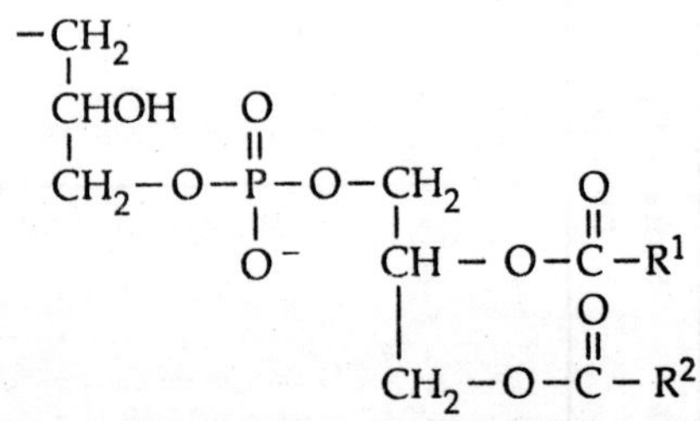

Fig. 1.19: Glycerophospholipids : The common glycerophospholipids are diacylglycerols linked to head-group alcohols through a phosphodiester bond. Phosphatidic acid, a phosphomonoester, is the parent compound. Each derivative is named for the head-group alcohol (X), with the prefix "phosphatidyl–". In cardiolipin, two phosphatidic acids share a single glycerol.

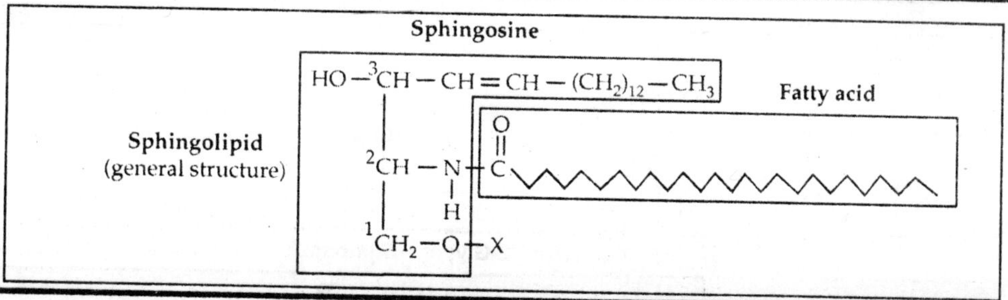

Fig. 1.20: **Sphingolipids :** The first three carbons at the polar end of sphingosine are analogous to the three carbons of glycerol in glycerophospholipids. The amino group at C-2 bears a fatty acid in amide linkage. The fatty acid is usually saturated or monounsaturated with 16, 18, 22 or 24 carbon atoms.

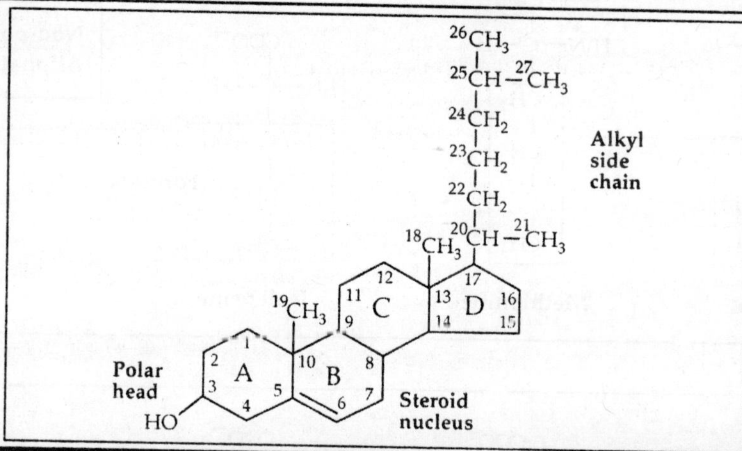

Fig. 1.21: **Cholesterol :** The rings are labeled A to D of steroid nucleus, the hydroxyl group on C-3 represents the polar head group. For storage and transport of the sterol, this hydroxyl group condenses with a fatty acid to form a sterol ester.

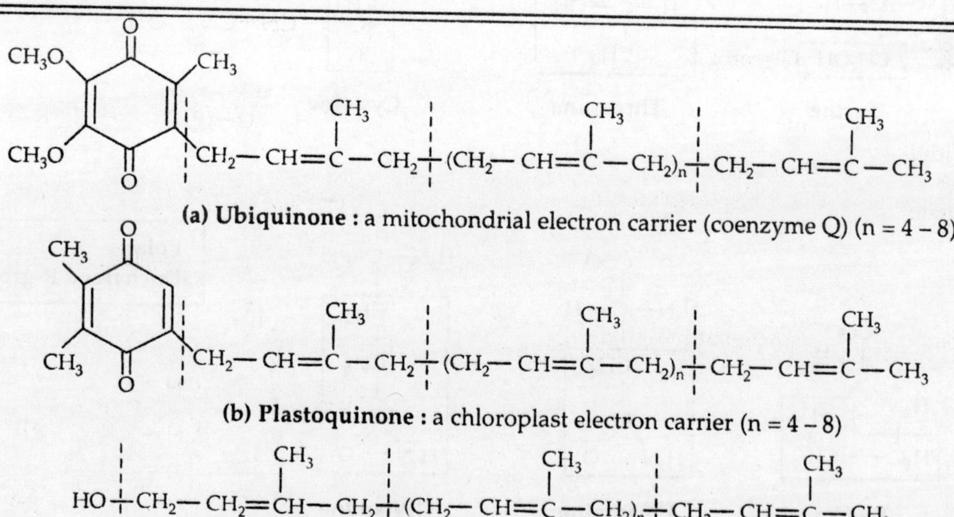

(a) Ubiquinone : a mitochondrial electron carrier (coenzyme Q) (n = 4 – 8)

(b) Plastoquinone : a chloroplast electron carrier (n = 4 – 8)

(c) Dolichol : a sugar carrier (n = 9 – 22)

Fig. 1.22: **Some biologically active isoprenoid compounds or derivatives :** Isoprene structural units are set off by dashed lines. In ubiquinone (also called coenzyme Q) has ten isoprene units. Bacterial dolichols have 11, and those of plants and fungi have 14 to 24.

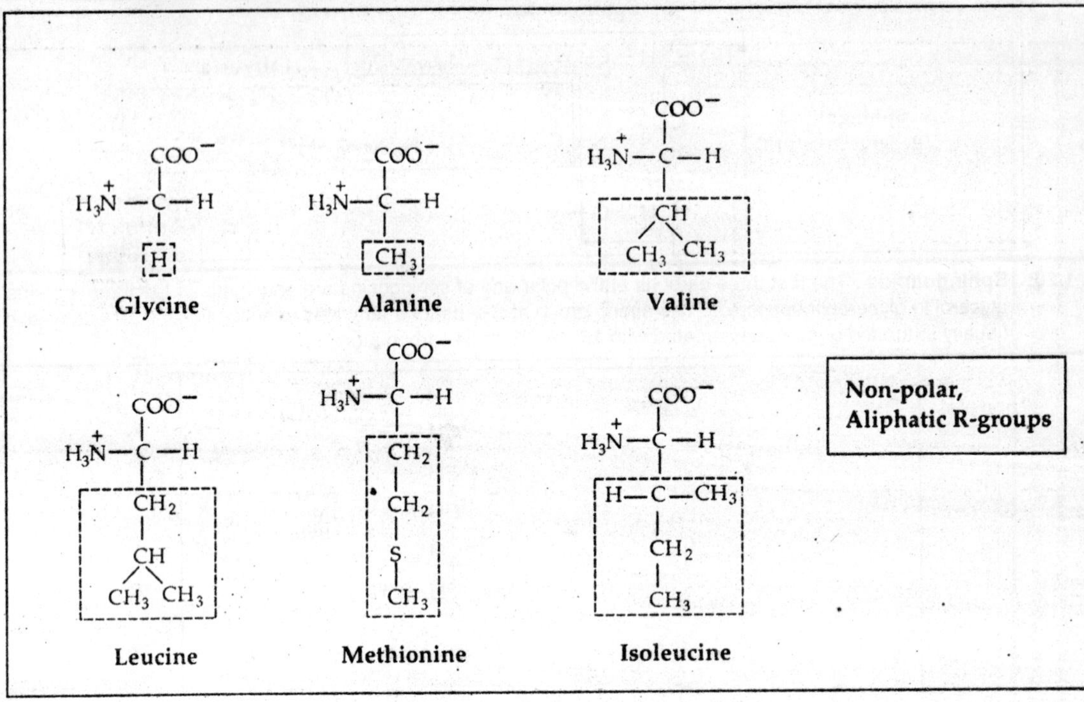
Glycine
Alanine
Valine
Non-polar,
Aliphatic R-groups
Leucine
Methionine
Isoleucine

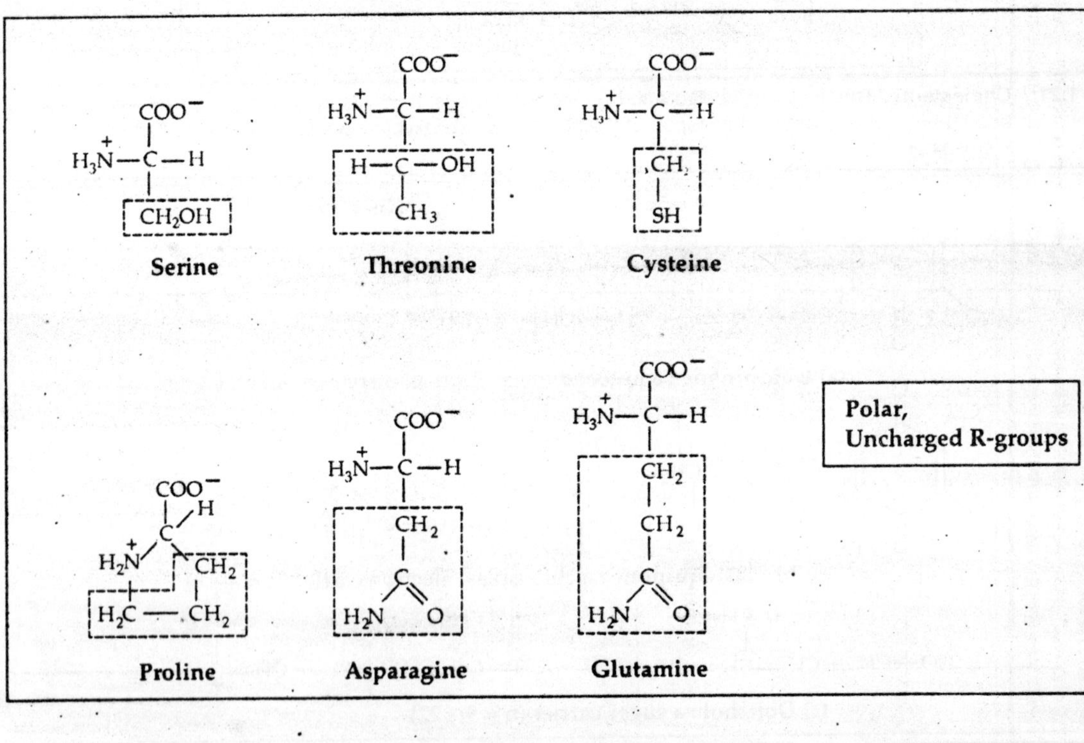
Serine
Threonine
Cysteine
Polar,
Uncharged R-groups
Proline
Asparagine
Glutamine

contd.

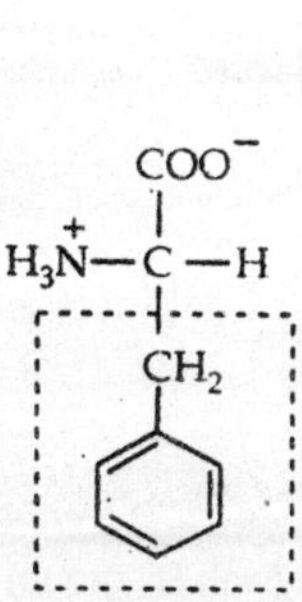

Phenylalanine

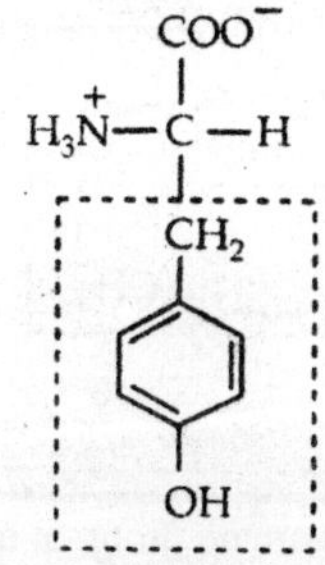

Tyrosine

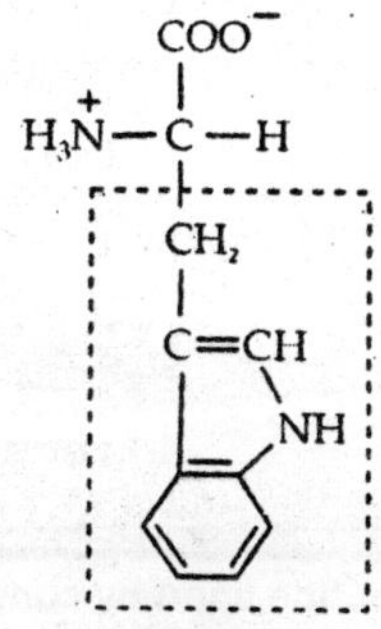

Tryptophan

Aromatic R-groups

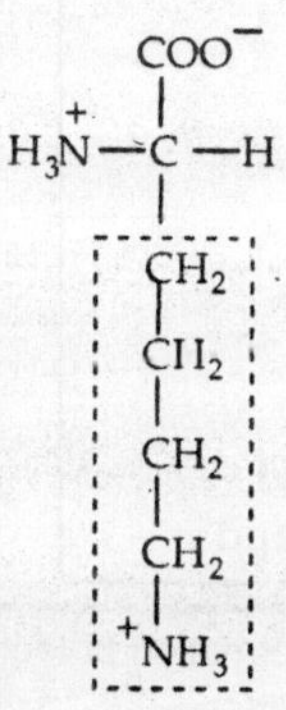

Lysine

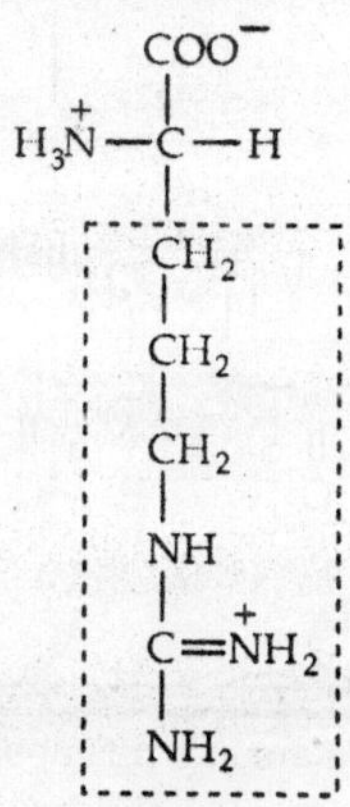

Arginine

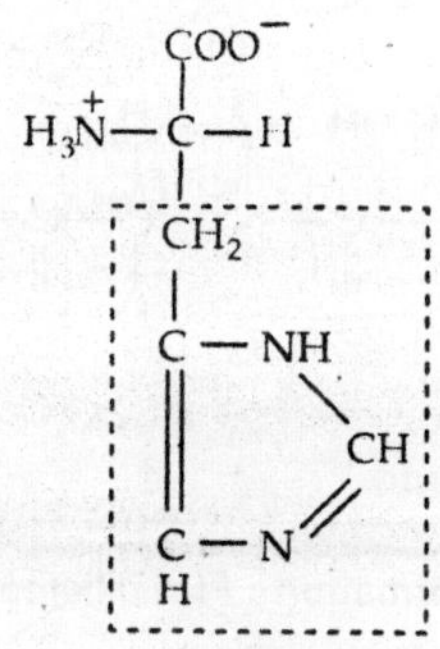

Histidine

Positively charged R-groups

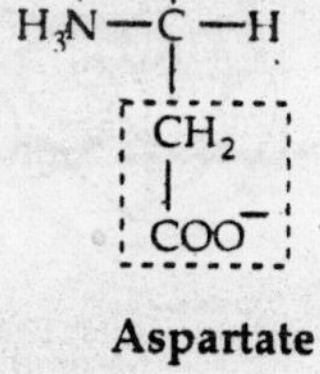

Aspartate

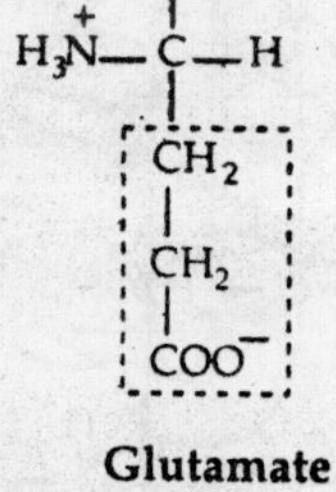

Glutamate

Negatively charged R-groups

Fig. 1.23: The 20 standard amino acids of proteins : The structural formulas show the state of ionization that would predominate at pH 7·0. The portions outside the box are those common to all the amino acids, the boxed portions are the R-groups.

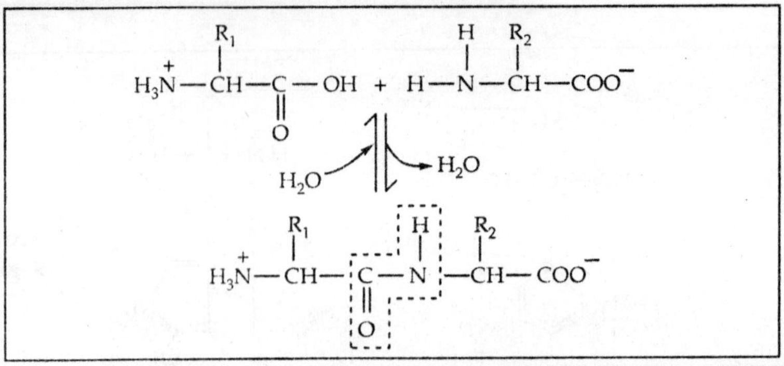

Fig. 1.24: Formation of peptide bond by condensation : The α-amino group of one amino acid (with R_2-group) acts as a nucleophile to displace the hydroxyl group of another amino acid (with R_1-group), forming a peptide bond (box). Amino groups are good nucleophiles, but the hydroxyl group is a poor leaving group and is not readily displaced.

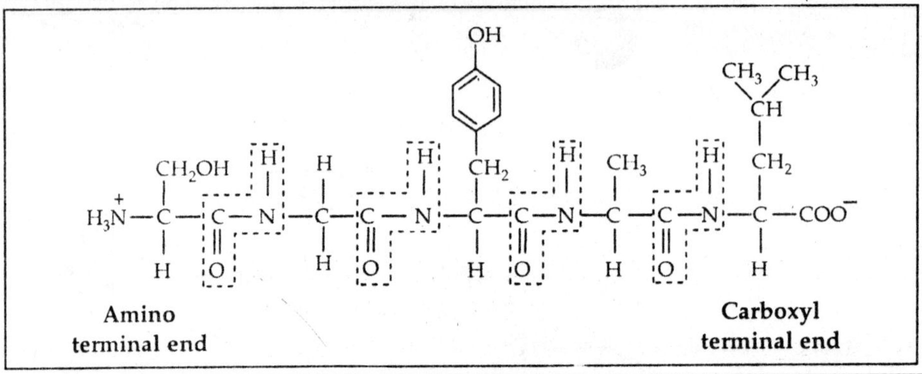

Fig. 1.25: A pentapeptide formation : Five different amino acids are in a pentapeptide chain, formation of four peptide bonds.

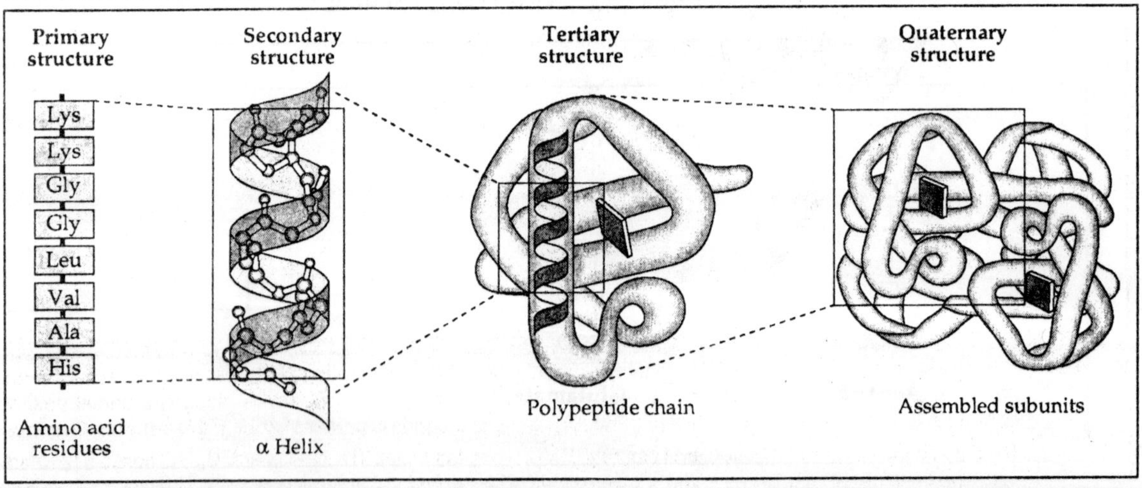

Fig. 1.26: Levels of structure in proteins : The **primary structure** consists of a sequence of amino acids linked together by peptide bonds and any disulphide bonds forming the polypeptide chain only. The **secondary structure** consists of α-helix which is the result of coiling of polypeptide. The **tertiary structure** is formed by the folding of the polypeptide where helix is a part. The **quaternary structure** is formed by several tertiary structures in case of a multisubunit protein.

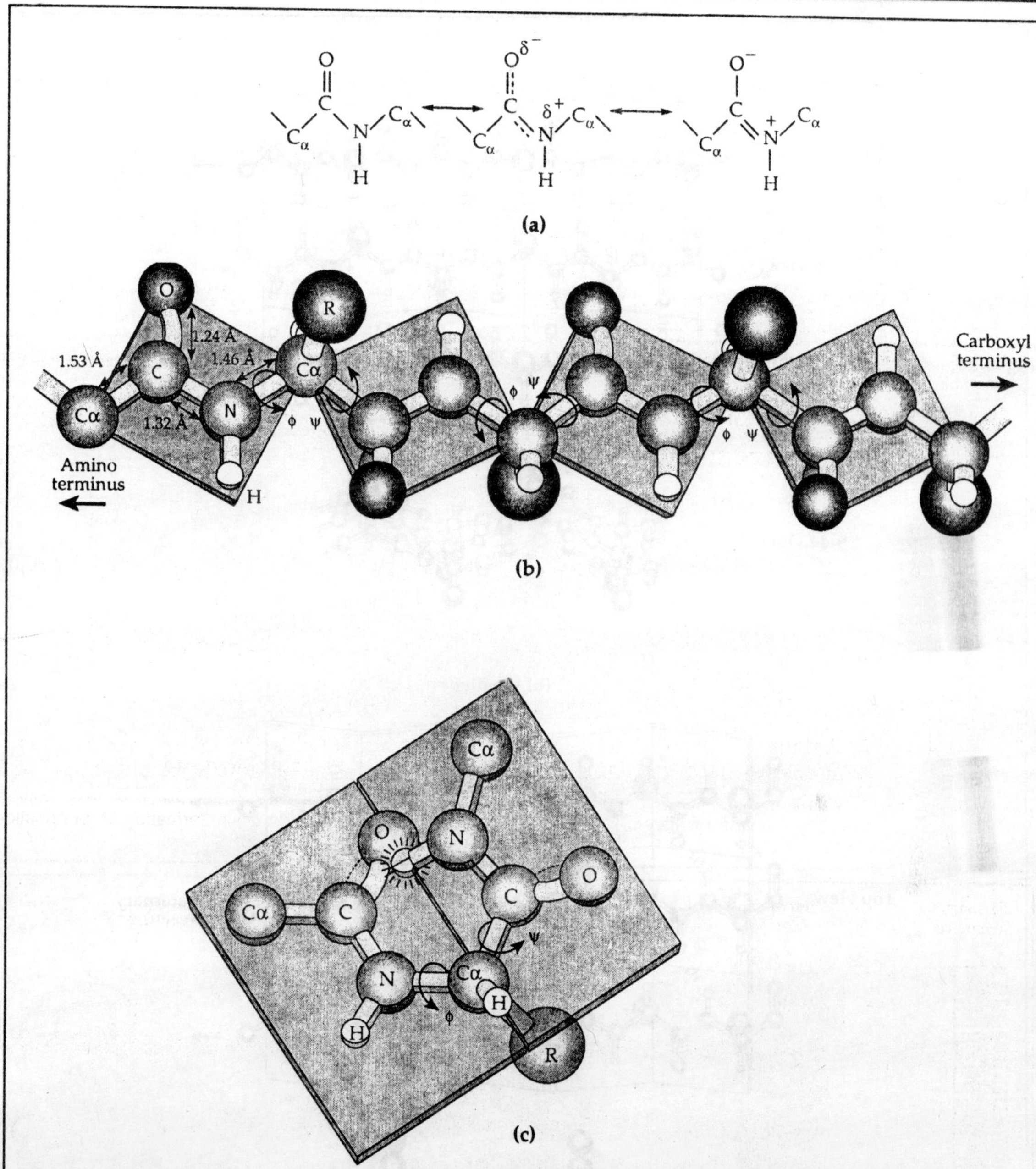

Fig. 1.27: The planar peptide group : (a) Each peptide bond has some double bond character due to resonance and cannot rotate. The carboxyl oxygen has a partial negative charge and the amide nitrogen a partial positive charge, setting up a small electric dipole. Virtually, all peptide bonds in proteins occur in this *trans* configuration. **(b)** Three bonds separate sequential α carbons in a polypeptide chain. The N – C_α and C_α – C bonds can rotate, with bond angles designated as φ and ψ respectively. The peptide C – N bond is not free to rotate. Other single bonds in the backbone may also be rotationally hindered, depending on the size and charge of the R-groups. **(c)** φ and ψ are both defined as 0° by convention, when the two peptide bonds flanking that α carbon are in the same plane and positioned as shown. In a protein, this conformation is prohibited by steric overlap between an α-carbonyl oxygen and α-amino hydrogen atom. To illustrate the bonds between atoms, the balls representing each atom are smaller than the van der Waals radii for this scale. 1 Å = 0·1 nm.

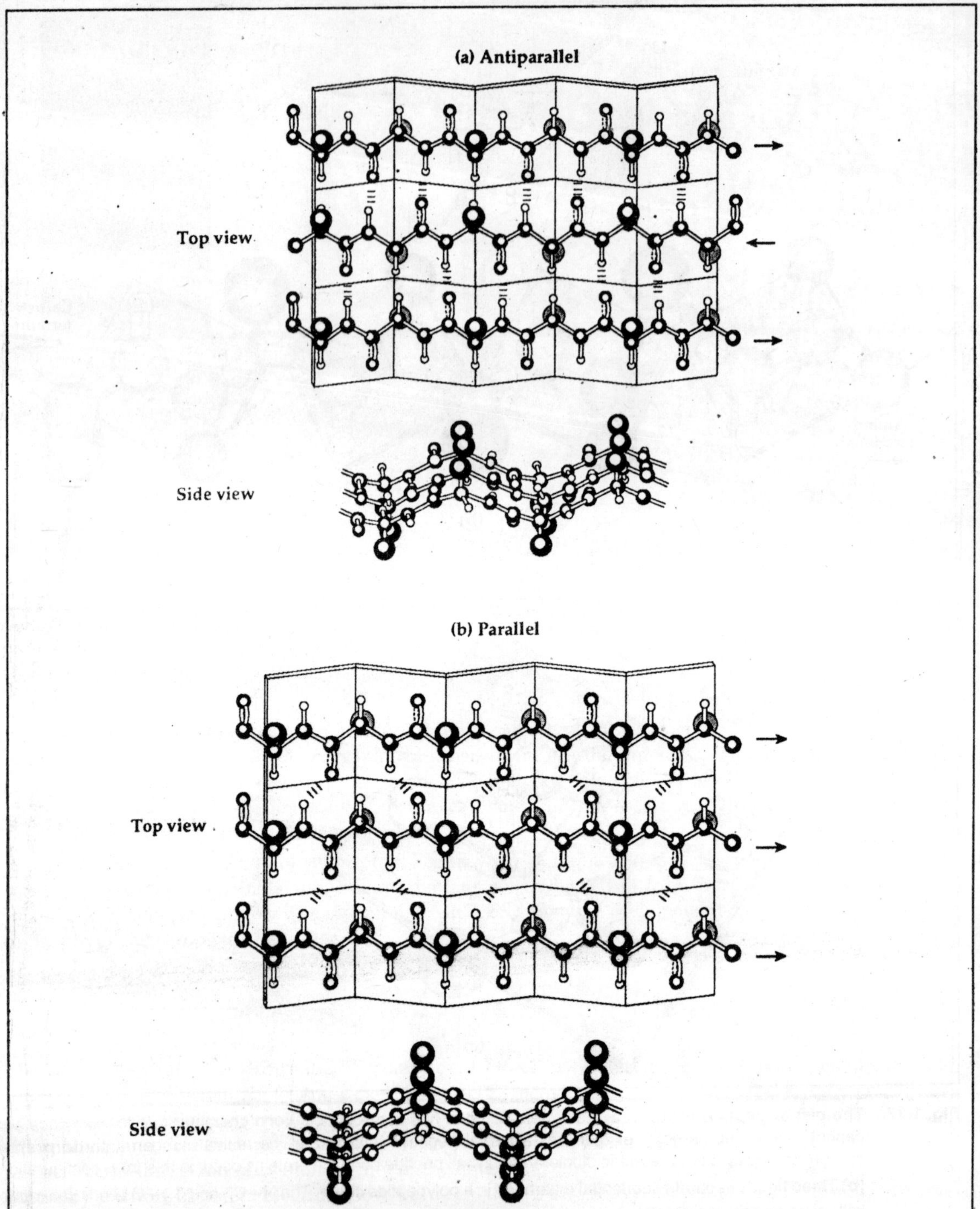

Fig. 1.28: The β conformation of polypeptide chains : These top and side views reveal the R-groups extending out from the β-sheet and emphasize the pleated shape described by the planes of the peptide bonds. Hydrogen bond cross links between adjacent chains are also shown. **(a)** Antiparallel β-sheet, in which the amino-terminal to carboxyl-terminal orientation of adjacent chains (arrows) is inverse. **(b)** Parallel β-sheet structure.

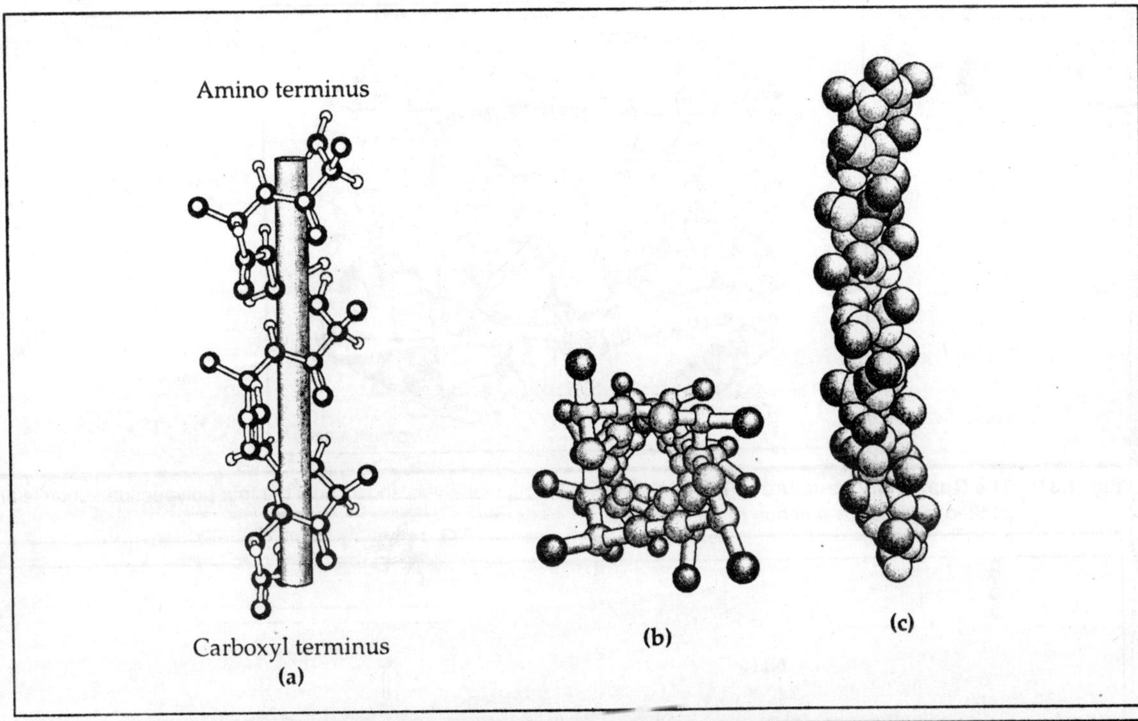

Fig. 1.29: Models of the α-helix, showing different aspects of its structure : (a) Formation of a right-handed α-helix. The planes of the rigid peptide bonds are parallel to the long axis of the helix. **(b)** The α-helix as viewed from one end, looking down the longitudinal axis. The model gives the false impression that the α-helix is hollow, it is due to the representation of the structure by ball-and-stick. **(c)** In the space-filling model it is shown that the molecules at center are in very close contact.

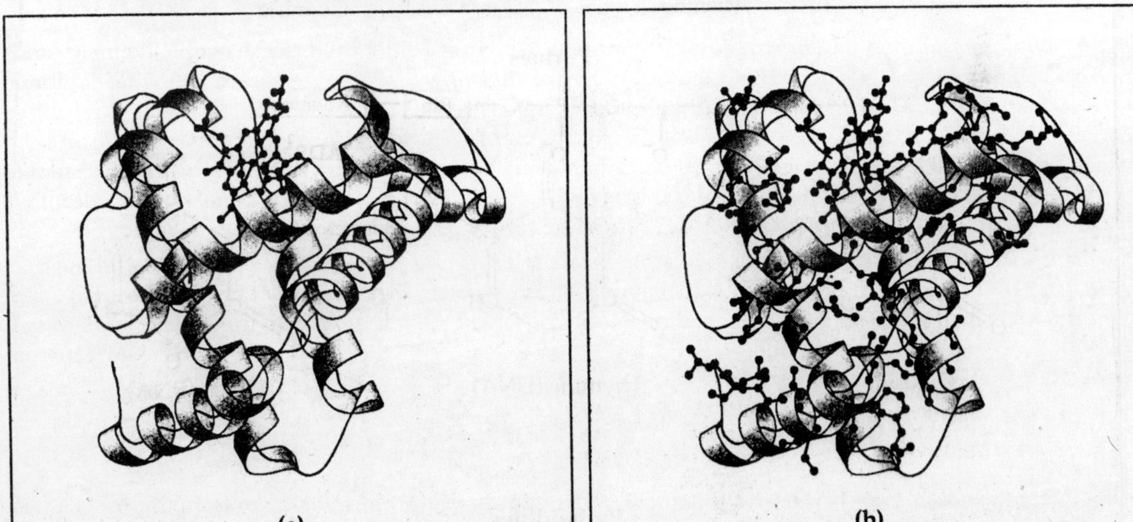

Fig. 1.30: The tertiary structure of protein : (a) The polypeptide backbone shown in a ribbon representation, which shows the regions of secondary structure, the α-helical regions are evident. **(b)** The ribbon representation, including side chains for the hydrophobic residues Leu, Ile, Val, Phe.

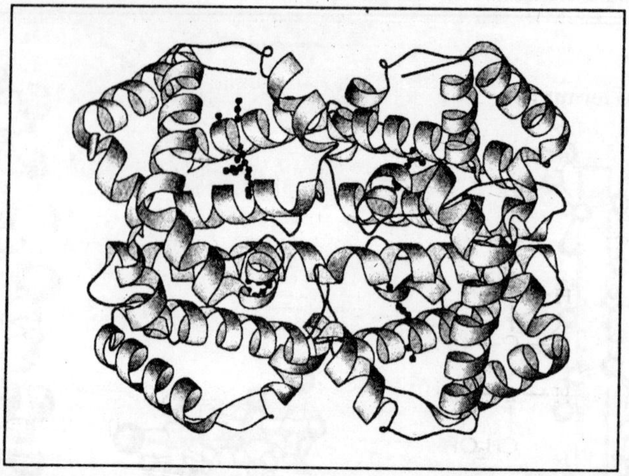

Fig. 1.31: The Quaternary structure of Protein : X-ray diffraction analysis shows how the four polypeptide subunits are packed together by a ribbon representation.

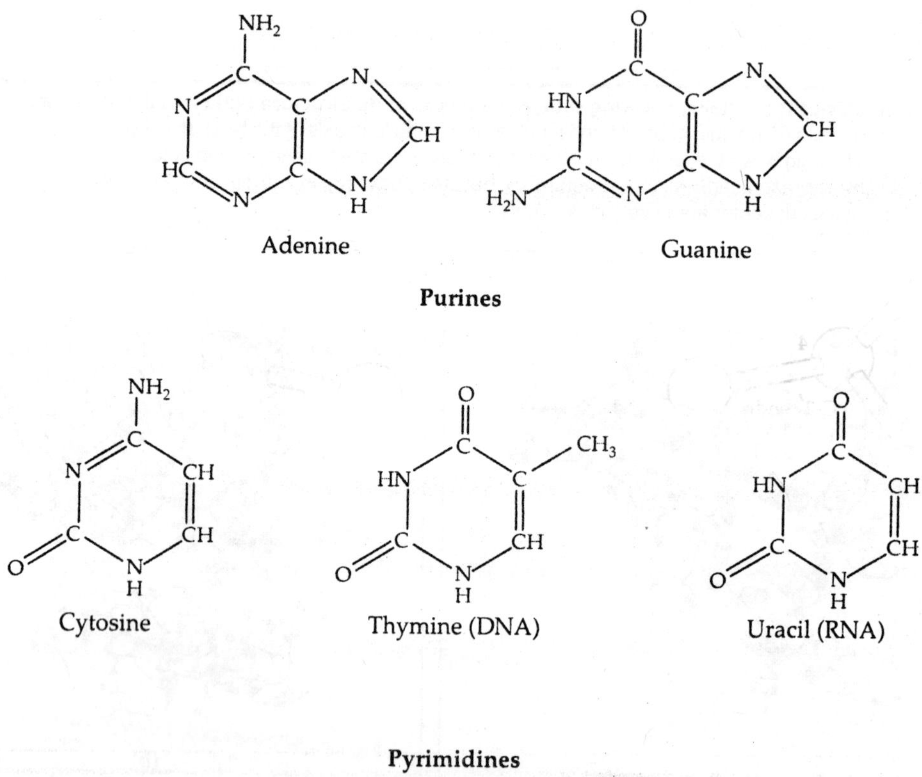

Fig. 1.32 : Major purine and pyrimidine bases of nucleic acids.

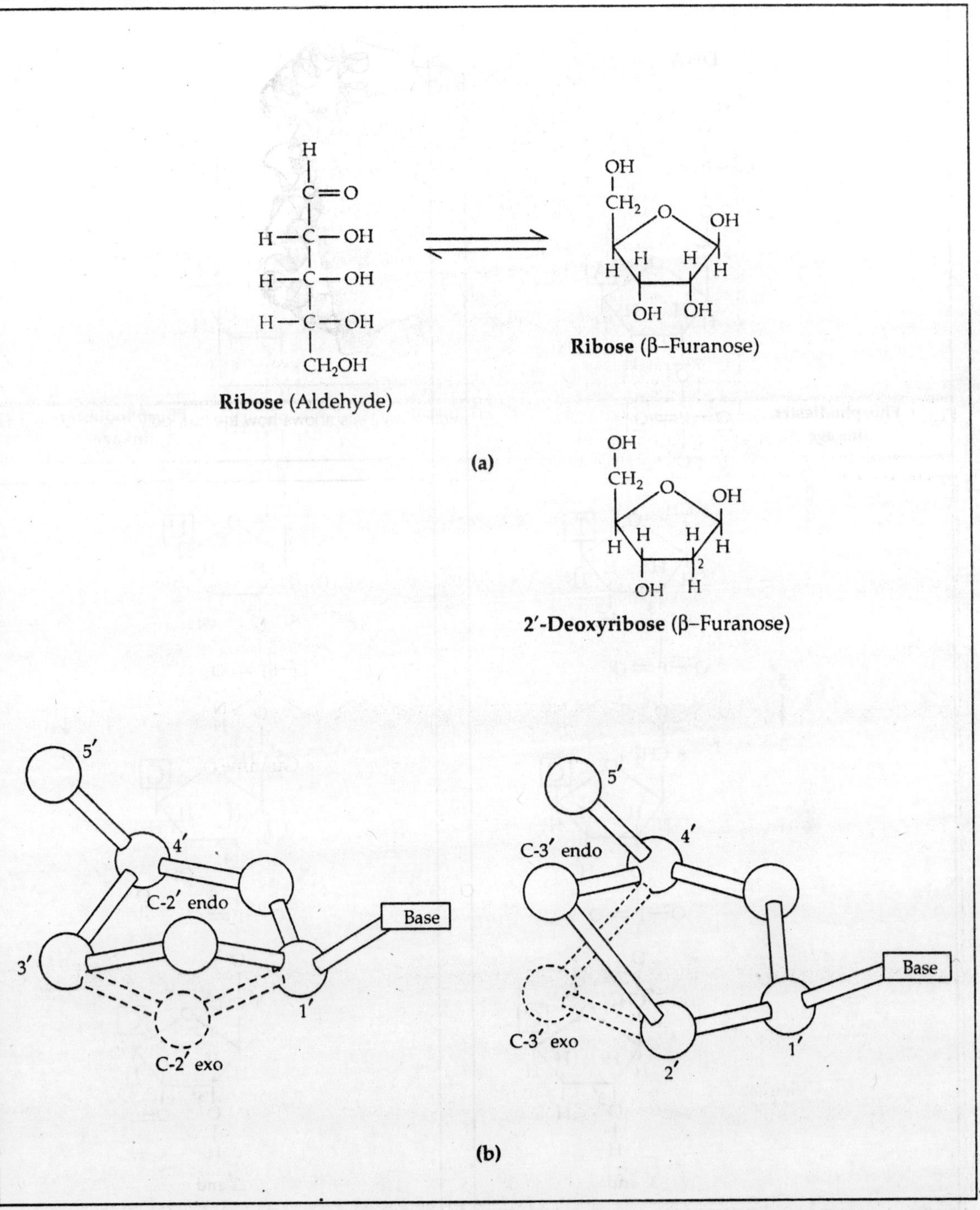

Fig. 1.33: Conformations of ribose : (a) In solution, the straight chain (aldehyde) and ring (β-furanose) forms of free ribose are in equilibrium. RNA contains only the ring form, β-D-ribofuranose. Deoxyribose undergoes a similar interconversion in solution, but in DNA exists solely as β-2′-deoxy-D-ribofuranose. **(b)** Ribofuranose rings in nucleotides can exist in four different puckered conformations. In all cases, four of the five atoms are in a single plane. The fifth atom (C-2′ or C-3′) is on either the same (endo) or the opposite (exo) side of the plane relative to the C-5′ atom.

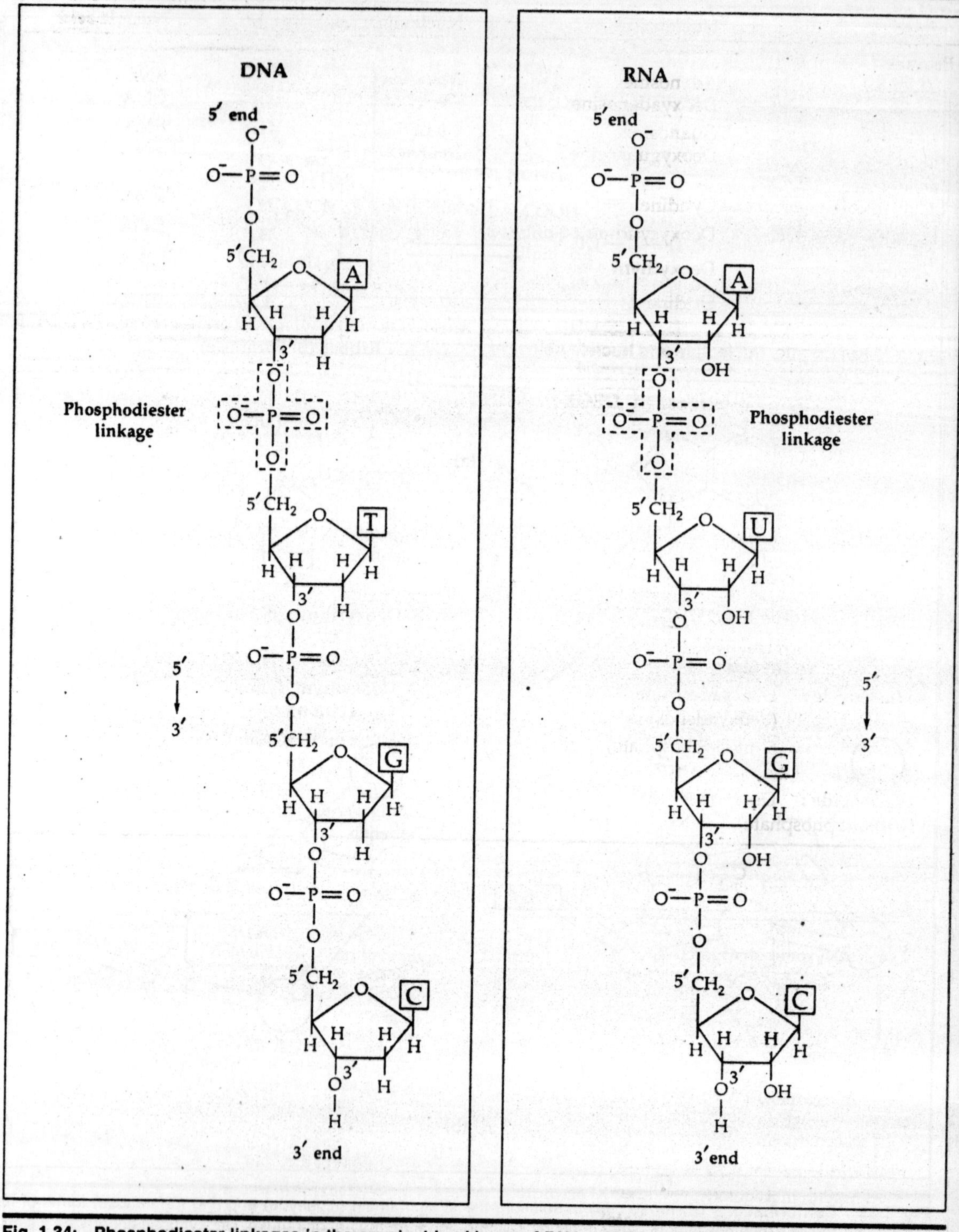

Fig. 1.34: Phosphodiester linkages in the covalent backbone of DNA and RNA : The phosphodiester bonds (one is shown in box) link successive nucleotide units. The backbone of alternating pentose and phosphate groups in both types of nucleic acid is highly polar. The 5′ end of the macromolecule lacks a nucleotide at the 5′ position, and the 3′ end lacks a nucleotide at the 3′ position.

Base	Nucleoside	Nucleotide	Nucleic acid
Purines			
Adenine	Adenosine	Adenylate	RNA
	Deoxyadenosine	Deoxyadenylate	DNA
Guanine	Guanosine	Guanylate	RNA
	Deoxyguanosine	Deoxyguanylate	DNA
Pyrimidines			
Cytosine	Cytidine	Cytidylate	RNA
	Deoxycytidine	Deoxycytidylate	DNA
Thymine	Deoxythymidine	Deoxythymidylate	DNA
Uracil	Uridine	Uridylate	RNA

Fig. 1.35: Nucleoside, Nucleotide and Nucleic Acid nomenclature.

DEOXYRIBONUCLEOTIDES

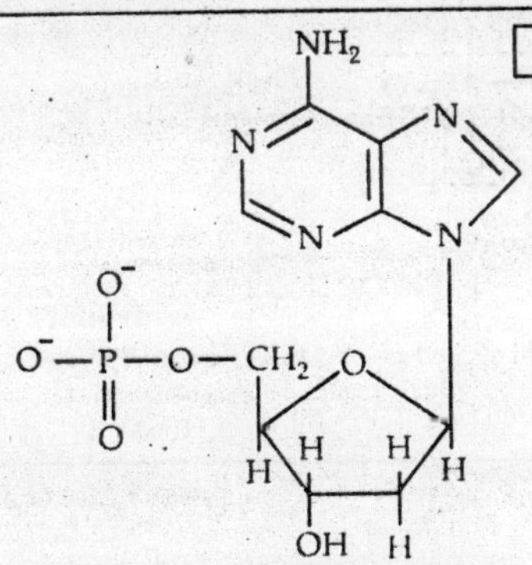

Nucleotide : Deoxyadenylate (deoxyadenosine 5′-monophosphate)
Symbols : A, dA, dAMP
Nucleoside : Deoxyadenosine (without phosphate)

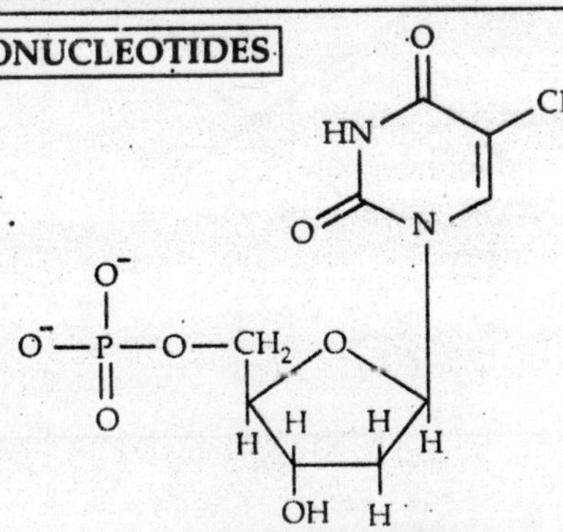

Deoxythymidylate (deoxythymidine 5′-monophosphate)
T, dT, dTMP
Deoxythymidine

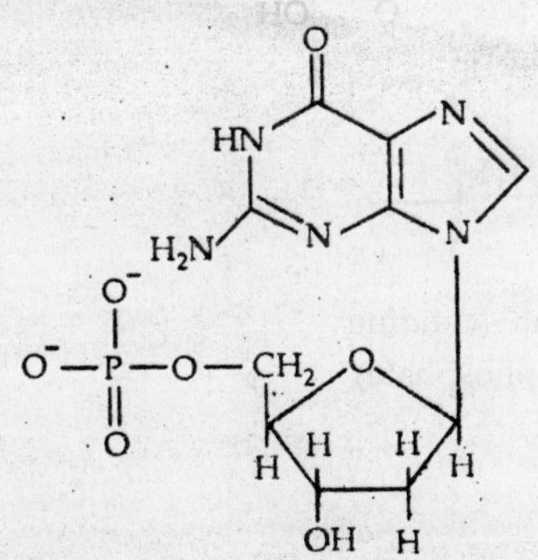

Nucleotide : Deoxyguanylate (deoxyguanosine 5′-monophosphate)
Symbols : G, dG, dGMP
Nucleoside : Deoxyguanosine (without phosphate)

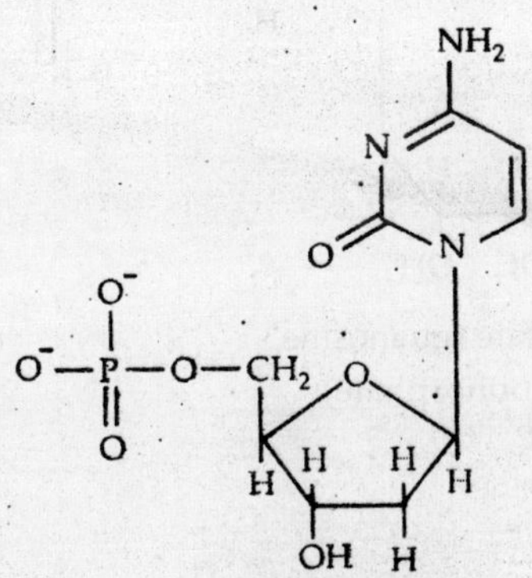

Deoxycytidylate (deoxycytidine 5′ monophosphate)
C, dC, dCMP
Deoxycytidine

RIBONUCLEOTIDES

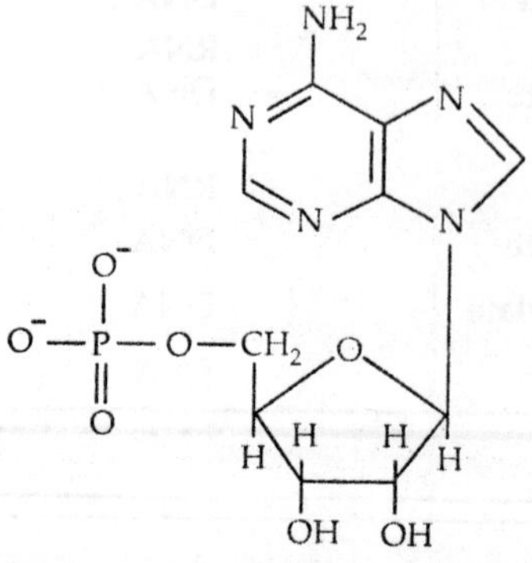

Nucleotide : Adenylate (adenosine 5′-monophosphate)
Symbols : A, AMP
Nucleoside : Adenosine
(without phosphate)

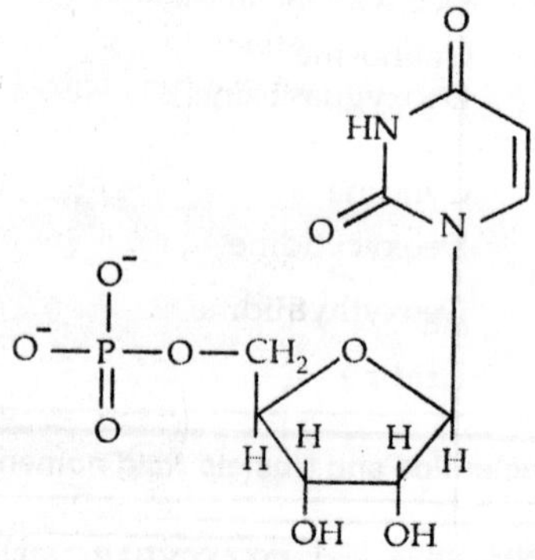

Uridylate (uridine 5′-monophosphate)
U, UMP
Uridine

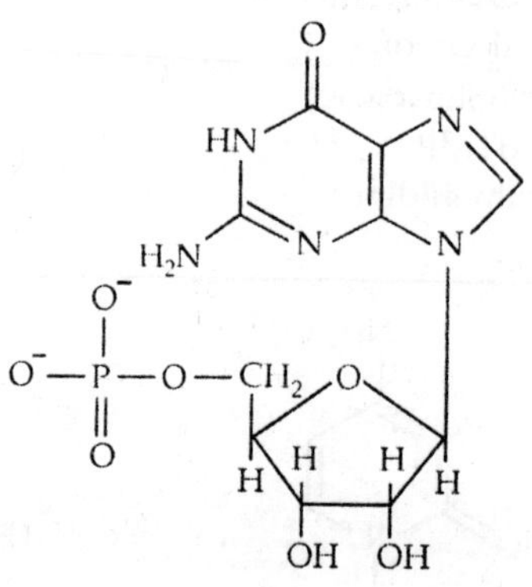

Nucleotide : Guanylate (guanosine 5′-monophosphate)
Symbols : G, GMP
Nucleoside : Guanosine
(without phosphate)

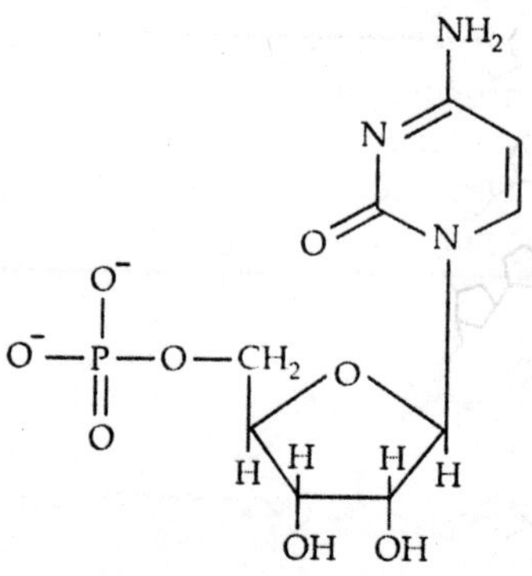

Cytidylate (cytidine 5′-monophosphate)
C, CMP
Cytidine

Fig. 1.36: Deoxyribonucleotides and ribonucleotides of nucleic acids : All nucleotides are shown in their free form at pH 7·0. The nucleotide units of DNA are usually symbolised as A, T, G, C or dA, dT, dG and dC and those of RNA as A, U, G, and C. The deoxyribonucleotides are commonly abbreviated dAMP, dTMP, dGMP and dCMP, whereas the ribonucleotides are named as AMP, UMP, GMP and CMP. Complete name is in parentheses which describes the attachment of phosphate group at 5′ position.

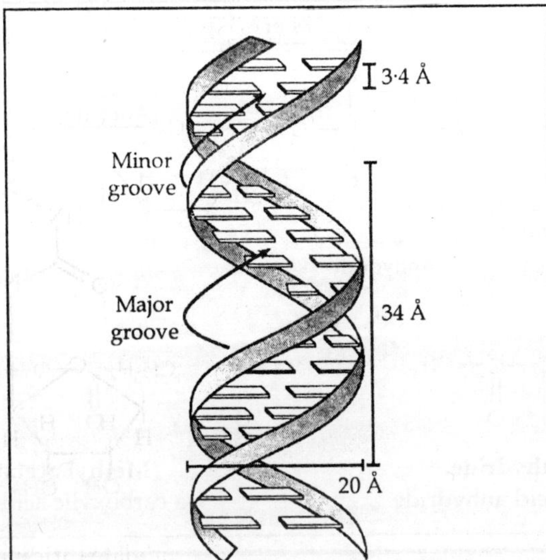

Fig. 1.37: Watson-Crick model for the structure of DNA : Schematic representation showing dimensions of the helix, 10 base pairs per turn of the helix measuring 34 Å (3·4 nm).

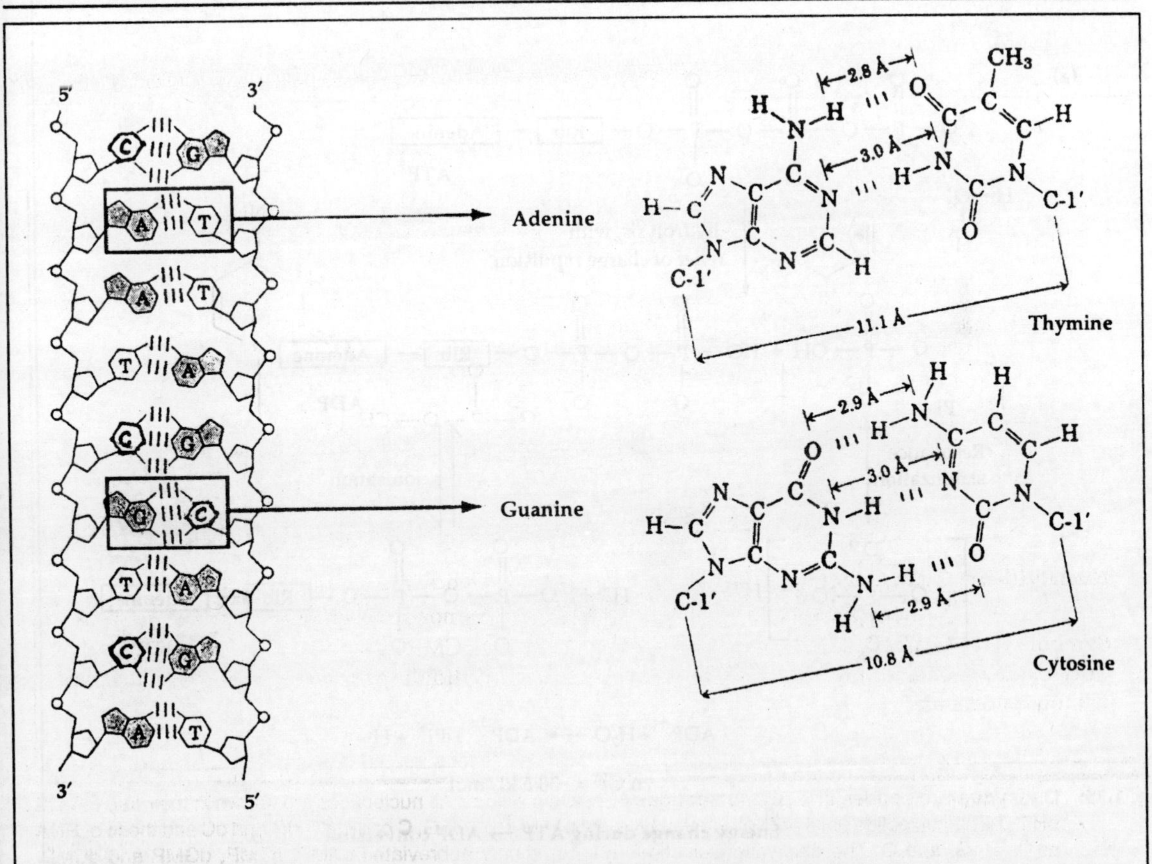

Fig. 1.38: Hydrogen-bonding patterns in the base pairs defined by Watson and Crick.

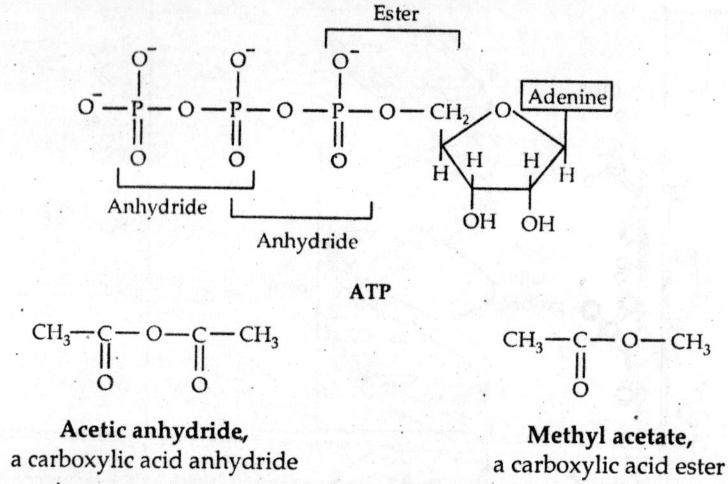

Fig. 1.39: The phosphate ester and phosphoanhydride bonds of ATP : Hydrolysis of an anhydride bond yields more energy than hydrolysis of the ester. A carboxylic acid anhydride and carboxylic acid ester are shown for comparison.

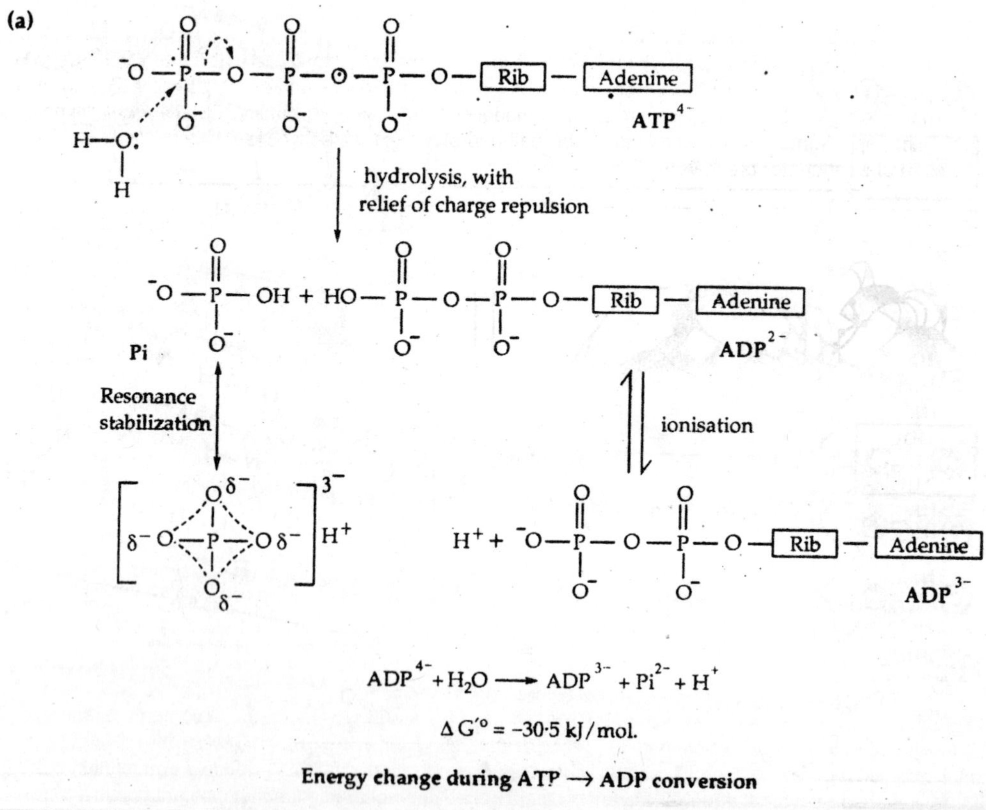

contd.

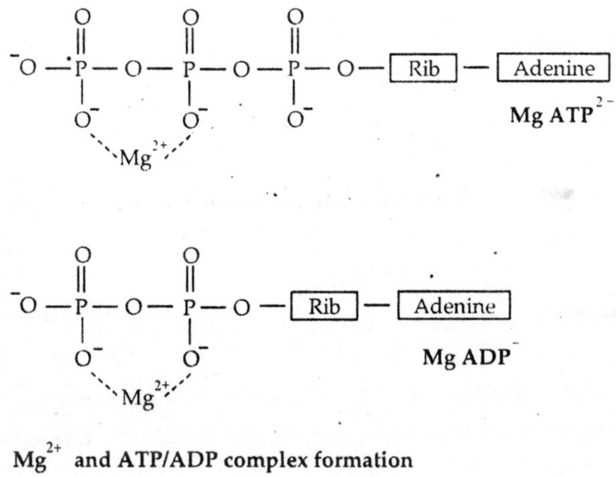

Fig. 1.40: (a) Chemical basis for the large free-energy change associated with ATP hydrolysis : Hydrolysis, by causing charge separation, relieves electrostatic repulsion among the four negative charges on ATP. Inorganic phosphate (Pi) released by hydrolysis is stabilised by formation of a resonance hybrid, in which each of the four P – O bonds has the same degree of double-bond character and the hydrogen ion is not permanently associated with any one of the oxygens. Some degree of resonance stabilization also occurs in phosphates involved in ester or anhydride linkages, but fewer resonance forms are possible than for Pi. ADP^{2-} produced by the hydrolysis immediately ionizes, releasing a proton into medium of very low [H^+] (pH – 7). There is another factor favours ATP hydrolysis, that is the greater degree of solvation (hydration) of the products Pi and ADP relative to ATP, which further stabilizes the products relative to the reactants. **(b) Mg^{2+} and ATP/ADP complex formation :** Formation of Mg^{2+} complexes partially shields the negative charges and influences the conformation of the phosphate groups in nucleotides such as ATP and ADP.

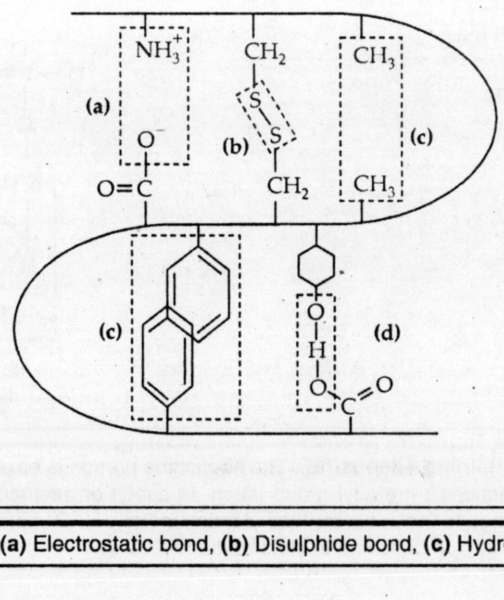

Fig. : Types of bond in Protein : (a) Electrostatic bond, **(b)** Disulphide bond, **(c)** Hydrophobic bond, **(d)** Hydrogen bond.

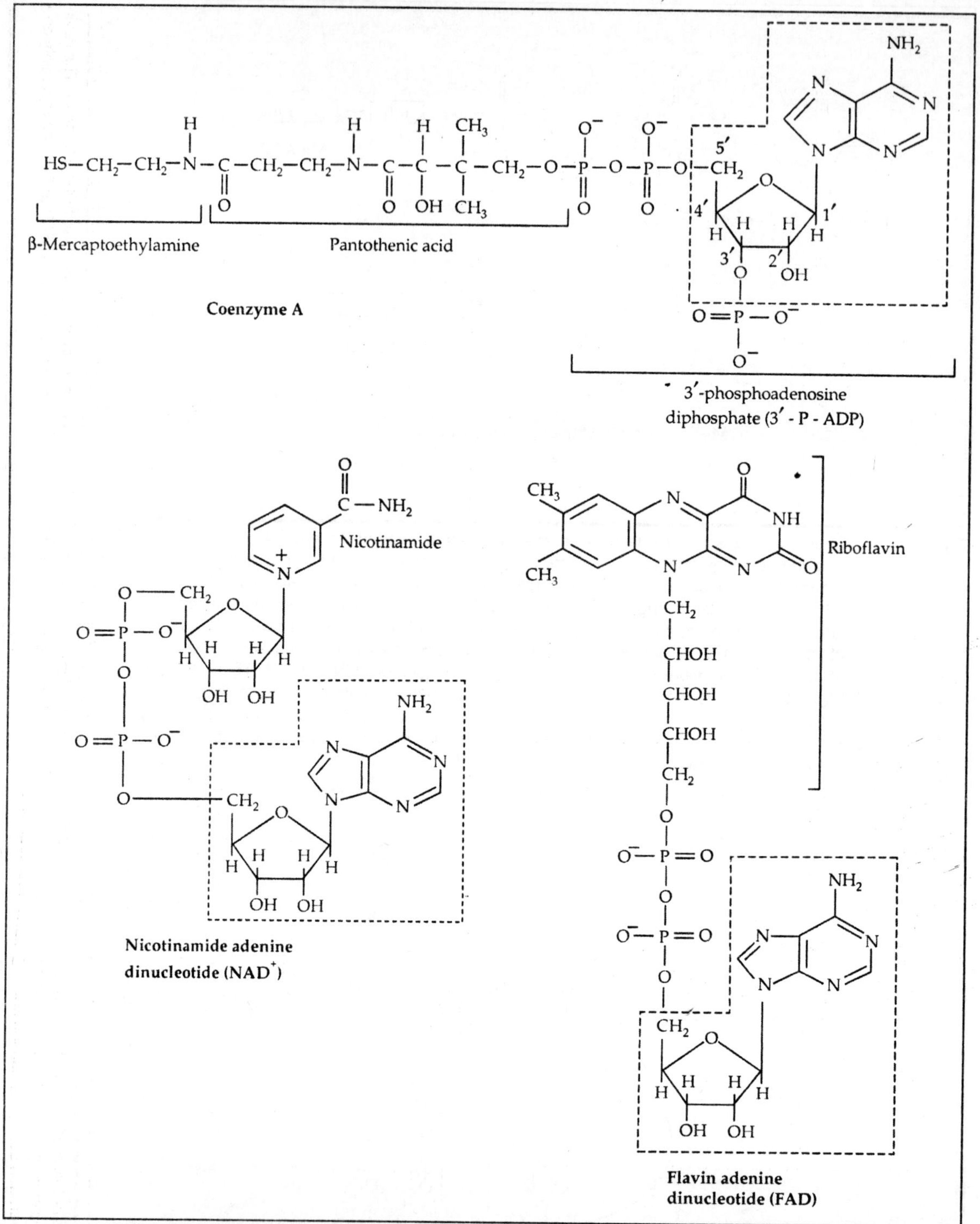

Fig. 1.41: Some coenzymes containing adenosine : The adenosine portion is boxed. **Coenzyme A** (CoA) functions in acyl group transfer reactions ; the acyl group (such as acetyl or acetoacetyl group) is attached to the CoA through a thioester linkage to the β-mercaptoethylamine moiety. NAD+ functions in hydride transfers, and FAD, the active form of vitamin B_2 (riboflavin), in electron transfers.

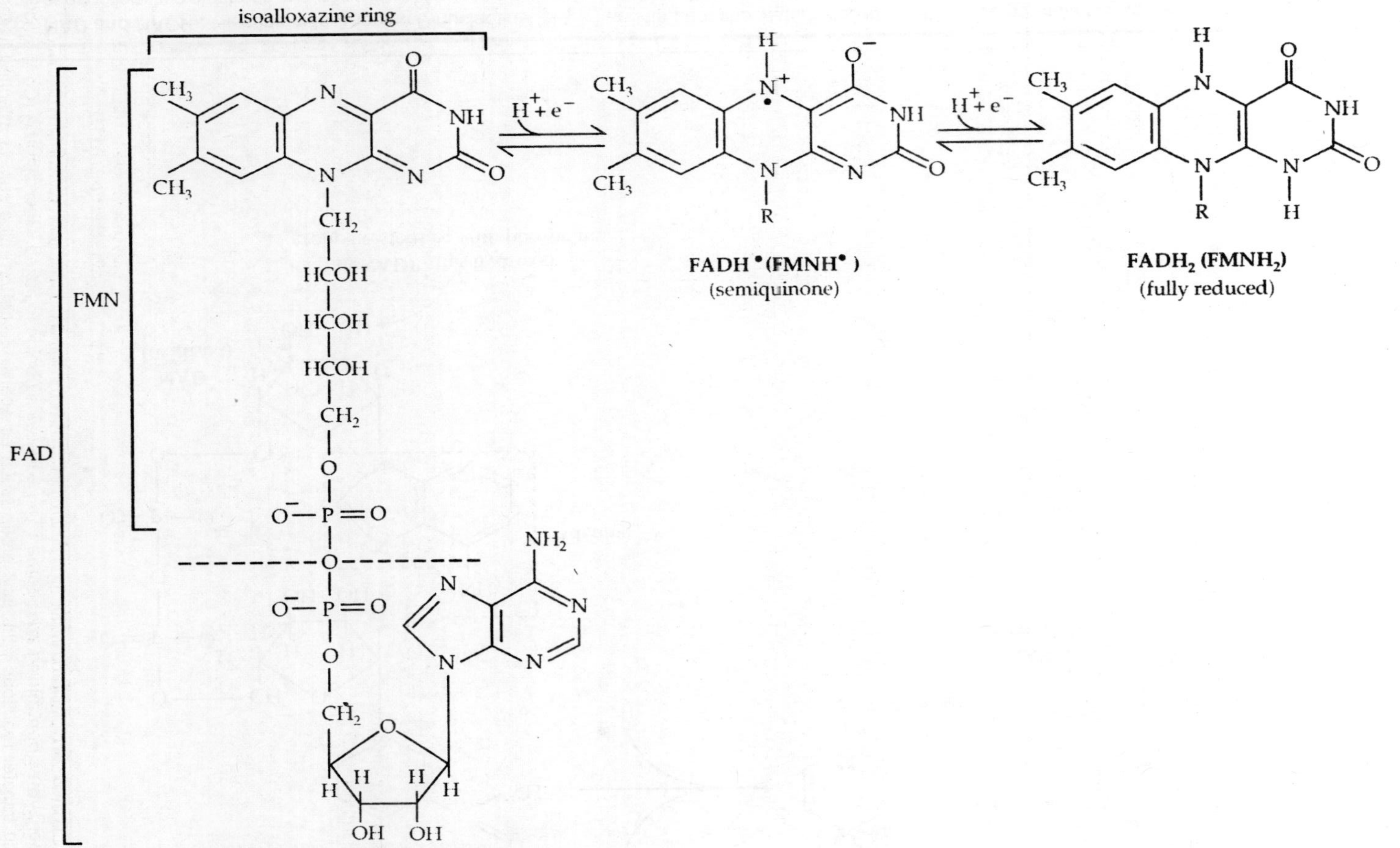

Fig. 1.42: Structures of oxidised and reduced FAD and FMN : FMN consists of the structure above the dashed line shown on the oxidised (FAD) structure. The flavin nucleotides accept two hydrogen atoms (two electrons and two protons), both of which appear in the flavin ring system. When FAD or FMN accepts only one hydrogen atom, the semiquinone, a stable free radical, forms.

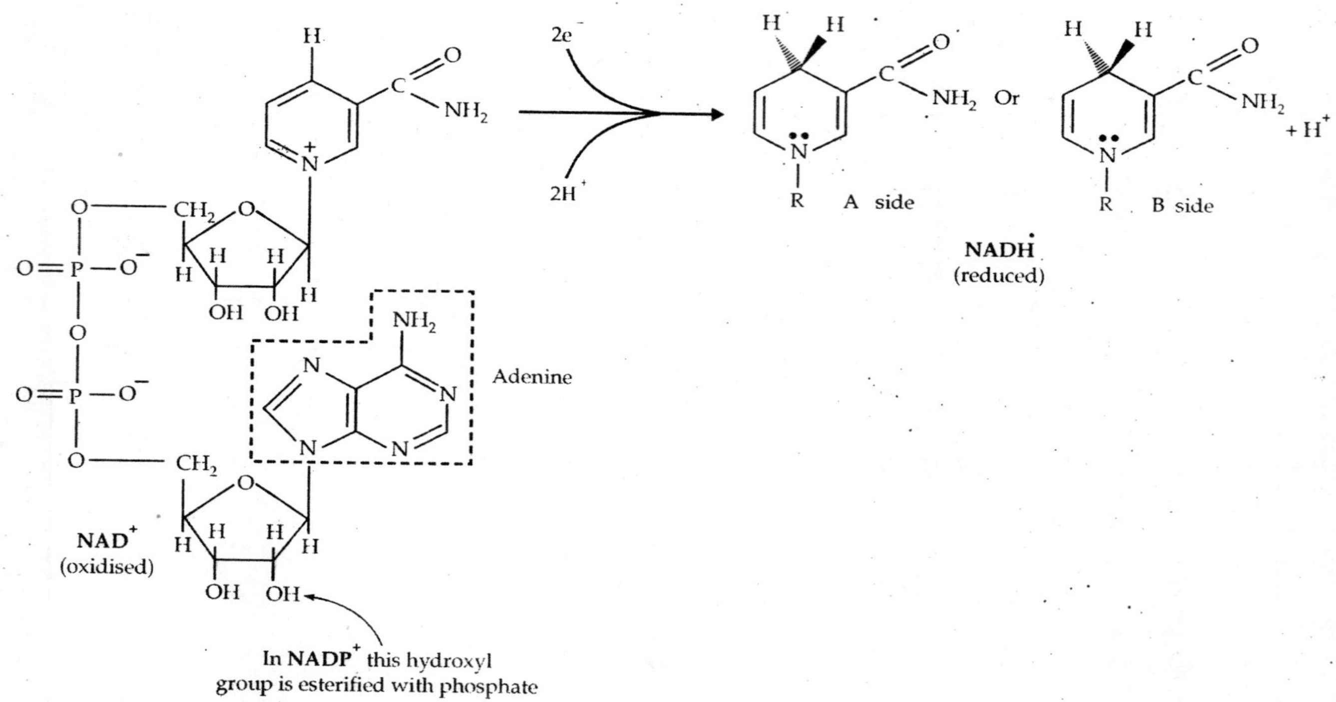

Fig. 1.43: **NAD and NADP :** Nicotinamide Adenine Dinucleotide (NAD^+) and its phosphorylated analog $NADP^+$ undergo reduction to NADH and NADPH, accepting a hydride ion (two electrons and one proton) from an oxidisable substrate. The hydride ion is added to either the front (the A side) or the back (the B side) of the planar nicotinamide ring.

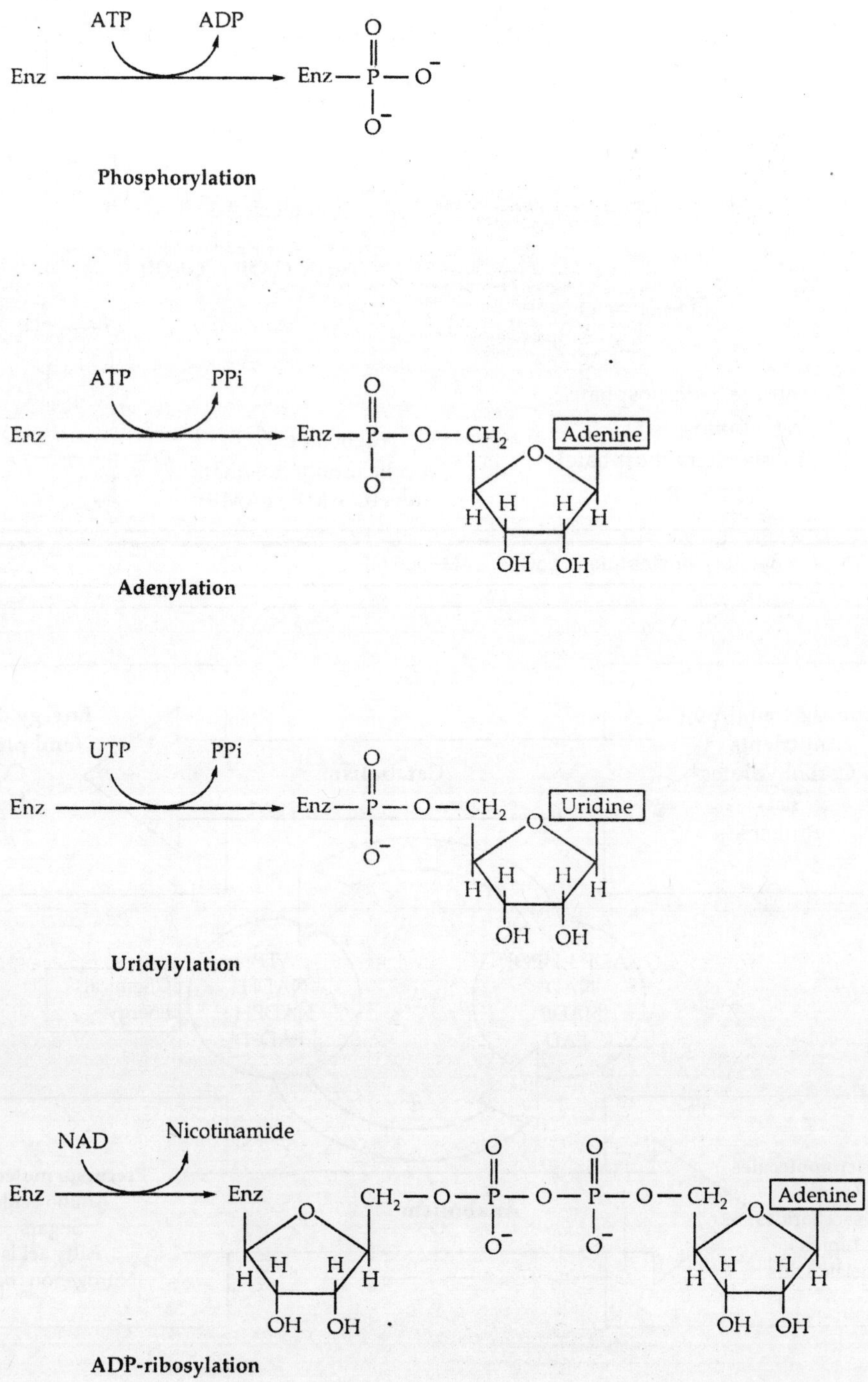

Fig. 1.44: Examples of enzyme modification reactions involving nucleotides.

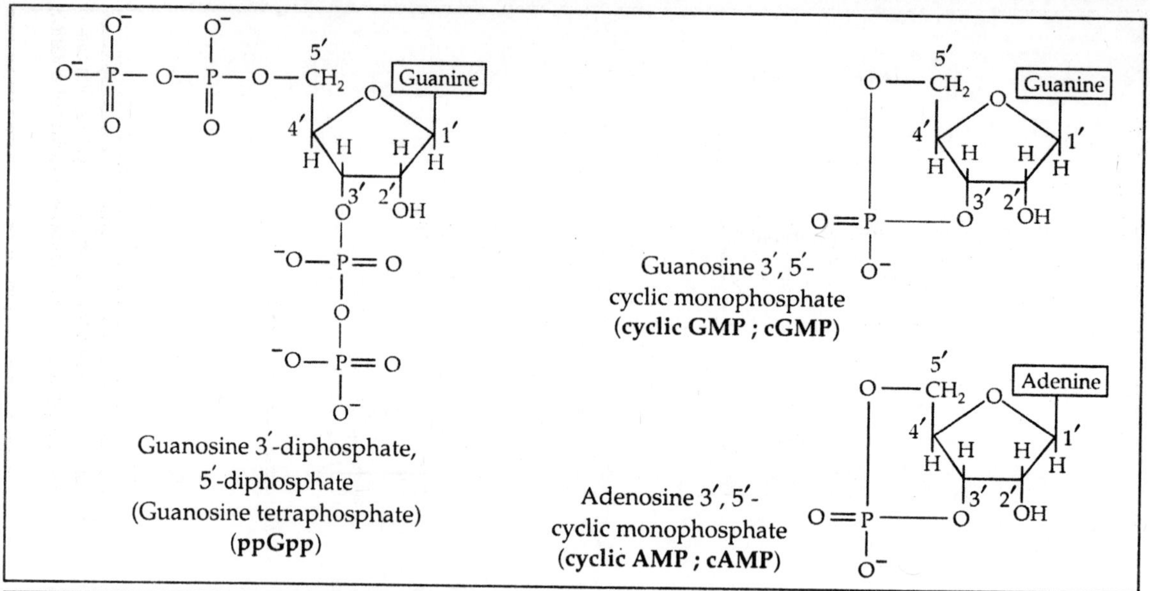

Fig. 1.45 : Three regulatory nucleotides : ppGpp, cAMP, cGMP.

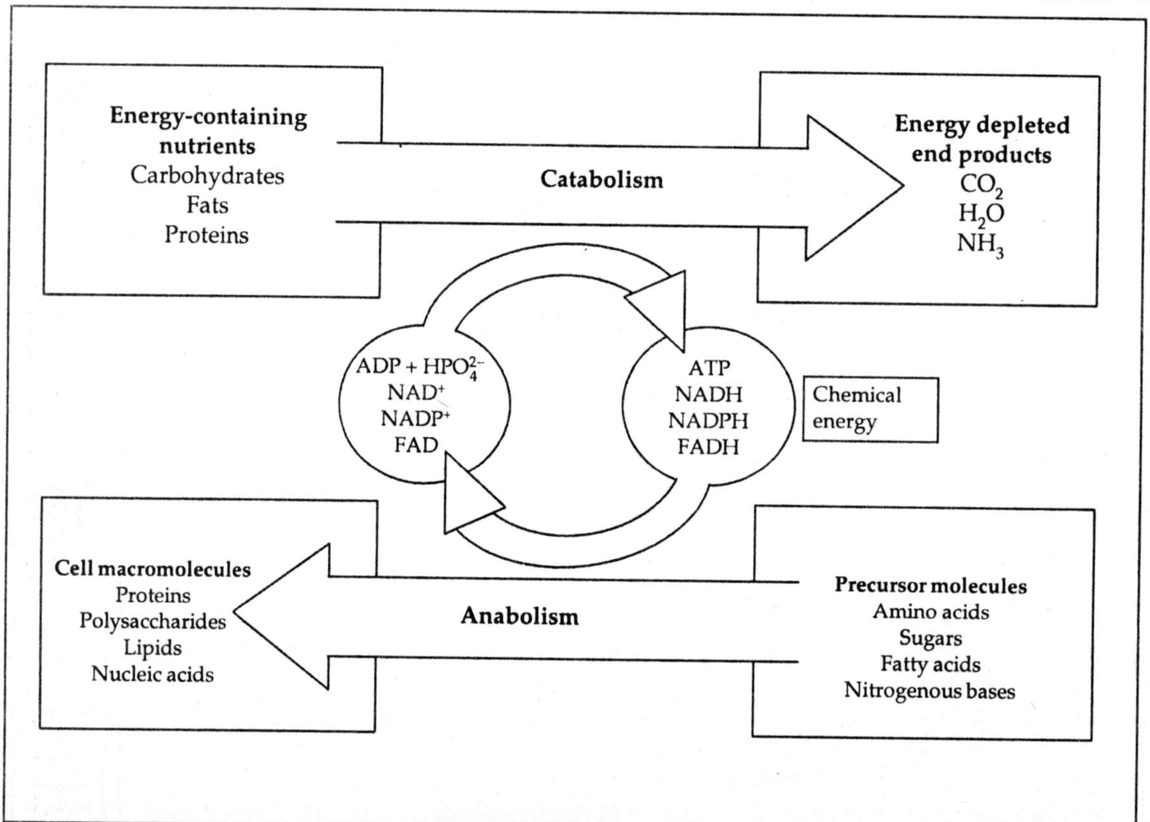

Fig. 1.46: Energy relationships between catabolic and anabolic pathways : Catabolic pathways deliver chemical energy in the form of ATP, NADH, NADPH and $FADH_2$. These energy carriers are used in anabolic pathways to convert small precursor molecules into cell macromolecules.

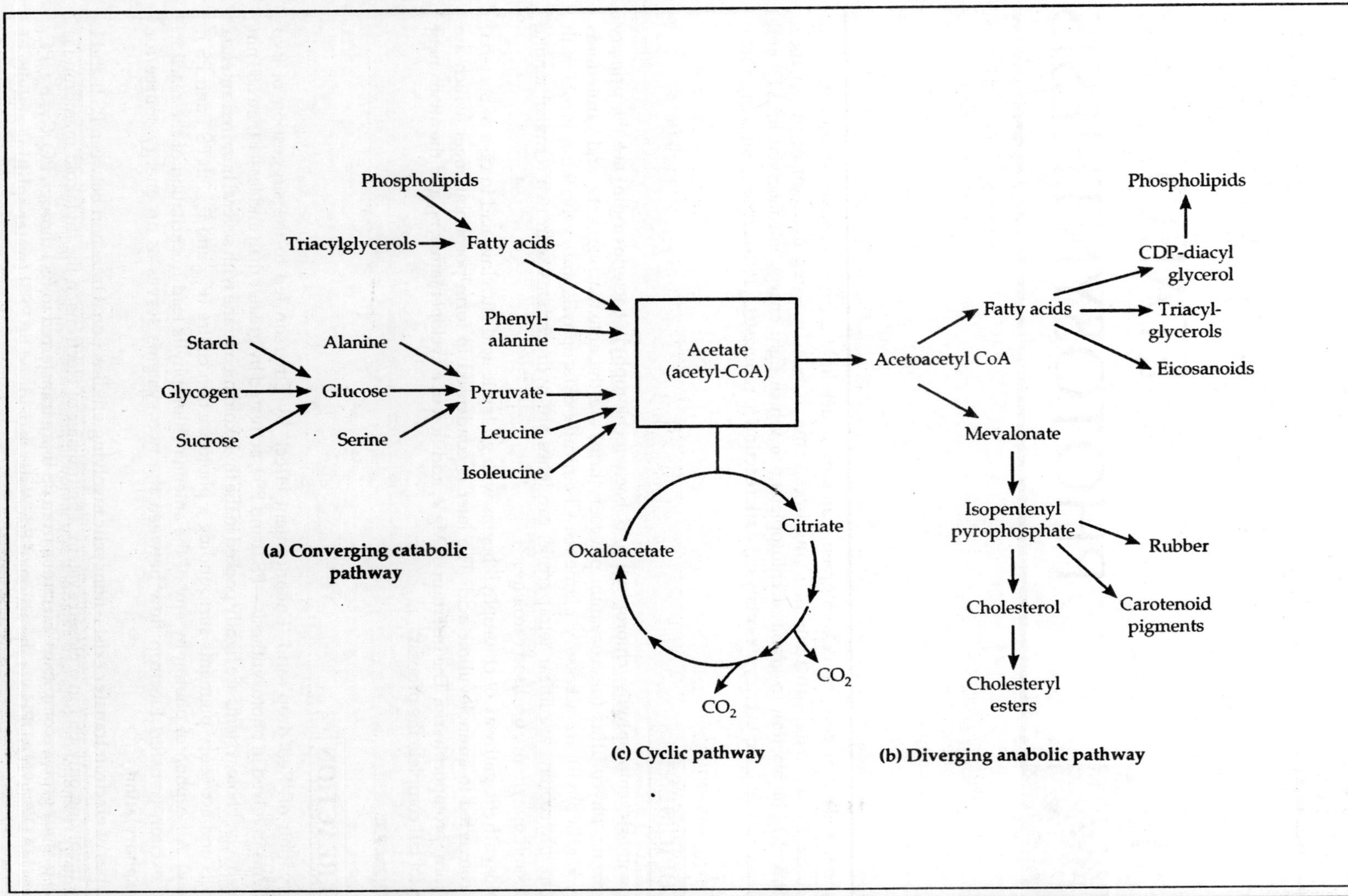

Fig. 1.47: Three types of nonlinear metabolic pathways : (a) Converging catabolic pathway, **(b)** Diverging anabolic pathway and **(c)** Cyclic pathway, in which one of the starting materials (oxaloacetate) is regenerated and re-enters the pathway. Acetate, a key metabolic intermediate, is the breakdown product of a variety of fuels (a), serves as precursor of a number of products (b), consumed in (c).

CHAPTER

2

CARBOHYDRATE METABOLISM: PHOTOSYNTHESIS

Photosynthesis is an anabolic process in green plants by which light energy is converted into chemical energy, necessary for vital functions of living organisms. It is a two-step mechanism involving a photochemical **light reaction** in grana of chloroplastid in which light energy is conserved in ATP and NADPH; these are utilised in enzyme-catalysed **dark reaction** to reduce CO_2 into carbohydrates in stroma.

Figure: 2.1

PHOTOPIGMENTS

In photosynthesis, **chlorophylls** (chlorophyll a, b, bacteriochlorophyll, bacterioviridin) are the primary photopigments, **phycobilins** (phycocyanin, phycoerythrin) are the antenna pigments, and **carotenoids** (carotene, xanthophyll) are accessory pigments. Chemically chlorophylls are cyclic tetrapyrrole with magnesium, phycobilins are linear tetrapyrrole, carotenes are hydrocarbons with two terminal ionone rings and xanthophylls are oxidised carotenes.

The **biosynthetic pathway** of chlorophyll begins with glutamic acid (succinic acid in case of bacteria) which is converted to aminolevulinic acid. The later is condensed to form porphobilinogen which are linked to form protoporphyrin. The insertion of Mg^{++}, addition of a cyclopentane ring and the attachment of the phytol tail complete the process.

Figures: 2.2 – 2.4

LIGHT REACTION

The observation of 'Red drop' and 'Enhancement effect' by Emerson led to the suggestion of two **photosystems** involved in photosynthesis—PS II and PS I, activated by greater than and lesser than 680 nm light respectively. Photosystems are tightly packed in the thylakoid membrane with several hundred antenna chlorophylls and accessory pigments surrounding a photoreaction centre (P_{700} and P_{680} in PS I and PS II respectively). Absorption of photon by any of the antenna chlorophylls lead to excitation of the reaction centre by exciton transfer. Electron flow between the two systems serves as an energy source for **photophosphorylation**.

Two types of **electron transfer** exist—**noncyclic** involving unidirectional transport between PS II and I and **cyclic** involving only PS I in a closed system. The transfer of electrons in the thylakoid membrane is carried out by four protein complexes. Electrons from PS II are transferred to PS I through PQ, Cyt b_6f, PC. PS I reduces $NADP^+$ to NADPH via the action of Fd. Water is oxidised by PS II (**water-splitting activity** of the oxygen-evolving complex) and protons are released in the lumen ; protons are also pumped into lumen by PQ and Cyt b_6f and their diffusion down the electrochemical energy gradient is used to synthesize ATP.

Figures: 2.5 – 2.9

DARK REACTION

Different pathways exist for CO_2 assimilation—**C_3 pathway** (Calvin cycle), **C_4 pathway** and **CAM.**

CALVIN CYCLE

Calvin cycle is called **C_3 cycle** since the early product in this pathway is a 3-carbon compound (3PGA). The cycle proceeds in three phases—**carboxylation, reduction** and **regeneration.** In the first phase, CO_2 is fixed by RuBP (C_5) to yield two molecules of 3PGA (C_3) with the help of **RUBISCO.** The second phase involves the reduction of 3PGA to 3PGAld with the help of photochemically derived ATP and NADPH. In the third phase 3PGAld through a series of intermediates (C_3 = DHAP ; C_4 = Erythrose-4-P; C_5 = Xylulose-5-P, Ribose-5-P, Ribulose-5-P ; C_6 = Fructose-1-6-P, Fructose-6-P ; C_7 = Sedoheptulose-1-7-P, Sedoheptulose-7-P) regenerate RuBP. For each molecule of CO_2 fixation, three molecules ATP and two molecules NADPH are consumed which are produced in light reaction. Several enzymes of Calvin cycle are also activated by light, mediated by thioredoxin.

From Calvin cycle, triose phosphate (C_3) is drained to **synthesize sucrose and starch** through hexose-phosphate in cytosol and chloroplast respectively, determined partly by Pi-concentration. The Pi-triose **antiport system** of the inner chloroplast membrane facilitates the transfer of stromal DHAP into cytosol to synthesize sucrose. The same antiporter can transport ATP and NADPH from chloroplast to cytosol through 3-PGA formation from 3PGAld. Triose-phosphate may be converted to sugar nucleotide (UDP-glucose), glucuronate and vitamin-C.

Figures: 2.10 – 2.33

C_4 PATHWAY

Some plants like maize, sorghum with **Kranz leaf anatomy** (chloroplast present both in mesophyll and bundle sheath cells) possess C_4 pathway where the early product of CO_2 fixation is C_4 acid, also called **HSK pathway.** In mesophyll chloroplast, CO_2 is accepted by PEP with the help of efficient **PEP carboxylase** to produce oxaloacetate and then to malate or aspartate which will diffuse to bundle sheath cell. Decarboxylation of four-carbon acid generates three-carbon acid and a high conc. of CO_2 which is utilised in Calvin cycle to synthesize sucrose and starch. The residual three-carbon acid transports back to the mesophyll cell where PEP is regenerated. The stoichiometry of C_4 path is 5ATP and 2NADPH per molecule of CO_2 fixed.

The **three variants** (NAD-malate type, NADP-malate type, PCK type) of the C_4 path differ principally in the nature of C_4 acid (Malate or Aspartate), C_3 acid (Pyruvate or Alanine) and the nature of the enzyme that catalyzes the decarboxylation step. C_4 plants are **photosynthetically more efficient** than C_3 plants in term of crop productivity.

Figures: 2.34 – 2.42

CAM

CAM is an adaptation of **succulants** to xerophytic environment where primary carboxylation and carbohydrate synthesis are separated with respect to time, i.e., day and night. As stomata remain closed at daytime to reduce water loss, CO_2 is fixed and stored as malic acid at night (**Dark acidification**) ; during daytime decarboxylation (**Light deacidification**) and refixation of internally released CO_2 via Calvin cycle occur to synthesize carbohydrate.

Figures: 2.43 – 2.45

PHOTORESPIRATION

Oxygenase activity of RUBISCO leads to the existence of photorespiration process in higher plants where glycolate, a C_2 acid, is an early product, therefore, called **C_2-cycle.** The operation of this cycle involves co-operative interaction among **three organelles** and oxygen is consumed in each organelle—chloroplast (RuBP-oxygenase), peroxisome (glycolate oxidase) and mitochondrion (cytochrome-oxidase). Glycolate is salvaged by conversion to glycine and serine and then to glycerate which re-enters the Calvin cycle. Large amount of fixed carbon is lost by reoxidation to CO_2, thus it is an **wasteful process.**

Figures: 2.46 – 2.49

BACTERIAL PHOTOSYNTHESIS

In some autotrophic bacteria (purple and green), photosynthesis occurs in chromatophores involving the pigment bacteriochlorophyll/bacterioviridin using reductant other than H_2O (e.g., H_2S) and do not evolve O_2, therefore, called **anoxygenic photosynthesis**. In **purple bacteria**, light energy drives electrons from reaction centre P_{870} and is cycled back through pheophytin, quinone and cytochromes. **Green bacteria** have two routes for electrons driven by excitation of P_{840}– cyclic involving quinone-cytochromes and a noncyclic through Fd^--NAD^+. Apart from Calvin cycle of **CO_2 fixation**, a reductive carboxylic acid cycle exists in bacteria.

Figures: 2.50 – 2.55

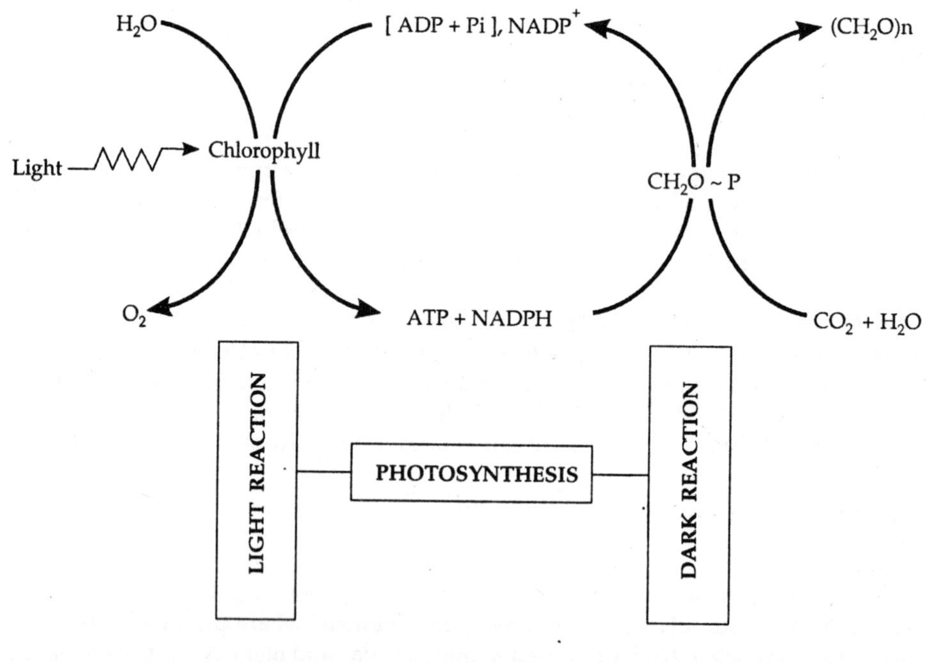

Fig. 2.1: Light and Dark reactions of Photosynthesis: Light is required to generate ATP and NADPH. The ATP and NADPH are consumed in dark reactions which reduce carbon dioxide to carbohydrate.

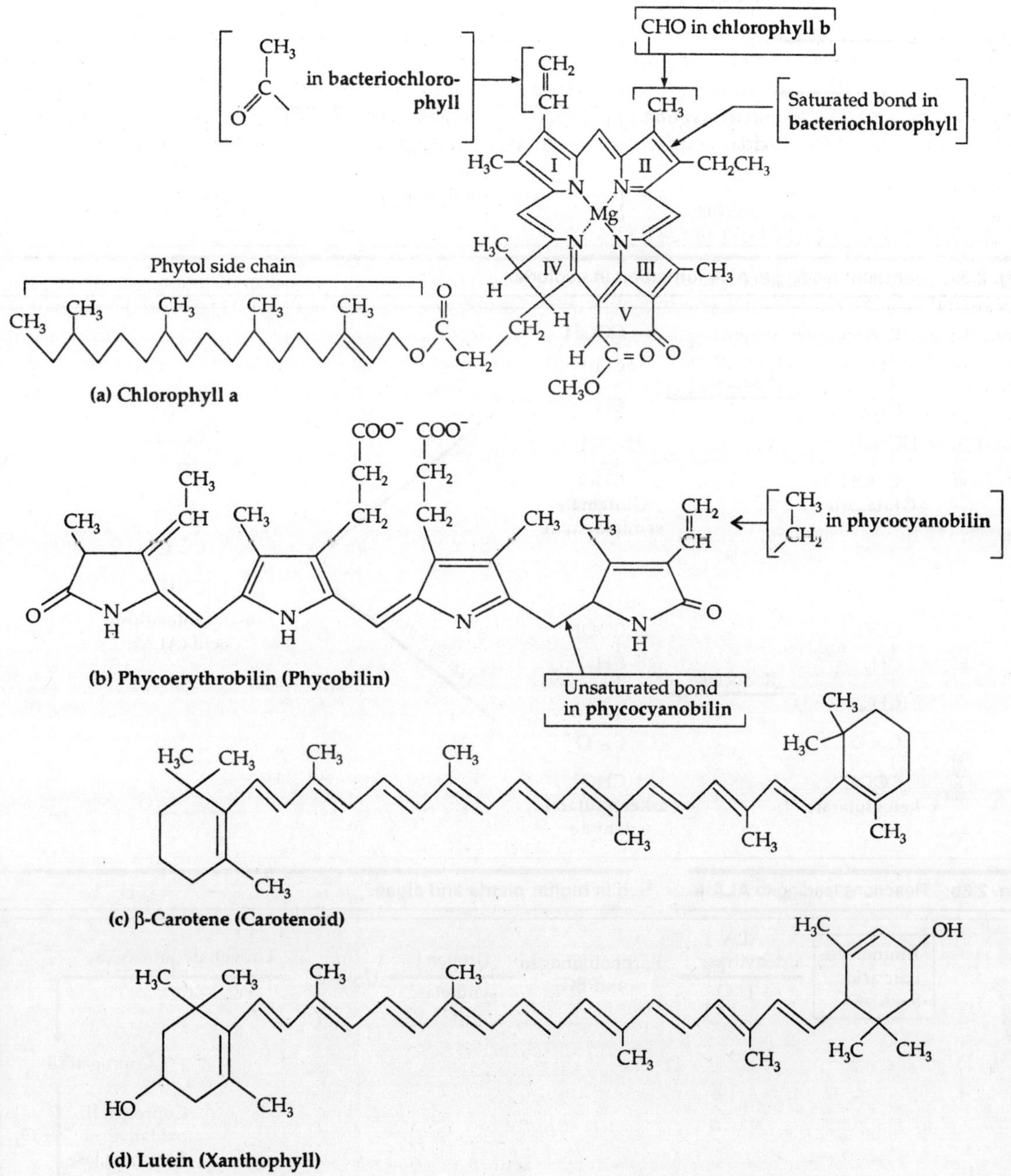

Fig. 2.2: Primary and Secondary photopigments : (a) Chlorophyll – Chlorophylls a, b and bacteriochlorophyll are the primary pigments capturing light energy. **(b) Phycobilin** – Phycoerythrobilin and phycocyanobilin are the antenna pigments in cyanobacteria and red algae. **(c) Carotenoid** – β-carotene is the accessory pigment. **(d) Xanthophyll** – Lutein is another accessory photopigment.

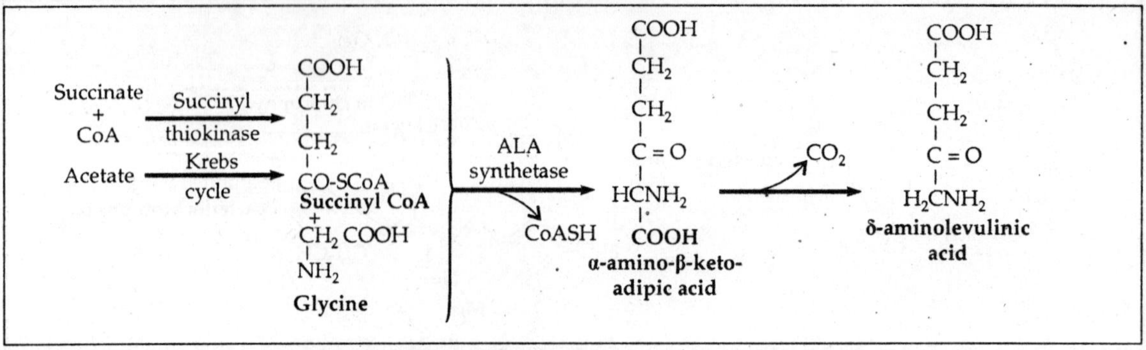

Fig. 2.3a: Reactions leading to **ALA formation in bacteria.**

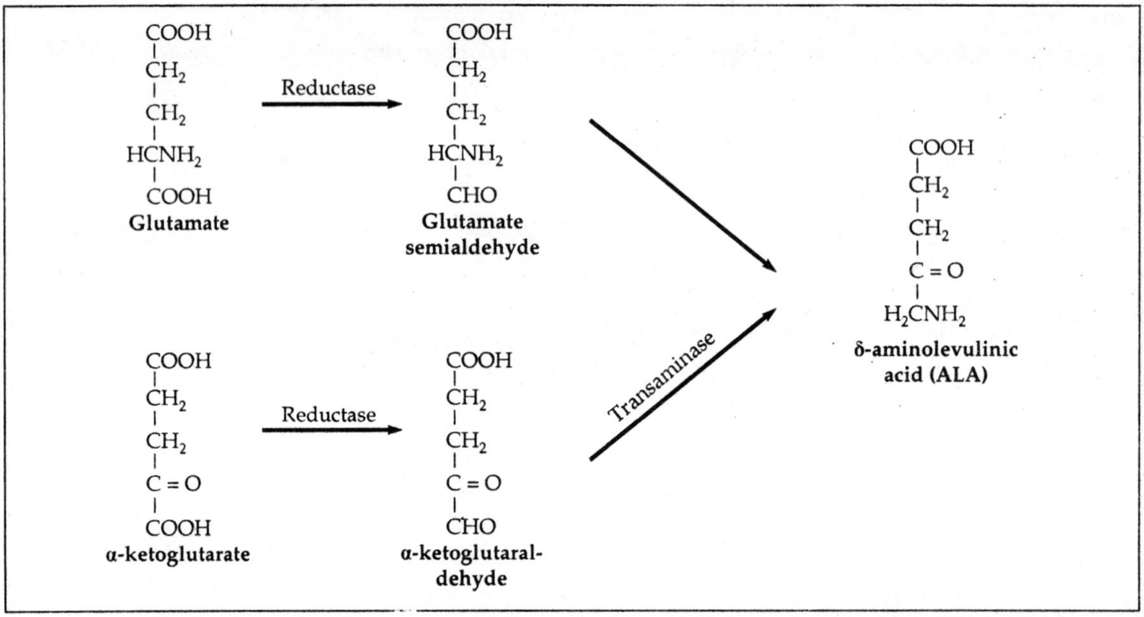

Fig. 2.3b: Reactions leading to **ALA fo… n in higher plants and algae.**

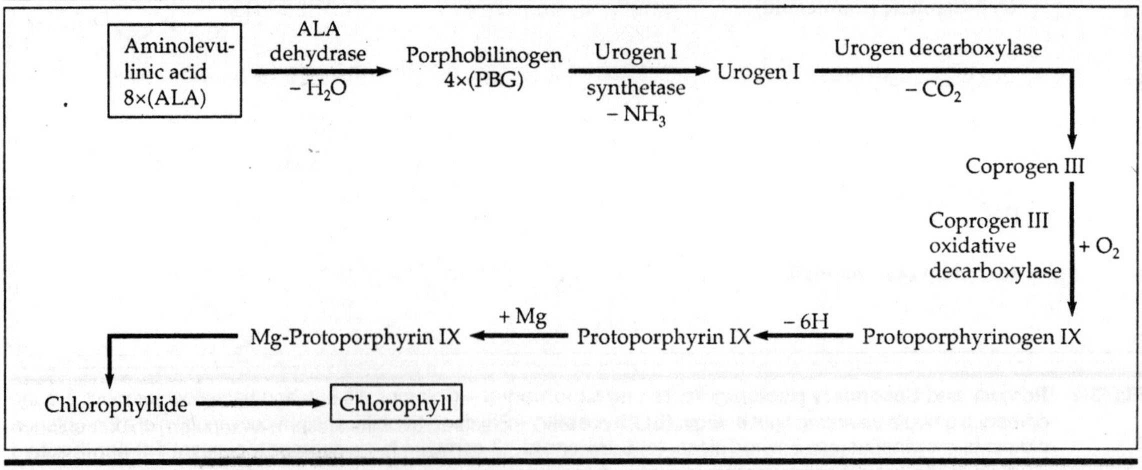

Fig. 2.3c: Reactions leading to **Chlorophyll synthesis from ALA.**

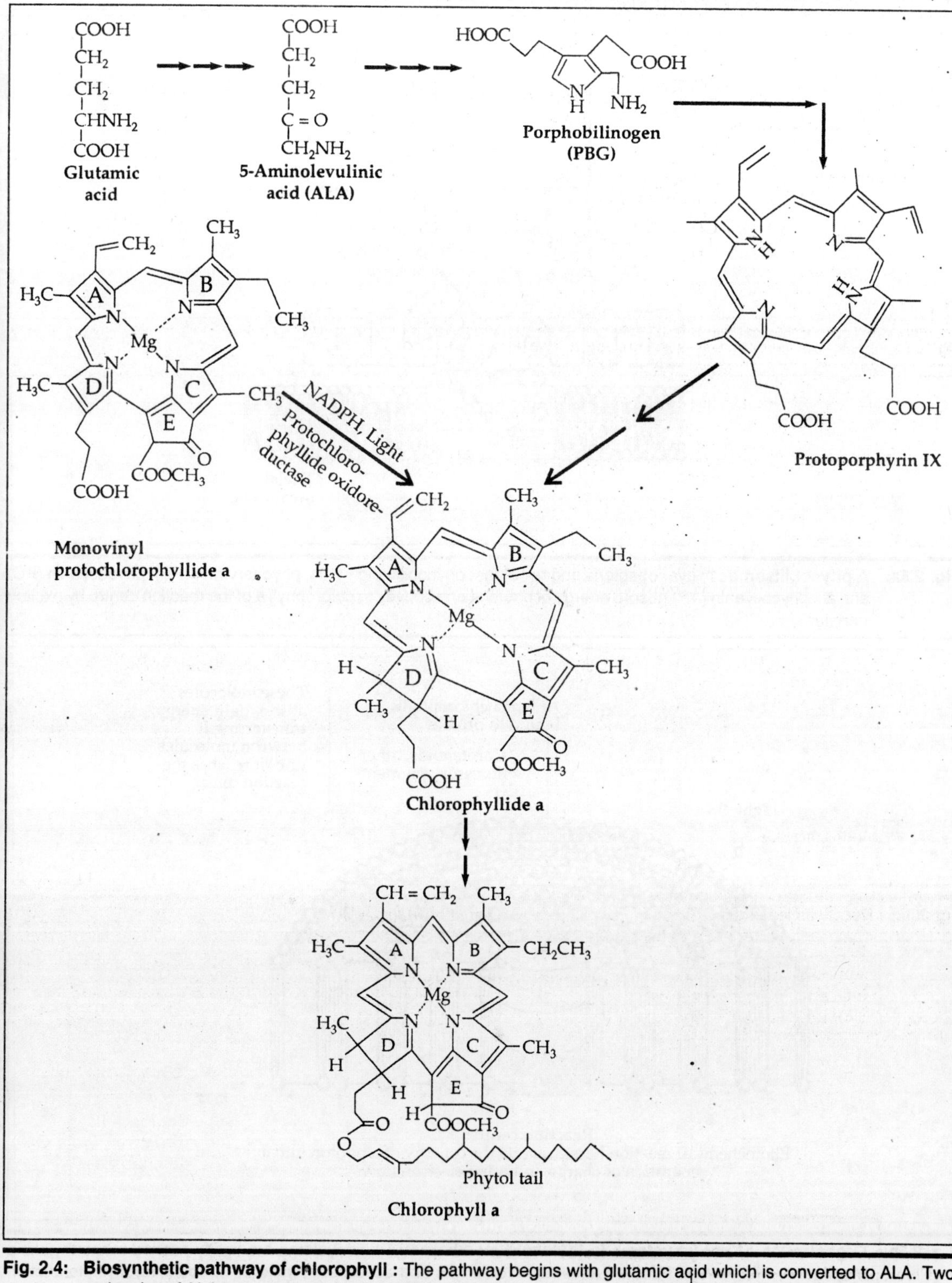

Fig. 2.4: Biosynthetic pathway of chlorophyll : The pathway begins with glutamic acid which is converted to ALA. Two molecules of ALA are condensed to form PBG. Four PBG molecules are linked to form protoporphyrin IX. Insertion of Mg and light-dependent cyclization of ring E, the reduction of ring D and phytol tail attachment complete the process.

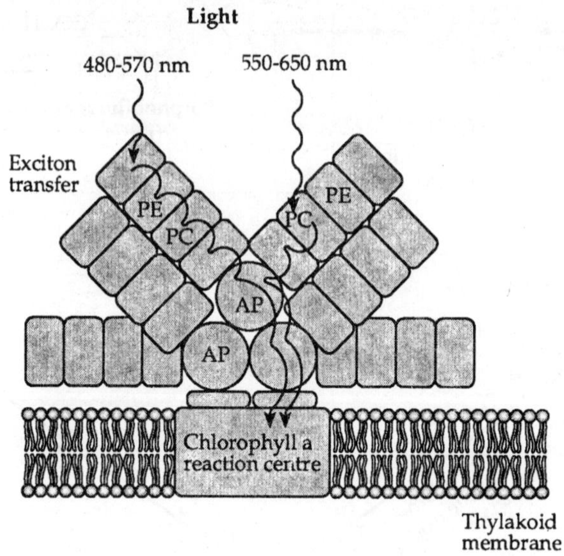

Fig. 2.5a: A phycobilisome : In cyanobacteria and red algae, phycobilin pigments, phycoerythrin (PE), phycocyanin (PC) and allophycocyanin (AP) absorb energy of photons and convey to chlorophyll a of the reaction centre by exciton transfer.

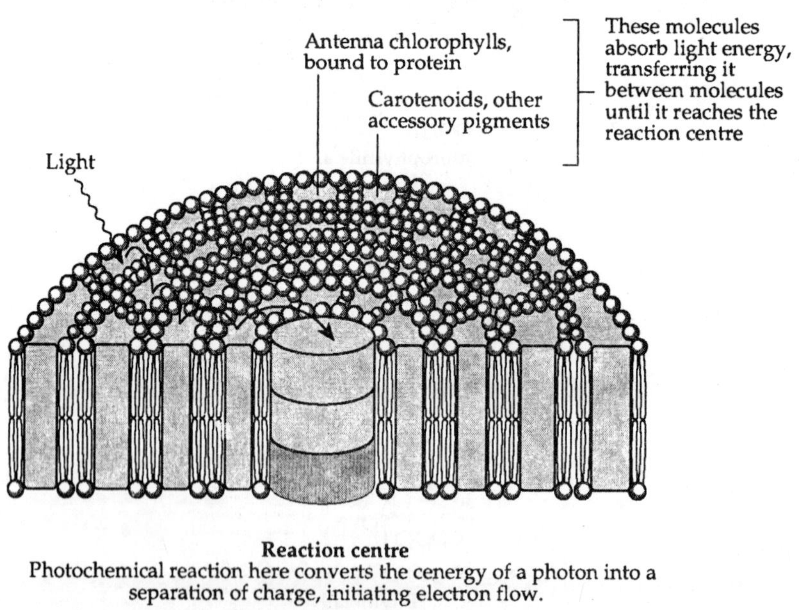

Fig. 2.5b: Organization of photosystems in thylakoid membrane : Photosystems are tightly packed in the thylakoid membrane, with several hundred antenna chlorophylls and accessory pigments surrounding a photoreaction centre. Absorption of a photon by any of the antenna chlorophylls leads to the excitation of the reaction centre by exciton transfer.

(1) Light excites an antenna molecule of chlorophyll or accessory pigment, raising an electron to a higher energy level.

(2) The excited antenna molecule passes energy to adjacent chlorophyll molecule by resonance energy transfer and excites it.

(3) The energy is transferred to a reaction-centre chlorophyll and excites it.

(4) The excited reaction-centre chlorophyll passes an electron to an electron acceptor.

(5) The electron hole in the reaction centre is filled by an electron from an electron donor.

The absorption of a photon has caused separation of charge in the reaction centre

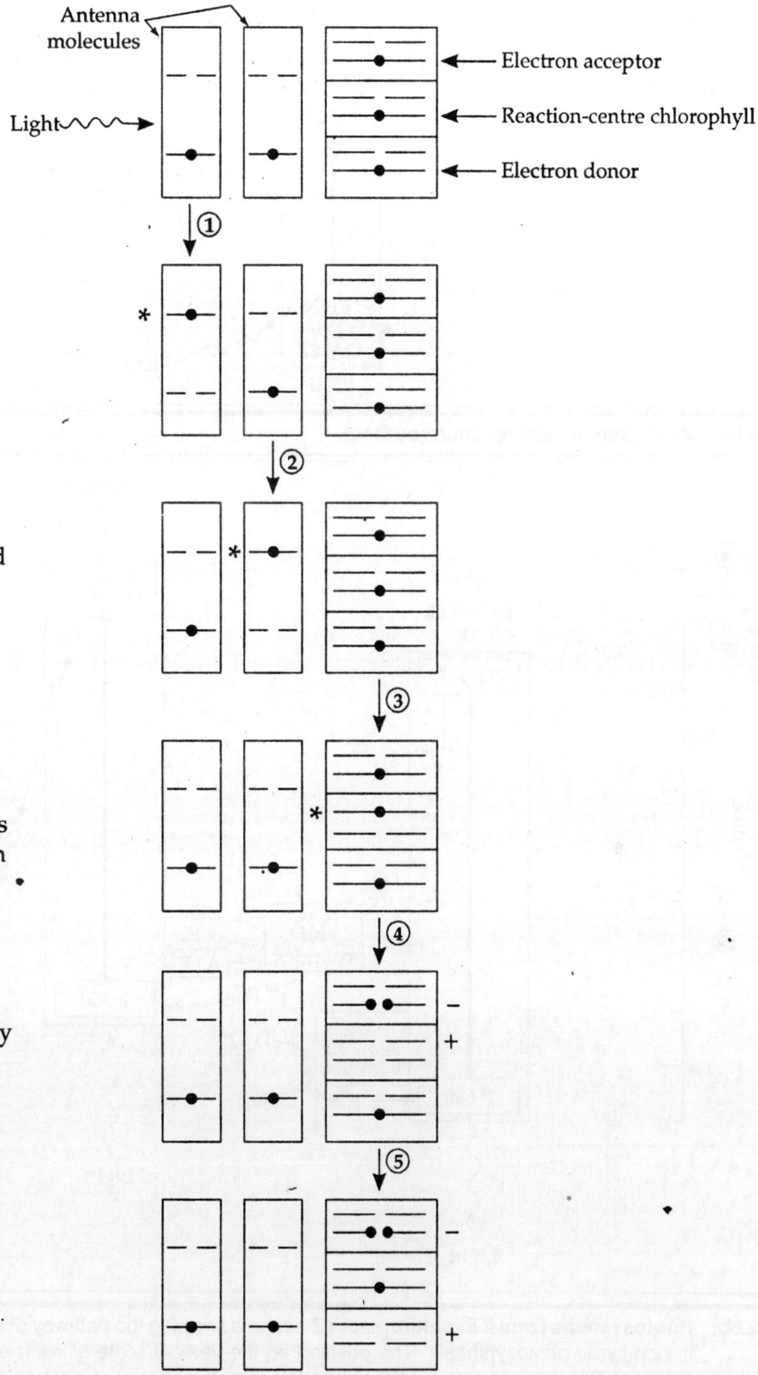

Fig. 2.5c: Exciton transfer : Conversion of energy from an absorbed photon into separation of charges of the reaction centre. Step one may be repeated a number of times between successive antenna molecules until a reaction-centre chlorophyll is reached. The asterisk (*) represents the excited state of an antenna molecule.

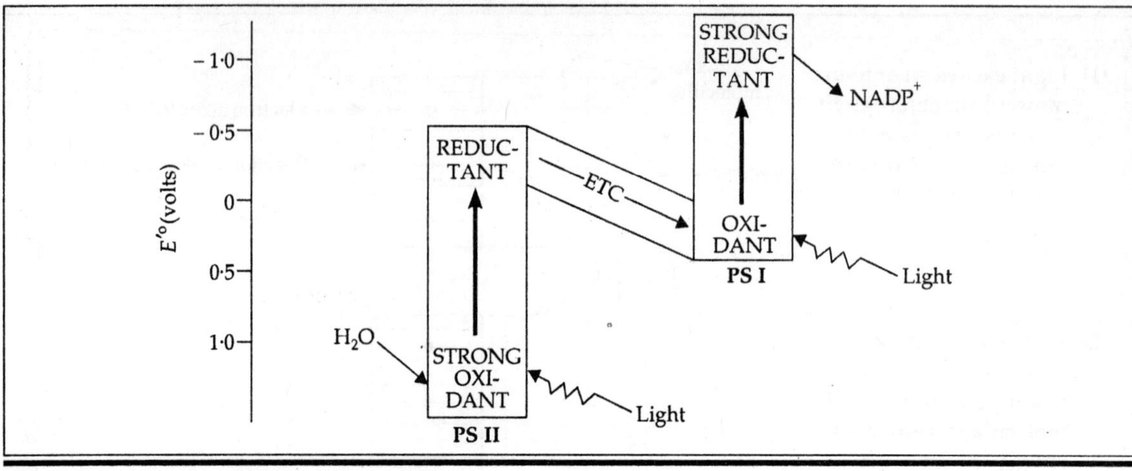

Fig. 2.6a: Z-scheme of light reaction (outline).

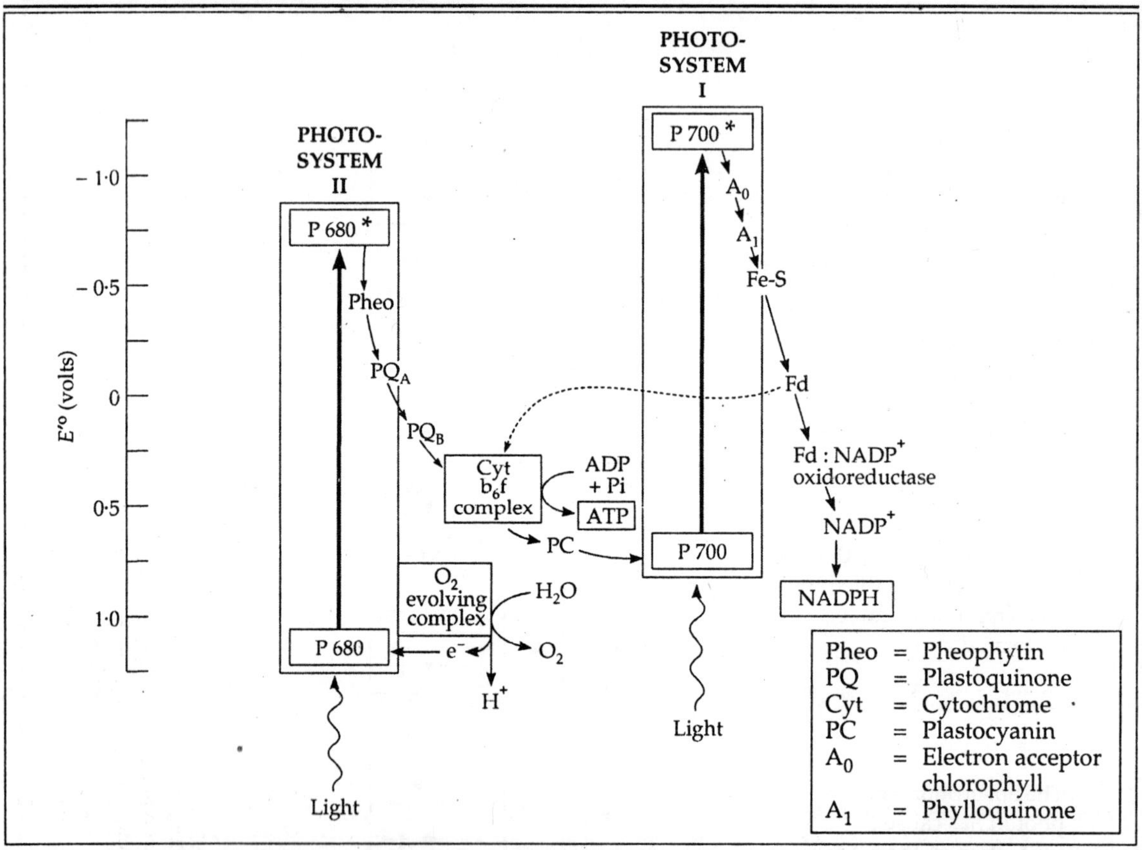

Fig. 2.6b: Photosystems I and II in chloroplast : Z-scheme showing the pathway of electron transfer from H_2O to $NADP^+$ in non-cyclic photosynthesis. The position on the vertical scale of each electron carrier reflects its reduction potential. Photons absorbed in PS I and PS II raise the energy of electron derived from H_2O to the energy level required to reduce $NADP^+$ to NADPH. One photon is required per electron in each photosystem. After excitation, the high energy electrons flow "downhill" through the carrier chains. Protons move across the thylakoid membrane during the water-splitting reaction and during electron transfer through the cyt b_6f complex, producing the proton gradient that is central to ATP formation. The dashed arrow is the path of cyclic electron transfer in which only PS I is involved; electrons return via the cyclic pathway to PS I, instead of reducing $NADP^+$ to NADPH.

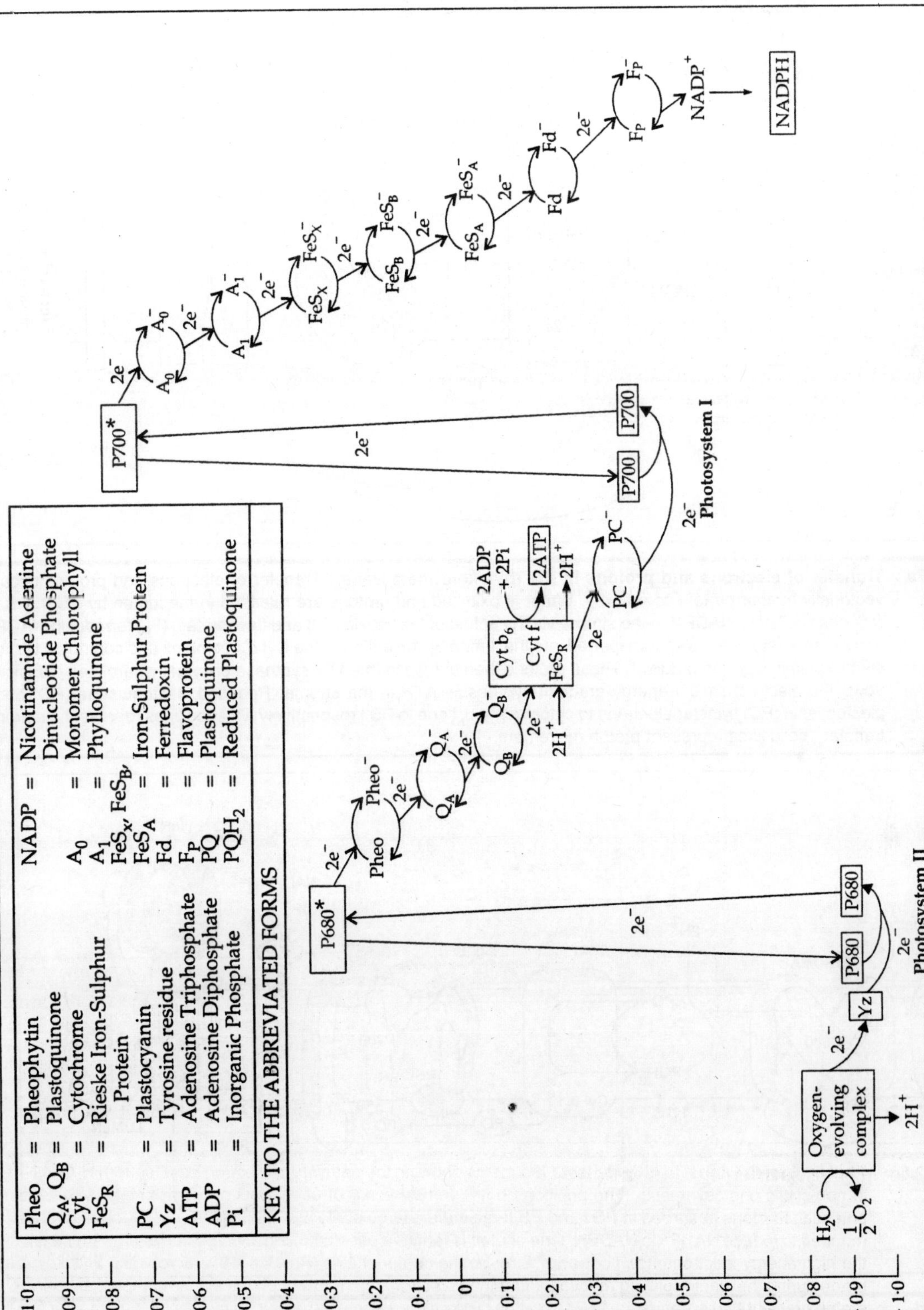

Fig. 2.6c: Detailed Z-scheme for oxygen evolving photosynthetic organisms.

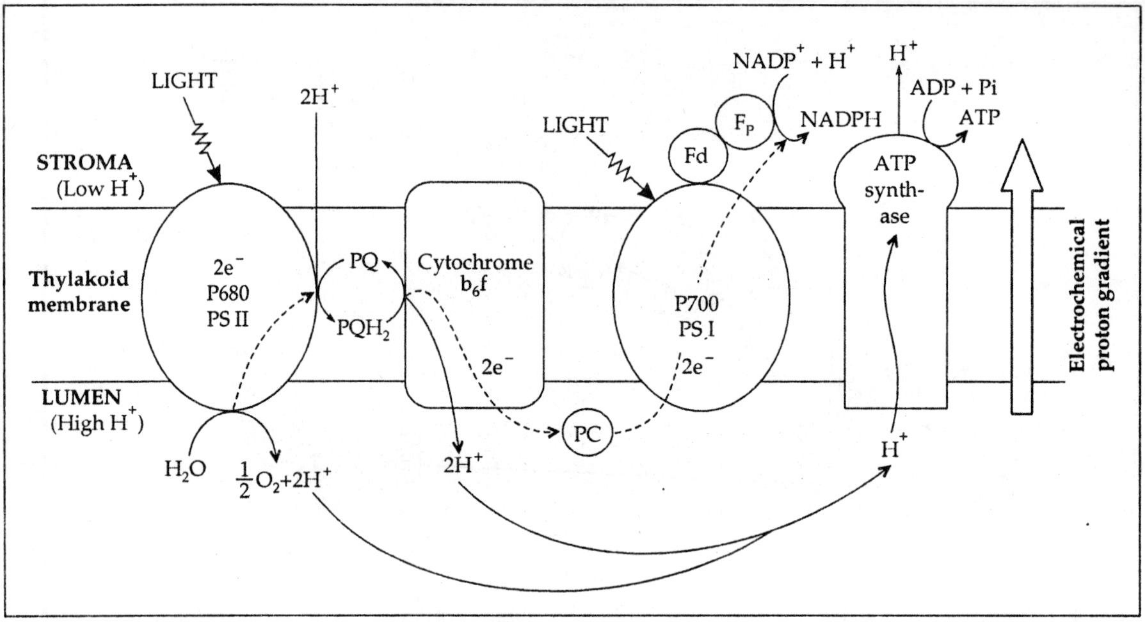

Fig. 2.7a : Transfer of electrons and protons in the thylakoid membrane : Transfer of electrons and protons occur vectorially by four protein complexes. Water is oxidized and protons are released in the lumen by PS II. PS I reduces $NADP^+$ to NADPH in the stroma via the action of ferredoxin (Fd) and flavoprotein (F_P) ferredoxin-NADP reductase. Protons are also transported into the lumen by the action of the Cyt b_6f complex and contribute to the electrochemical proton gradient. These protons then diffuse to the ATP synthase enzyme, where their diffusion down the electrochemical energy gradient synthesize ATP in the stroma. Reduced plastoquinone (PQ) and plastocyanin (PC) transfer electrons to cytochrome b_6f and to PS I respectively. Dashed lines represent electron transfer ; solid lines represent proton movement.

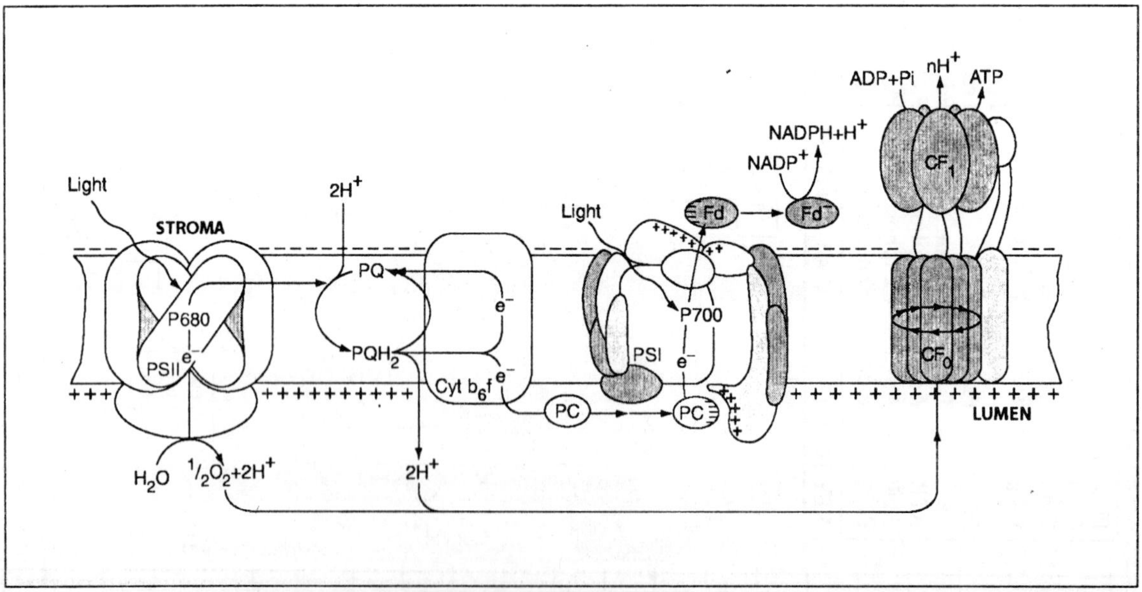

Fig. 2.7b: Components of photosynthetic electron transport system : PS II, PS I, cytochrome b_6f complex, ATP synthase (CF_1 and CF_0) in the thylakoid membrane. Due to electron transport a proton gradient is established across the thylakoid membrane which is utilised in the formation of ATP by ATP synthase.

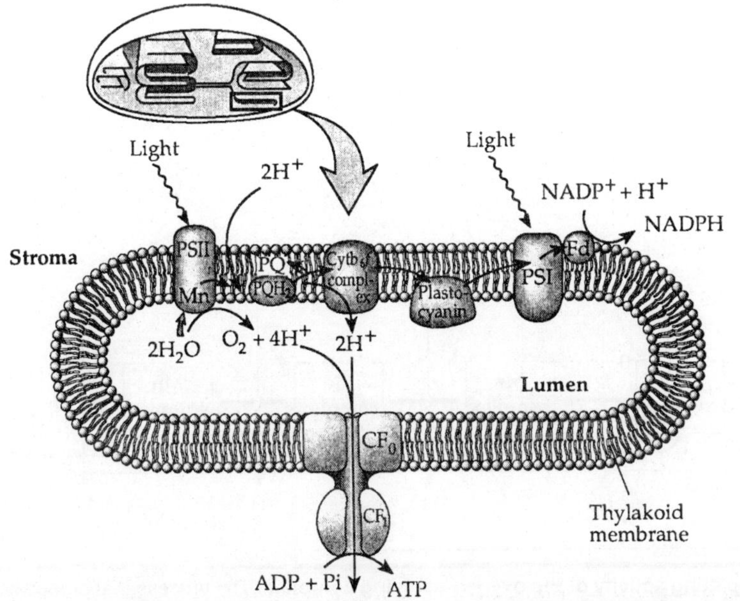

Fig. 2.8a: Proton and electron circuits in thylakoids : Electrons move from H_2O through PS II, the intermediate chain of carriers, PS I and finally to $NADP^+$. Protons are pumped into the thylakoid lumen by the flow of electrons through the chain of carriers between PS II and PS I, and re-enter the stroma through proton channels formed by F_o portion of the ATP synthase, designated CF_o in the chloroplast enzyme. The F_1 subunit, CF_1, catalyzes synthesis of ATP.

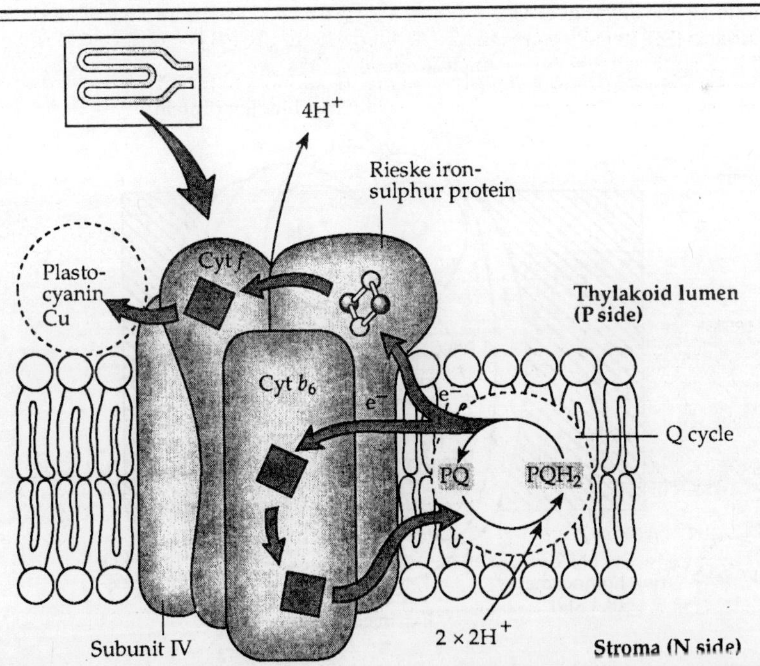

Fig. 2.8b: Flow of electron and proton through cytochrome b_6f complex : Reduced plastoquinone formed in PS II is oxidized by the cytochrome b_6f in a series of steps. One electron passes to the Fe-S centre of the Rieske protein and the other to the hemes of cytochrome b_6. The net effect is passage of electrons from PQH_2 to the soluble protein plastocyanin, which carries them to PS I.

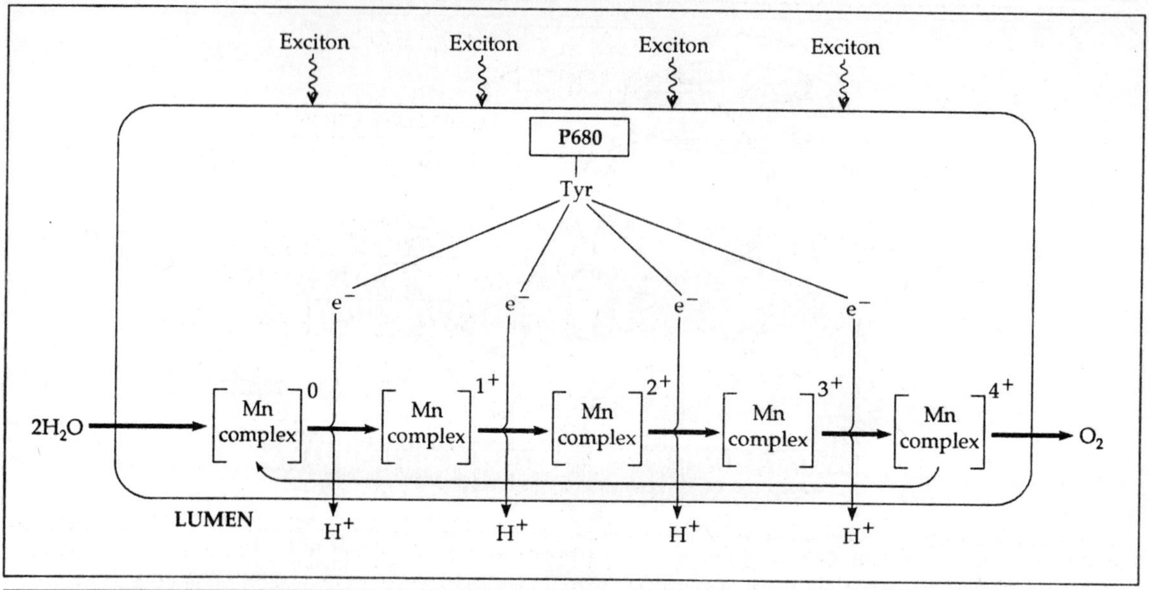

Fig. 2.9a: **Water-splitting activity of the oxygen-evolving complex :** The process that produces a four-electron oxidizing agent, a multinuclear centre with several Mn ions, in the water-splitting complex of PS II. The sequential absorption of four photons, each causing the loss of one electron from the Mn centre, produces an oxidizing agent that can take four electrons from two molecules of water, producing O_2. The electrons lost from the Mn centre pass one at a time to an oxidized Tyr residue in a PS II protein.

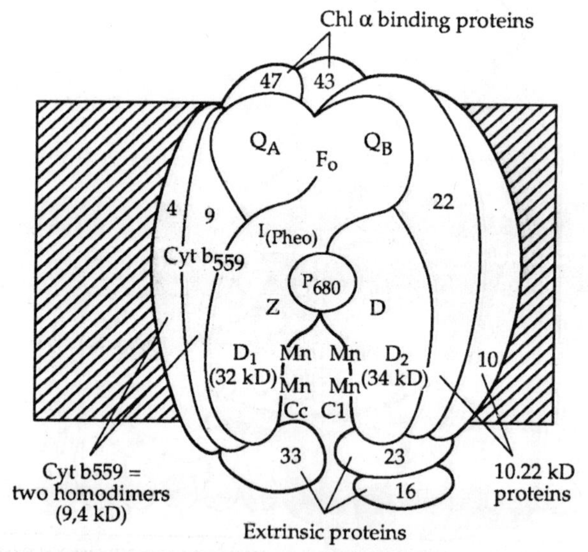

Fig. 2.9b: A model explaining the organization of **PS II polypeptides** according to Anderson (1988).

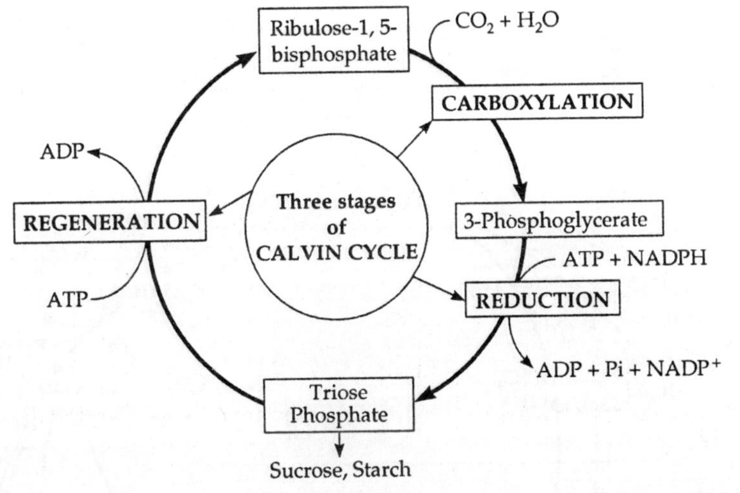

Fig. 2.10a: Three stages of Calvin cycle : (1) Carboxylation – CO_2 is covalently linked to a carbon skeleton. **(2) Reduction** – With the help of photochemically derived ATP and NADPH, glyceraldehyde and carbohydrate formed. **(3) Regeneration** – CO_2 acceptor ribulose–1, 5–bisphosphate resynthesised.

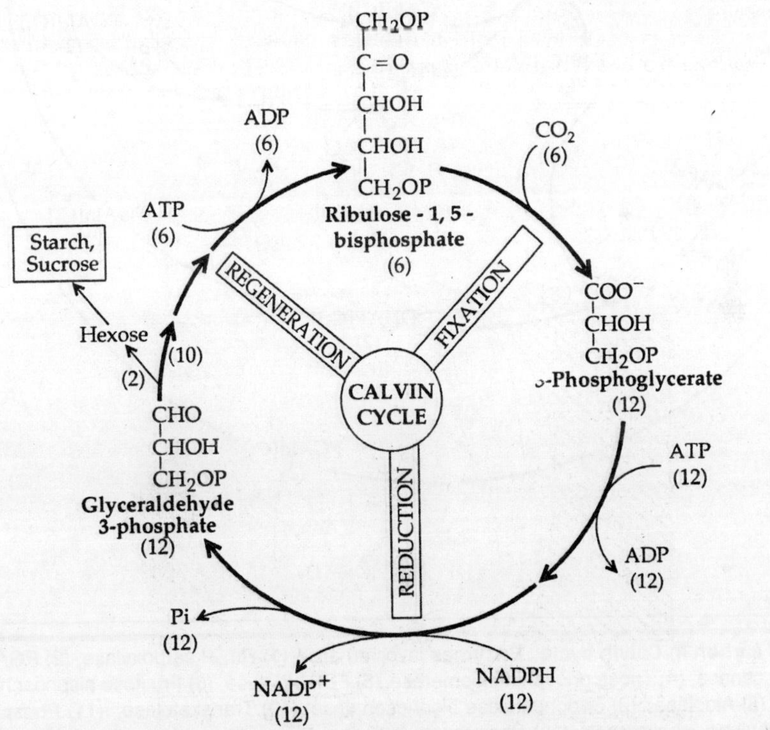

Fig. 2.10b: Three stages of CO_2 assimilation in photosynthetic organisms : Six molecules of CO_2 are fixed to permit net synthesis of two molecules of glyceraldehyde 3-phosphate which synthesize one molecule sucrose or enter into other carbohydrates like starch. Number of molecules in parenthesis indicate stoichiometries. The cycle is also called photosynthetic carbon reduction cycle. To synthesize one hexose molecule, 18 molecules ATP and 12 molecules NADPH are required.

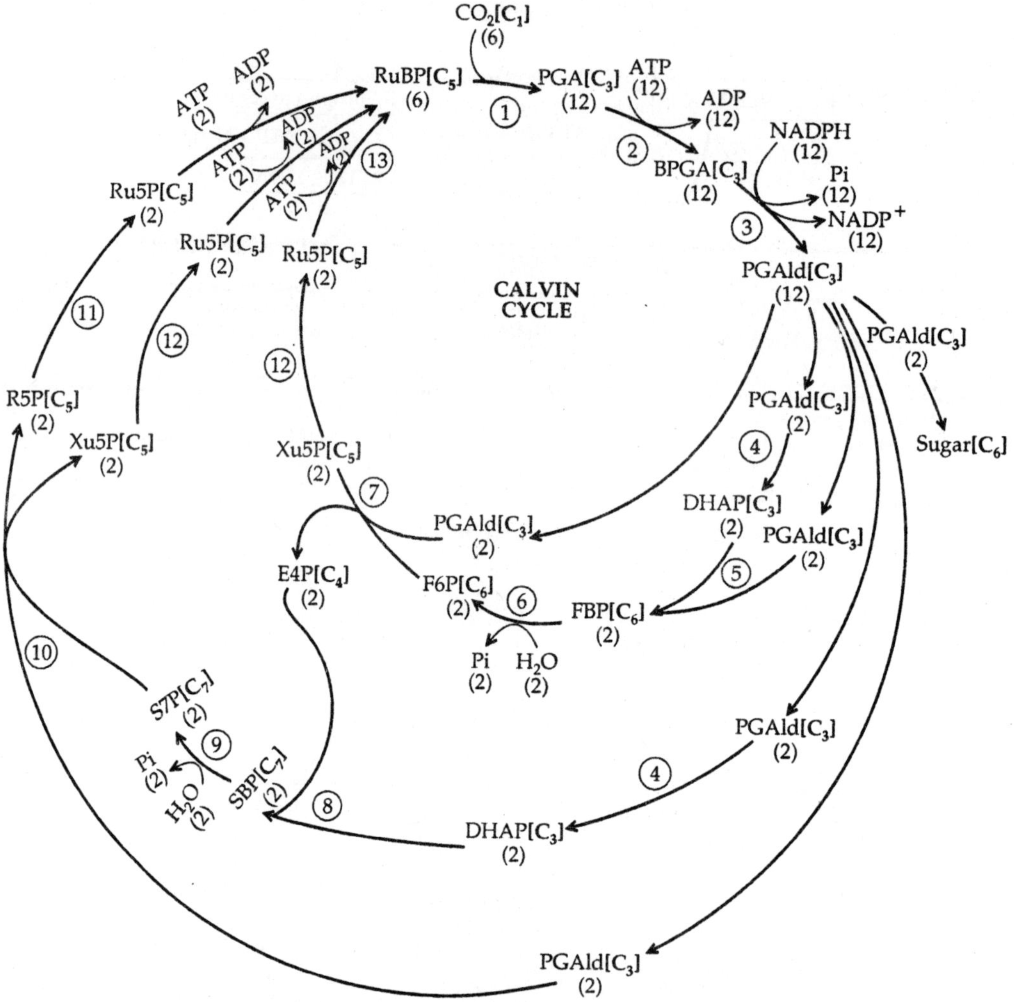

Fig. 2.10c: Path of carbon in Calvin cycle : Enzymes involved are—**(1)** RuBP carboxylase, (2) PGA kinase, (3) PGAld dehydrogenase, (4) Triose phosphate isomerase, **(5)** FBP aldolase, (6) Fructose bisphosphatase, (7) Transketolase, (8) Aldolase, (9) Sedoheptulose bisphosphatase, (10) Transketolase, (11) Phosphoribuloisomerase, (12) (Ribulose 3-epimerase, (13) Phosphoribulokinase. Number of carbon atom in the compound has been indicated within [ ]. Number of molecules involved in one molecule sugar (Glucose) synthesis has been indicated within ( ). **Abbreviations :** RuBP—Ribulose-1, 5-bisphosphate ; PGA—3-Phosphoglyceric acid ; **BPGA**—1, 3-Bisphosphoglycerate ; PGAld—3-Phosphoglyceraldehyde ; DHAP—Dihydroxyacetone phosphate ; **FBP**—Fructose-1, 6-bisphosphate ; F6P—Fructose-6-phosphate ; E4P—Erythrose-4-phosphate ; Xu5P—Xylulose-5-phosphate; SBP—Sedoheptulose-1, 7-bisphosphate ; S7P—Sedoheptulose-7-phosphate ; Ru5P—Ribulose-5-phosphate ; R5P—Ribose-5-phosphate.

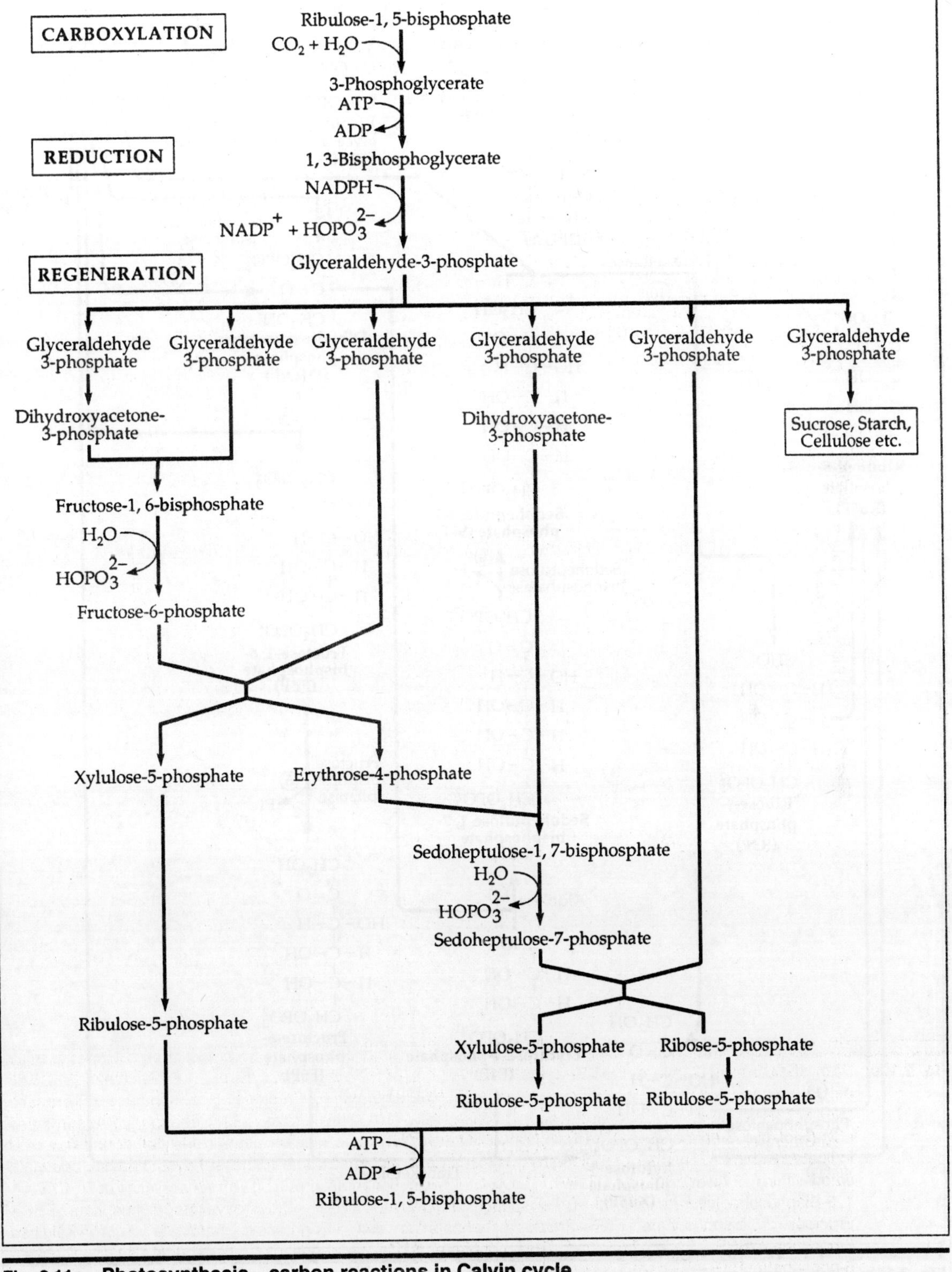

Fig. 2.11a: Photosynthesis—carbon reactions in Calvin cycle.

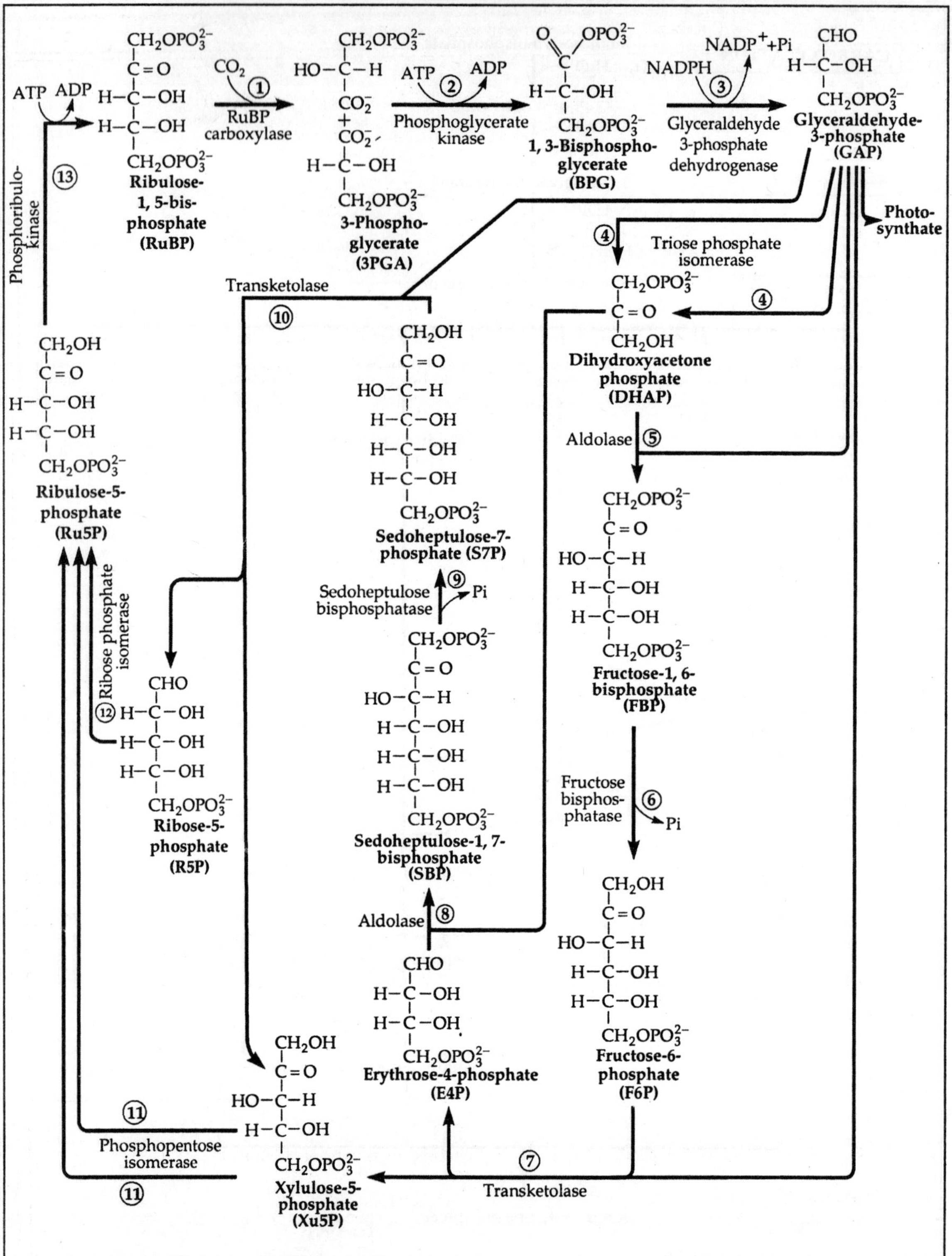

Fig. 2.11b: Dark Reactions (Calvin cycle).

Fig. 2.12: Reactions of the Calvin cycle and the corresponding enzymes.

(1) ***Ribulose-1, 5-bisphosphate carboxylase :***

Ribulose-1, 5-bisphosphate + CO_2 + H_2O ⟶ **2(3-Phosphoglycerate)** + $2H^+$

(2) ***3-phosphoglycerate kinase :***

3-Phosphoglycerate + ATP ⟶ **1, 3-Bisphosphoglycerate** + ADP

(3) ***NADP-glyceraldehyde-3-phosphate dehydrogenase :***

1, 3-Bisphosphoglycerate + NADPH + H^+ ⟶ **Glyceraldehyde-3-phosphate** + $NADP^+$ + $HOPO_3^{2-}$

(4) ***Triose phosphate isomerase :***

Glyceraldehyde-3-phosphate ⟶ **Dihydroxyacetone-3-phosphate**

(5) ***Aldolase :***

Glyceraldehyde-3-phosphate + **Dihydroxyacetone-3-phosphate** ⟶ **Fructose-1, 6-bisphosphate**

(6) ***Fructose-1, 6-bisphosphate phosphatase :***

Fructose-1, 6-bisphosphate + H_2O ⟶ **Fructose-6-phosphate** + $HOPO_3^{2-}$

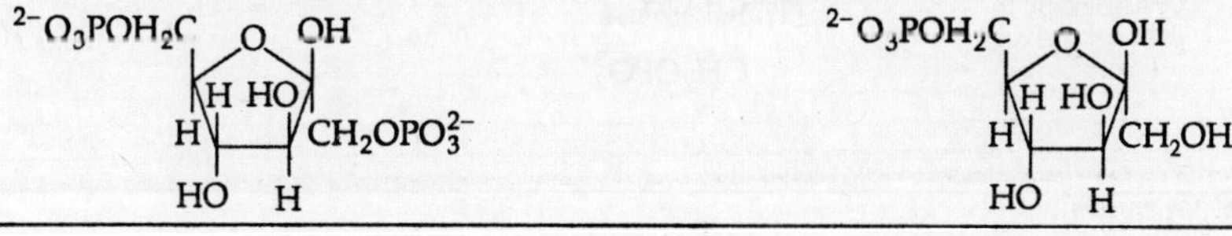

contd.

(7) ***Transketolase*** **:**

Fructose-6-phosphate + **Glyceraldehyde-3-phosphate** ⟶ **Erythrose-4-phosphate** + **Xylulose-5-phosphate**

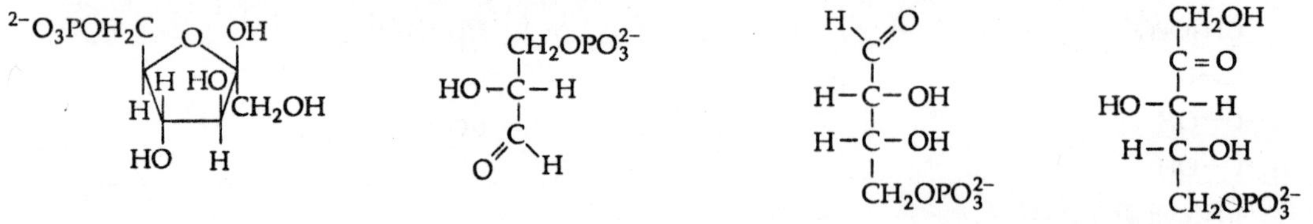

(8) ***Aldolase*** **:**

Erythrose-4-phosphate + **Dihydroxyacetone-3-phosphate** ⟶ **Sedoheptulose-1, 7-bisphosphate**

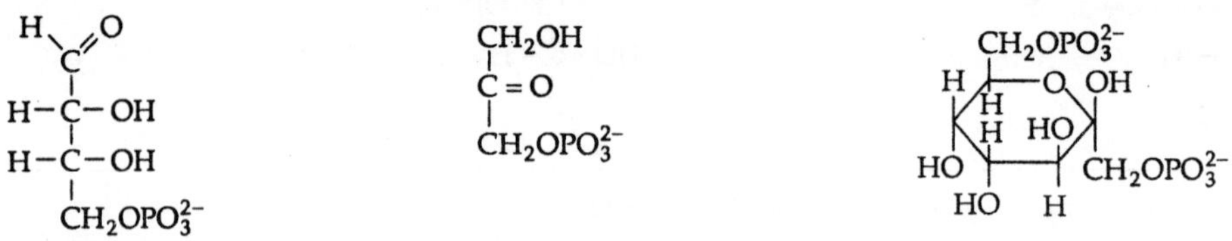

(9) ***Sedoheptulose-1, 7-bisphosphate phosphatase*** **:**

Sedoheptulose-1, 7-bisphosphate + H_2O ⟶ **Sedoheptulose-7-phosphate** + $HOPO_3^{2-}$

(10) ***Transketolase*** **:**

Sedoheptulose-7-Phosphate + **Glyceraldehyde-3-phosphate** ⟶ **Ribose-5-phosphate** + **Xylulose-5-phosphate**

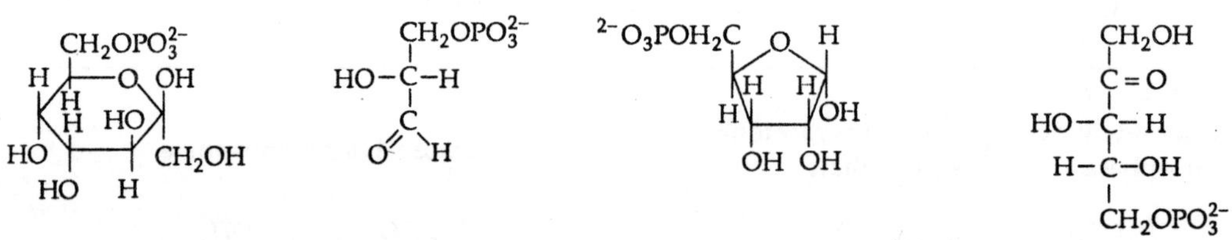

(11) ***Ribulose-5-phosphate epimerase*** **:**

Xylulose-5-phosphate ⟶ **Ribulose-5-phosphate**

contd.

(12) ***Ribose-5-phosphate isomerase :***

Ribose-5-phosphate ⟶ **Ribulose-5-phosphate**

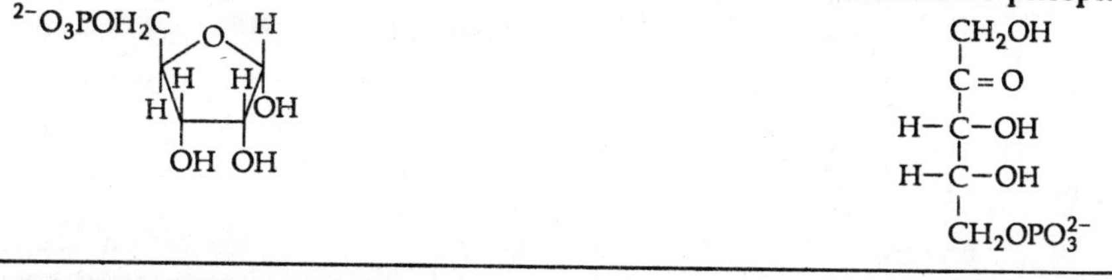

(13) ***Ribulose-5-phosphate kinase :***

Ribulose-5-phosphate + ATP ⟶ **Ribulose-1, 5-bisphosphate** + ADP

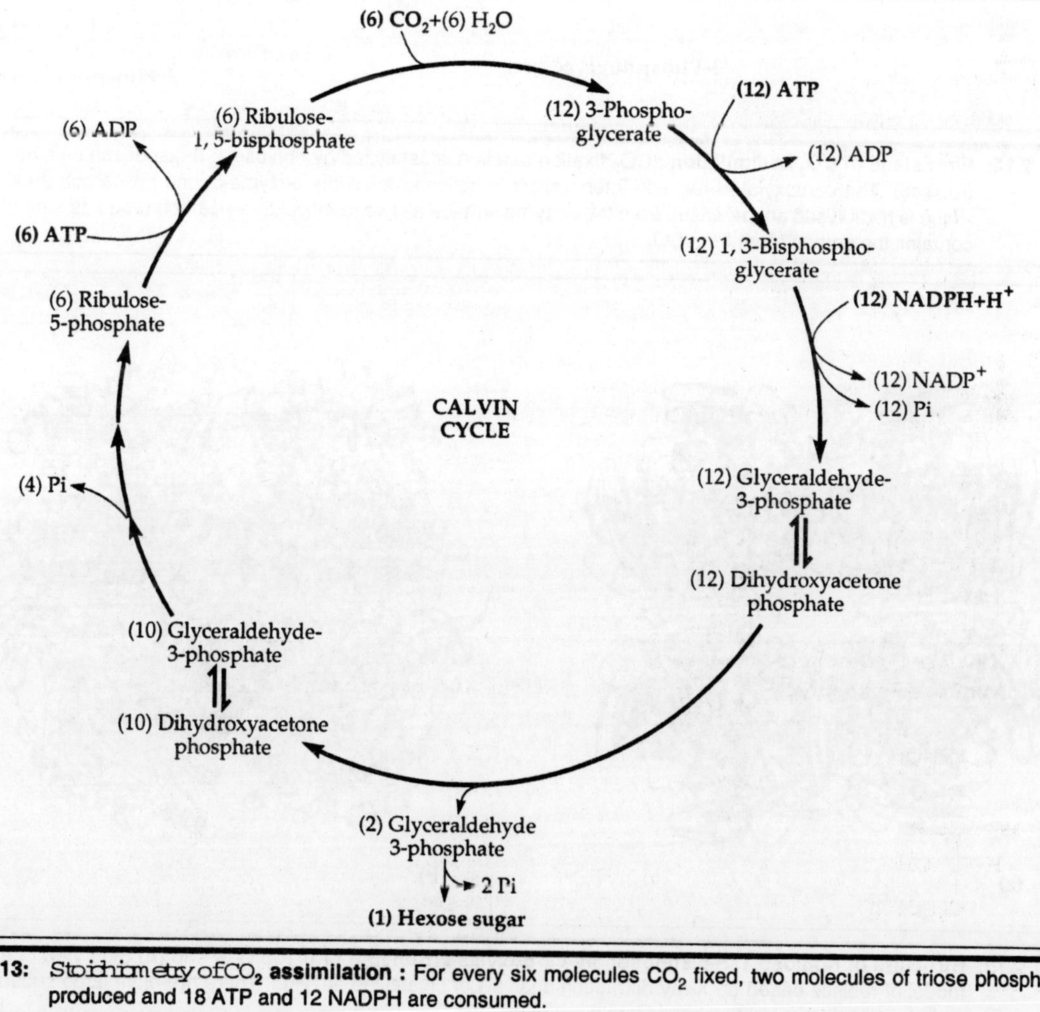

Fig. 2.13: **Stoichiometry of CO_2 assimilation :** For every six molecules CO_2 fixed, two molecules of triose phosphate produced and 18 ATP and 12 NADPH are consumed.

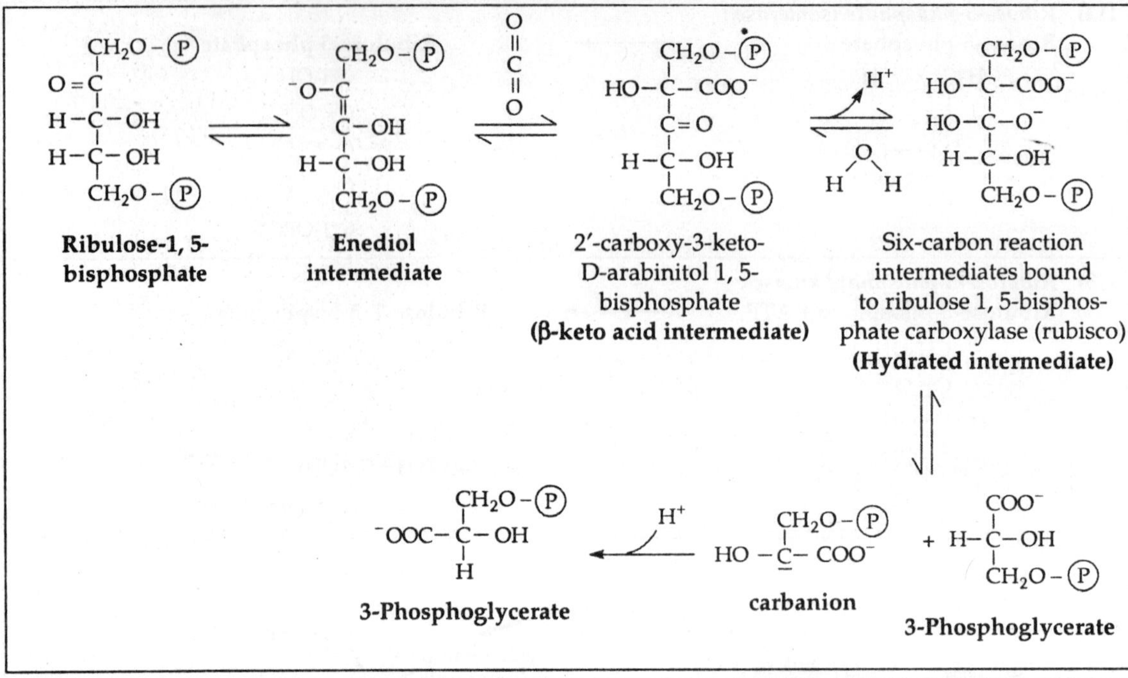

Fig. 2.14: First stage of CO_2 assimilation : CO_2 fixation reaction is catalyzed by ribulose 1, 5-bisphosphate carboxylase (rubisco). The carboxylated reaction intermediate is believed to be the enzyme-bound six-carbon β-keto acid, which is hydrolysed and released from the enzyme surface as two identical three-carbon products, one of which contains the carbon atom from CO_2.

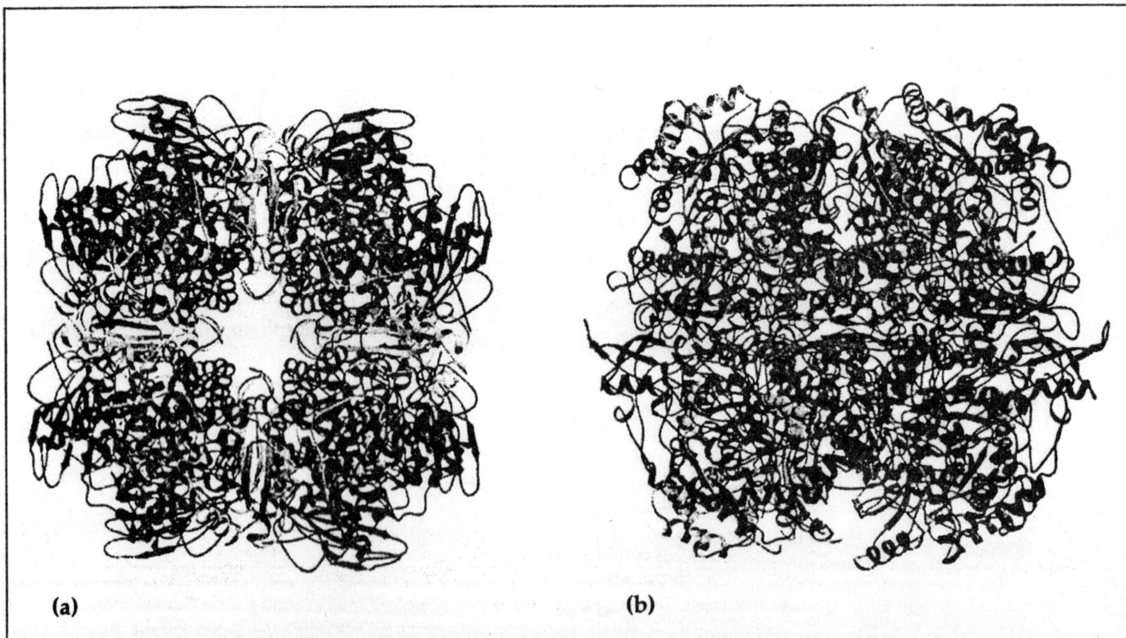

Fig. 2.15: Structure of ribulose 1, 5-bisphosphate carboxylase (rubisco) : (a) Top view and **(b)** Side view of a ribbon model of rubisco based on X-ray diffraction analysis of crystalline enzyme. There are eight large subunits and eight small subunits tightly packed into a structure.

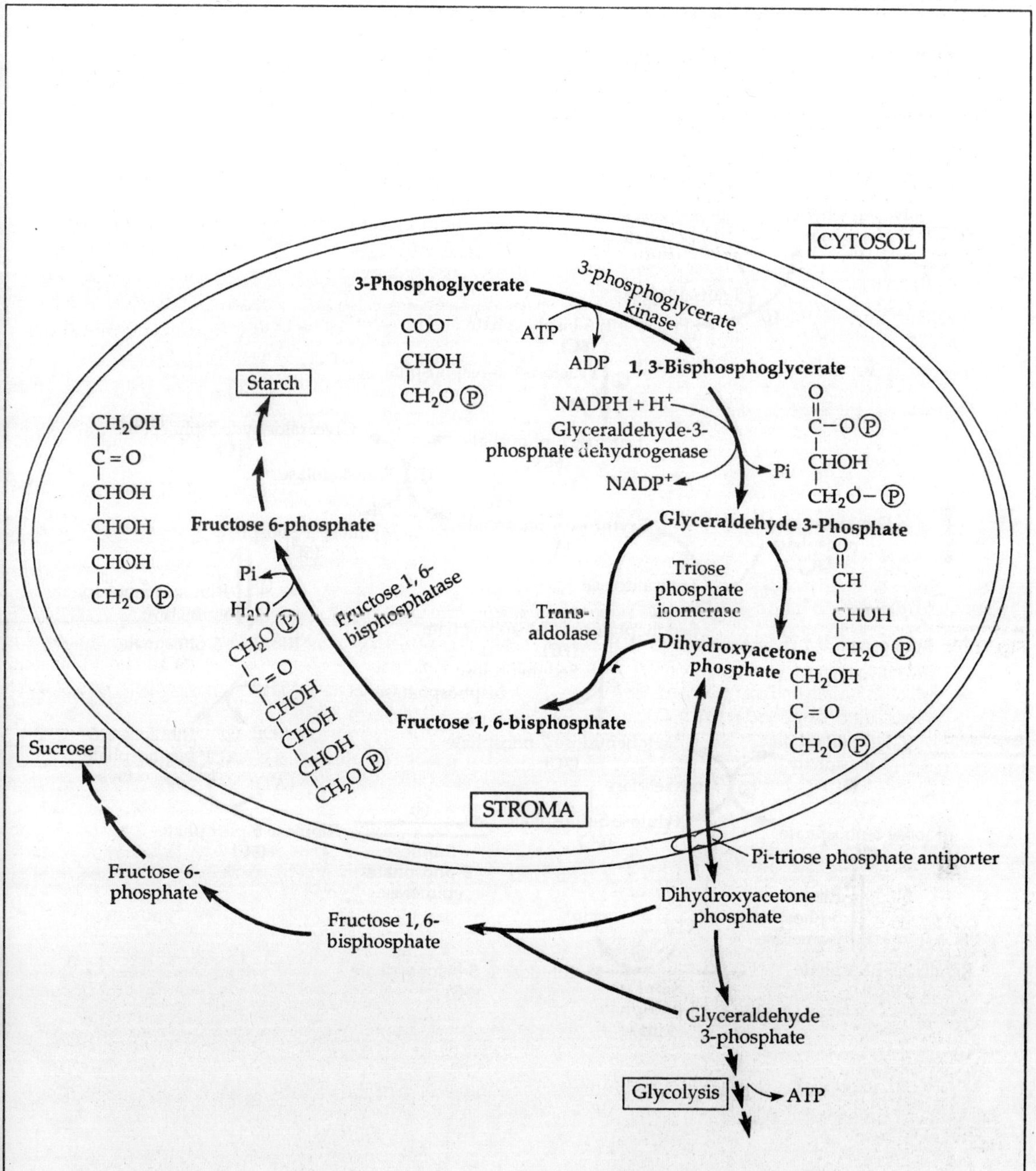

Fig. 2.16: Second stage of CO_2 assimilation ; 3-phosphoglycerate is converted to glyceraldehyde 3-phosphate. Most is recycled to ribulose 1, 5-bisphosphate. A small fraction of the extra glyceraldehyde 3-phosphate may be used immediately as a source of energy, but most is converted to sucrose for transport or is stored as starch for future use. If glyceraldehyde 3-phosphate is needed for starch synthesis, it condenses with dihydroxyacetone phosphate, which leaves the chloroplast via a specific antiporter. In the cytosol, dihydroxyacetone phosphate can be degraded glycolytically to provide energy or can be used to form fructose 6-phosphate and hence sucrose.

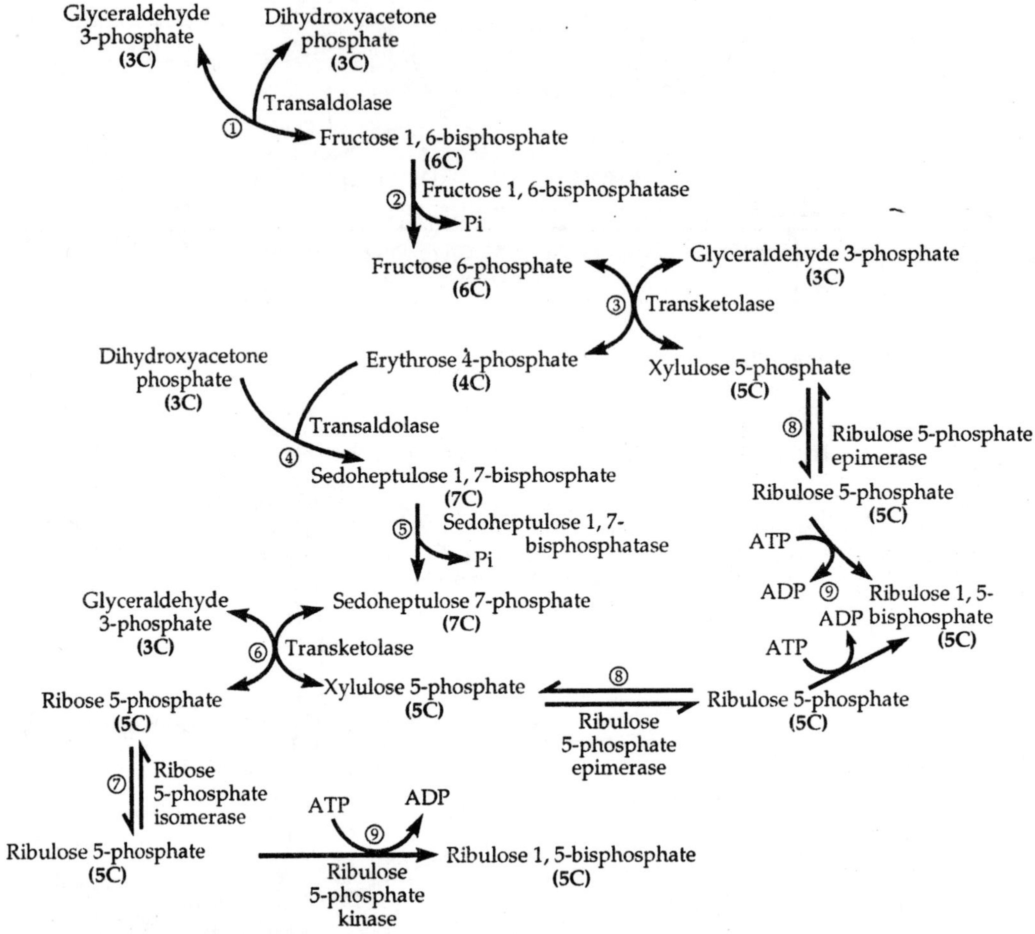

Fig. 2.17: Third stage of CO_2 assimilation : The diagram shows the interconversions of triose phosphates and pentose phosphates. The starting materials are glyceraldehyde 3-phosphate (3C) and dihydroxyacetone phosphate (3C). Reactions catalysed by transaldolase [① and ④] and transketolase [③ and ⑥] produce pentose phosphates that are converted to ribulose 5-phosphate–ribose 5-phosphate by ribose 5-phosphate isomerase ⑦ and xylulose 5-phosphate by ribulose 5-phosphate epimerase ⑧. In step ⑨, ribulose 5-phosphate is phosphorylated by ribulose 5-phosphate kinase regenerating ribulose 1, 5 bisphosphate. Dephosphorylation occurs in steps ② (fructose 1, 6-bisphosphatase) and ⑤ (sedoheptulose 1, 7-bisphosphatase).

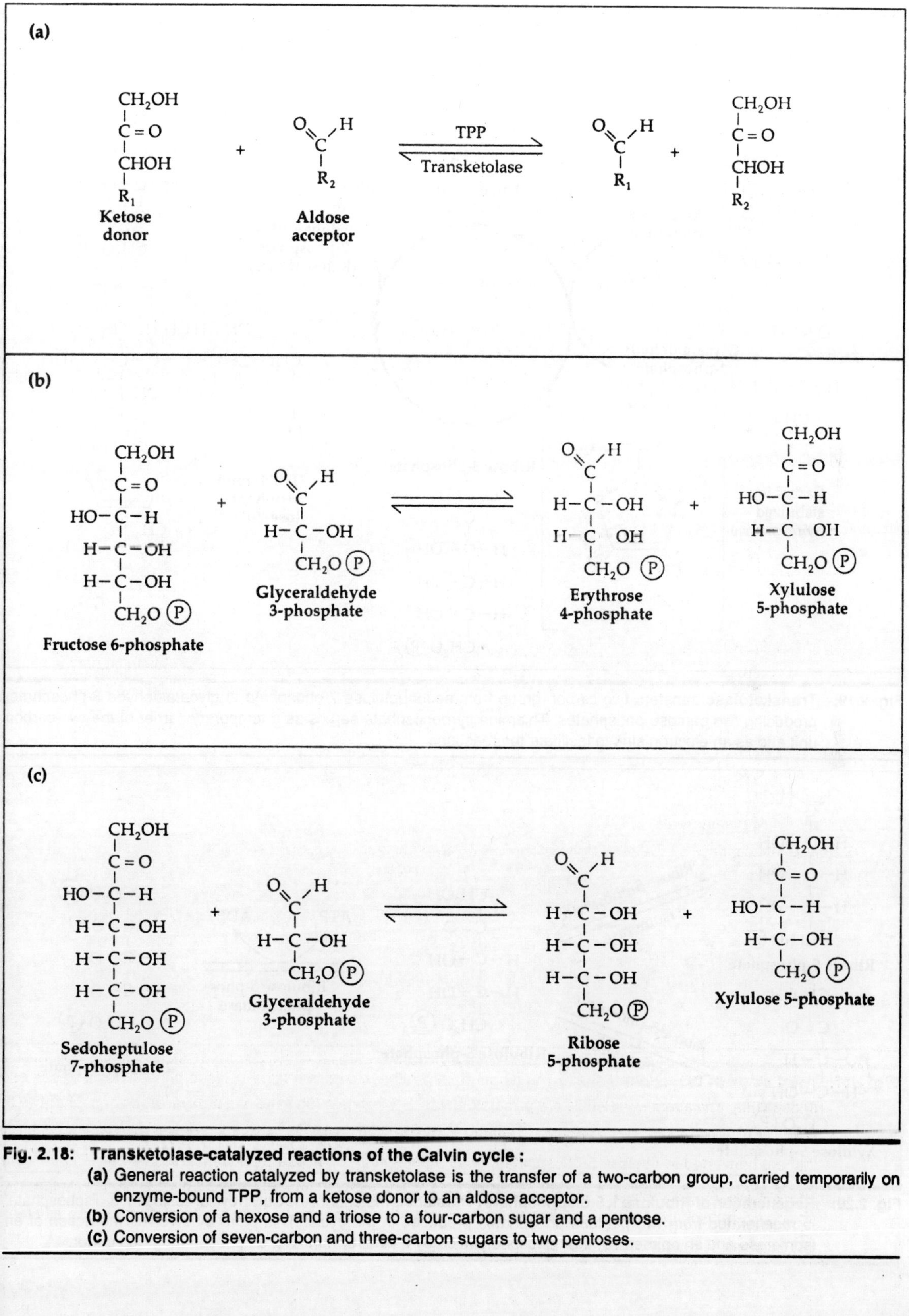

Fig. 2.18: Transketolase-catalyzed reactions of the Calvin cycle :

(a) General reaction catalyzed by transketolase is the transfer of a two-carbon group, carried temporarily on enzyme-bound TPP, from a ketose donor to an aldose acceptor.

(b) Conversion of a hexose and a triose to a four-carbon sugar and a pentose.

(c) Conversion of seven-carbon and three-carbon sugars to two pentoses.

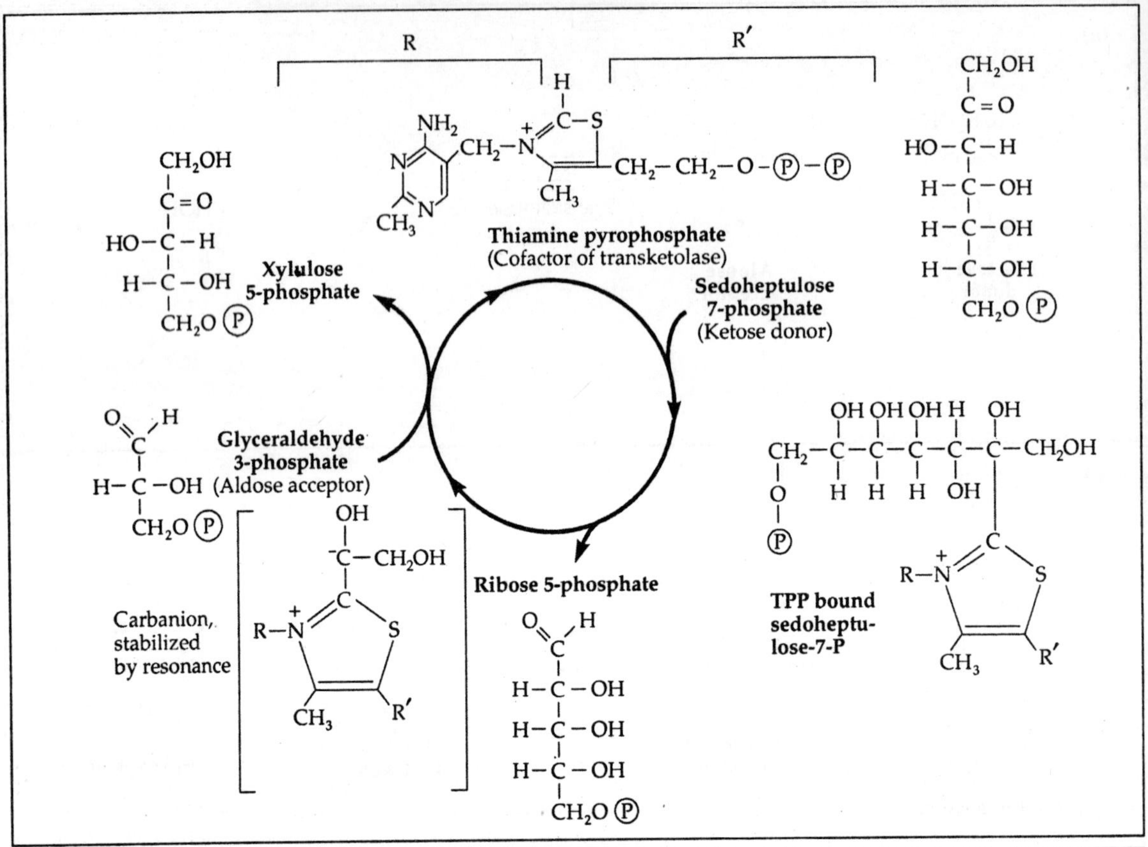

Fig. 2.19: **Transketolase** transfers two-carbon group from sedoheptulose 7-phosphate to glyceraldehyde 3-phosphate, producing two pentose phosphates. Thiamine pyrophosphate serves as a temporary carrier of the two-carbon unit and as an electron sink to facilitate the reactions.

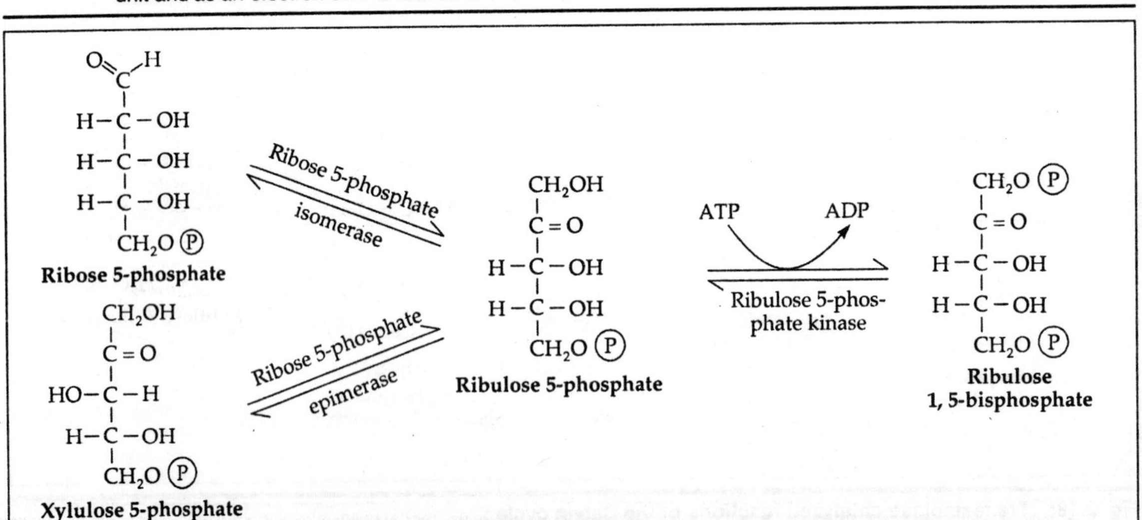

Fig. 2.20: **Regeneration of Ribulose 1,5-bisphosphate:** The starting material for the Calvin cycle, ribulose 1,5-bisphosphate, is regenerated from two pentose phosphates produced in the cycle. This pathway involves the action of an isomerase and an epimerase, then phosphorylation by a kinase, with ATP as phosphate group donor.

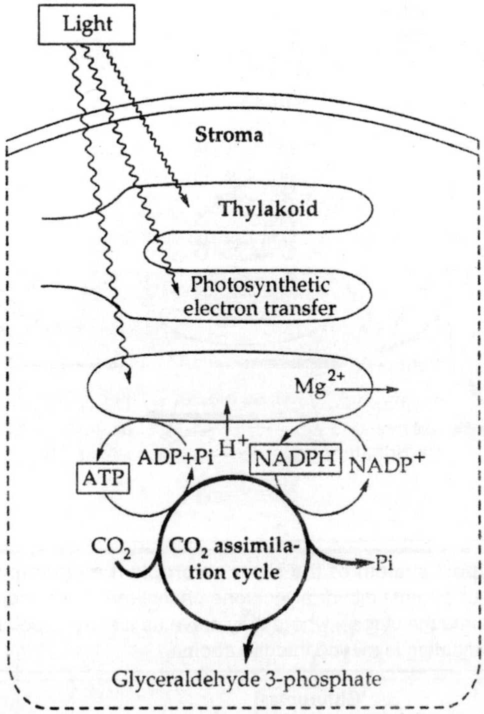

Fig. 2.21: ATP and NADPH produced by the light reactions are essential substrates for the reduction of CO_2. The photosynthetic reactions that produce ATP and NADPH are accompanied by movement of protons from the stroma into the thylakoid, creating alkaline conditions in the stroma. Magnesium ions pass from the thylakoid into the stroma, increasing the stromal [Mg^{2+}].

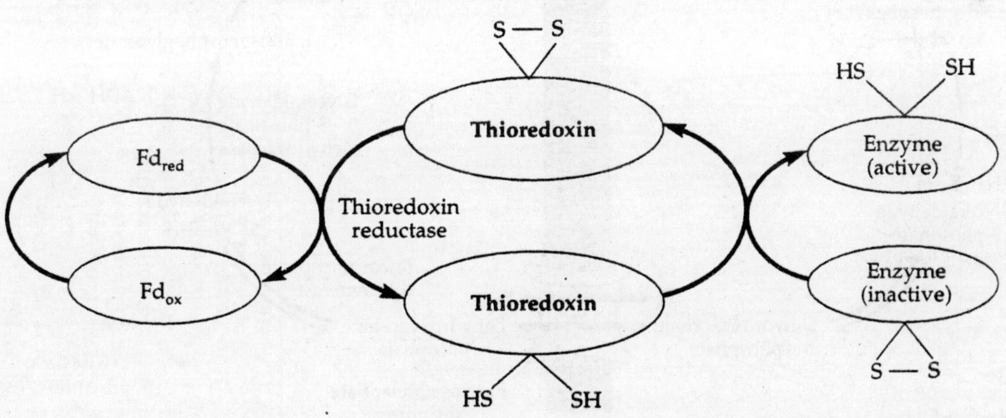

Fig. 2.22: Thioredoxin : Light activation of several enzymes of the Calvin cycle is mediated by a small, disulphide-containing protein, thioredoxin. It is reduced by electron from ferredoxin (Fd) in light, which then reduces the disulphide bonds of sedoheptulose 1, 7-bisphosphatase, fructose 1, 6-bisphosphatase, ribulose 5-phosphate kinase and glyceraldehyde 3-phosphate dehydrogenase, thereby activating these enzymes.

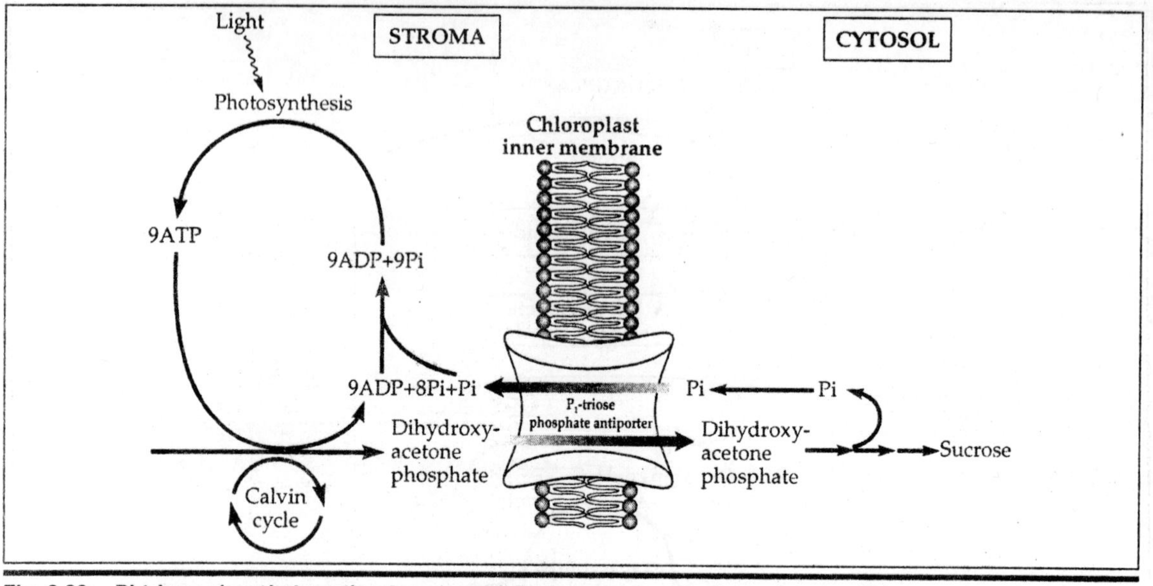

Fig. 2.23: Pi-triose phosphate antiport system of the inner chloroplast membrane : This transporter facilitates the exchange of cytosolic Pi for stromal dihydroxyacetone phosphate. The products of photosynthetic carbon assimilation are thus moved into the cytosol where they serve as a starting point for sucrose biosynthesis, and Pi required for photophosphorylation is moved into the stroma.

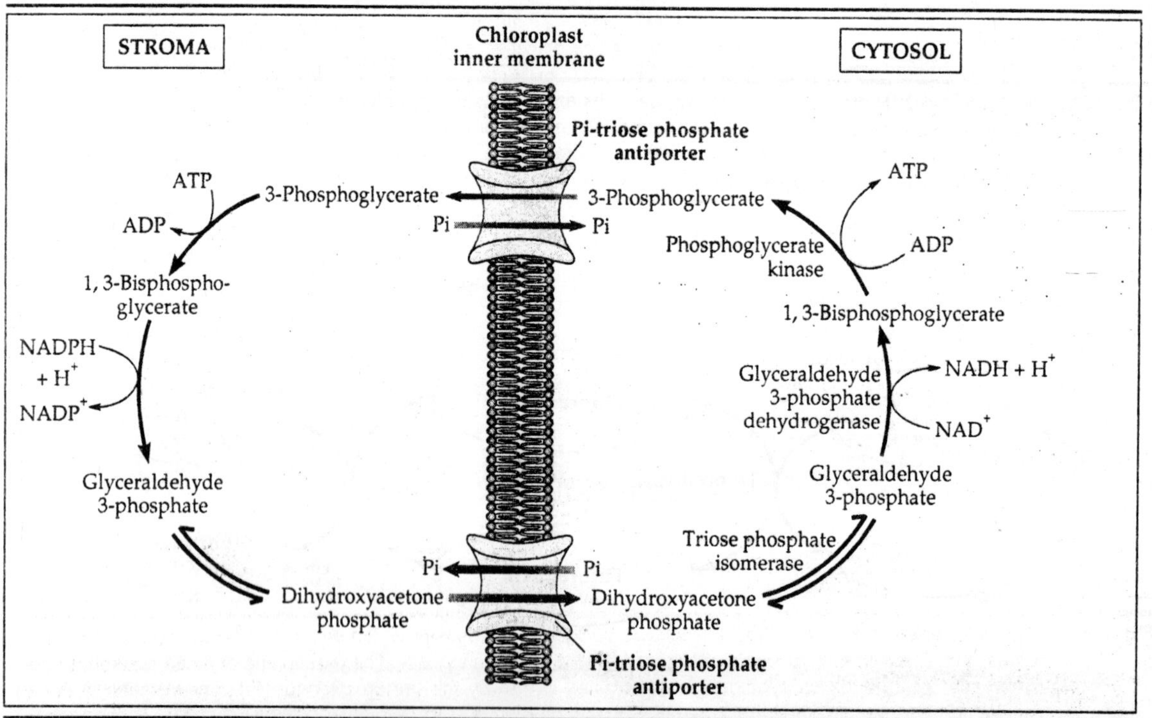

Fig. 2.24: Role of the Pi-triose phosphate antiporter in the transport of ATP and reducing equivalents : Dihydroxyacetone phosphate leaves the chloroplast and is converted to glyceraldehyde 3-phosphate in the cytosol. The cytosolic glyceraldehyde 3-phosphate dehydrogenase and phosphoglycerate kinase reactions then produce NADH, ATP and 3-phosphoglycerate. The latter re-enters the chloroplast and is reduced to dihydroxy-acetone phosphate, completing a cycle that effectively moves ATP and reducing equivalents (NADPH/NADH) from chloroplast to cytosol.

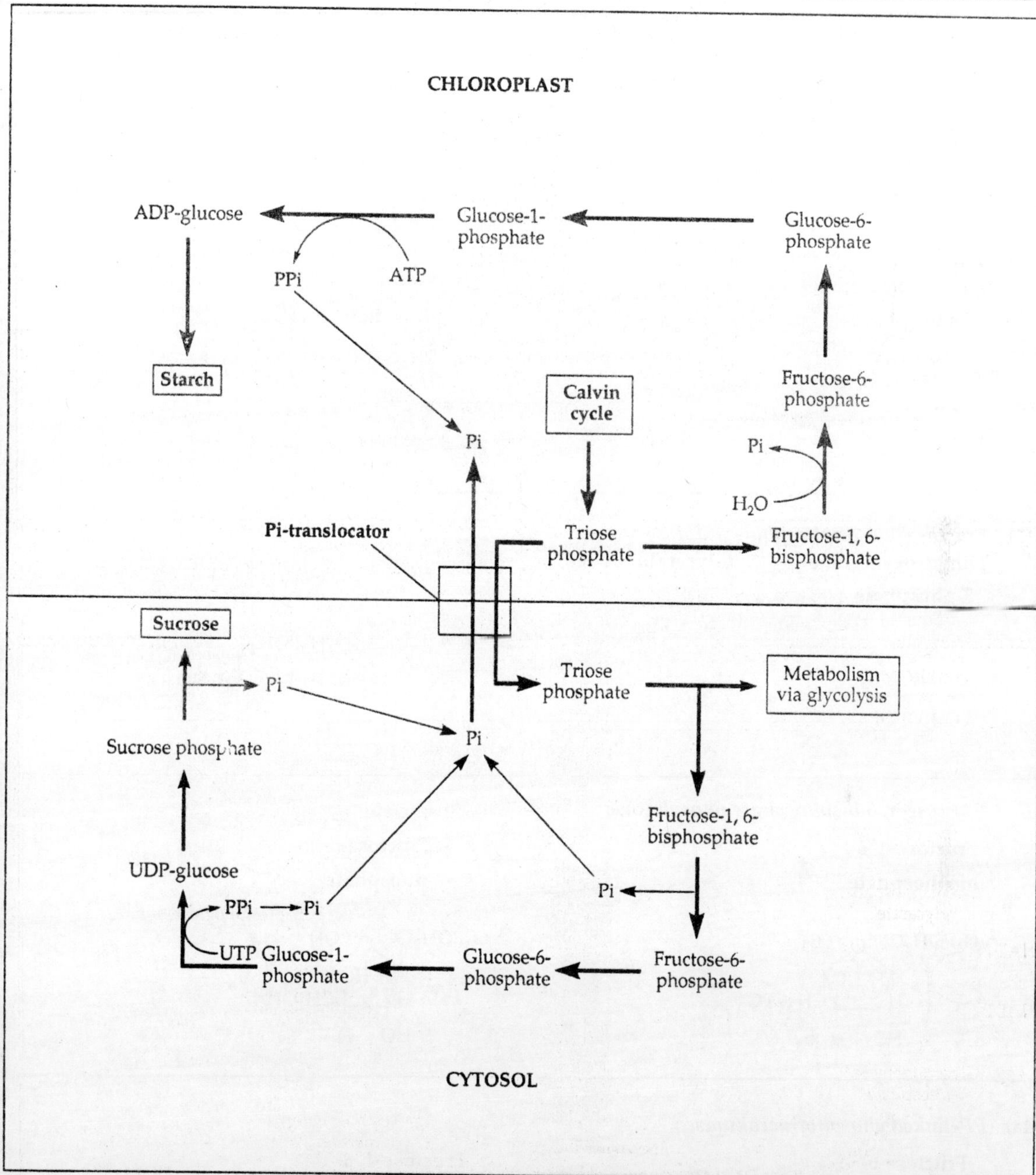

Fig. 2.25: **The syntheses of Sucrose and Starch** are competing processes which occur in different cellular compartments. Starch synthesis occurs in chloroplast and sucrose in cytosol. The partitioning of triose phosphate between sucrose and starch synthesis is determined in part by the orthophosphate (Pi) concentration in these cellular compartments. When the cytosolic Pi concentration is high, chloroplast triose-phosphate is exported to the cytosol in exchange of Pi and sucrose is synthesized. When the cytosolic Pi concentration is low, triose phosphate is retained within the chloroplast and starch is synthesized. Reversibility of reaction is effected by the Pi translocator. Transport of hexoses from the chloroplast to the cytosol is also important especially at night, when the starch is broken down by amylases. This hexose transport mechanism bypasses the cytosolic fructose-1, 6-bisphosphate phosphatase, which would be largely inactive because of the high level of fructose 2, 6-bisphosphate occurring at night.

Fig. 2.26: Reactions of Sucrose synthesis from Triose phosphate in the cytosol and the enzymes involved.

(1) ***Phosphate/Triose phosphate translocator*** **:**

Triose phosphate (chloroplast) + $HOPO_3^{2-}$ (cytosol) ⟶ Triose phosphate (cytosol) + $HOPO_3^{2-}$ (chloroplast)

(2) ***Triose phosphate isomerase*** **:**

Dihydroxyacetone-3-phosphate ⟶ **Glyceraldehyde-3-phosphate**

(3) ***Fructose-1, 6-bisphosphate aldolase*** **:**

Dihydroxyacetone-3-phosphate + **Glyceraldehyde-3-phosphate** ⟶ **Fructose-1, 6-bisphosphate**

(4) ***Fructose-1, 6-bisphosphate phosphatase*** **:**

Fructose-1, 6-bisphosphate + H_2O ⟶ **Fructose-6-phosphate** + $HOPO_3^{2-}$

(4a) ***PP-linked phosphofructokinase*** **:**

Fructose-6-phosphate + $H_2P_2O_7^{2-}$ ⟶ **Fructose-1, 6-bisphosphate** + $HOPO_3^{2-}$

contd.

(5) ***Hexose-phosphate isomerase :***

Fructose-6-phosphate ⟶ **Glucose-6-phosphate**

(6) ***Phosphoglucomutase :***

Glucose-6-phosphate ⟶ **Glucose-1-phosphate**

(7) ***UDP-glucose pyrophosphorylase :***

Glucose-1-phosphate + UTP ⟶ **UDP-glucose +** $H_2P_2O_7^{2-}$

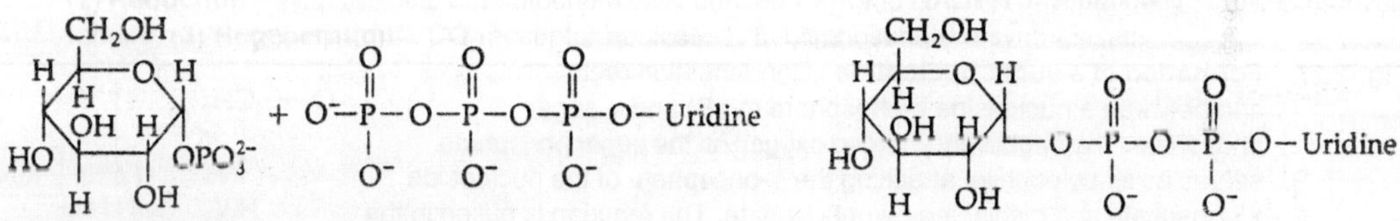

(8) ***Sucrose phosphate synthase :***

UDP-glucose + Fructose-6-phosphate ⟶ **UDP + Sucrose-6-phosphate**

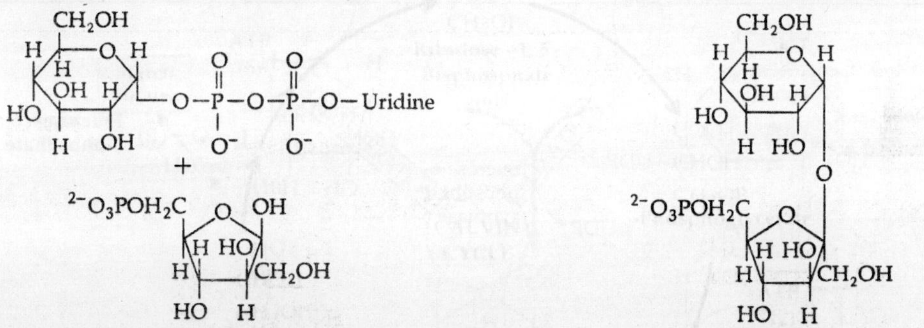

(9) ***Sucrose phosphate phosphatase :***

Sucrose-6-phosphate + H_2O ⟶ **Sucrose +** $HOPO_3^{2-}$

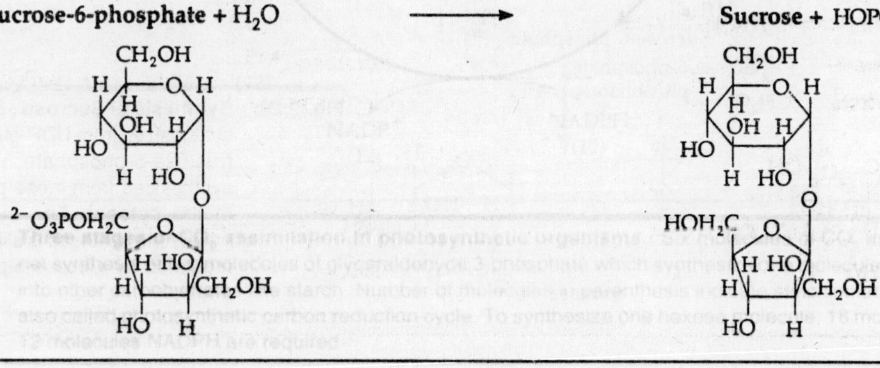

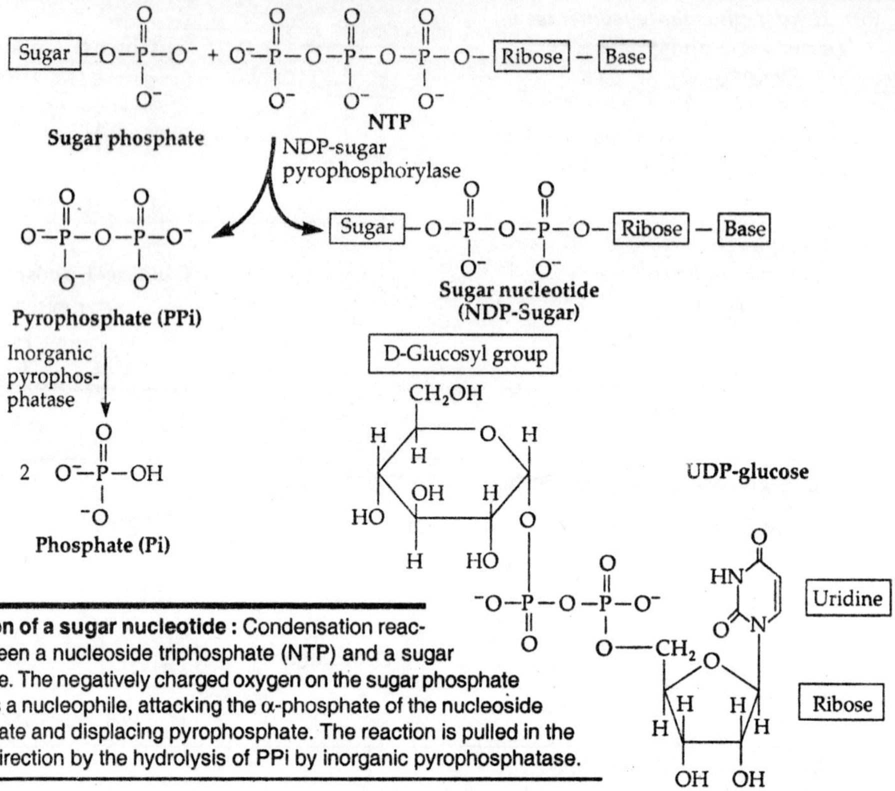

Fig. 2.27: Formation of a sugar nucleotide : Condensation reaction between a nucleoside triphosphate (NTP) and a sugar phosphate. The negatively charged oxygen on the sugar phosphate serves as a nucleophile, attacking the α-phosphate of the nucleoside triphosphate and displacing pyrophosphate. The reaction is pulled in the forward direction by the hydrolysis of PPi by inorganic pyrophosphatase.

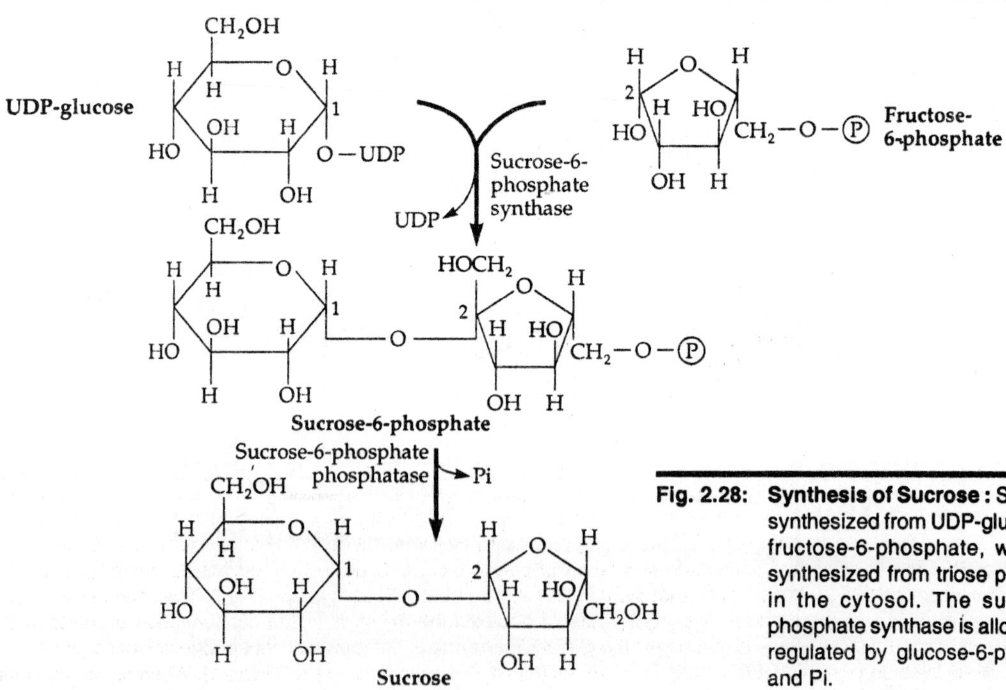

Fig. 2.28: Synthesis of Sucrose : Sucrose is synthesized from UDP-glucose and fructose-6-phosphate, which are synthesized from triose phosphate in the cytosol. The sucrose-6-phosphate synthase is allosterically regulated by glucose-6-phosphate and Pi.

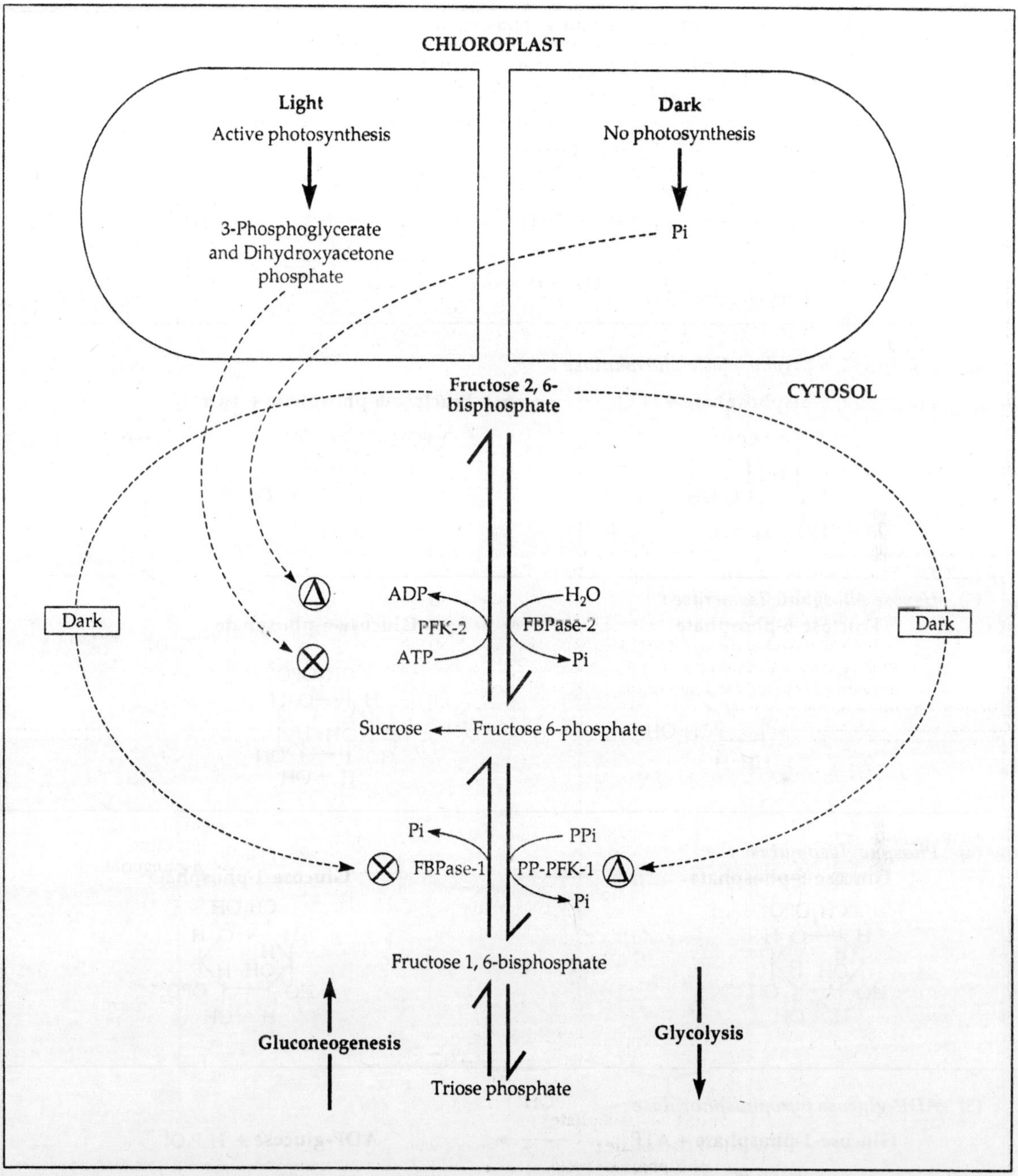

Fig. 2.29: Fructose 2, 6-bisphosphate as regulator of sucrose synthesis : The concentration of allosteric regulator fructose 2, 6-bisphosphatase (FBPase-2) is regulated by the products of photosynthetic carbon assimilation and by Pi. Dihydroxyacetone phosphate and 3-phosphoglycerate produced by CO_2 assimilation inhibit phosphofructokinase-2 (PFK-2), the enzyme that synthesizes the regulator; Pi stimulates PFK-2. The concentration of the regulator is inversely proportional to the rate of photosynthesis. In dark, the concentration of fructose 2, 6-bisphosphate increases and stimulates the glycolytic enzyme PPi-dependent phosphofructokinase (PP-PFK-1), while inhibiting the gluconeogenetic enzyme fructose 1, 6-bisphosphatase (FBPase-1). When photosynthesis is active in presence of light, the concentration of the regulator drops and the synthesis of fructose 6-phosphate and sucrose is favoured.

Fig. 2.30: Reactions of starch synthesis from triose phosphate in chloroplasts.

(1) *Fructose-1, 6-bisphosphate aldolase* :

Dihydroxyacetone-3-phosphate + Glyceraldehyde-3-phosphate ⟶ Fructose-1, 6-bisphosphate

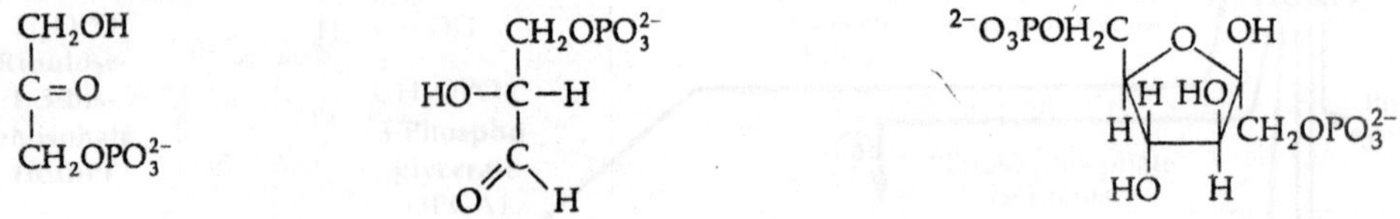

(2) *Fructose-1, 6-bisphosphate phosphatase* :

Fructose-1, 6-bisphosphate + H_2O ⟶ Fructose-6-phosphate + $HOPO_3^{2-}$

(3) *Hexose phosphate isomerase* :

Fructose-6-phosphate ⟶ Glucose-6-phosphate

(4) *Phosphoglucomutase* :

Glucose-6-phosphate ⟶ Glucose-1-phosphate

(5) *ADP-glucose pyrophosphorylase* :

Glucose-1-phosphate + ATP ⟶ ADP-glucose + $H_2P_2O_7^{2-}$

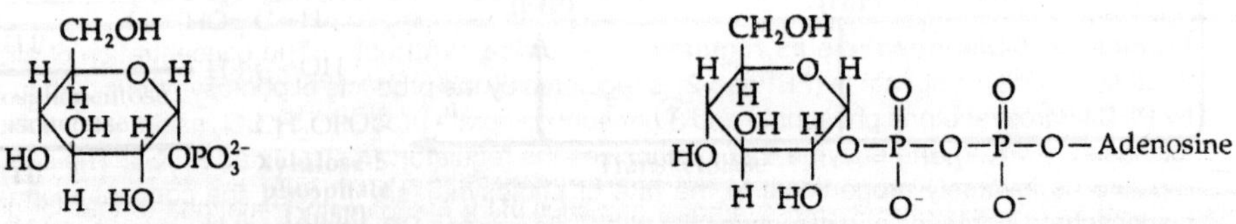

contd.

(6) ***Pyrophosphatase*** **:**

$$H_2P_2O_7^{2-} + H_2O \longrightarrow 2HOPO_3^{2-} + 2H^+$$

(7) ***Starch synthase*** **:**

ADP glucose + (1, 4-α-D-glucosyl)$_n$ ⟶ ADP + (1, 4-α-D-glucosyl)$_{n+1}$

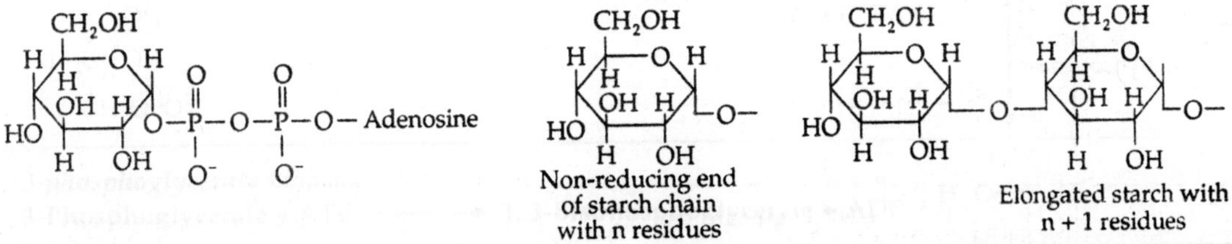

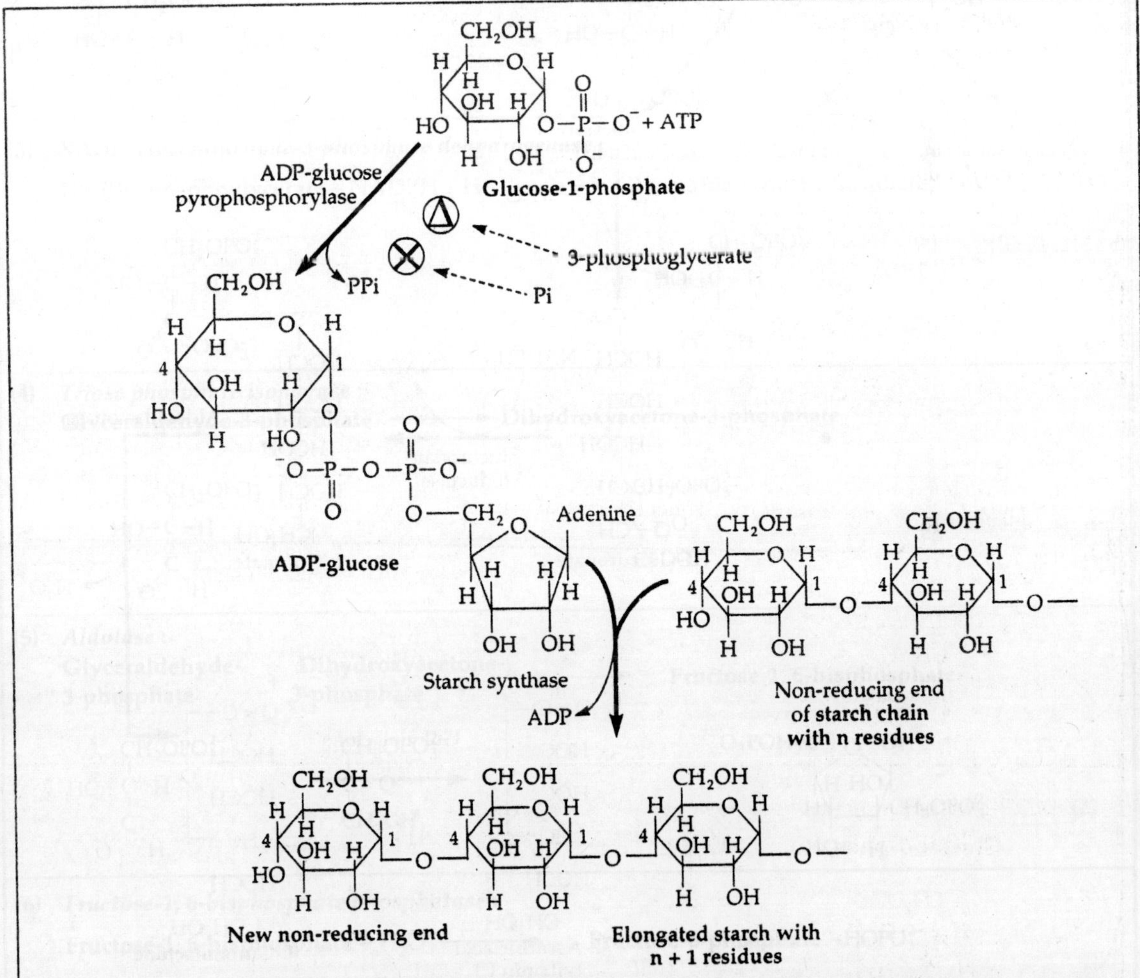

Fig. 2.31: Synthesis of starch : Starch synthesis proceeds by a mechanism analogous to that for glycogen synthesis except that the activated substrate is ADP-glucose instead of UDP-glucose. Glucose is transferred to the non-reducing end of an existing starch molecule, in (α1→4) linkage. ADP-glucose pyrophosphorylase is a regulatory enzyme.

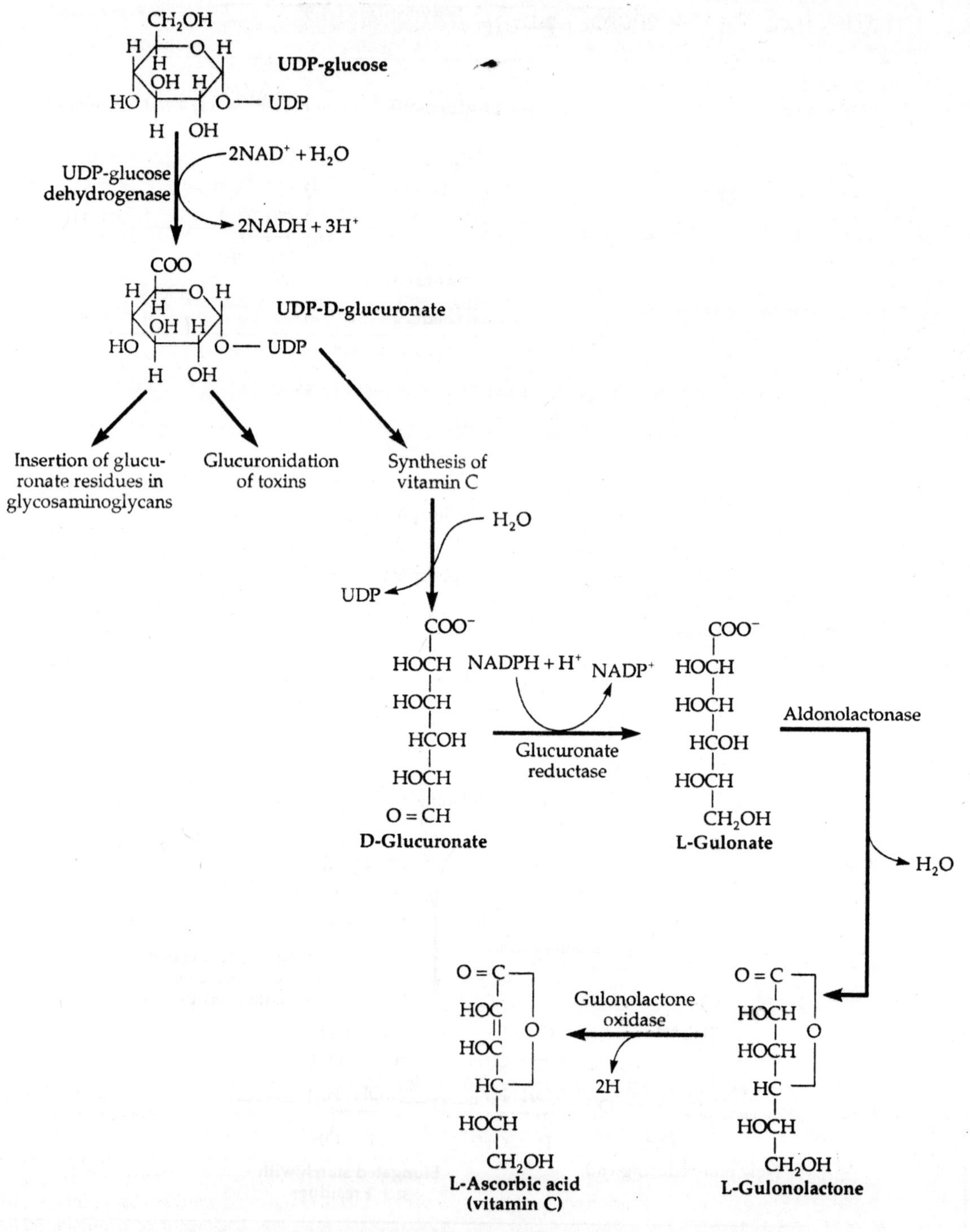

Fig. 2.32: Synthesis of Glucuronate and Vitamin C : UDP-glucuronate formed from glucose is the precursor of glucuronate (GlcA) residues in glycosaminoglycans and of vitamin C (ascorbic acid).

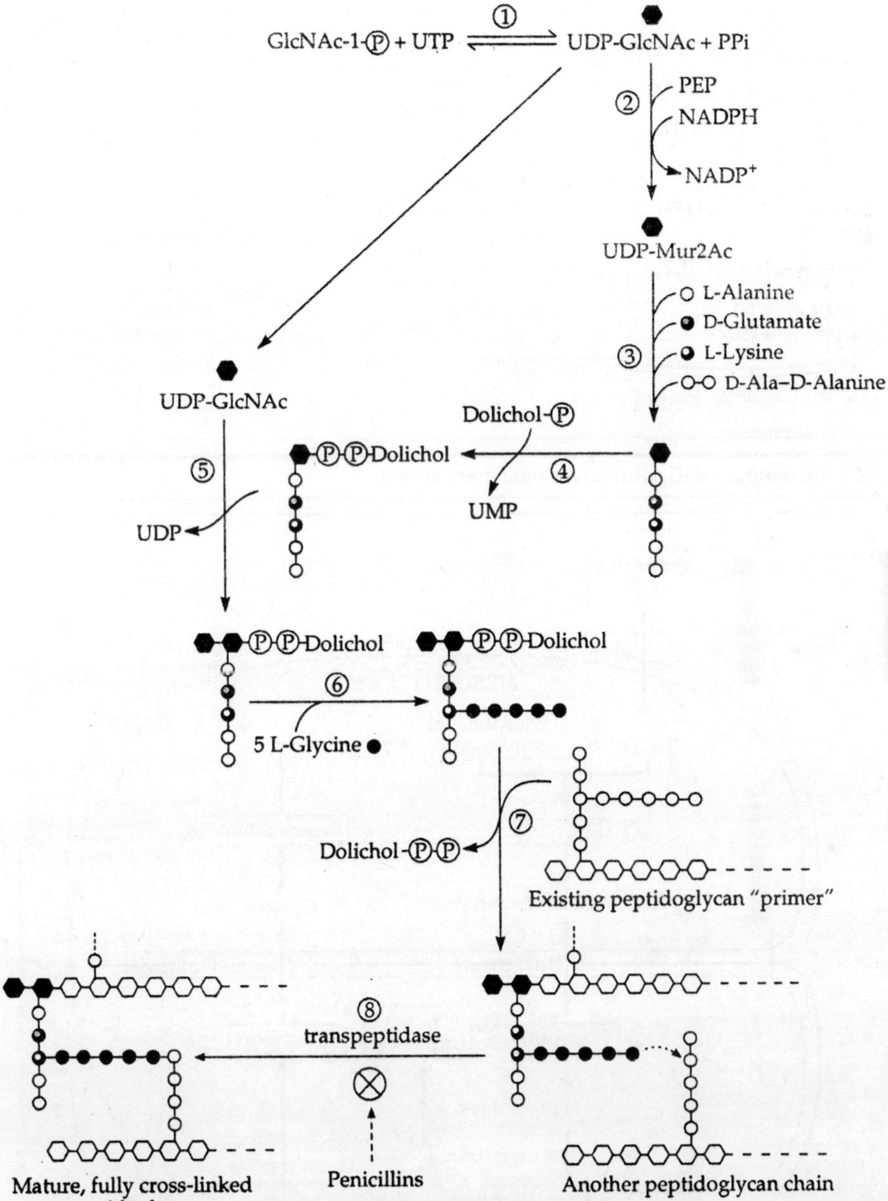

Fig. 2.33: Synthesis of bacterial peptidoglycan : In steps 1 through 4, N-acetyl glucosamine (GlcNAc) and N-acetylmuramic acid (Mur2Ac) are activated by attachment of a uridine nucleotide (UDP) to their anomeric carbons and in case of Mur2Ac, of a long chain isoprenyl alcohol (dolichol) through a phosphodiester bond. These activating groups participate in the formation of glycosidic linkages ; they serve as excellent leaving groups. After assembly (steps 5 and 6) of a disaccharide with a peptide side chain (10 amino acid residues), this precursor is transferred (step 7) to the non-reducing end of an existing peptidoglycan chain, which serves as a primer for the polymerization reaction. Finally, a transpeptidation reaction (step 8) occurs between the peptide side chains on two different peptidoglycan molecules. A Gly residue at the end of one peptide chain displaces a terminal D-Ala in the other peptide chain, forming a cross-link. This transpeptidation reaction is inhibited by the penicillins, which kill bacteria by weakening their cell walls.

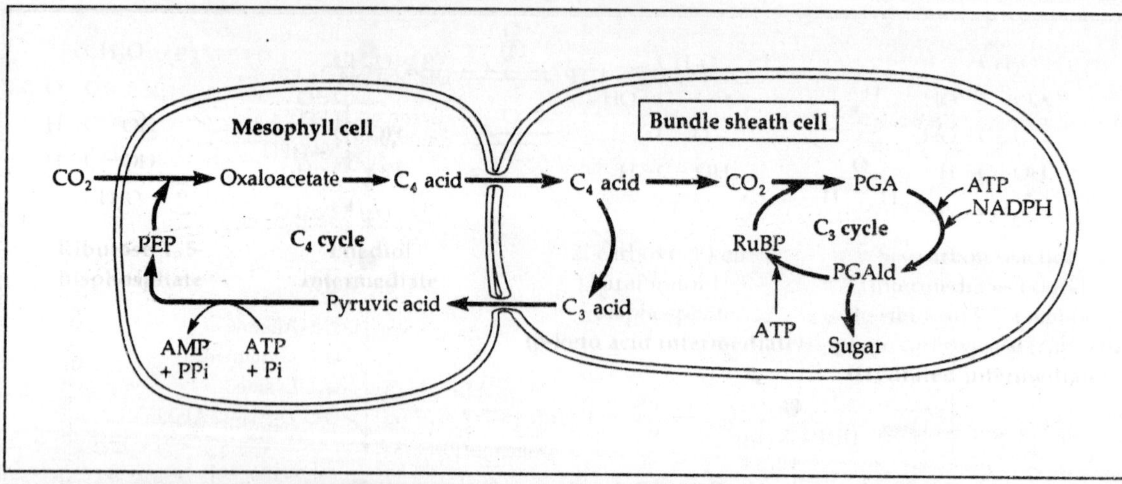

Fig. 2.34: Schematic diagram of C_4 photosynthetic mechanism.

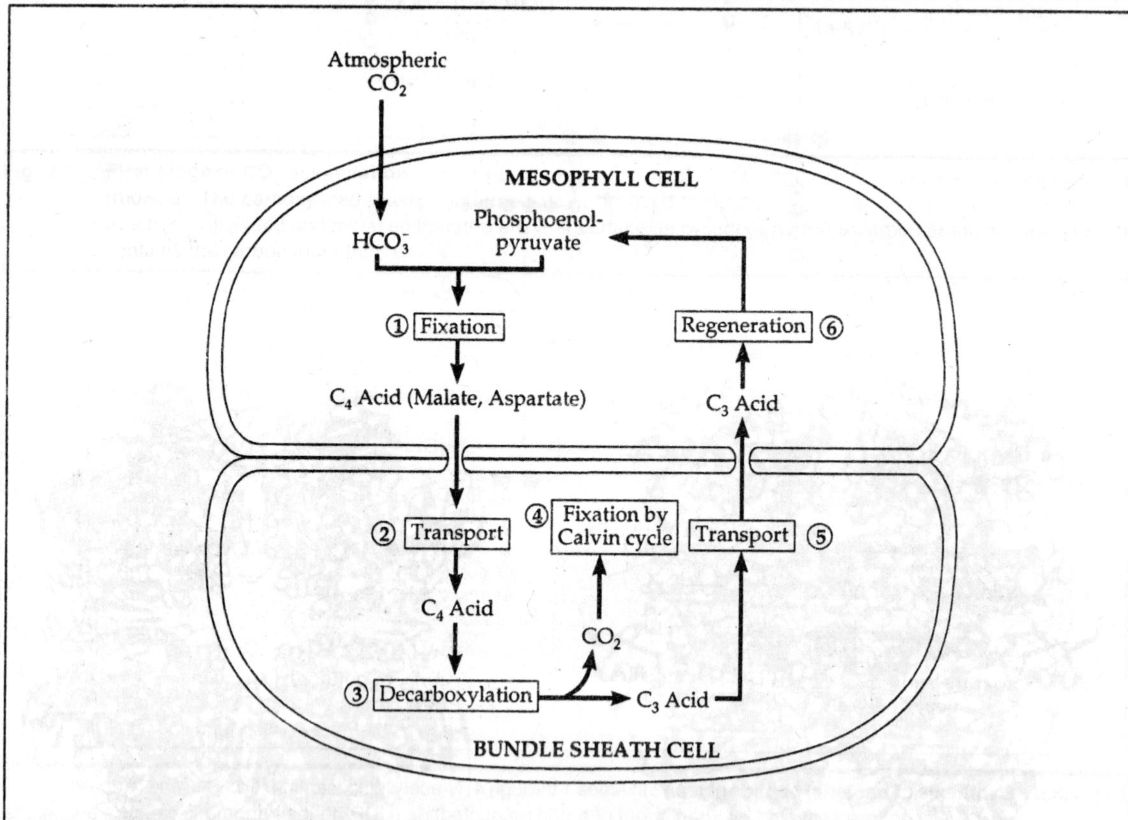

Fig. 2.35: C_4 photosynthetic carbon cycle : It involves two types of cells and proceeds in several stages—**(1)** Fixation of atmospheric CO_2 into a four-carbon acid in mesophyll cell. **(2)** Transport of the four-carbon acid from mesophyll cell to bundle sheath cell. **(3)** Decarboxylation of the four-carbon acid generating a high concentration of CO_2 in the bundle sheath cell. **(4)** Fixation of CO_2 by Calvin cycle in bundle sheath cell to synthesize sugars. **(5)** Transport of the residual three-carbon acid back to the mesophyll cells. **(6)** Regeneration of original CO_2 acceptor, phosphoenol pyruvate from C_3 acid in mesophyll cell.

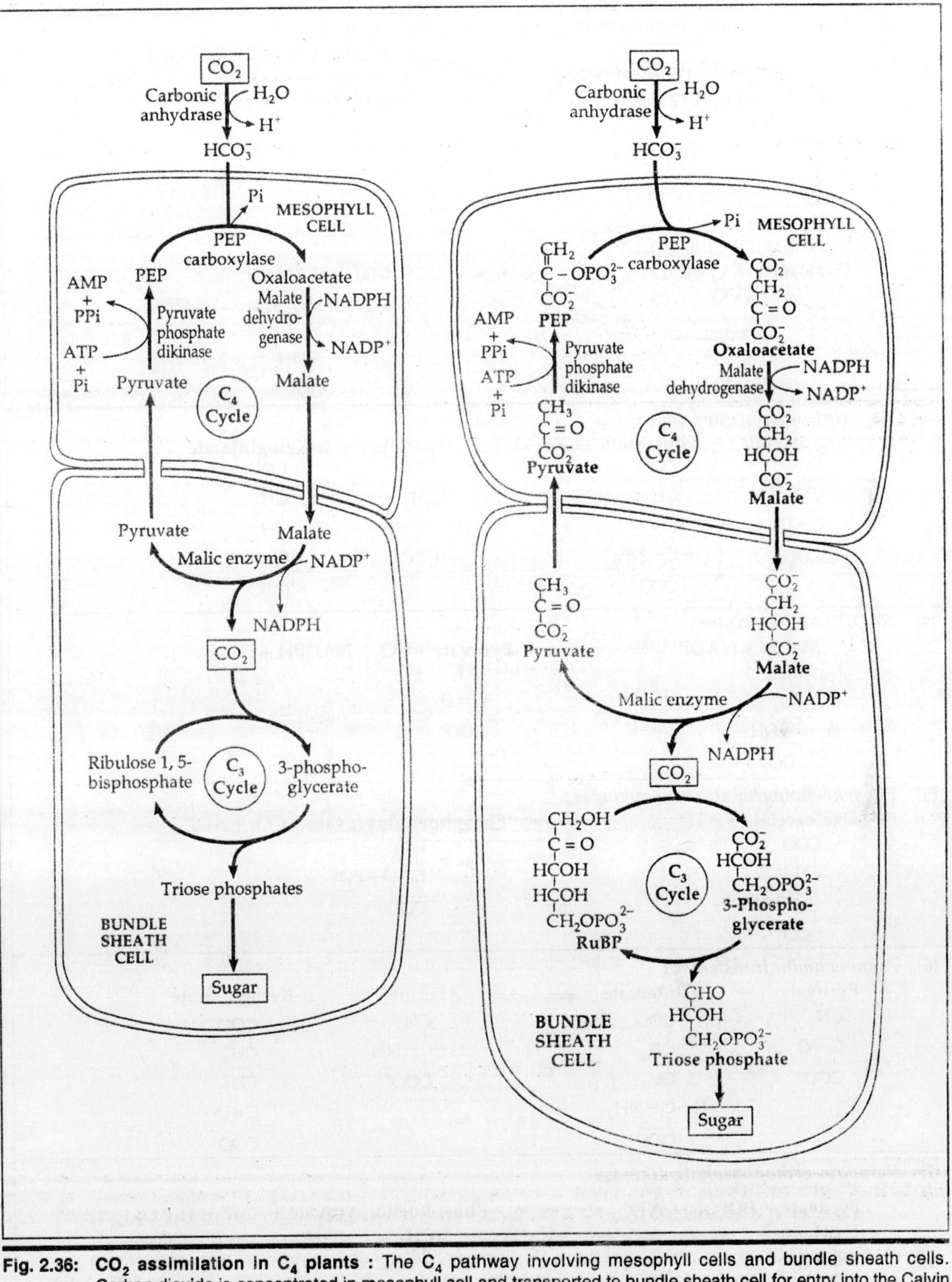

Fig. 2.36: CO_2 assimilation in C_4 plants : The C_4 pathway involving mesophyll cells and bundle sheath cells. Carbon dioxide is concentrated in mesophyll cell and transported to bundle sheath cell for entry into the Calvin cycle.

Fig. 2.37: Reactions of the C_4 photosynthetic carbon cycle and the enzymes.

(1) *Phosphoenolpyruvate carboxylase :*

Phosphoenolpyruvate + HCO_3^- ⟶ Oxaloacetate + $HOPO_3^{2-}$

(2) *NADP-malate dehydrogenase :*

Oxaloacetate + NADPH + H^+ ⟶ Malate + $NADP^+$

(3) *Aspartate aminotransferase :*

Oxaloacetate + Glutamate ⇌ Aspartate + α-Ketoglutarate

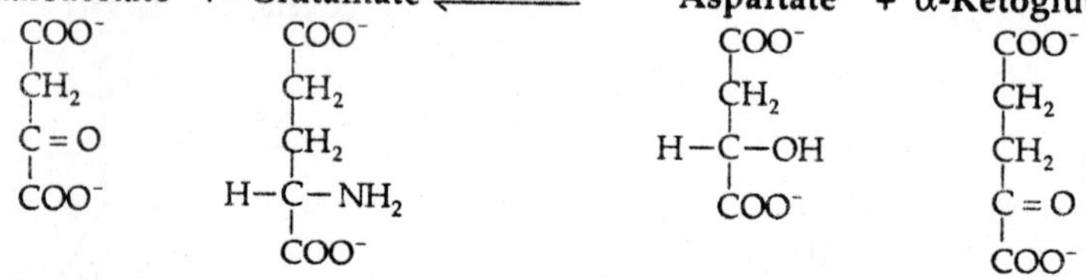

(4) *NAD(P) Malic enzyme :*

Malate + $NADP^+$ ⟶ Pyruvate + CO_2 + NADPH + H^+

(5) *Phosphoenolpyruvate carboxykinase :*

Oxaloacetate + ATP ⟶ Phosphoenolpyruvate + CO_2 + ADP

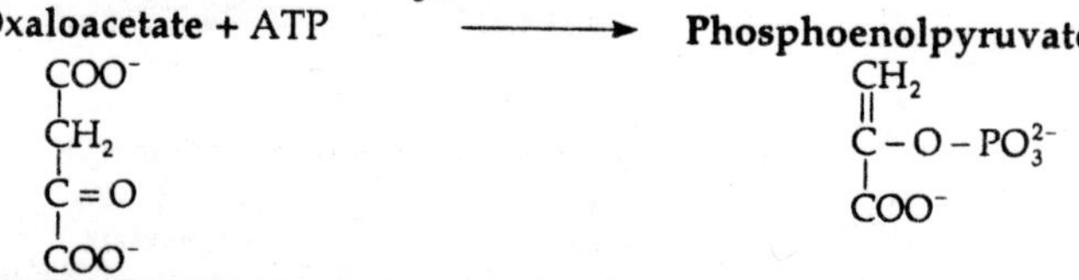

(6) *Alanine aminotransferase :*

Pyruvate + Glutamate ⇌ Alanine + α-Ketoglutarate

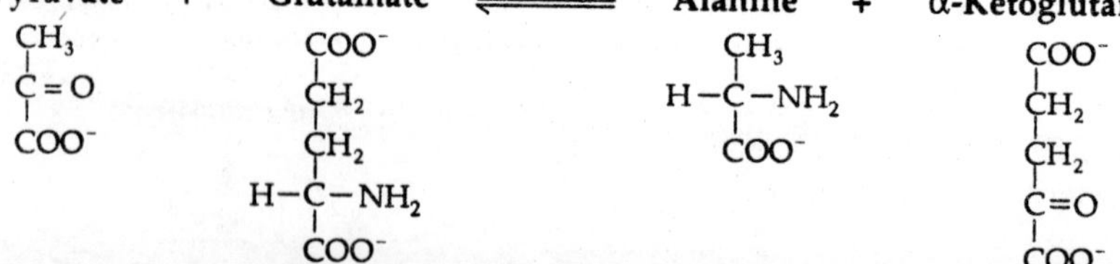

(7) *Pyruvate-orthophosphate dikinase :*

Pyruvate + $HOPO_3^{2-}$ + ATP ⟶ Phosphoenolpyruvate + AMP + $H_2P_2O_7^{2-}$

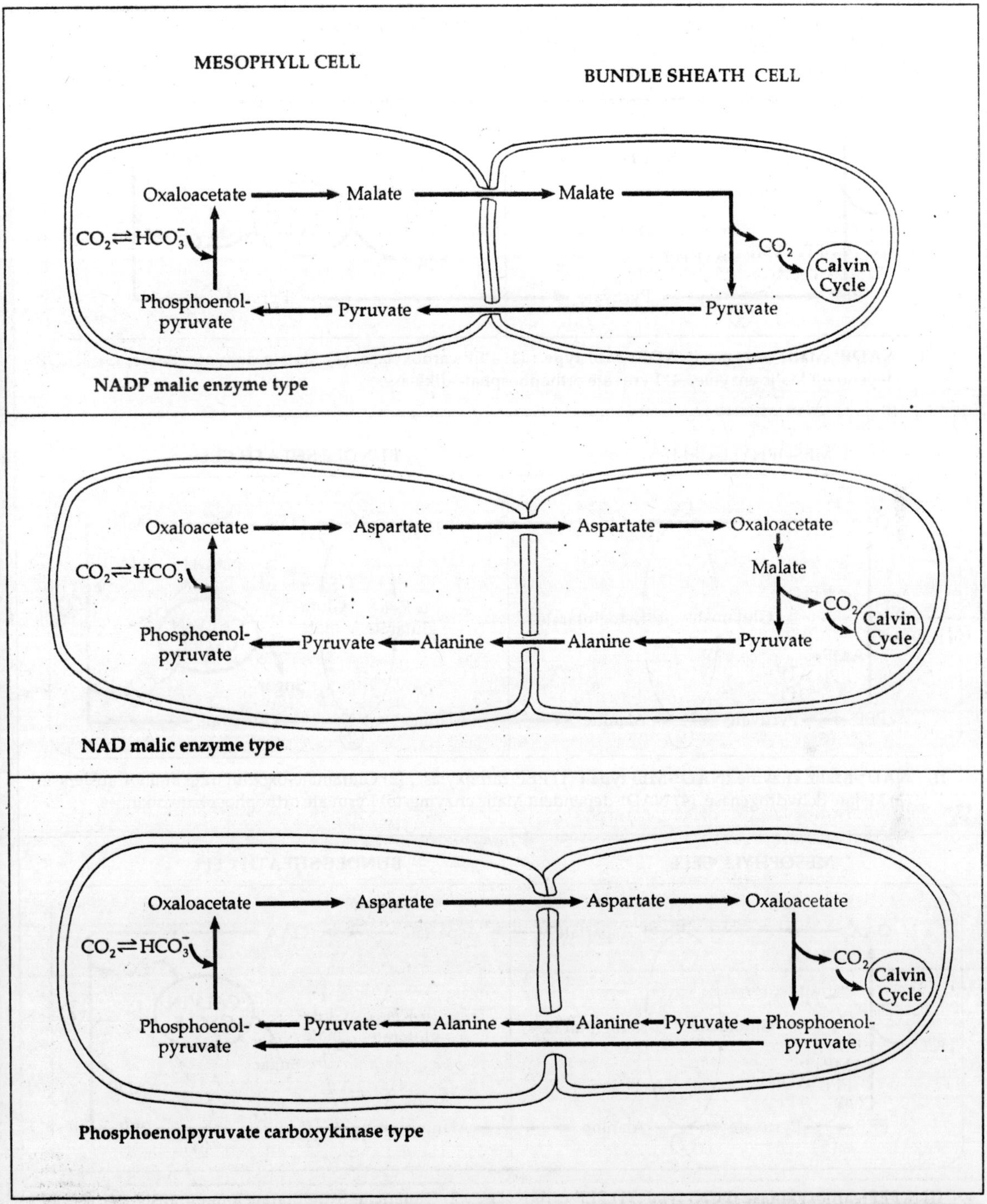

Fig. 2.38: Three variants of the C_4 photosynthetic carbon cycle : The variants differ mainly in **(a)** the nature of the four-carbon acid (malate / aspartate) transported into the bundle sheath cell and of the three-carbon acid (pyruvate / alanine) returned to the mesophyll cell ; **(b)** the nature of the enzyme that catalyzes the decarboxylation step in the bundle sheath cell. The three variants are named after the enzymes that catalyze the decarboxylation reactions. Representatives of each variant include maize, sugarcane (NADP malic enzyme), millet (NAD malic enzyme), guineagrass (phosphoenolpyruvate carboxykinase).

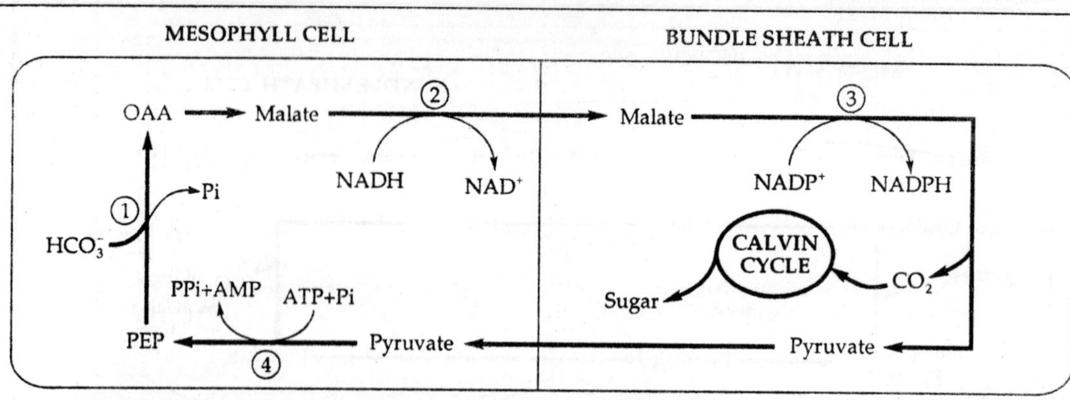

I. NADP⁺-Malic enzyme (NADP⁺-ME) type : (1) PEP-carboxylase, (2) Malate dehydrogenase, (3) NADP⁺ dependent Malic enzyme, (4) Pyruvate orthophosphate dikinase.

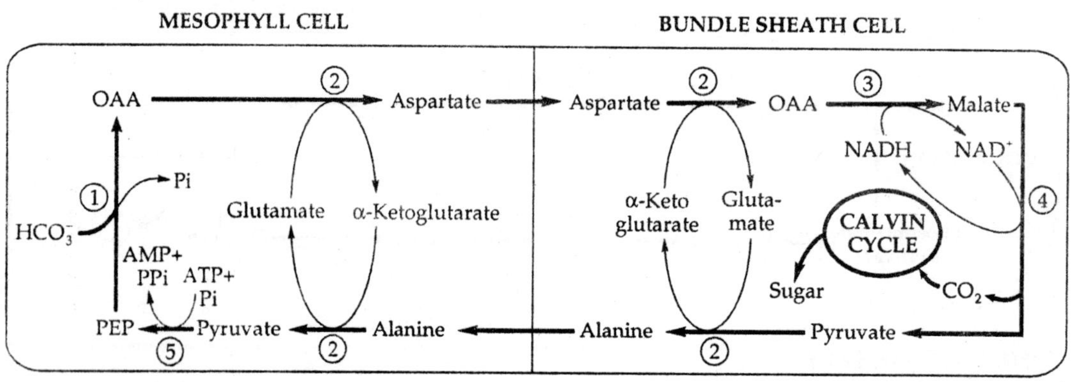

II. NAD⁺-Malic enzyme (NAD⁺-ME) type : (1) PEP-carboxylase, (2) Glutamate-oxoglutarate aminotransferase, (3) Malate dehydrogenase, (4) NAD⁺-dependent Malic enzyme, (5) Pyruvate orthophosphate dikinase.

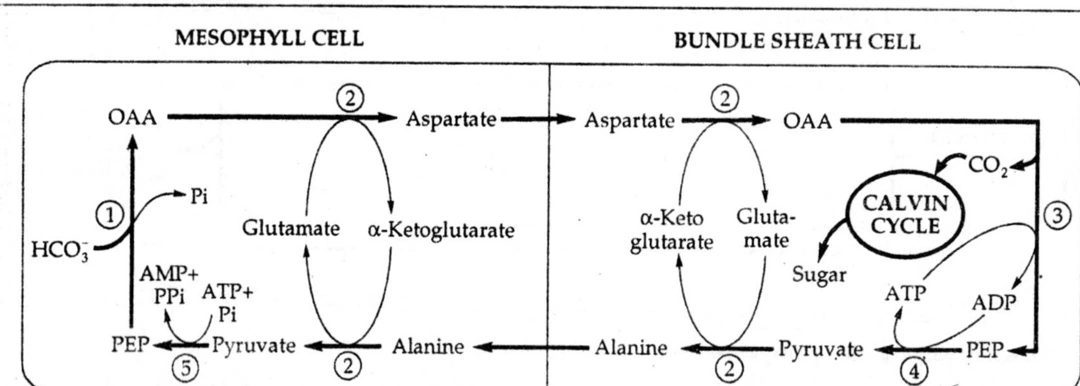

III. PEP-carboxykinase (PCK) type : (1) PEP-carboxylase, (2) Glutamate-oxoglutarate aminotransferase, (3) PEP-carboxykinase, (4) Pyruvate kinase, (5) Pyruvate orthophosphate dikinase.

Fig. 2.39: Three types of C_4 cycle :
I. NADP⁺-Malic enzyme (NADP⁺-ME) type;
II. NAD⁺-Malic enzyme (NAD⁺-ME) type;
III. PEP-carboxykinase (PCK) type.

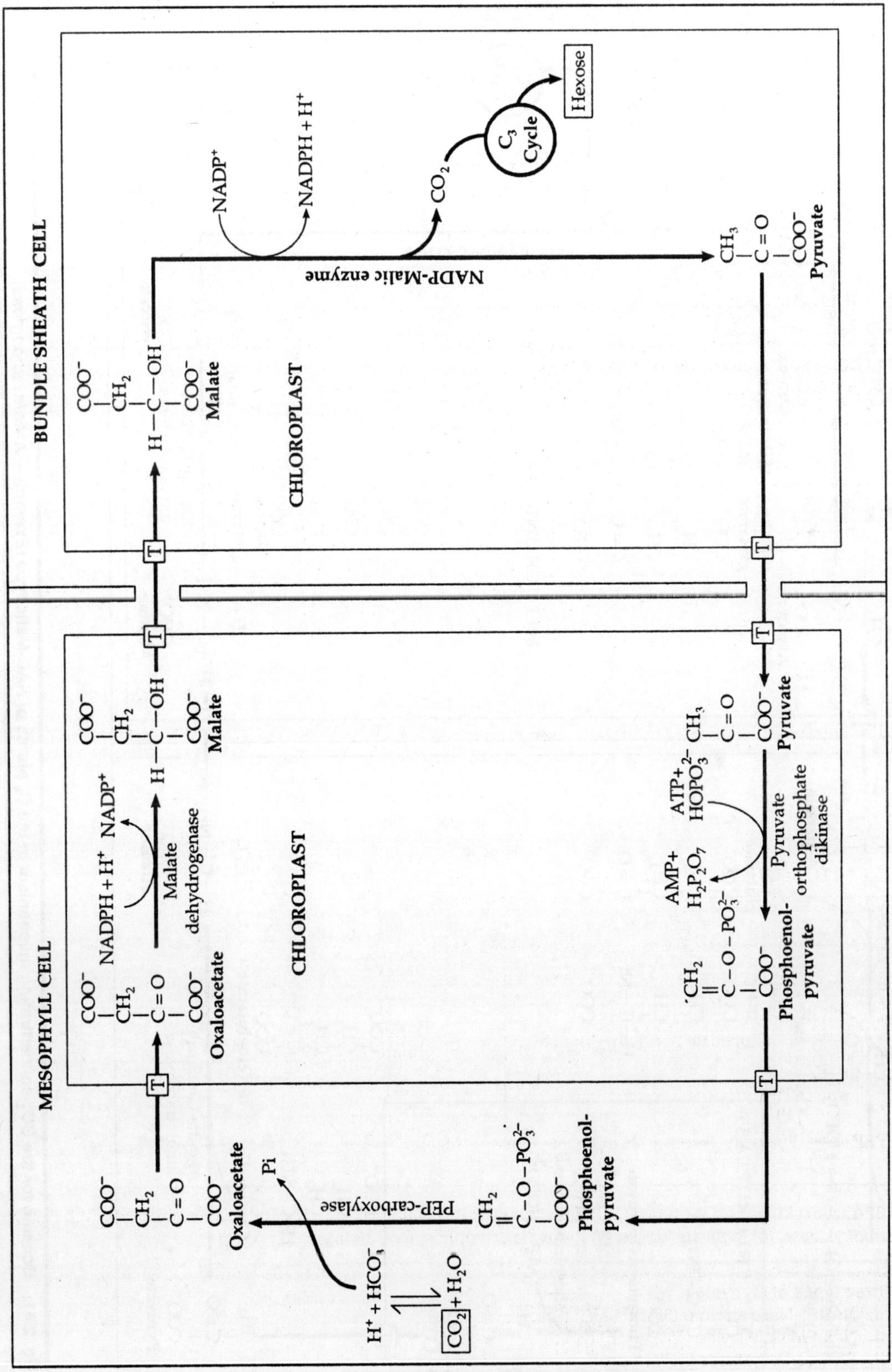

Fig. 2.40 : Scheme for CO_2-concentrating mechanism in the C_4-plants of NADP-Malic type (Example—*Zea mays*).

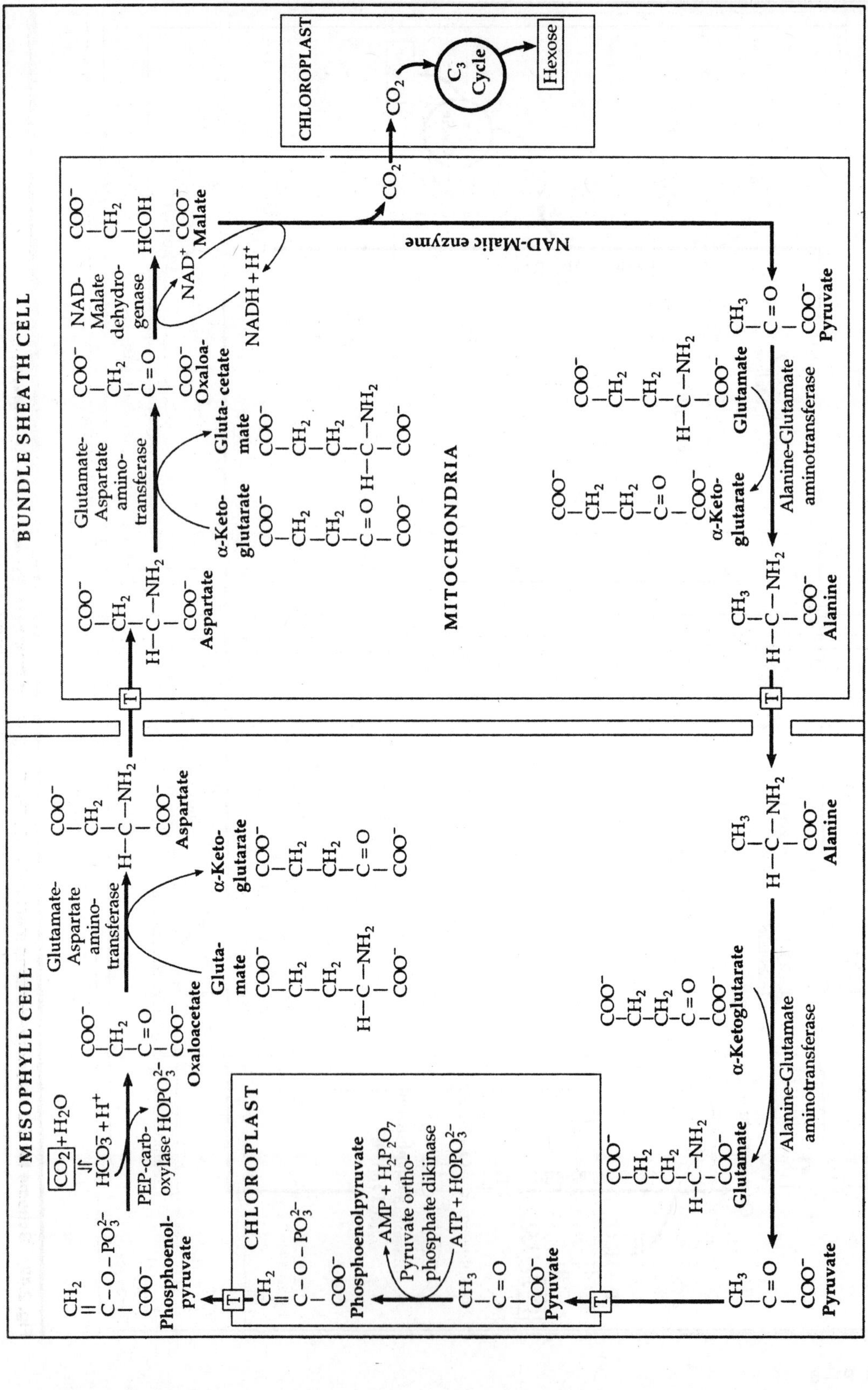

Fig. 2.41: Scheme for the CO_2-concentrating mechanism in the C_4 plants of NAD-Malic type (Example—*Atriplex spongiosa*)

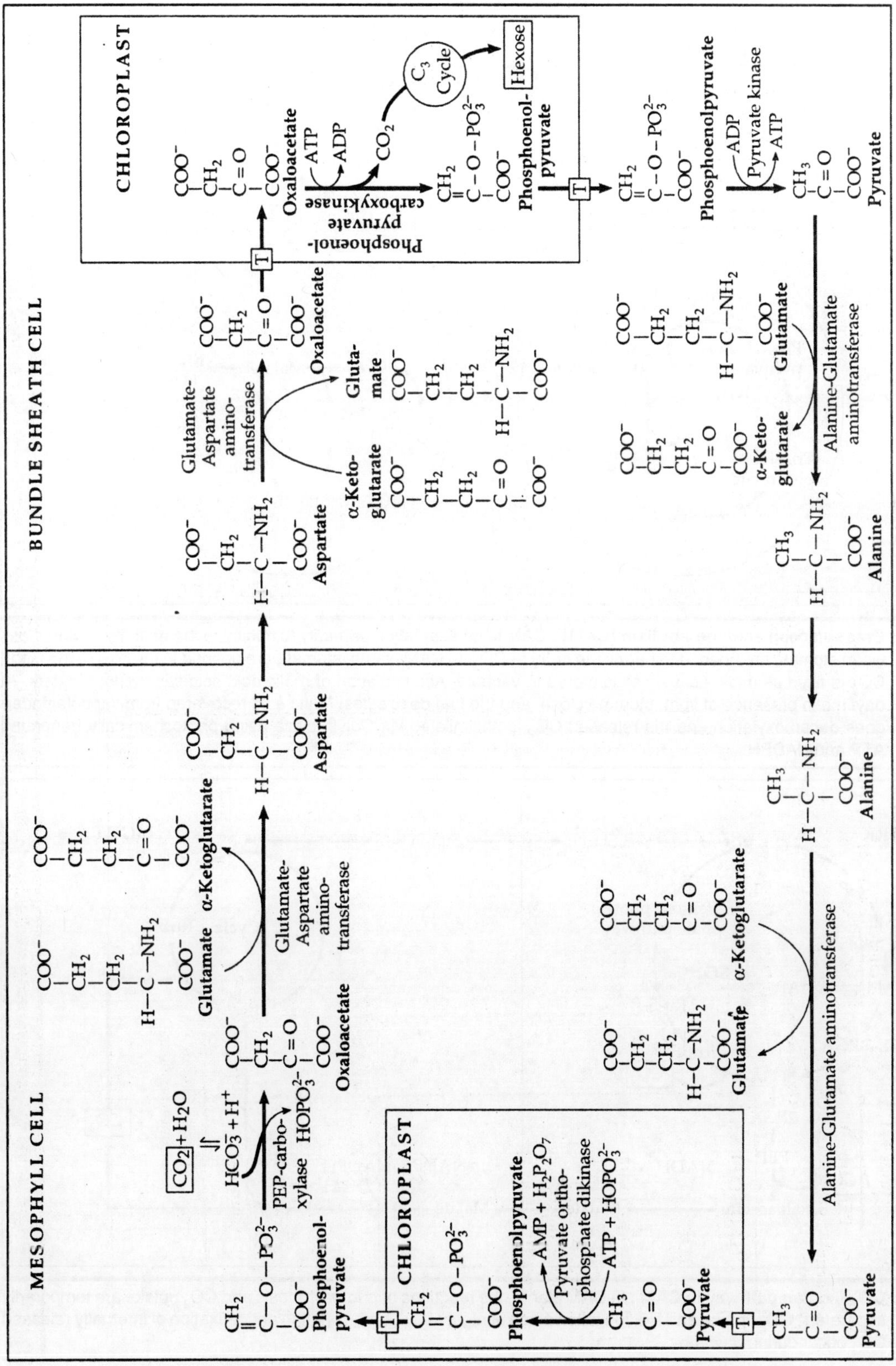

Fig. 2.42: Scheme for the CO_2-concentrating mechanism in the C_4 plants of PCK type (Example—*Panicum maximum*)

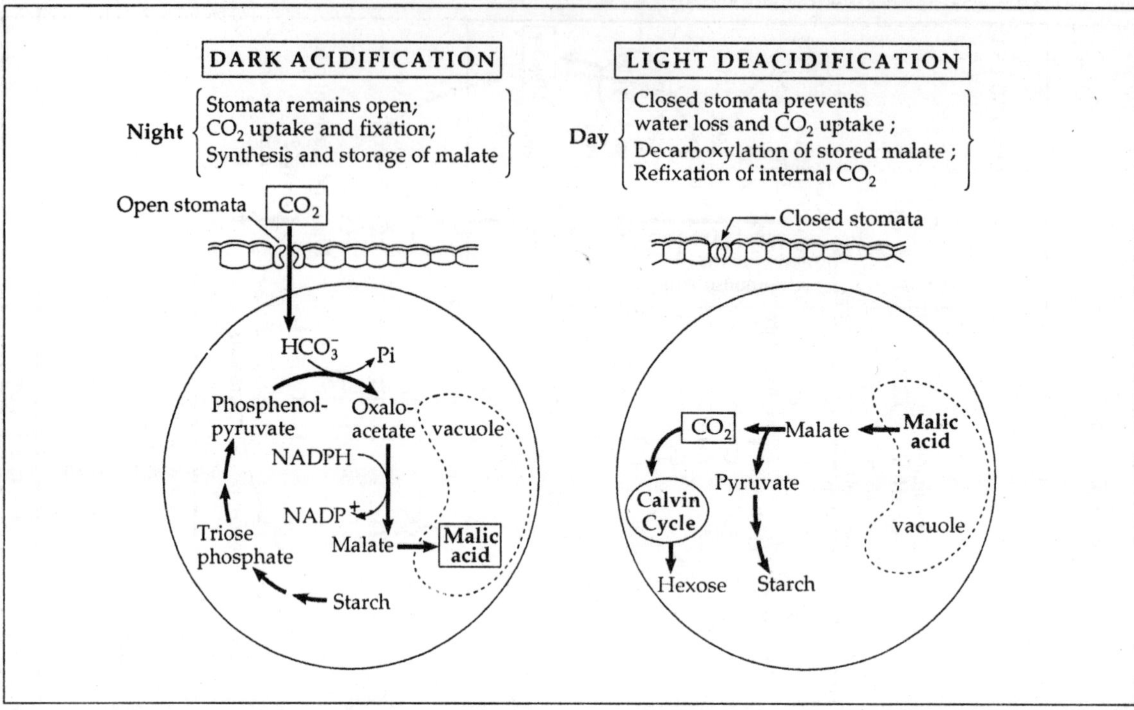

Fig. 2.43a: Crassulacean acid metabolism [CAM] : CAM is an adaptation primarily to minimize the quantity of water loss when stomata are opened to permit the entry of CO_2. In CAM plants, stomata are opened in the cool of the night. CO_2 is fixed as malic acid which is stored in vacuole. Accumulation of malic acid acidifies the leaf in dark. At daytime in presence of light, stomata close, and the leaf deacidifies. Malic acid recovered from vacuole undergoes decarboxylation and the released CO_2 is assimilated via Calvin cycle using photochemically generated ATP and NADPH.

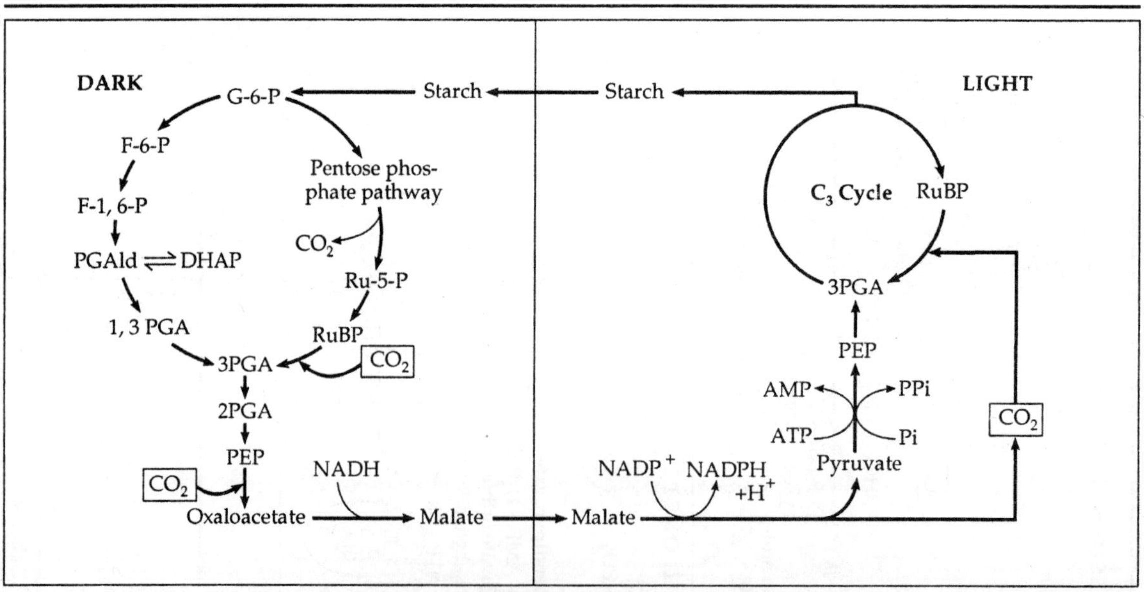

Fig. 2.43b: CO_2-fixation pathway in CAM : In CAM plants, the reactions of photosynthesis and CO_2 uptake are temporarily separated : CO_2 uptake and fixation takes place at night and decarboxylation and refixation of internally released CO_2 occur during daytime.

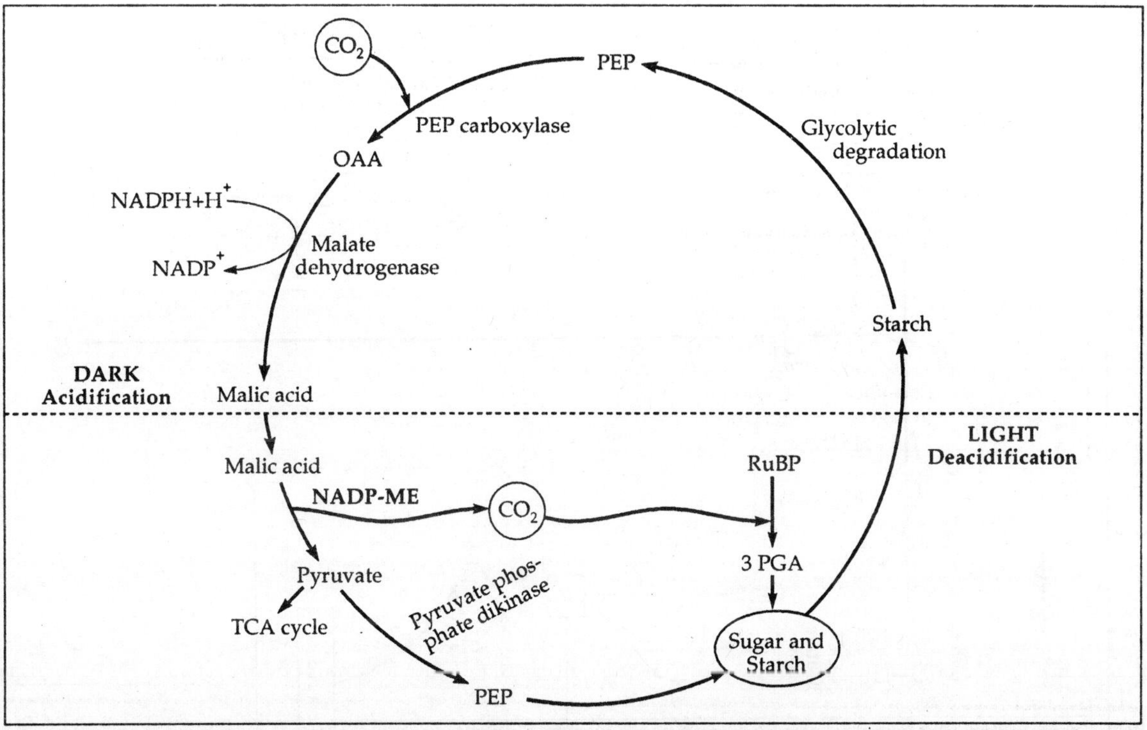

Fig. 2.44a: Schematic representation of dark acidification and light deacidification of CAM pathway (*ME-CAM Type*).

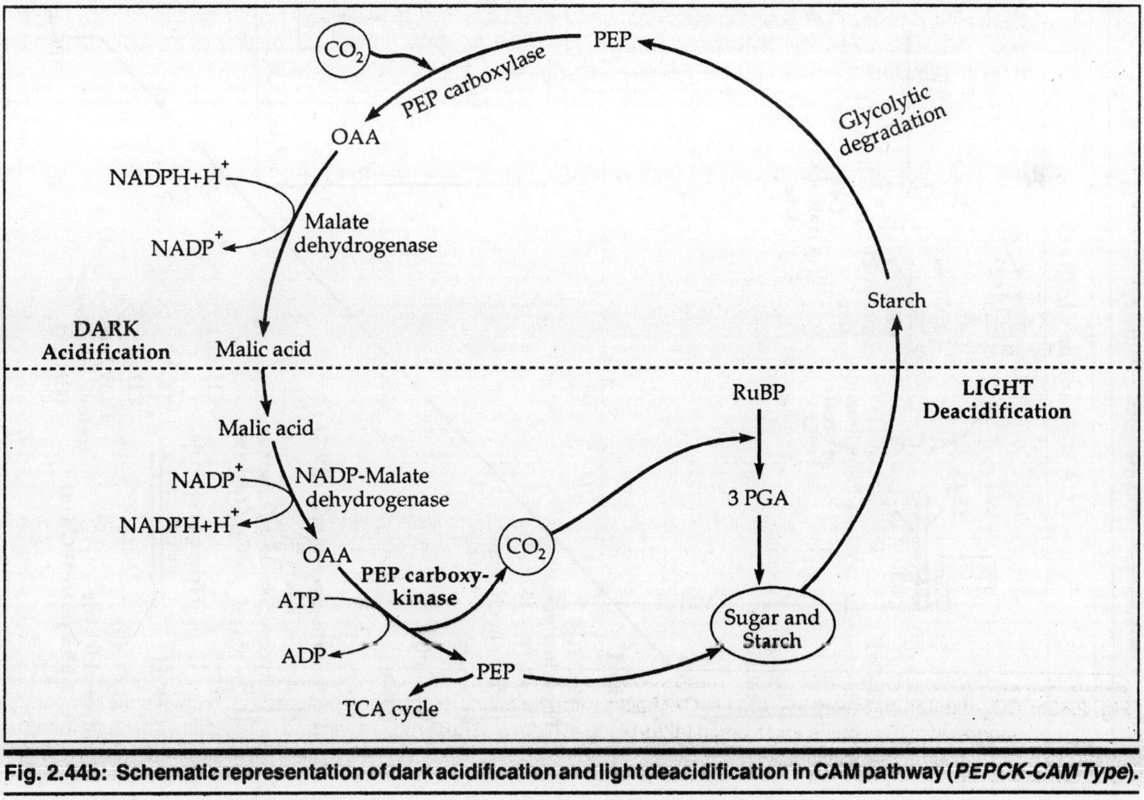

Fig. 2.44b: Schematic representation of dark acidification and light deacidification in CAM pathway (*PEPCK-CAM Type*).

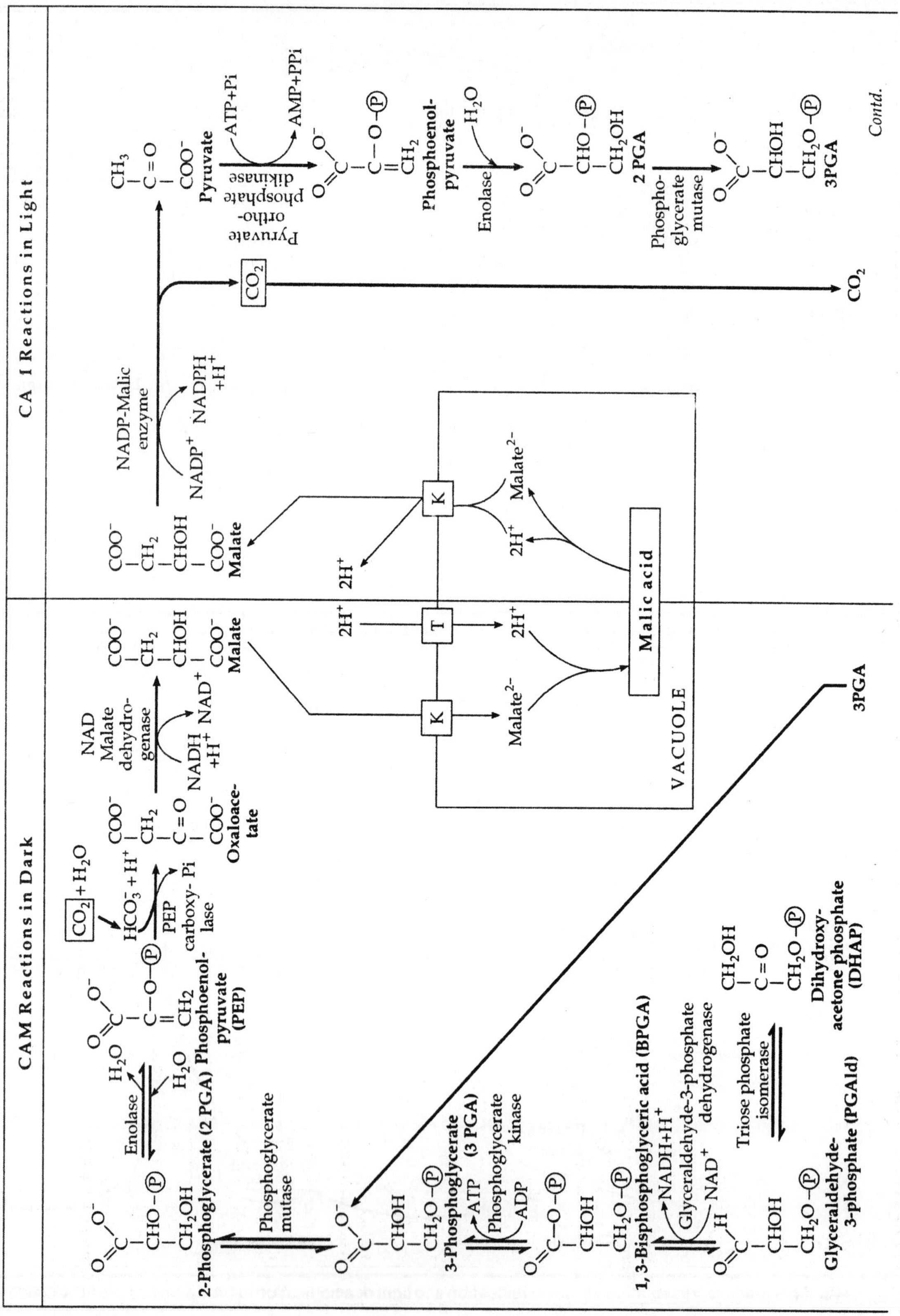

Fig. 2.45: Crassulacean Acid Metabolic Pathway.

Contd.

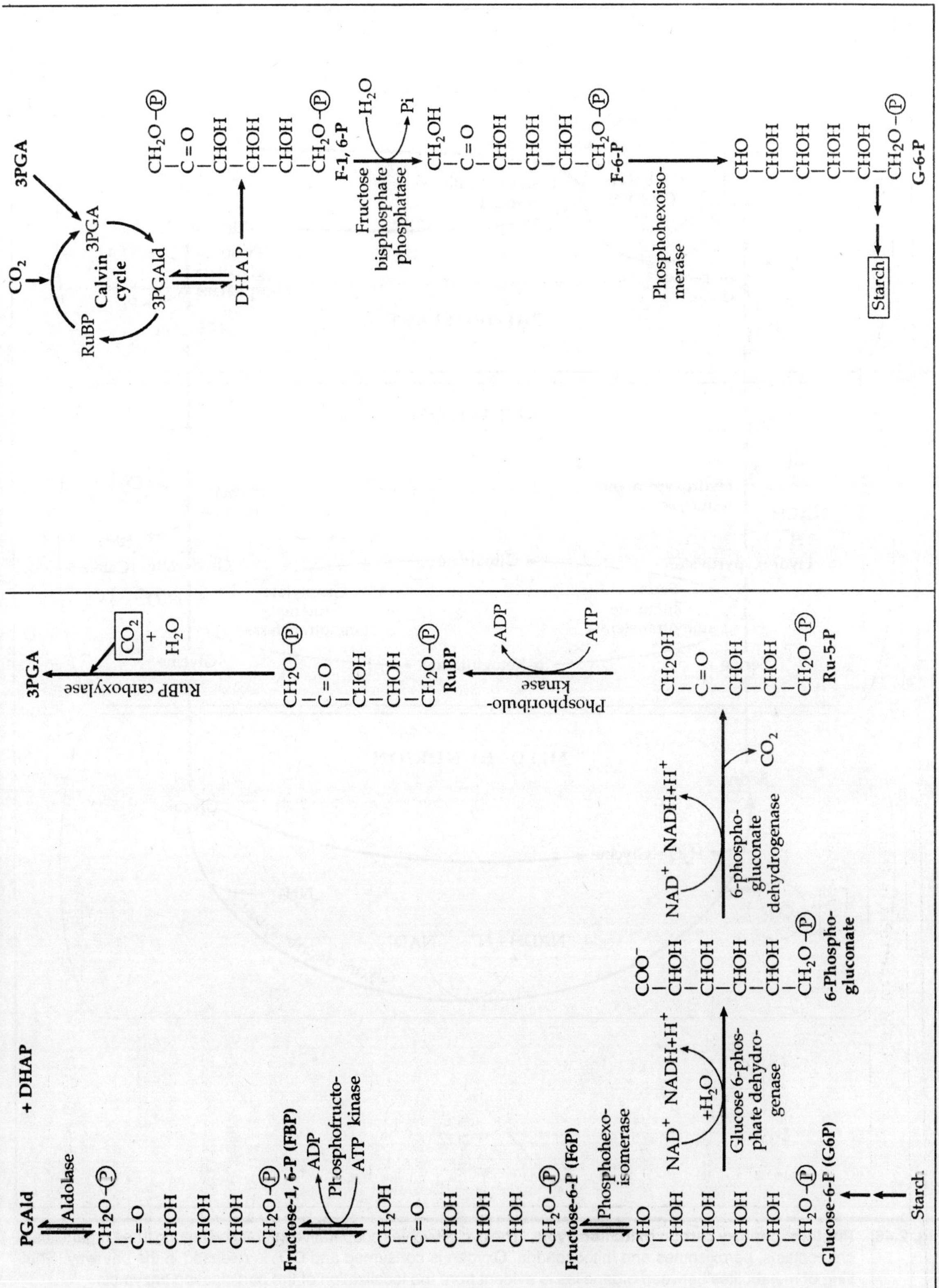
PGAld
+ DHAP
Aldolase
Fructose-1, 6-P (FBP)
ADP
Phosphofructo-
ATP kinase
Fructose-6-P (F6P)
Phosphohexo-
isomerase
Glucose-6-P (G6P)
Starch
NAD+ NADH+H+
+H2O
Glucose 6-phos-
phate dehydro-
genase
6-Phospho-
gluconate
6-phospho-
gluconate
dehydrogenase
CO2
Ru-5-P
Phosphoribulo-
kinase
ADP
ATP
RuBP
RuBP carboxylase
3PGA
CO2
+
H2O
RuBP
Calvin
cycle
3PGAld
DHAP
F-1, 6-P
Fructose
bisphosphate
phosphatase
H2O
Pi
F-6-P
Phosphohexoiso-
merase
G-6-P
Starch

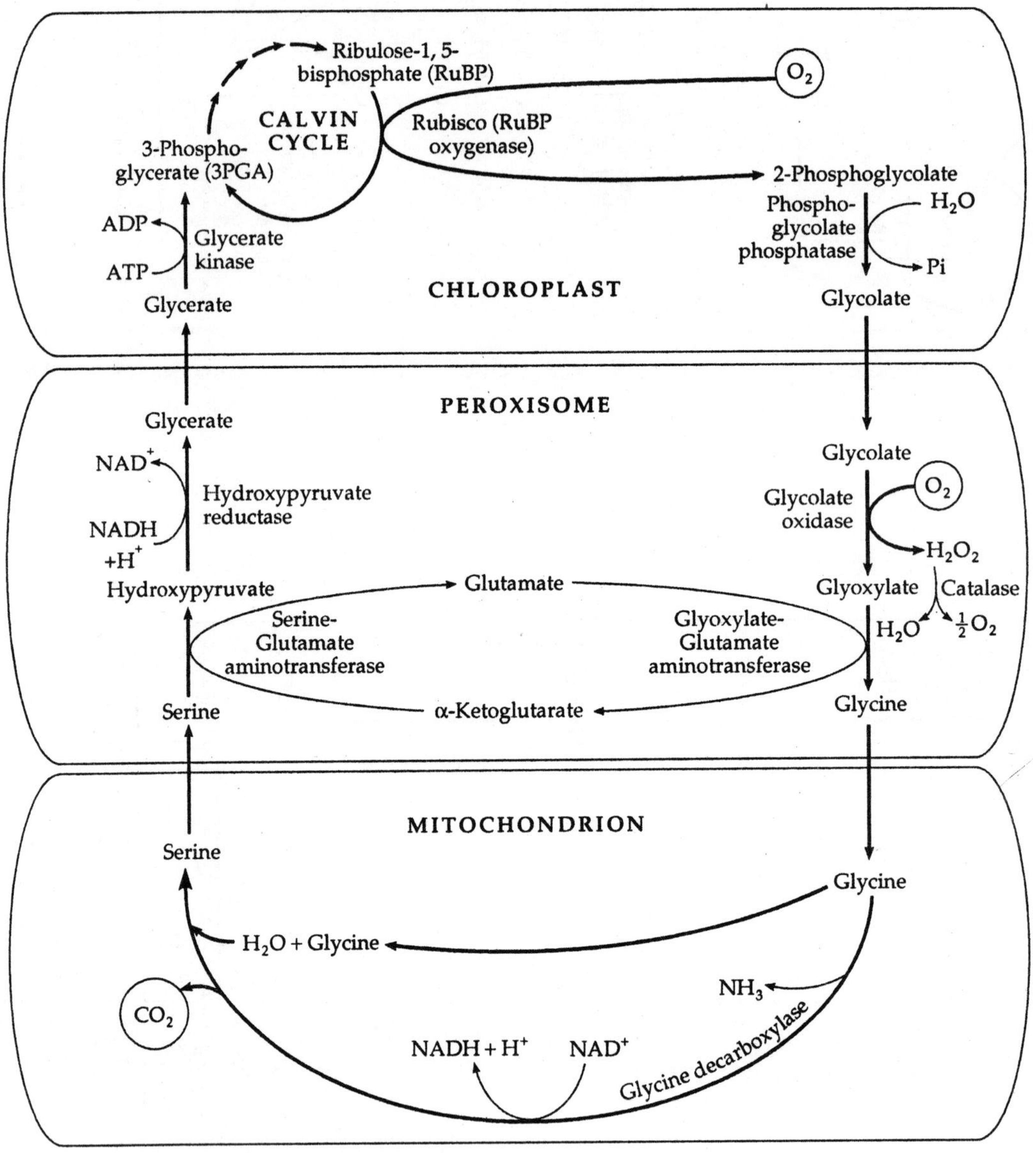

Fig. 2.46: Photorespiratory carbon oxidation cycle : The cycle involves cooperative interaction among three organelles—chloroplasts, peroxisomes and mitochondria. Oxygen is consumed and CO_2 is released in the pathway. Shuttling of metabolites between organelles is evident from the localization of the participating enzymes.

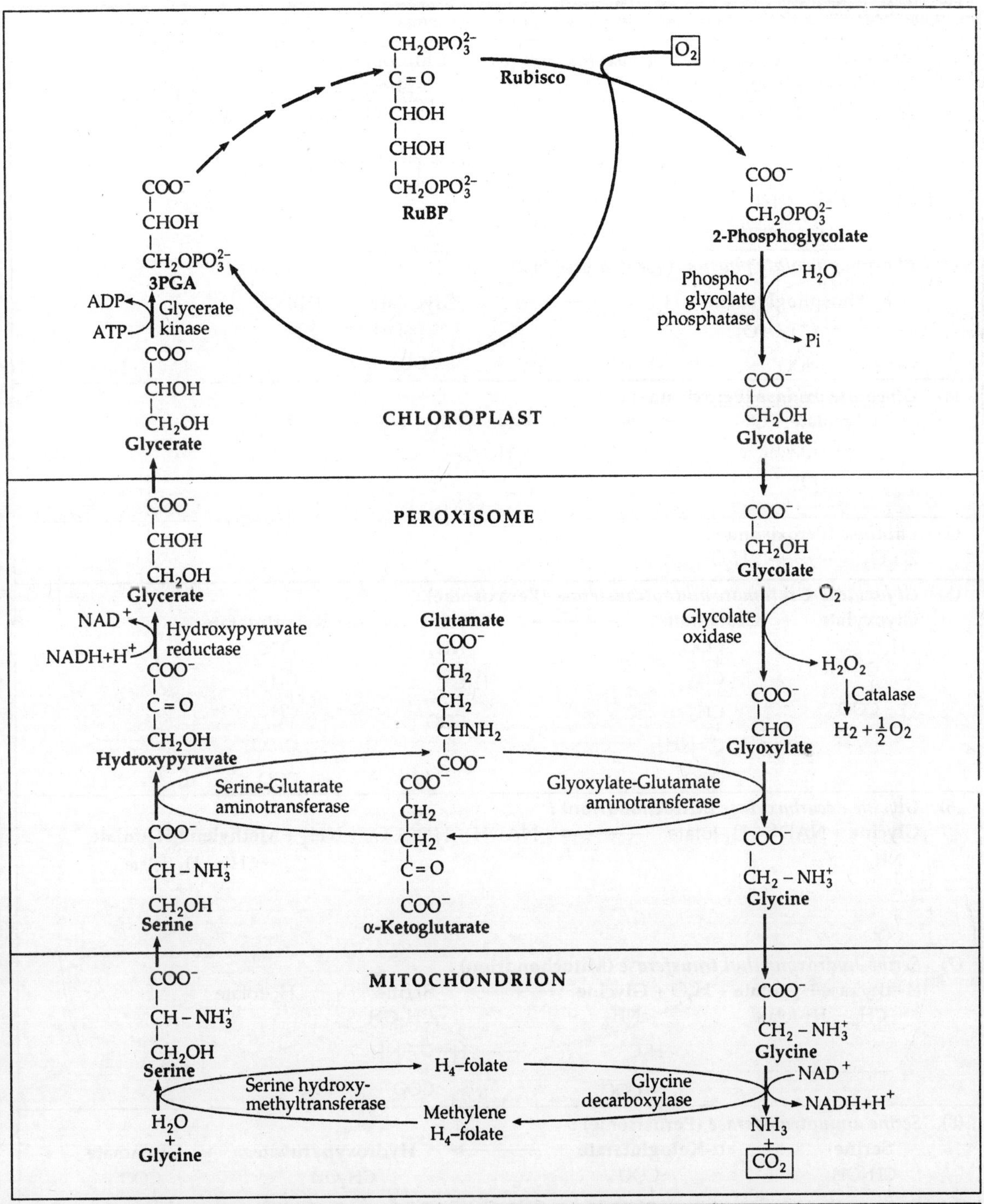

Fig. 2.47: Photorespiratory Glycolate Pathway : This pathway metabolizes the phosphoglycolate produced by RuBP carboxylase-catalyzed oxidation of RuBP. Phosphoglycolate by dephosphorylation forms glycolate in chloroplast which is converted to glyoxylate in peroxisome and subsequently transaminated to glycine. In mitochondria, two molecules of glycine condense to form serine and CO_2 released by the action of glycine decarboxylase. Serine is converted to hydroxypyruvate and then to glycerate in peroxisome. Glycerate re-enters chloroplast where it is phosphorylated to rejoin the Calvin cycle.

Fig. 2.48: Reactions of the photorespiratory carbon oxidation cycle and the corresponding enzymes.

(1) ***Ribulose-1, 5-bisphosphate carboxylase/oxygenase* (Chloroplast) :**

Ribulose-1, 5-bisphosphate + O_2 ⟶ **2-phosphoglycolate** + **3-phosphoglycerate** + $2H^+$

$CH_2OPO_3^{2-}$–C(=O)–CH(OH)–CH(OH)–$CH_2OPO_3^{2-}$ ⟶ $CH_2OPO_3^{2-}$–COO^- + $CH_2OPO_3^{2-}$–C(HO)(H)–COO^-

(2) ***Phosphoglycolate phosphatase* (Chloroplast) :**

Phosphoglycolate + H_2O ⟶ **Glycolate** + $HOPO_3^{2-}$

$CH_2OPO_3^{2-}$–COO^- ⟶ CH_2OH–COO^-

(3) ***Glycolate oxidase* (Peroxisome) :**

Glycolate + O_2 ⟶ **Glyoxylate** + H_2O_2

CH_2OH–COO^- ⟶ H–C(=O)–COO^-

(4) ***Catalase* (Peroxisome) :**

$2H_2O_2$ ⟶ $2H_2O + O_2$

(5) ***Glyoxylate-Glutamate aminotransferase* (Peroxisome) :**

Glyoxylate + **Glutamate** ⟶ **Glycine** + **α-Ketoglutarate**

H–C(=O)–COO^- + COO^-–CH_2–CH_2–CH(NH_2)–COO^- ⟶ NH_2–H_2C–COO^- + COO^-–CH_2–CH_2–C(=O)–COO^-

(6) ***Glycine decarboxylase* (Mitochondrion) :**

Glycine + NAD^+ + H_4-folate ⟶ NADH + H^+ + CO_2 + NH_3 + **Methylene-H_4-folate**

NH_2–H_2C–COO^- ⟶ $-CH_2-H_4$-folate

(7) ***Serine-hydroxymethyl transferase* (Mitochondrion) :**

Methylene-H_4-folate + H_2O + **Glycine** ⟶ **Serine** + H_4-folate

$-CH_2-H_4$-folate + NH_2–H_2C–COO^- ⟶ CH_2OH–CH(NH_2)–COO^-

(8) ***Serine aminotransferase* (Peroxisome) :**

Serine + **α-Ketoglutarate** ⟶ **Hydroxypyruvate** + **Glutamate**

CH_2OH–CH(NH_2)–COO^- + COO^-–CH_2–CH_2–C(=O)–COO^- ⟶ CH_2OH–C(=O)–COO^- + COO^-–CH_2–CH_2–CH(NH_2)–COO^-

contd.

(9) ***Hydroxypyruvate reductase*** **(Peroxisome) :**

Hydroxypyruvate + NADH + H^+ ⟶ **Glycerate** + NAD^+

(10) ***Glycerate kinase*** **(Chloroplast) :**

Glycerate + ATP ⟶ **3-Phosphoglycerate** + ADP + H^+

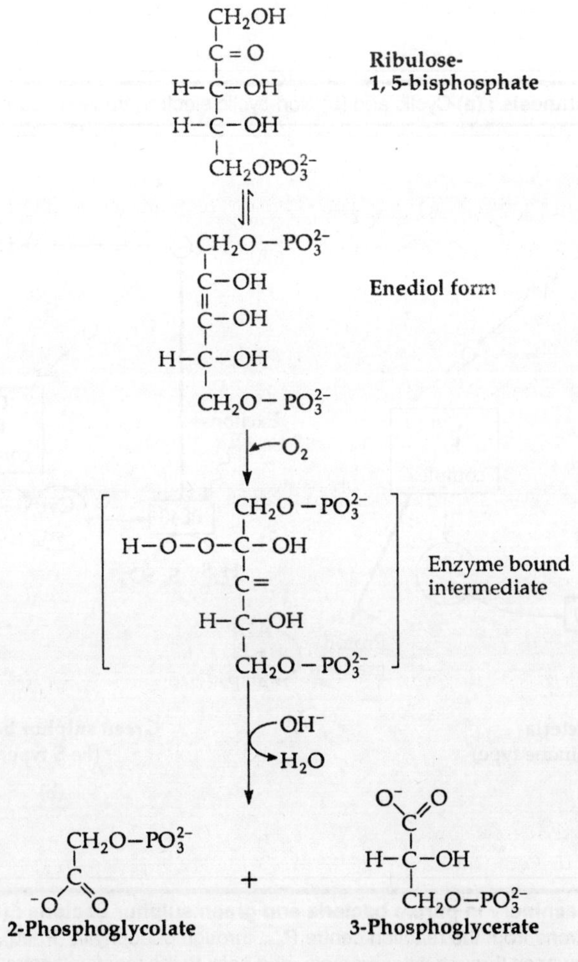

Fig. 2.49: Oxygenase activity of rubisco : Rubisco can incorporate O_2 rather than CO_2 into ribulose-1, 5-bisphosphate. The unstable intermediate thus formed splits into phosphoglycolate, which is recycled and 3-phosphoglycerate, which can re-enter the Calvin cycle.

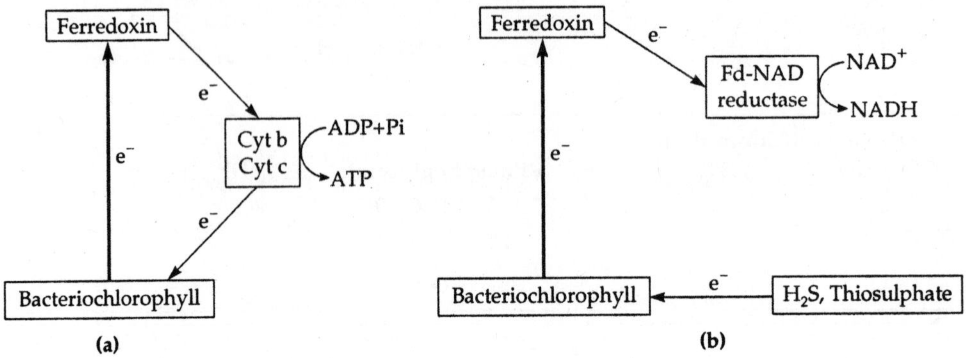

Fig. 2.50: Bacterial photosynthesis : (a) Cyclic and **(b)** Non-cyclic electron transport pathway.

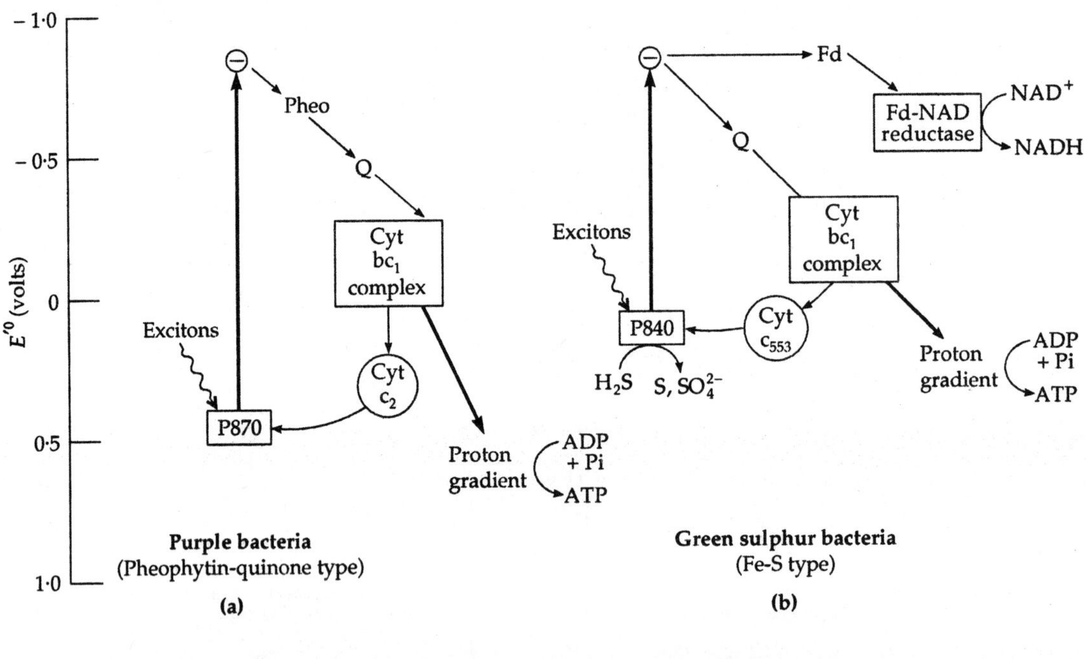

Fig. 2.51: Photosynthetic machinery in purple bacteria and green sulphur bacteria : (a) In the purple bacteria, light energy drives electrons from the reaction centre P_{870} through pheophytin (Pheo), a quinone (Q), and the cytochrome bc_1 complex, then through cytochrome c_2 and back to the reaction centre. Electron flow through the Cyt bc_1 complex causes proton pumping, creating an electrochemical potential that powers ATP synthesis. **(b)** Green sulphur bacteria have two routes for electrons driven by excitation of P_{840}. A cyclic route passes through a quinone (Q) to the cytochrome bc_1 complex and back to the reaction centre via cytochrome c. A non-cyclic route passes from the reaction centre through the iron-sulphur protein ferredoxin (Fd), then to NAD^+ in a reaction catalyzed by ferredoxin-NAD reductase.

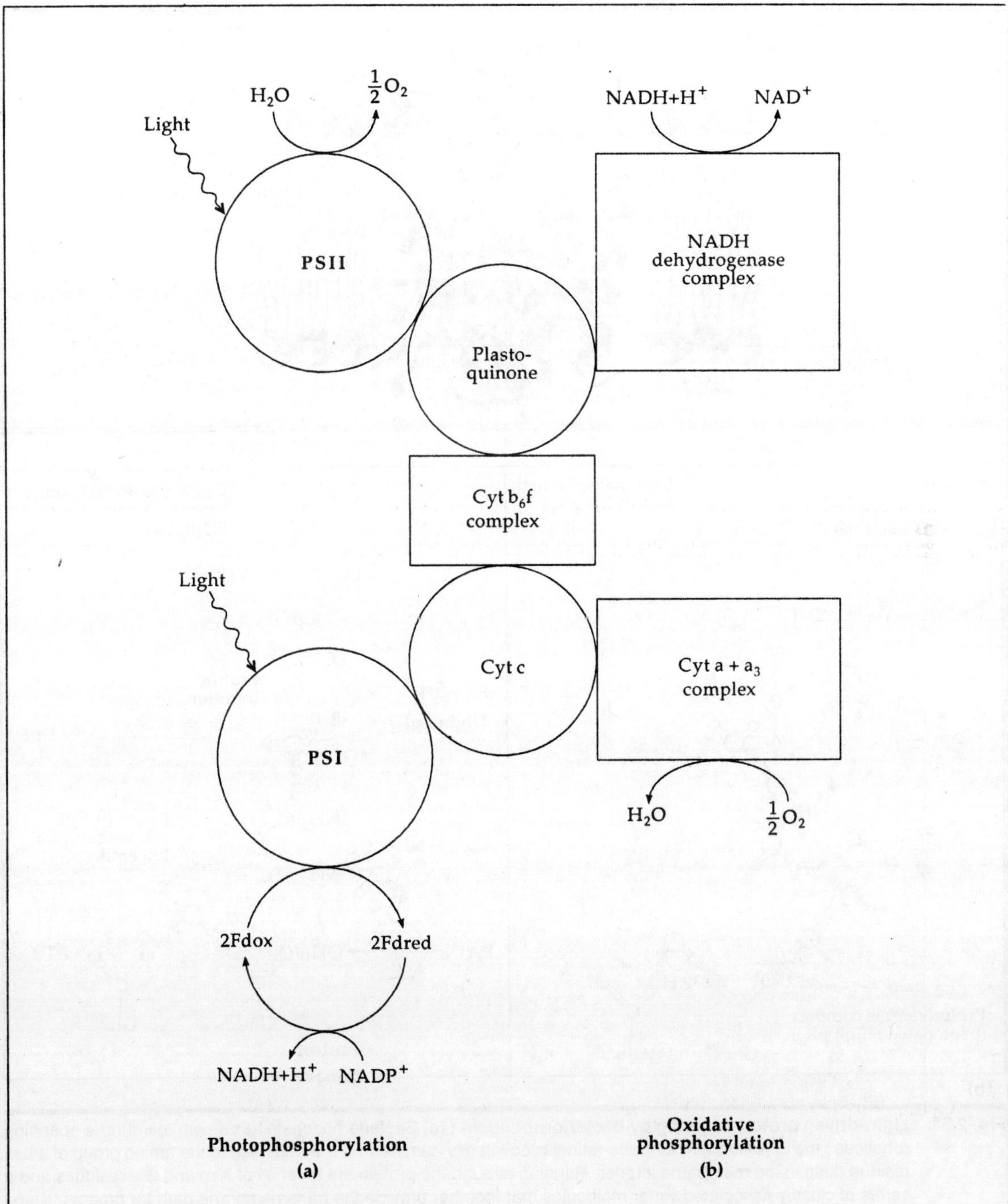

Fig. 2.52: Dual roles of Cyt b_6f and Cytochrome c in Cyanobacteria : Cyanobacteria use cytochrome b_6f, cytochrome c and plastoquinone for both oxidative phosphorylation and photophosphorylation. **(a)** In photophosphorylation, electrons flow from water to $NADP^+$. **(b)** In oxidative phosphorylation, electrons flow from NADH to O_2. Both processes are accompanied by proton movement across the membrane, accomplished by a Q cycle.

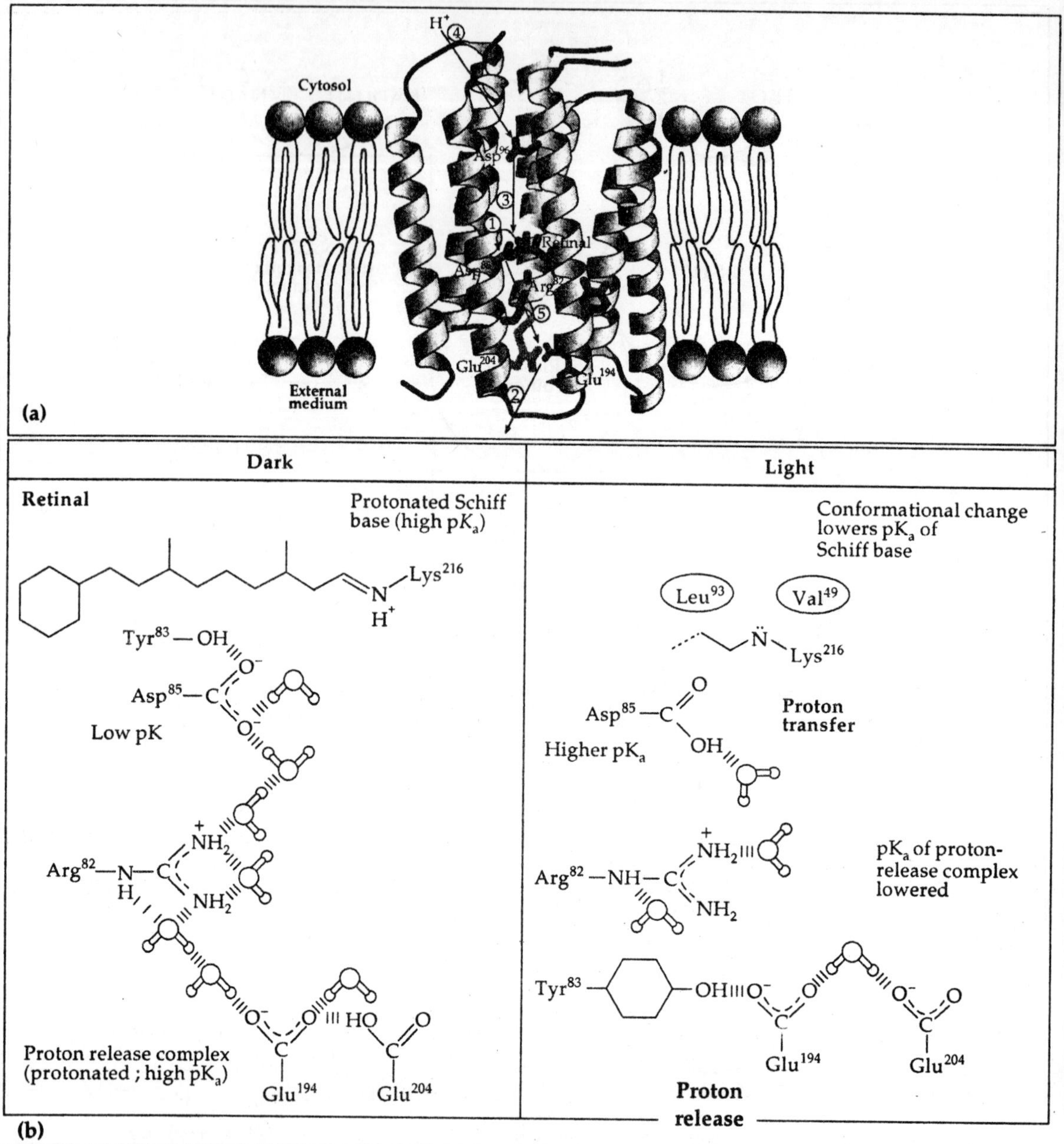

Fig. 2.53: Light-driven proton pumping by bacteriorhodopsin : (a) Bacteriorhodopsin has seven membrane-spanning α-helices ; the chromophore all-*trans*-retinal is covalently attached via a Schiff base to the amino group of a Lys residue deep in the membrane interior. Running through the protein are a series of Asp and Glu residues and a series of closely associated water molecules that together provide the transmembrane path for protons. Steps ① through ⑤ indicate proton movements. **(b)** In the dark, the Schiff base is protonated. Illumination photoisomerizes the retinal, forcing subtle conformational changes in the protein that alter the distance between the Schiff base and its neighbouring amino acid residues. Interaction with these neighbours lowers the pK_a of the protonated Schiff base, and the base gives up its proton to a nearby carboxyl group on Asp^{85} (step-**1** in **a**). This initiates a series of concerted proton hops between water molecules in the interior of the protein, which ends with step-**2**, the release of a proton that was shared by Glu^{194} and Glu^{204} near the extracellular surface. The Schiff base reacquires a proton from Asp^{96} (step-**3**), which in turn takes up a proton from the cytosol (step-**4**). Finally, Asp^{85} gives up its proton, leading to the reprotonation of the Glu^{204}-Glu^{194} pair (step-**5**).

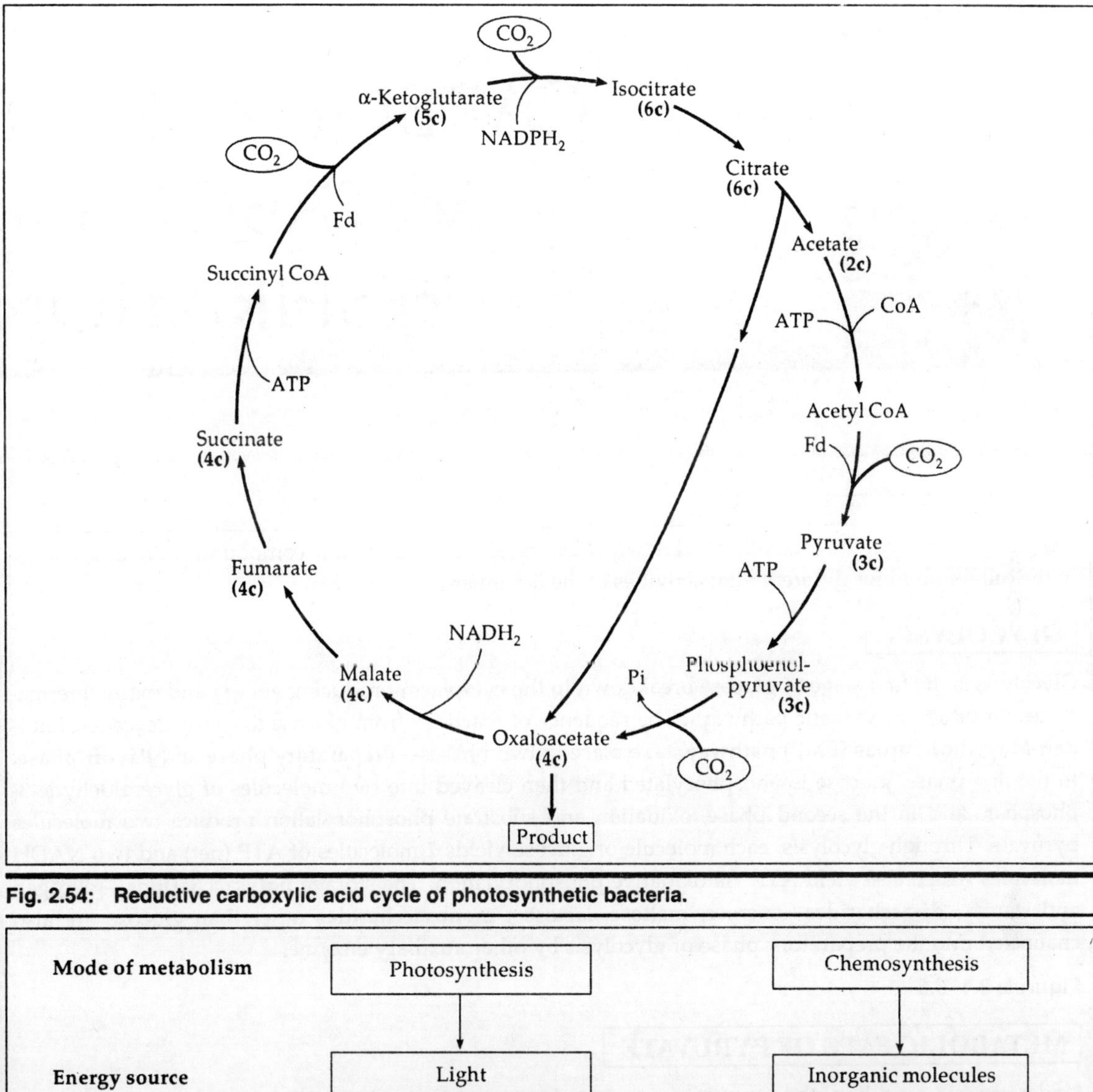

Fig. 2.54: Reductive carboxylic acid cycle of photosynthetic bacteria.

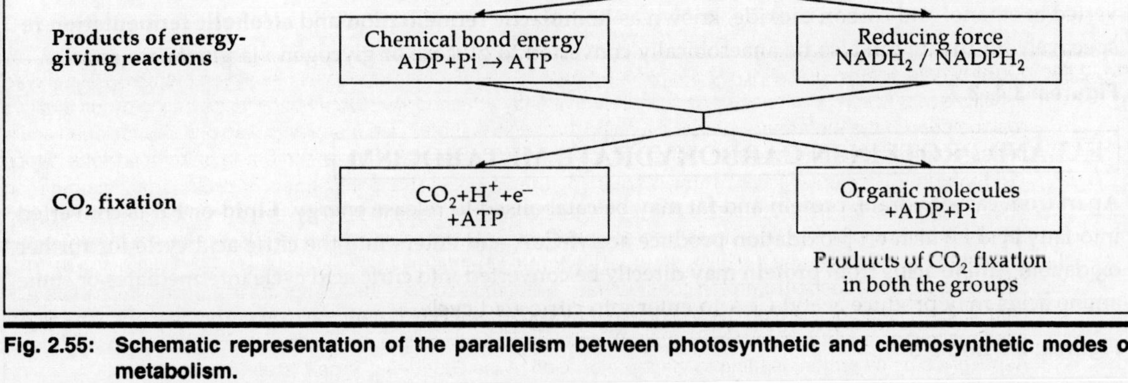

Fig. 2.55: Schematic representation of the parallelism between photosynthetic and chemosynthetic modes of metabolism.

CHAPTER

3

CARBOHYDRATE METABOLISM: RESPIRATION

Respiration involves the catabolic breakdown of carbon skeleton of organic compounds to release energy in the cell required for different vital activities of the organism.

GLYCOLYSIS

Glycolysis is the first stage of glucose breakdown in the cytoplasm producing energy and many intermediates for other biosynthetic pathways. The sequence of reactions from glucose to pyruvate, called **Embden-Meyerhof-Parnas** (EMP) **pathway**, take place in two phases—**Preparatory phase** and **Payoff phase.** In the first phase, glucose is phosphorylated and then cleaved into two molecules of glyceraldehyde-3-phosphate and in the second phase, oxidation and substrate phosphorylation produce two molecules pyruvate. Through glycolysis, each molecule of glucose yields 2 molecules of ATP (net) and two NADH molecules which also yield ATP via oxidative phosphorylation. The storage **polysaccharides**—glycogen and starch ; **disaccharides**—sucrose, maltose, lactose ; **monosaccharides** other than glucose are also channeled into the preparatory phase of glycolysis by other auxiliary enzymes.

Figures: 3.1–3.3

METABOLIC FATE OF PYRUVATE

Under **aerobic condition,** the pyruvate is converted to acetyl-CoA which will be completely oxidised to CO_2 through **citric acid cycle**. Whereas under **anaerobic condition,** pyruvate is reduced to lactate or converted to ethanol and carbon dioxide, known as **homolactic fermentation** and **alcoholic fermentation** respectively. Pyruvate can also be anaerobically converted to glucose or glycogen via **gluconeogenesis.**

Figures: 3.4–3.7

FAT AND PROTEIN IN CARBOHYDRATE METABOLISM

Apart from carbohydrate, protein and fat may be catabolised to release energy. **Lipid** or **Fat** is converted into fatty acids which on β-oxidation produce acetyl-CoA and enters into the citric acid cycle for further oxidation. Amino acids from **protein** may directly be converted into citric acid cycle intermediates or some amino acids may produce acetyl-CoA to enter into citric acid cycle.

Figures: 3.8 and 3.9

CITRIC ACID CYCLE

The **Krebs cycle** or citric acid cycle or tricarboxylic acid (TCA) cycle, occurs in the mitochondria, oxidises pyruvate to CO_2 and H_2O with concomitant production of energy. The cycle begins with the condensation of acetyl-CoA and oxaloacetate to form citric acid. Citrate through various **oxidation, decarboxylation** and other conversion steps regenerates oxaloacetate. Each step of oxidation requires coenzyme, either NAD^+ or FAD^+, which get reduced and upon reoxidation through electron transport chain produces energy. Each turn of Krebs cycle yields three NADH, one $FADH_2$, one GTP (ATP) and two CO_2 by oxidative decarboxylation reactions. Reoxidation of NADH and $FADH_2$ through electron transport system yields 11 ATP ; thus total **ATP yields** per turn of Krebs cycle are 12 molecules. Intermediates of TCA cycle are also used as precursors of many biosynthetic pathways.

Figures: 3.10–3.13

MITOCHONDRIAL ELECTRON TRANSPORT CHAIN

Electron transport and oxidative phosphorylation reoxidise NADH and $FADH_2$, and trap the energy released as ATP. The electron transport chain in the inner mitochondrial membrane consists of **4 enzyme complexes :** Complex I (NADH dehydrogenase), Complex II (Succinate dehydrogenase), Complex III (Ubiquinone-cytochrome c oxidoreductase), Complex IV (Cytochrome oxidase). During electron transfer to oxygen by each complex, the protons are translocated leading to the generation of proton electrochemical gradient that drives ATP synthesis. According to **chemiosmotic model** of **oxidative phosphorylation,** due to electron transfer the proton gradient is formed across the inner mitochondrial membrane, which creates a chemical gradient (ΔpH) across the membrane. The proton flow through the ATP synthase complex is coupled to the conversion of ADP + Pi to ATP. Approximately 3 and 2 molecules of ATP are synthesized per NADH and $FADH_2$ oxidised respectively. Some chemicals like DNP, FCCP are **uncoupling agents** which allow electron transport to proceed without ATP synthesis, they carry protons across the inner membrane thus dissipate the proton gradient and the energy released as heat.

Figures: 3.14–3.29

OXIDATION OF CYTOSOLIC NADH

Cytosolic NADH can be oxidised via the **glycerol-3-phosphate shuttle.** Cytosolic DHAP reduced to glycerol-3-phosphate by NADH which enters the mitochondria and converted back to DHAP by FAD dehydrogenase. $FADH_2$ produced is reoxidised by transferring electrons to ubiquinone in the electron transport chain and produces 2 molecules of ATP. Cytosolic NADH can also be oxidised via **malate-aspartate shuttle.** Oxaloacetate in cytosol is reduced to malate by NADH and enters mitochondria where it is reoxidised to oxaloacetate to produce NADH.

Bacterial respiratory chain, occurring in the inner plasma membrane, contains a minimal form of complex I which carries NADH to ubiquinone or menaquinone, pumping protons outwards and creating an electrochemical potential for ATP synthesis.

Figures: 3.30–3.33

ALTERNATIVE OXIDASE

Mitochondrial cytochrome pathway of electron transport is inhibited by cyanide, antimycin A, rotenone, amytal, azide, CO, etc. **Resistant pathways** exist in higher plants with the involvement of alternative oxidase system. Metabolic conditions resulting in the accumulation of pyruvate, intramitochondrial NADH or reduced ubiquinone have the potential to increase electron transport to alternative oxidase.

Figures: 3.34–3.36

REGULATION OF RESPIRATORY PATH

Regulation of **glycolysis** occurs in steps which are catalysed by phosphofructokinase (PFK), but hexokinase and pyruvate kinase are additional control sites. PFK is allosterically inhibited by ATP and citrate. The inhibition of hexokinase and pyruvate kinase occurs by glucose 6-P and ATP respectively. Conversion of pyruvate to acetyl-CoA is inhibited by acetyl-CoA and NADH.

The **citric acid cycle** is regulated at the steps catalysed by citrate synthase, isocitrate dehydrogenase and α-ketoglutarate dehydrogenase via feedback inhibition by ATP, citrate, NADH and succinyl-CoA.

Electron transport is under regulatory control of ADP ; if ADP is available, electron transport proceeds to produce ATP.

Figure: 3.37

ANABOLIC ROLE OF RESPIRATORY PATH

Glycolysis and the TCA cycle contribute intermediates to a variety of other metabolic pathways in higher plants. Glycolytic intermediates or end products may be reversed back to glucose synthesis, called **gluconeogenesis**, which may be converted to polysaccharides—starch, cellulose, etc. Entry through **shikimic acid pathway** may result in alkaloids, flavonoids, lignin, amino acids, proteins, IAA, etc. Entry through **pentose-phosphate pathway** may result in nucleotides, which can enter into nucleic acids, coenzymes, cytokinin, etc. Intermediates may enter into lipids also. Acetyl-CoA may be converted to fatty acids and lipids, terpenoids, isoprenoids, sterols, GA, ABA, etc.

Citric acid cycle has **amphibolic function**—intermediates are drawn off by cataplerotic reactions ; and anaplerotic reactions replenish cycle intermediates. Intermediates of the cycle are used as precursors in many biosynthetic pathways leading to the synthesis of fatty acids and sterols, amino acids and proteins, porphyrins and heme, pyrimidines and nucleotides, glucose and other carbohydrates.

Figures: 3.38–3.46

GLYOXYLATE CYCLE

Glyoxylate cycle involves the conversion of citric acid to succinate and glyoxylate in **glyoxysomes.** Succinate is transported to mitochondria, converted to malate and transported to cytosol for use in gluconeogenesis. Glyoxylate and acetyl-CoA condense to form malate which regenerates oxaloacetate, completing the cycle. In **germinating fatty seeds,** during conversion of fat to glucose, malate is the true product of glyoxylate cycle which is converted to sucrose in cytosol whereas succinate is used to replenish oxaloacetate to take part in citrate synthesis.

Figures: 3.47–3.50

PENTOSE PHOSPHATE PATHWAY

Pentose phosphate pathway also known as **hexose monophosphate shunt** or **phosphogluconate pathway,** is an alternative pathway of direct oxidation of glucose in cytoplasm. The pathway involves oxidative decarboxylation reactions generating NADPH for use in reductive reactions and pentose phosphate for nucleotide synthesis. Excess pentose phosphate is converted to glycolytic intermediates by a sequence of reactions and can regenerate hexose phosphate.

The **Enter-Doudoroff pathway** branches from the ketogluconate pathway of hexose oxidation in some bacteria.

Figures: 3.51–3.56

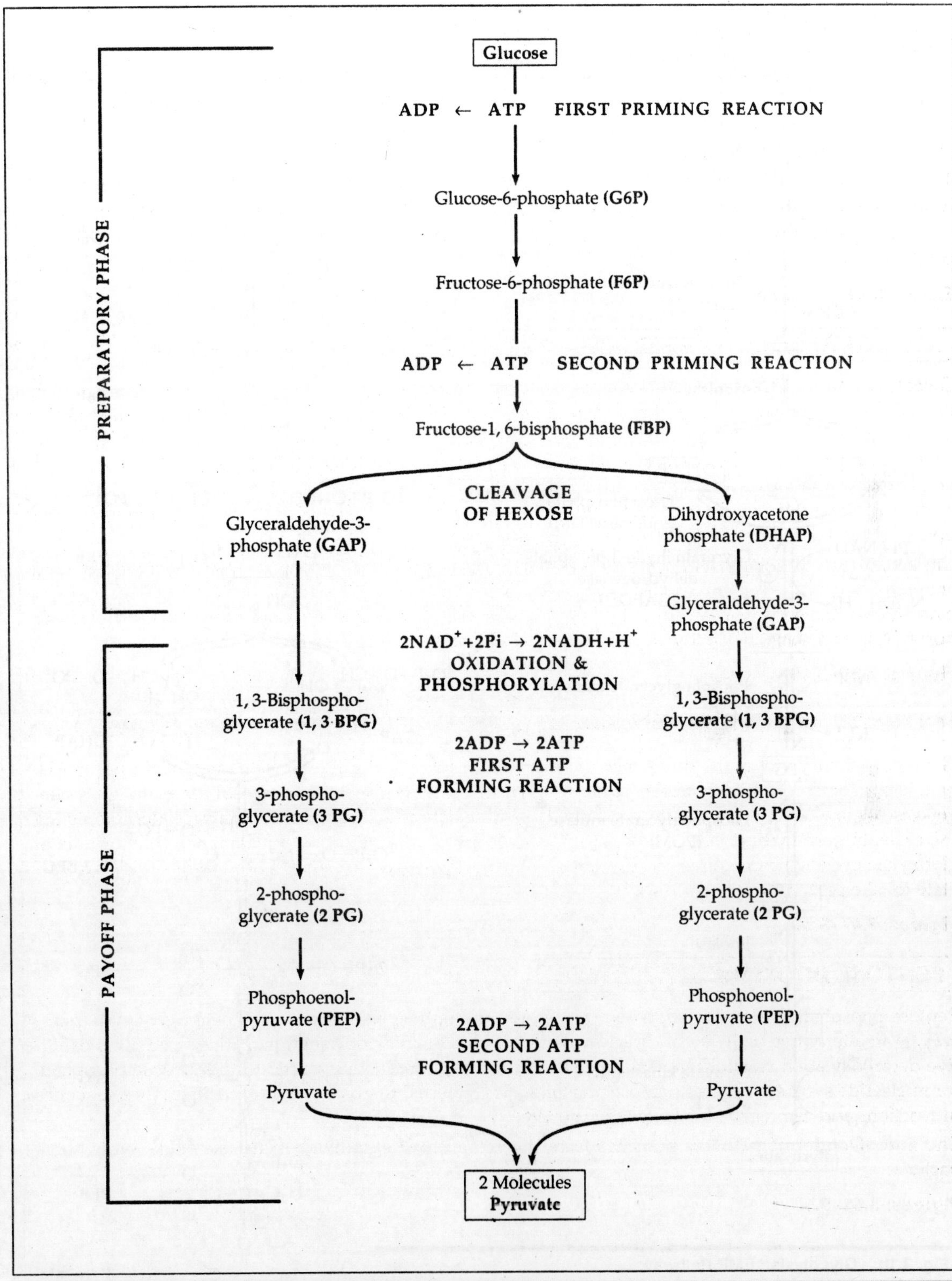

Fig. 3.1a: Glycolysis : Breakdown of glucose into pyruvate.

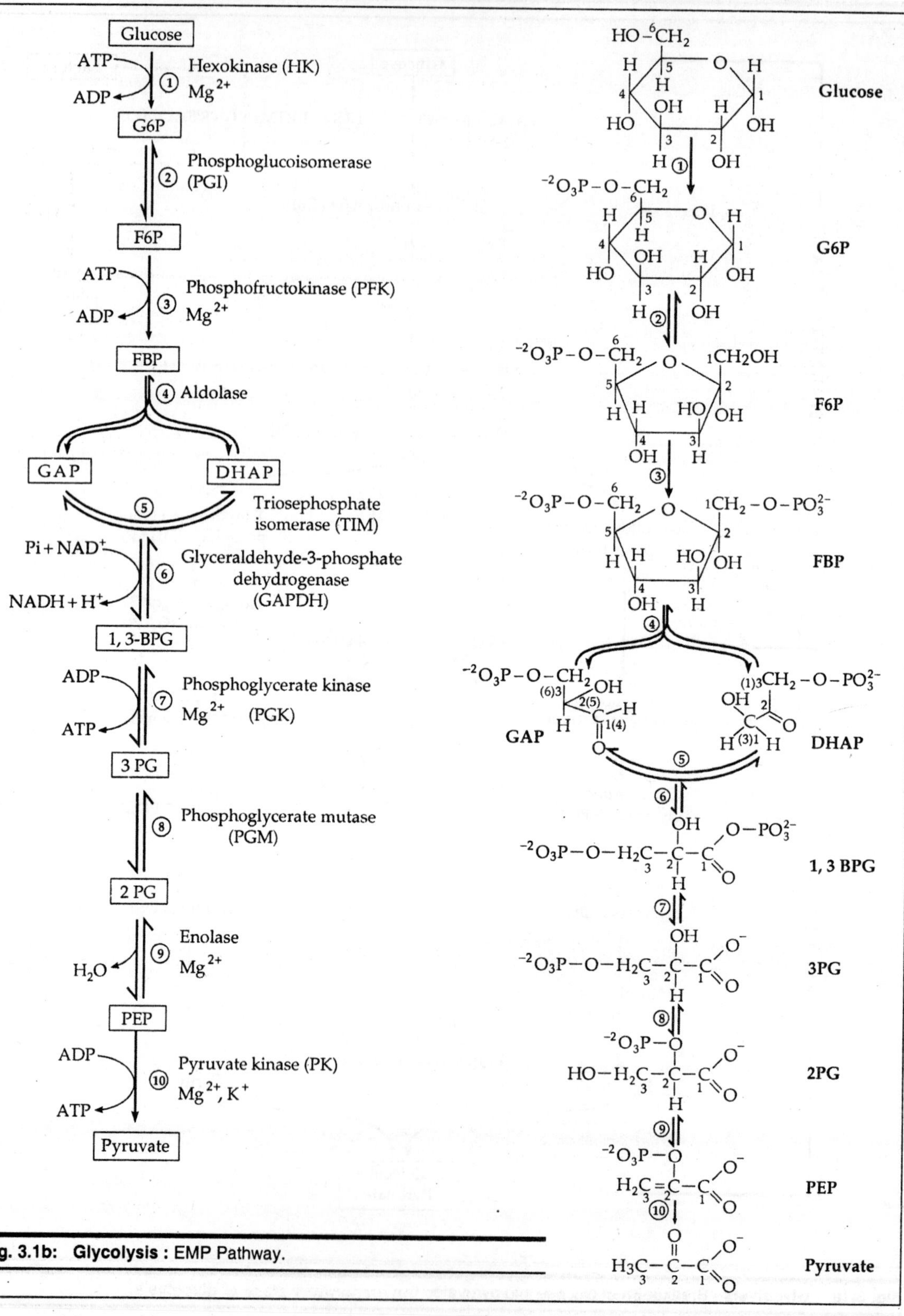

Fig. 3.1b: Glycolysis : EMP Pathway.

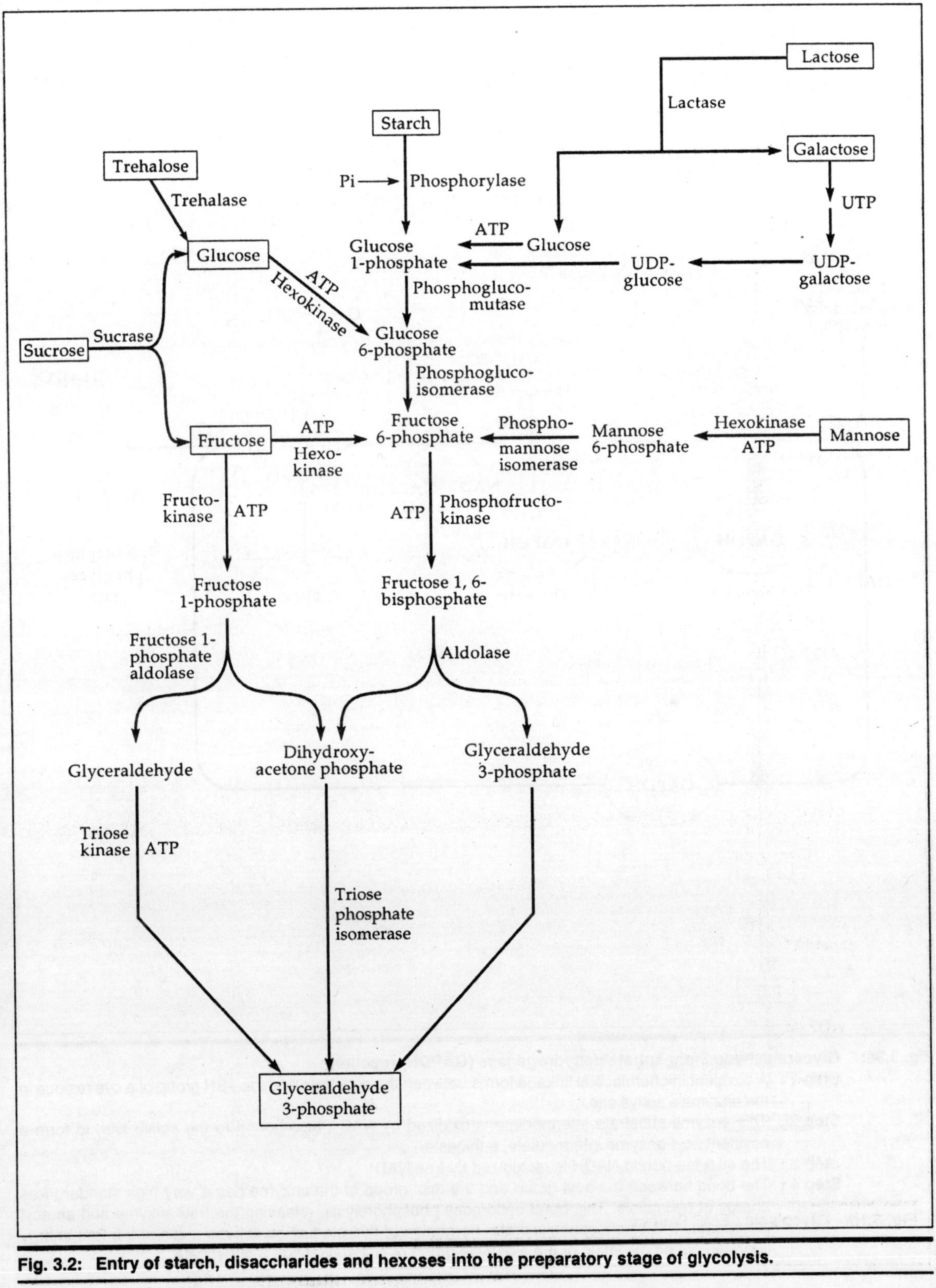

Fig. 3.2: Entry of starch, disaccharides and hexoses into the preparatory stage of glycolysis.

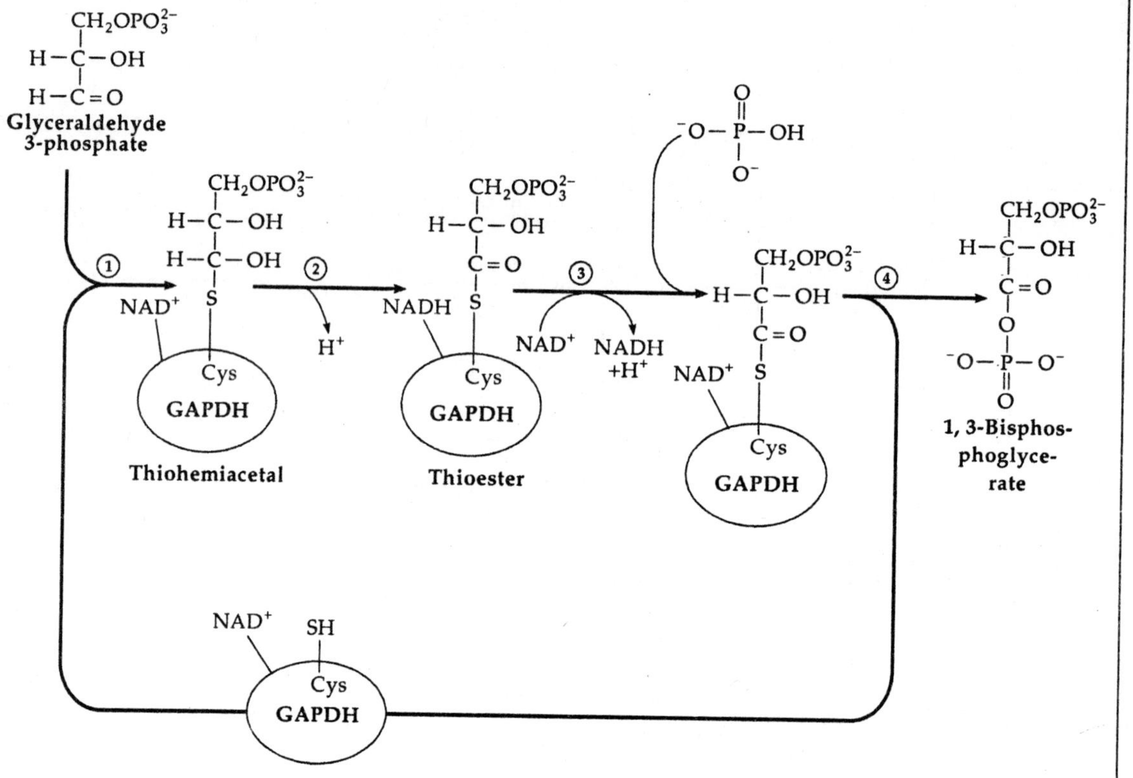

Fig. 3.3a: **Glyceraldehyde 3-phosphate dehydrogenase (GAPDH) reaction :**

Step 1 : A covalent thiohemiacetal linkage forms between the substrate and the –SH group of a cys residue in the enzyme's active site.

Step 2 : This enzyme-substrate intermediate is oxidized by NAD^+, also bound to the active site, to form a covalent acyl-enzyme intermediate, a thioester.

Step 3 : The enzyme-bound NADH is reoxidized by free NAD^+.

Step 4 : The bond between the acyl group and the thiol group of the enzyme has a very high standard free energy of hydrolysis. This bond undergoes phosphorolysis, releasing the free enzyme and an acyl phosphate, 1, 3-bisphosphoglycerate. Formation of this product conserves much of the free energy liberated during oxidation of the aldehyde group of glyceraldehyde 3-phosphate.

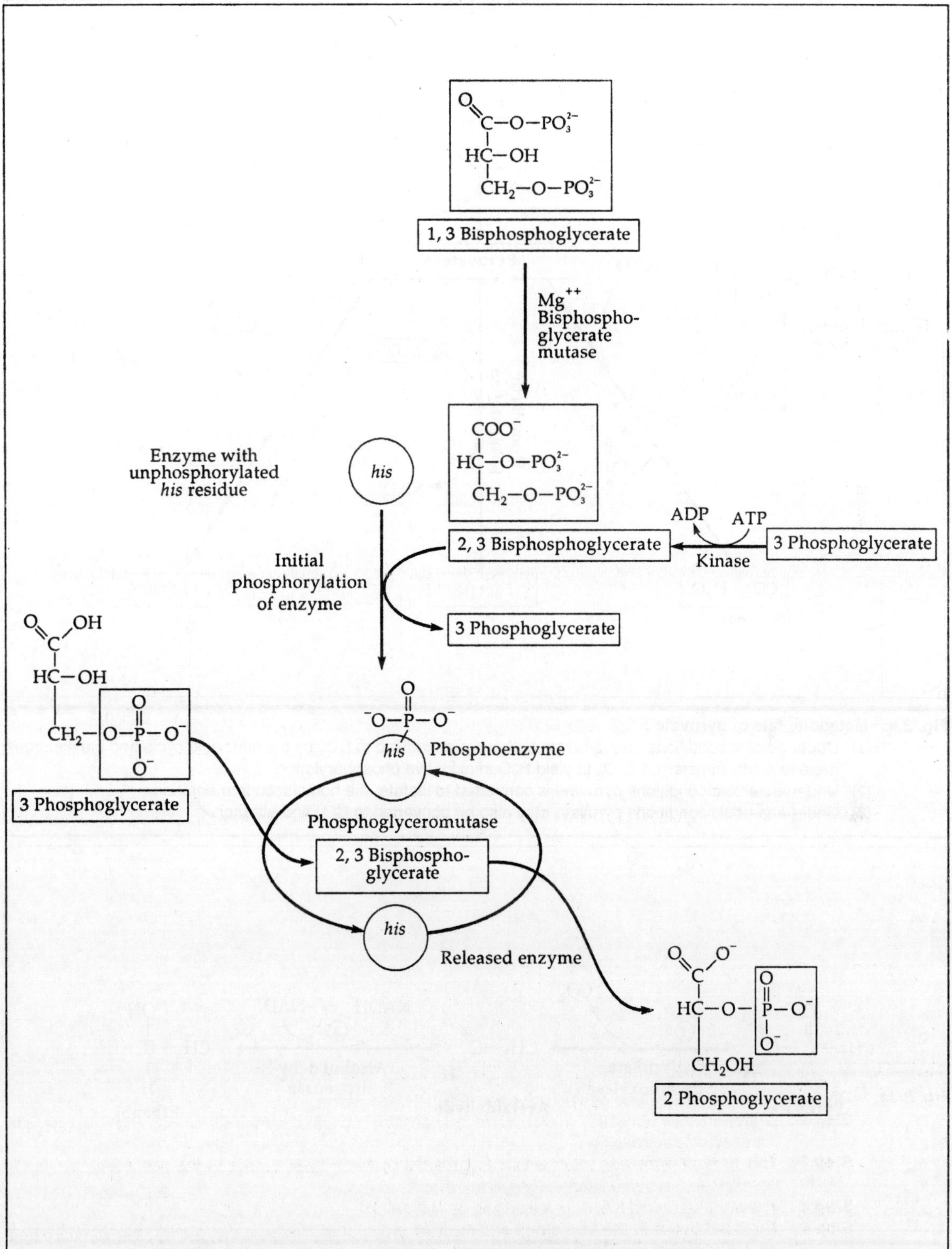

Fig. 3.3b: Formation of 2-Phosphoglycerate from 3-Phosphoglycerate.

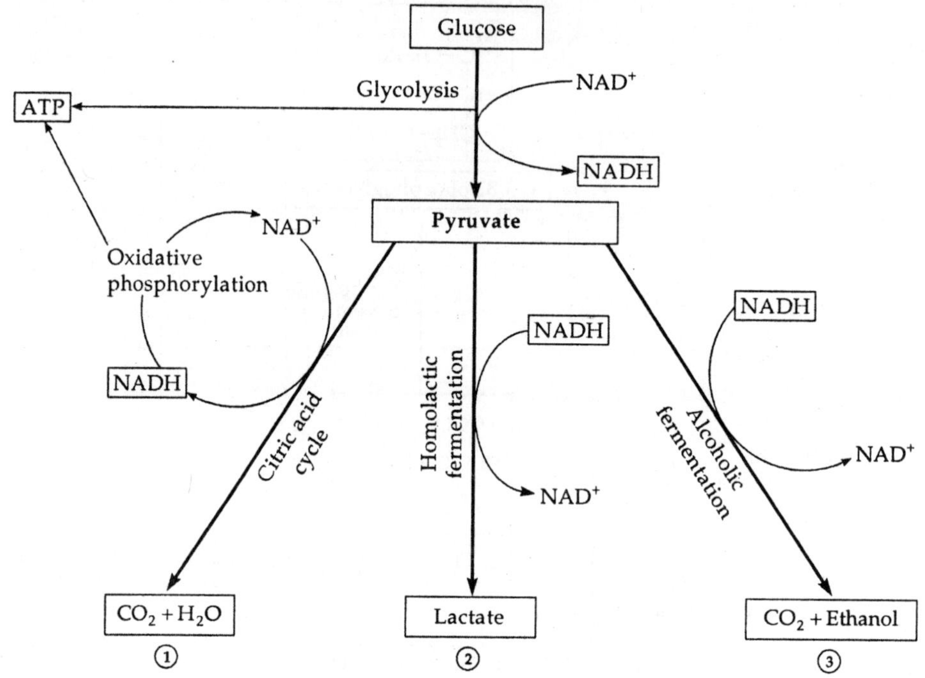

Fig. 3.4: Metabolic fate of pyruvate :

(1) Under aerobic conditions, the pyruvate carbons are oxidized to CO_2 by the citric acid cycle and the electrons are eventually transferred to O_2 to yield H_2O in oxidative phosphorylation.

(2) Under anaerobic conditions pyruvate is converted to lactate, the homolactic fermentation.

(3) Under anaerobic conditions pyruvate may also be converted to CO_2 and ethanol.

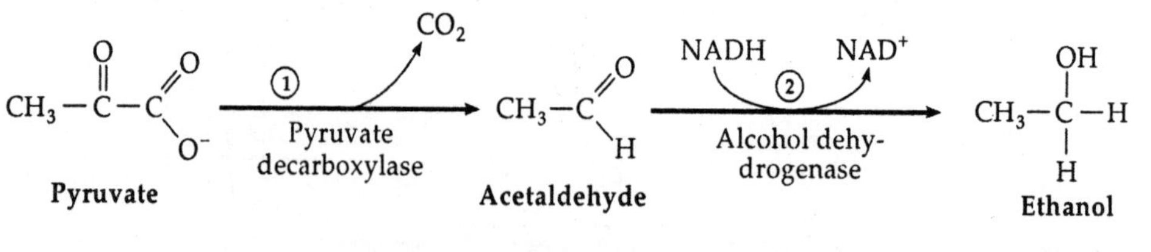

Fig. 3.5a: The two reactions of alcoholic fermentation :

(1) Decarboxylation of pyruvate to form acetaldehyde.

(2) Reduction of acetaldehyde to ethanol by NADH.

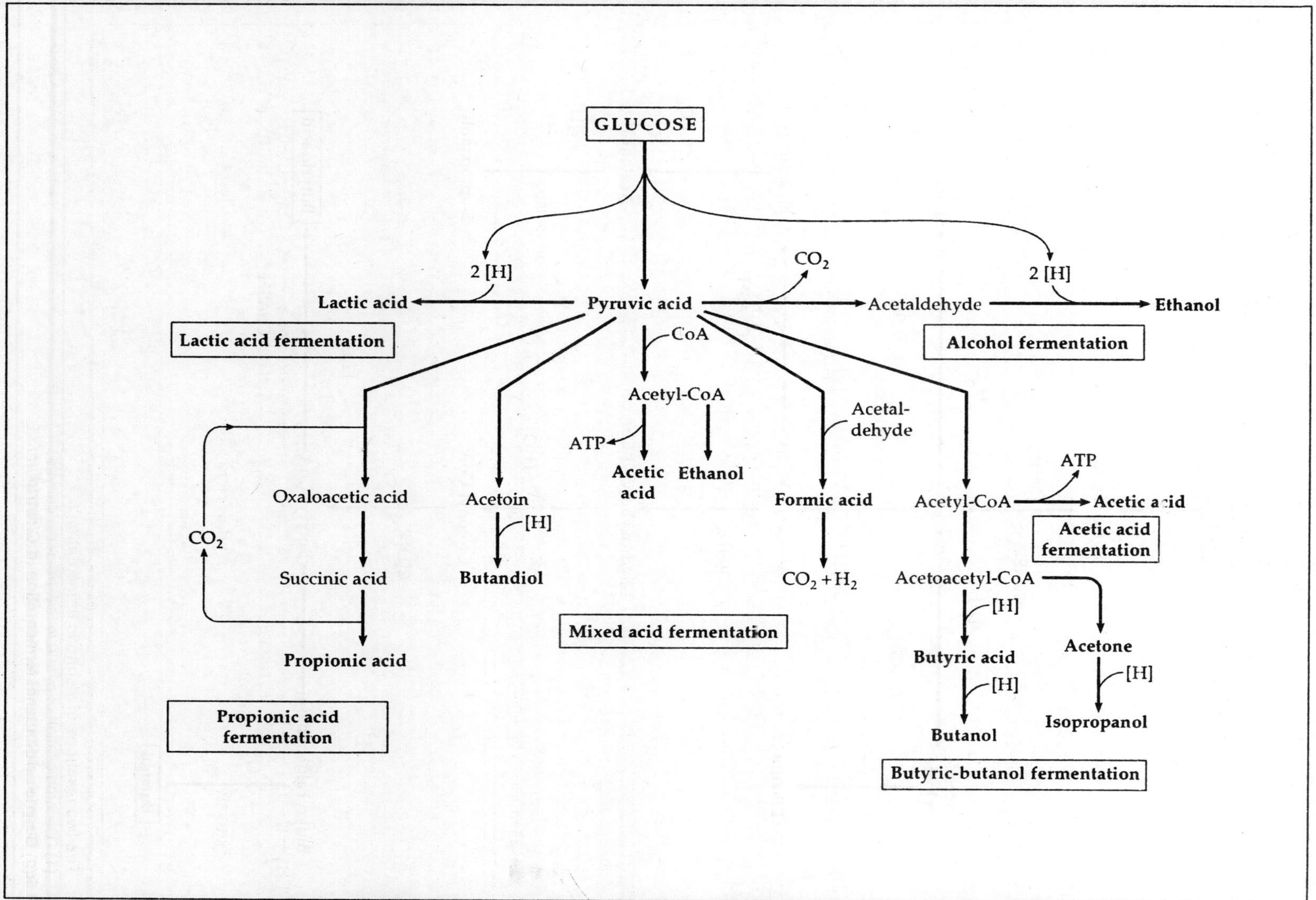

Fig. 3.5b: Different types of fermentations, their products and the role of the key metabolite, pyruvic acid produced by glucose breakdown in bacteria.

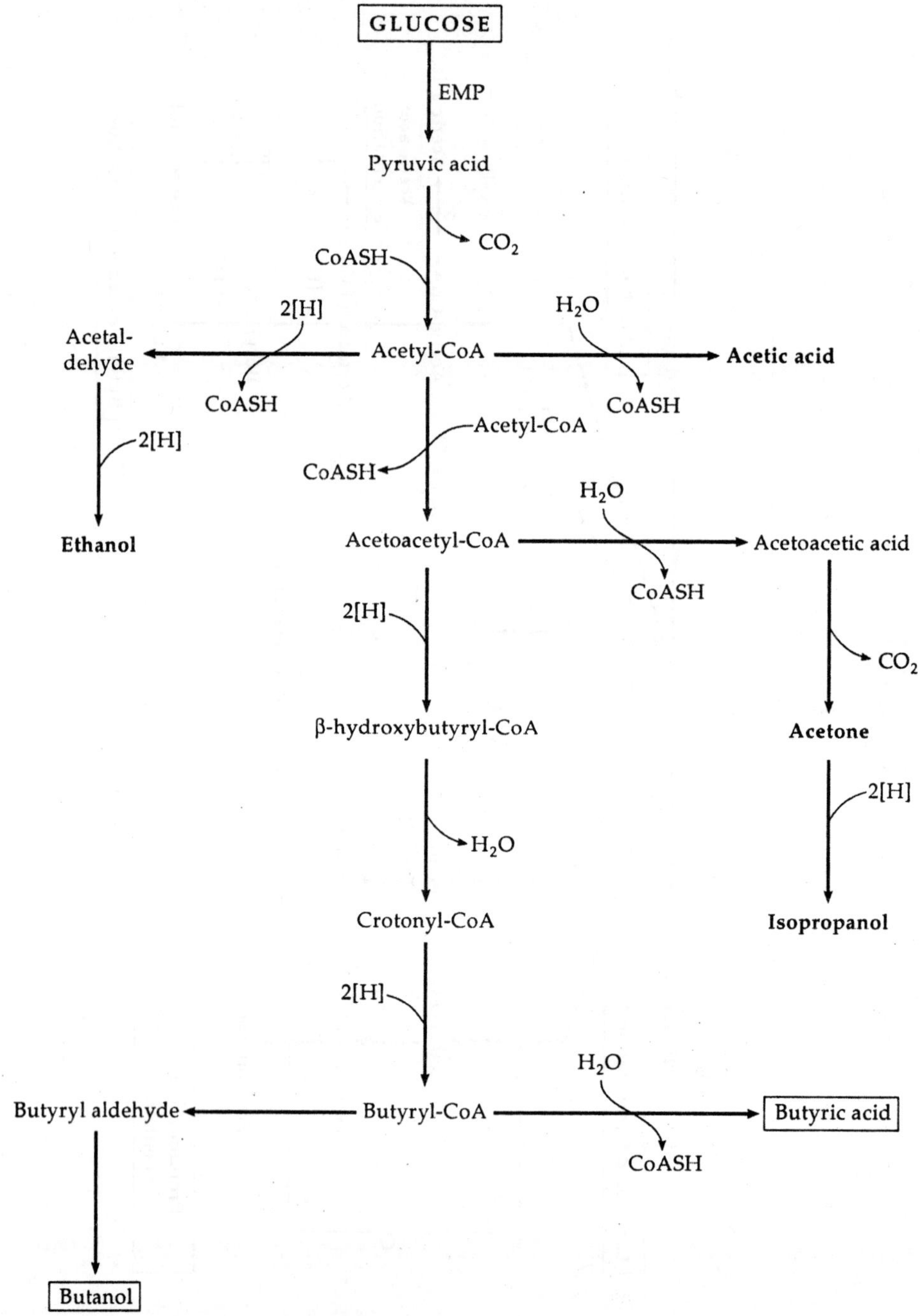

Fig. 3.5c: Butyric acid-butanol fermentation in *Clostridium*.

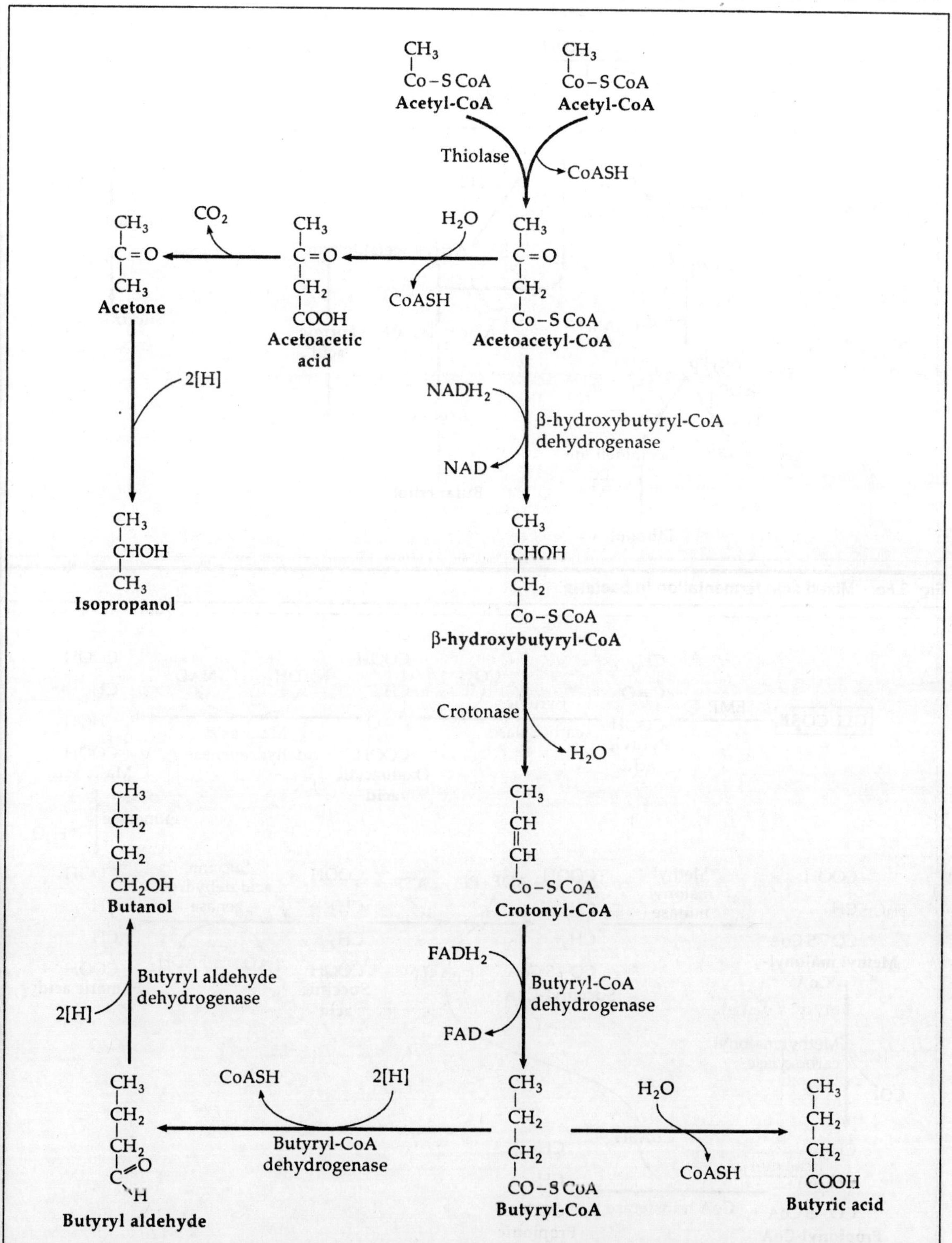

Fig. 3.5d: Synthesis of butyric acid and butanol, acetone and isopropanol in fermentation by *Clostridium*.

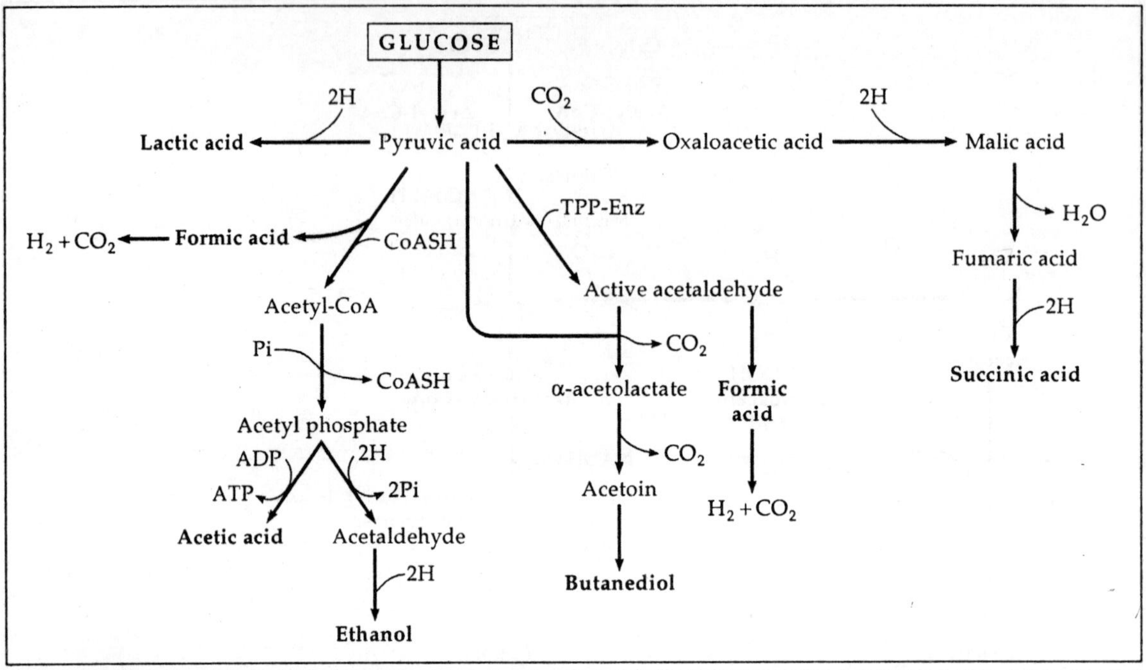

Fig. 3.5e: Mixed acid fermentation in bacteria.

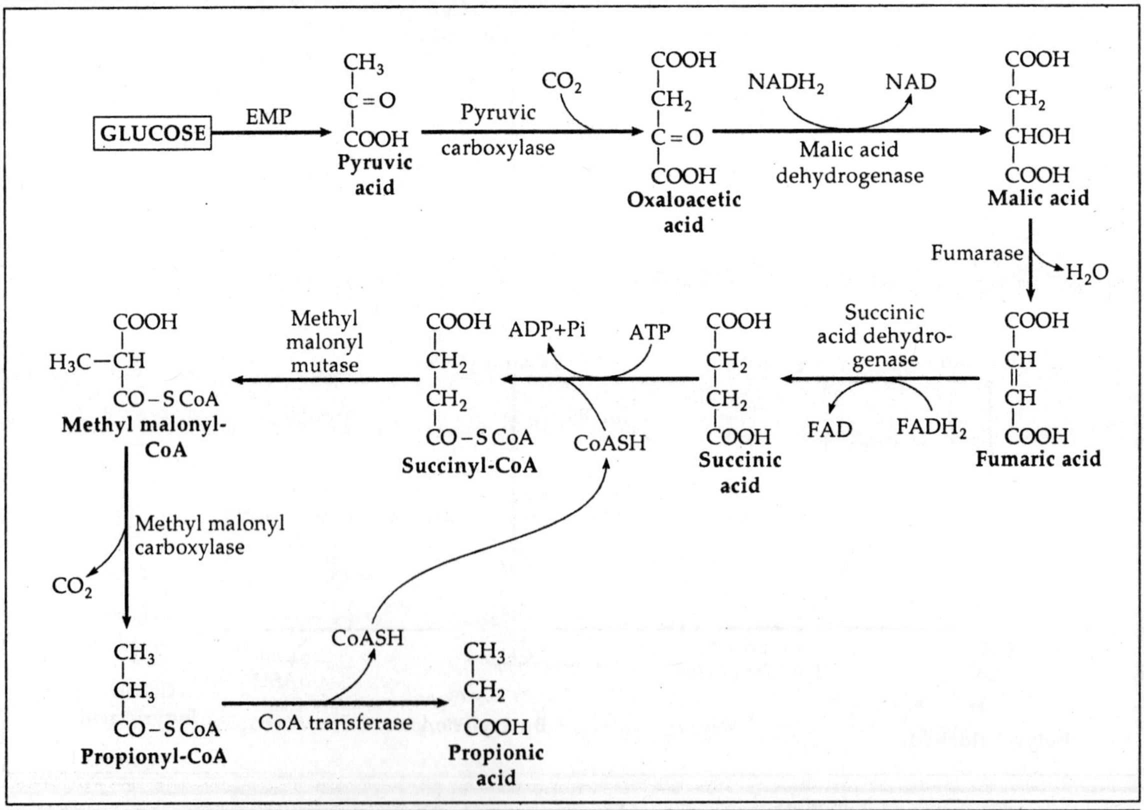

Fig. 3.5f: Propionic acid fermentation in bacteria.

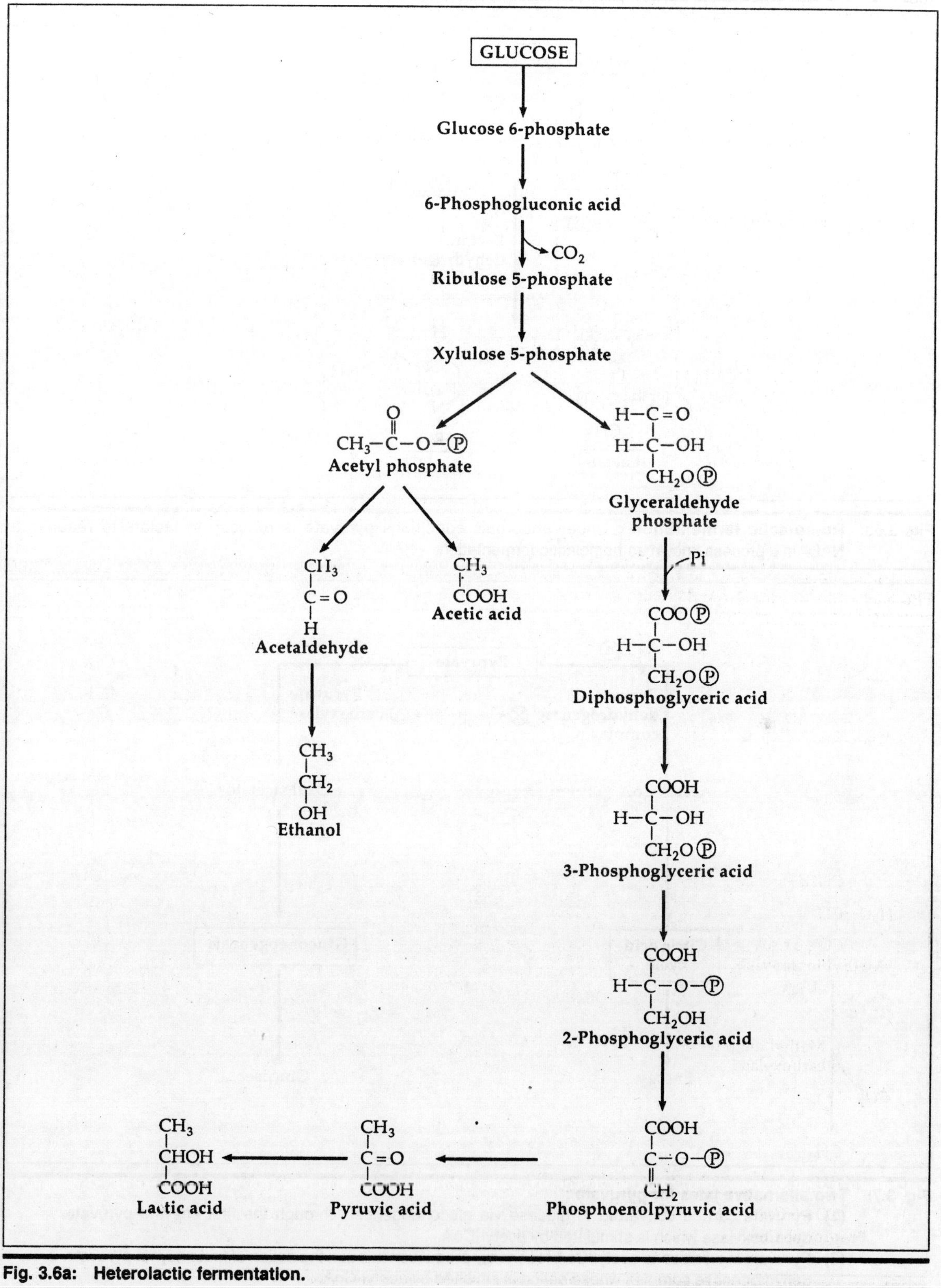

Fig. 3.6a: Heterolactic fermentation.

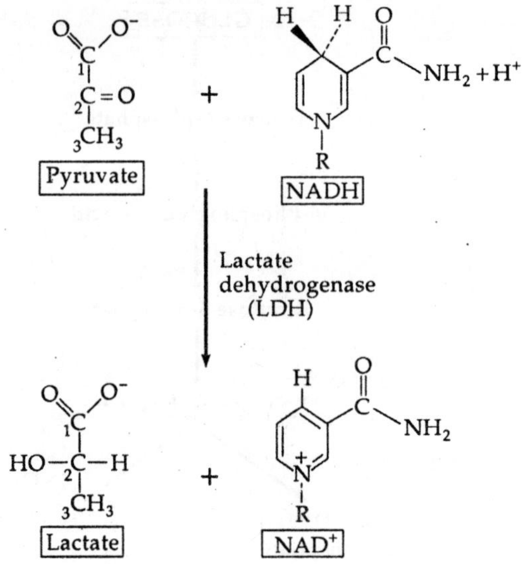

Fig. 3.6b: Homolactic fermentation : Under anaerobic conditions pyruvate is reduced to lactate to regenerate NAD^+ in a process known as homolactic fermentation.

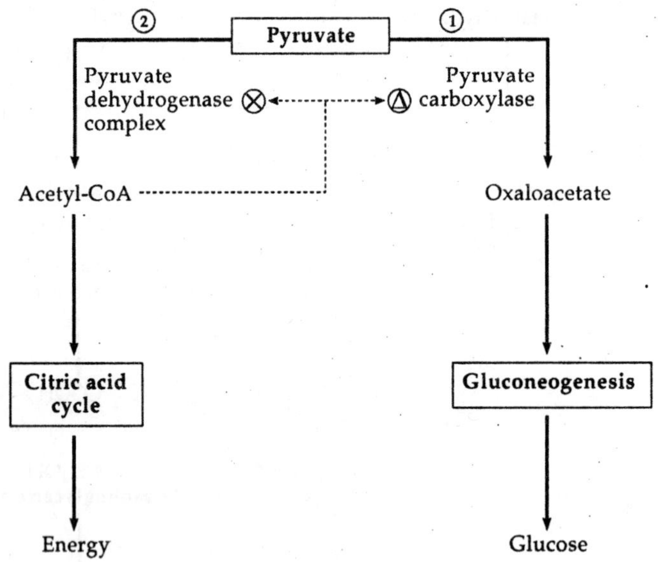

Fig. 3.7: Two alternative fates for pyruvate :

(1) Pyruvate can be converted to glucose via gluconeogenesis through the first enzyme pyruvate decarboxylase which is stimulated by acetyl-CoA.

(2) Pyruvate is oxidized to acetyl-CoA for energy production through the action of first enzyme pyruvate dehydrogenase complex whose action is inhibited by acetyl-CoA.

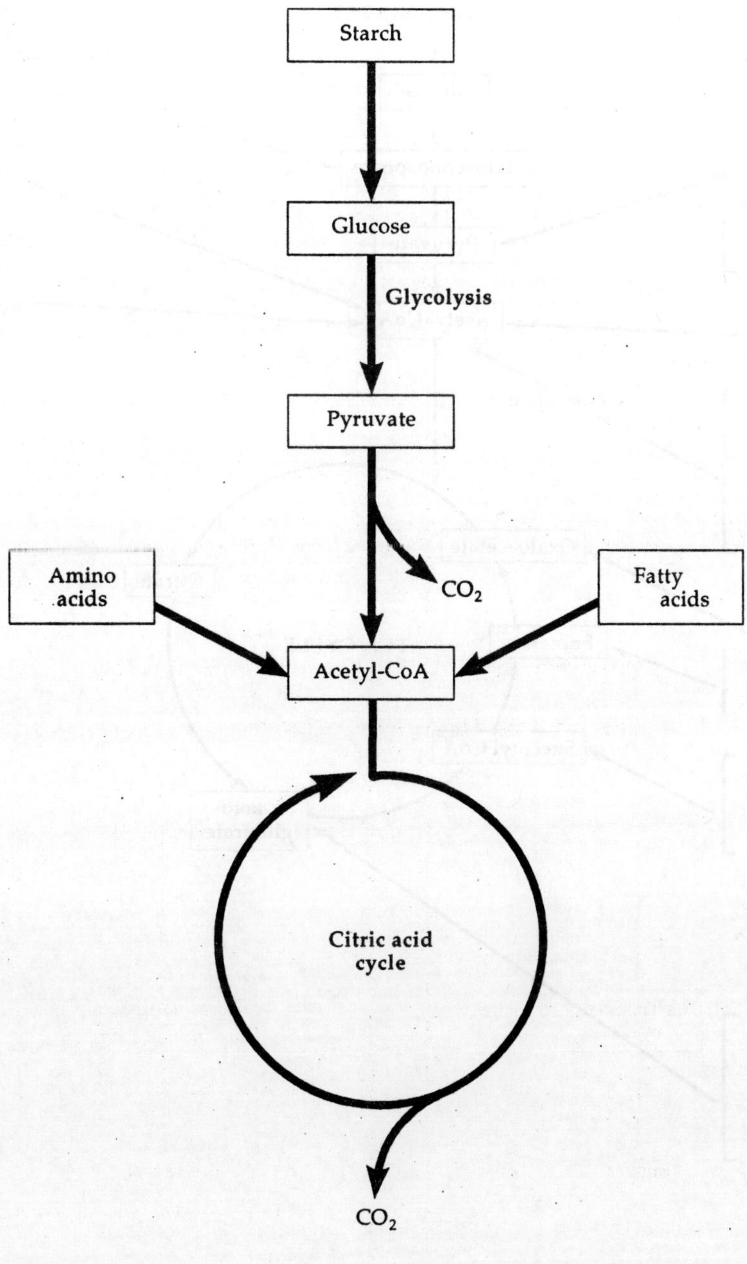

Fig. 3.8: An overview of oxidative fuel metabolism : Acetyl groups derived from carbohydrates, amino acids and fatty acids enter the citric acid cycle to be oxidized to CO_2.

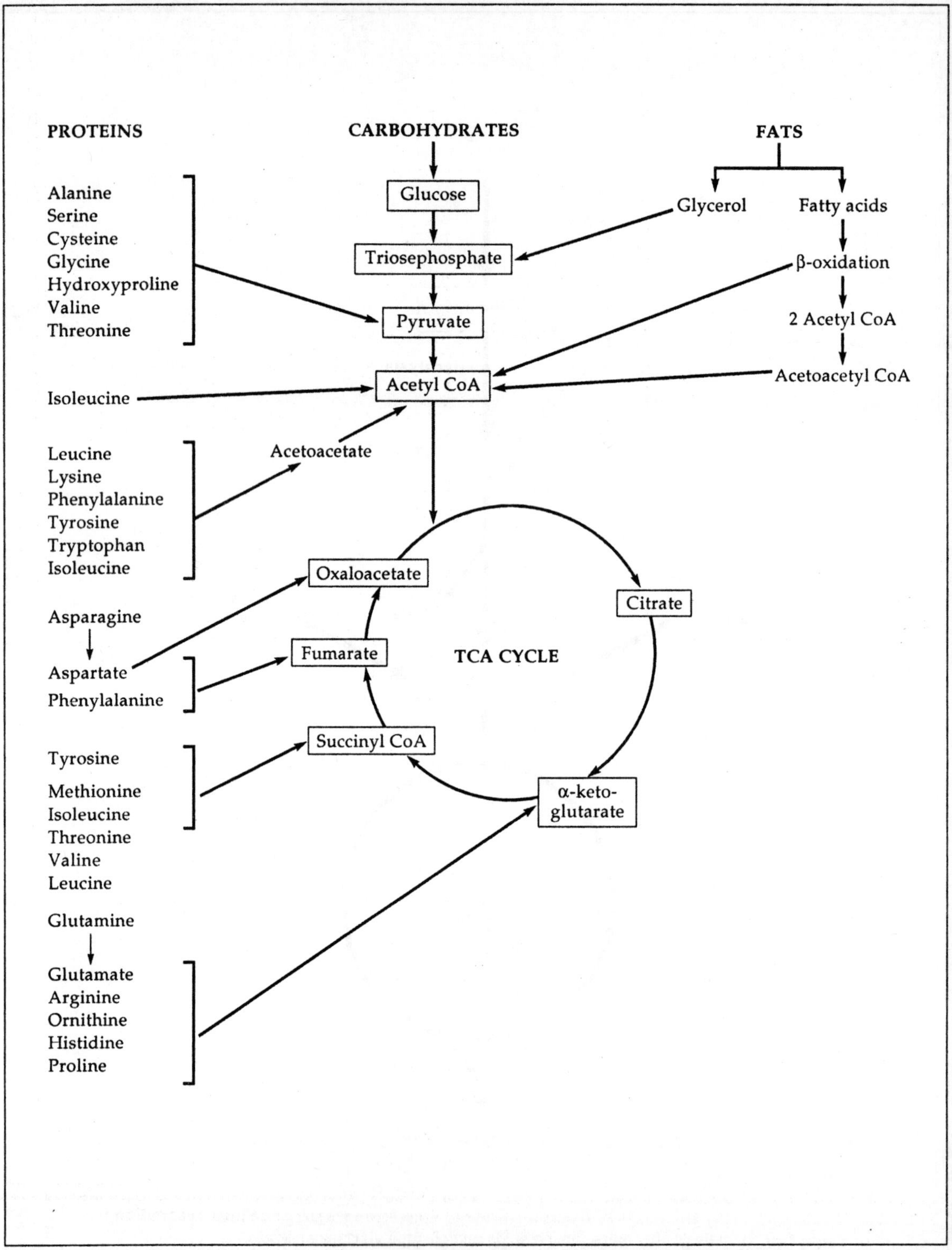

Fig. 3.9a: Metabolic mill showing the interrelation among protein, carbohydrate and fat metab[illegible]m and Krebs cycle.

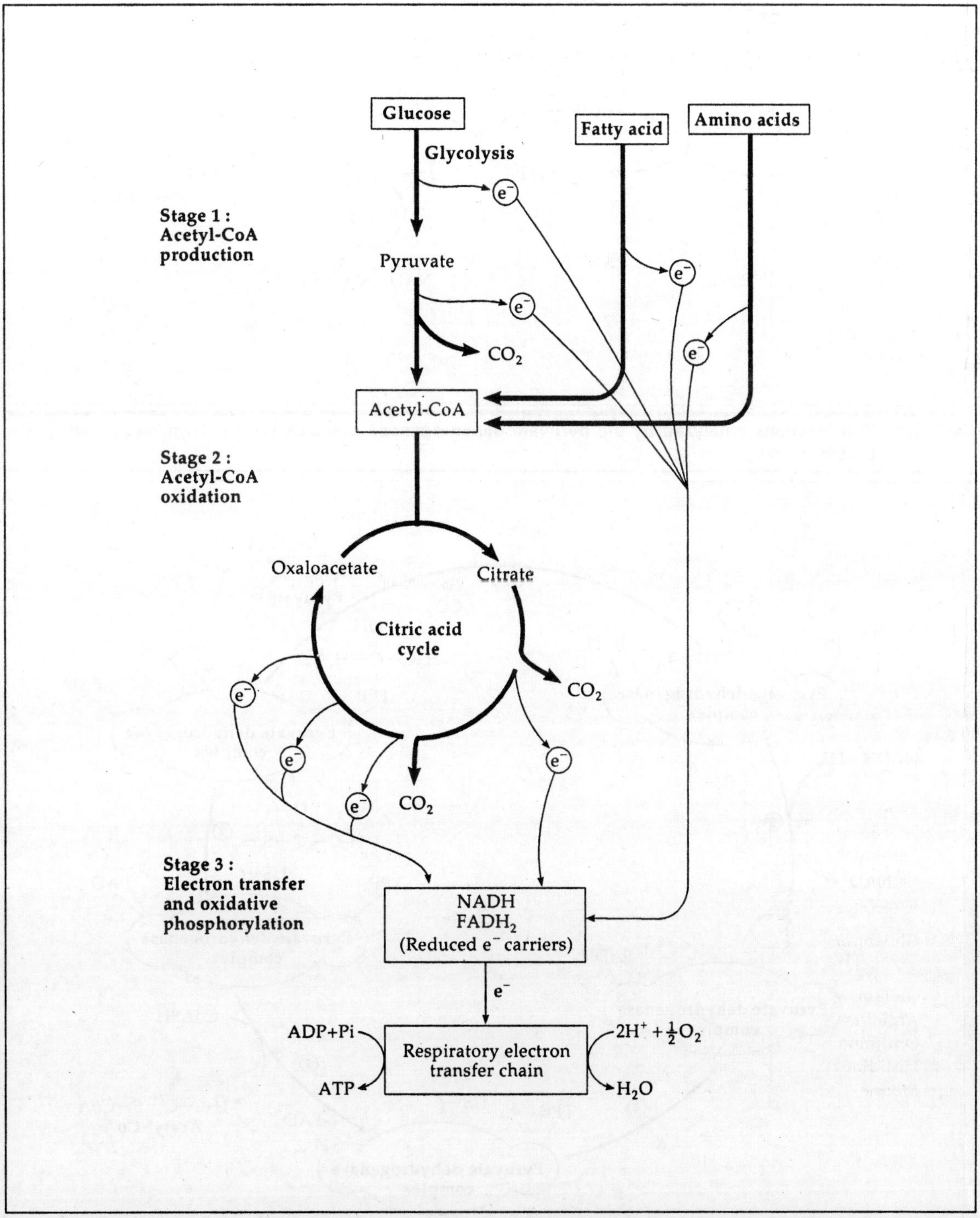

Fig. 3.9b: Catabolism of proteins, fats and carbohydrates in the three stages of cellular respiration :
Stage 1 : Oxidation of fatty acids, amino acids and glucose yields acetyl-CoA ;
Stage 2 : Oxidation of acetyl groups in the citric acid cycle includes 4 steps in which electrons are abstracted ;
Stage 3 : Electrons carried by NADH and $FADH_2$ are funneled into a chain of electron carriers, the respiratory chain, ultimately reducing O_2 to H_2O. This electron flow drives the production of ATP.

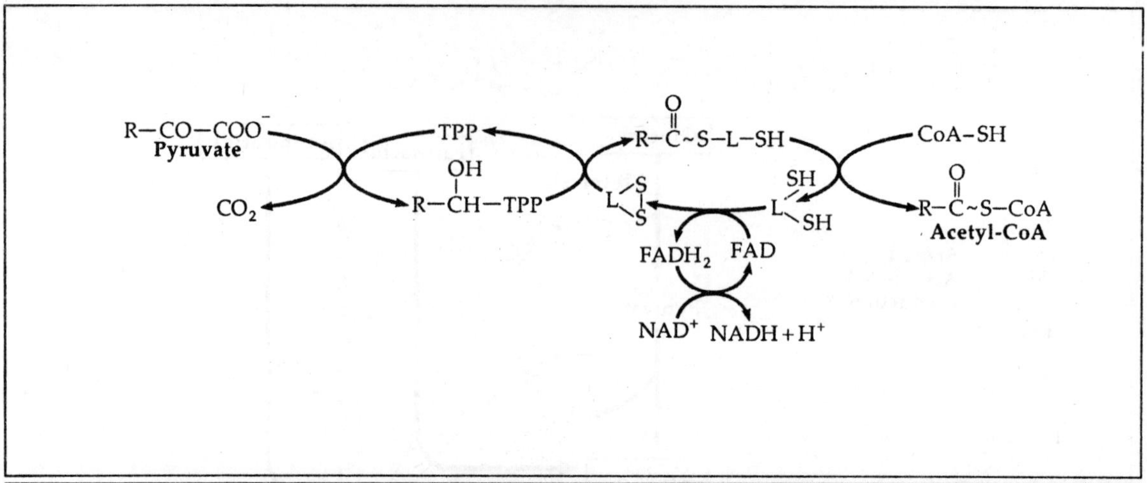

Fig. 3.10a: The reactions catalyzed by the pyruvate dehydrogenase complex [TPP = Thymine pyrophosphate, L = Lipoic acid].

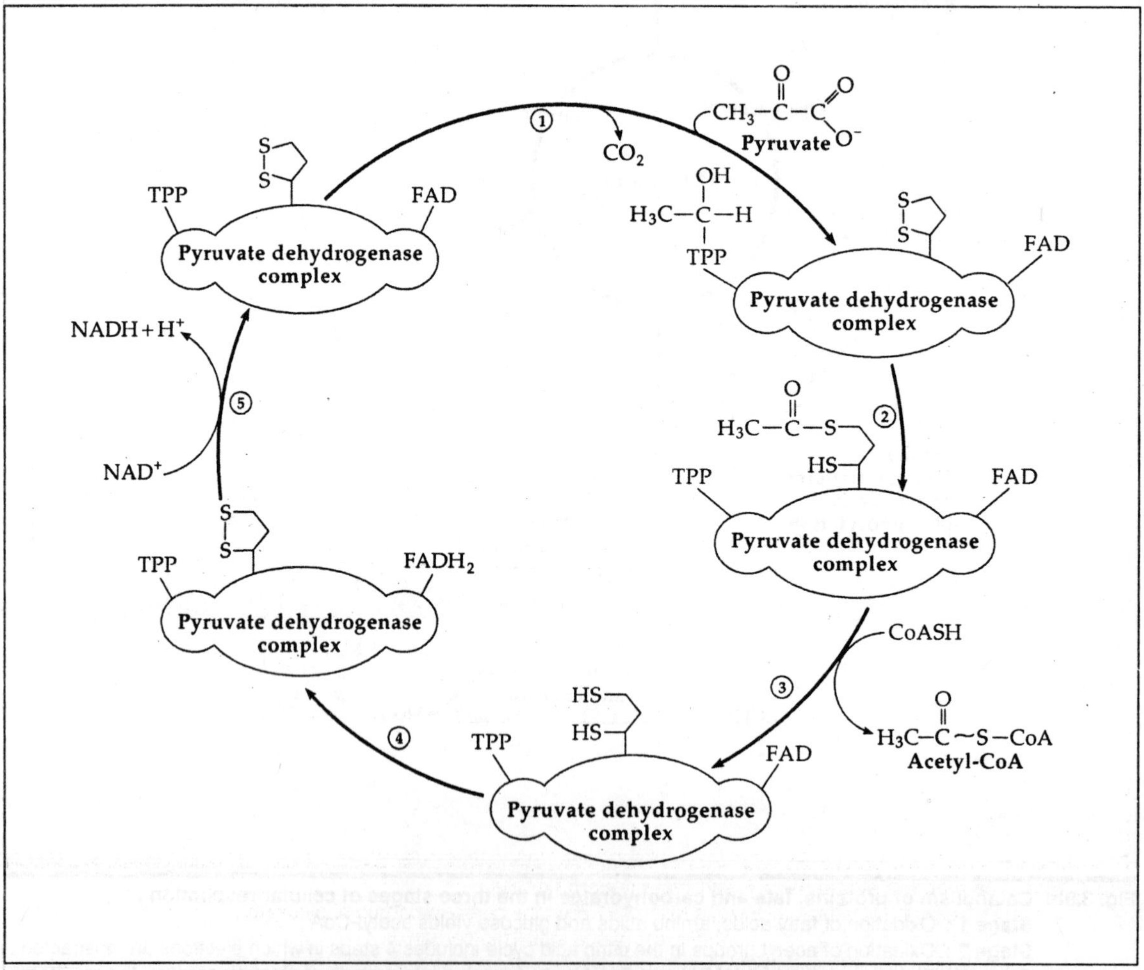

Fig. 3.10b: Conversion of pyruvate to acetyl-CoA by pyruvate dehydrogenase complex.

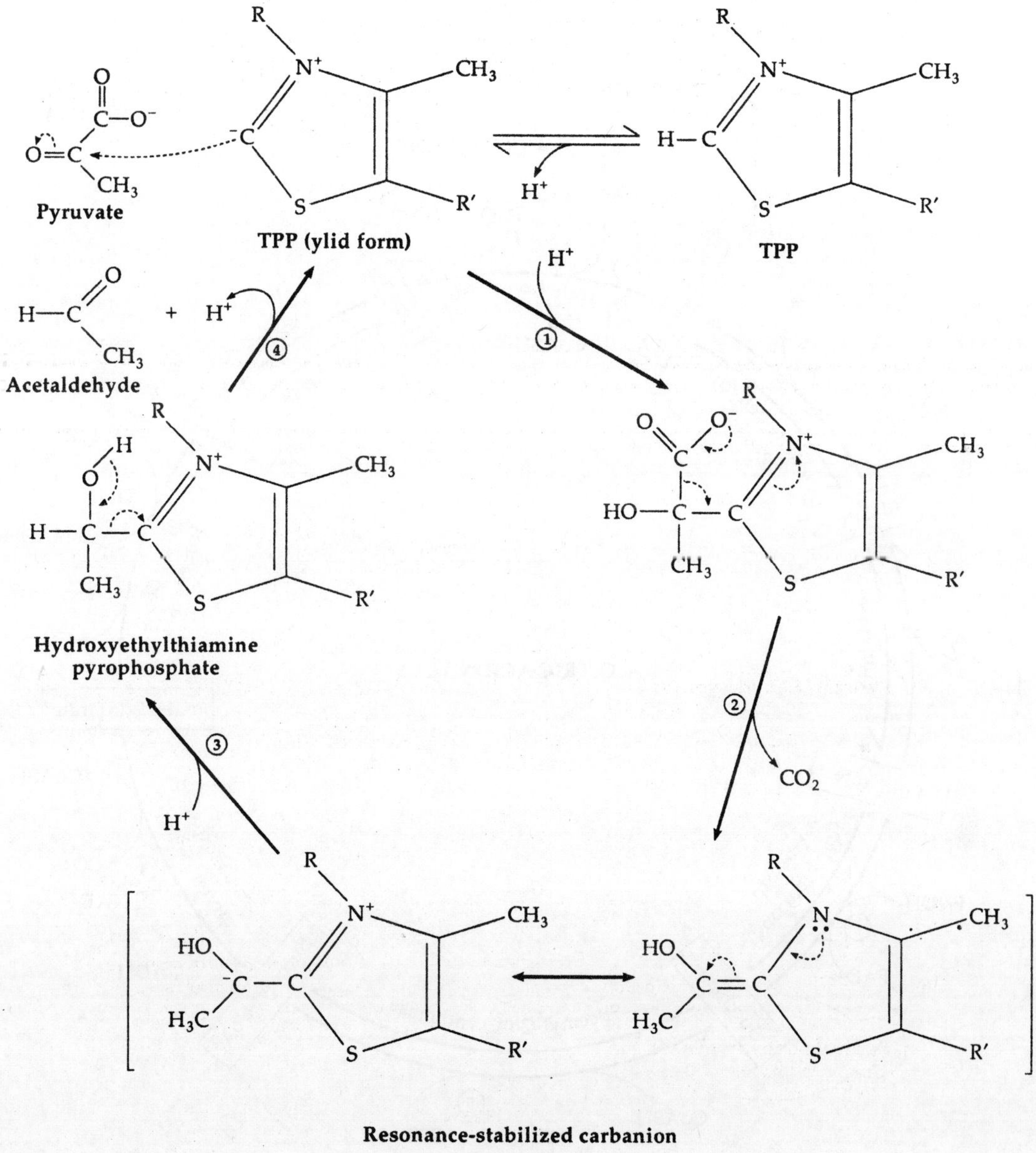

Fig. 3.10c: The reaction mechanism of pyruvate decarboxylase :

(1) Nucleophilic attack by the ylid form of TPP on the carbonyl carbon of pyruvate ;

(2) Departure of CO_2 to generate a resonance-stabilized carbanion ;

(3) Protonation of the carbanion ;

(4) Elimination of the TPP ylid and release of product.

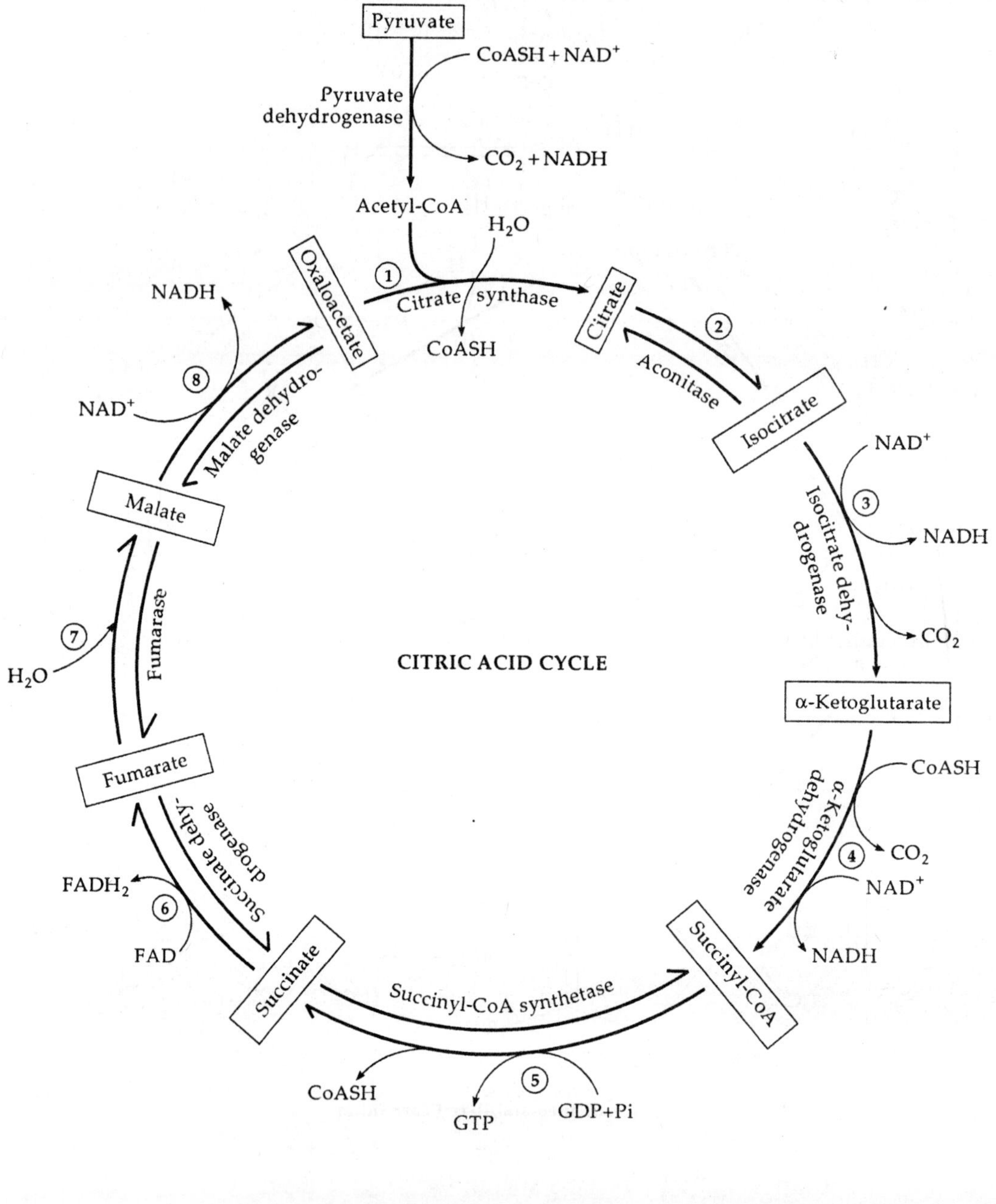

Fig. 3.11a: Citric acid cycle : Pyruvic acid produced through glycolysis under aerobic condition initiates the cycle through the synthesis of acetyl-CoA. The substrate acetyl-CoA enters the cycle by reaction with oxaloacetate forming the first compound citric acid. Citric acid is a tricarboxylic acid thus the cycle is also known as TCA cycle. The different intermediate acids are isocitrate (C6), α-ketoglutarate (C5), succinate (C4), fumarate (C4) and lastly regenerates oxaloacetate (C4) completing the cycle. Decarboxylation occurs at two steps [(3), (4)]. Oxidative steps are (3), (4), (6), (8).

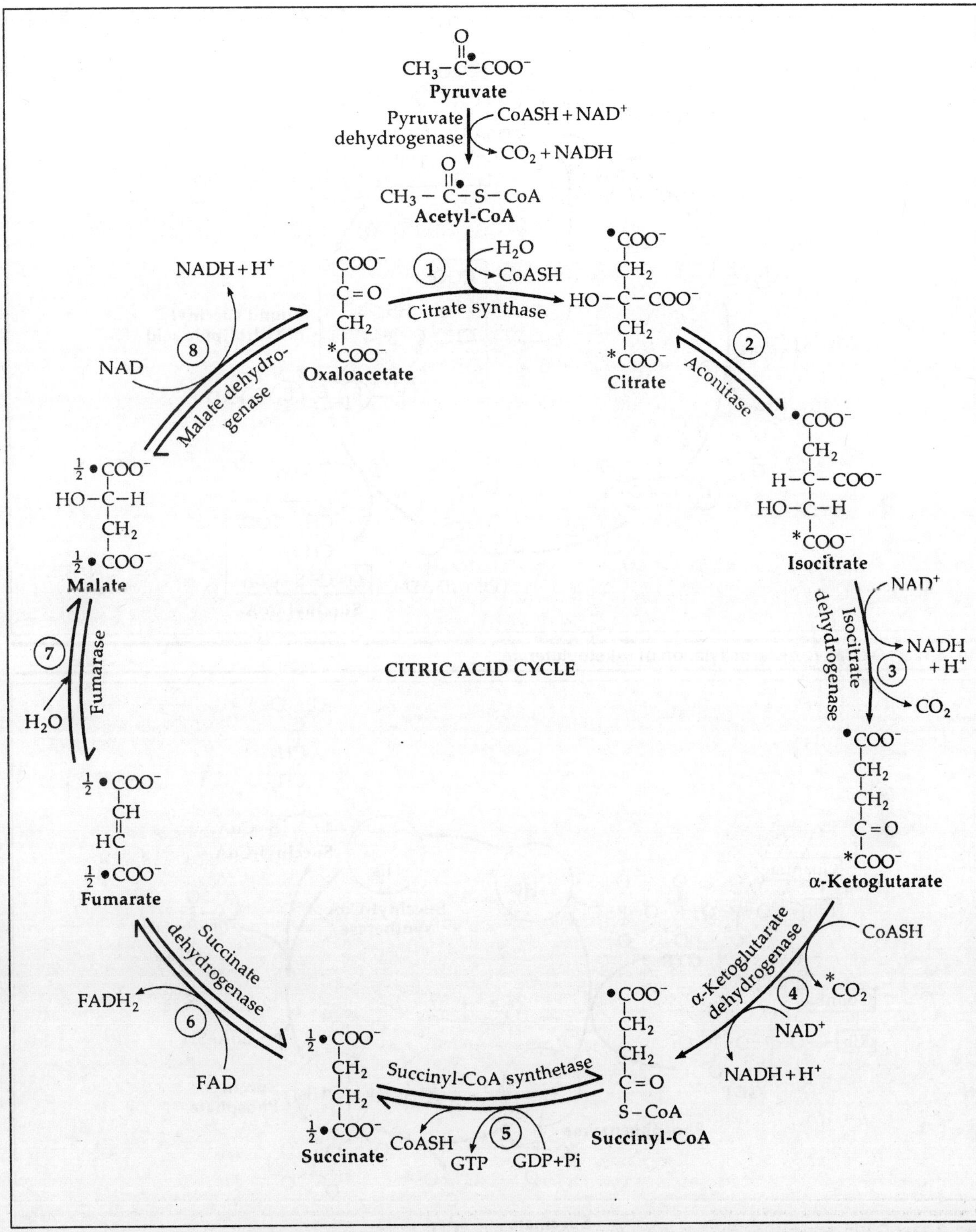

Fig. 3.11b: The reactions of citric acid cycle : The pyruvate → acetyl-CoA reaction supplies cycle's substrate but is not considered to be part of the cycle. An isotopic label at C_4 of oxaloacetate (*) becomes C1 of α-Ketoglutarate and is released as CO_2 in reaction (4). An isotopic label at C1 of acetyl-CoA (•) becomes C5 of α-Ketoglutarate and is scrambled in reaction (5) between C1 and C4 of succinate (½ •).

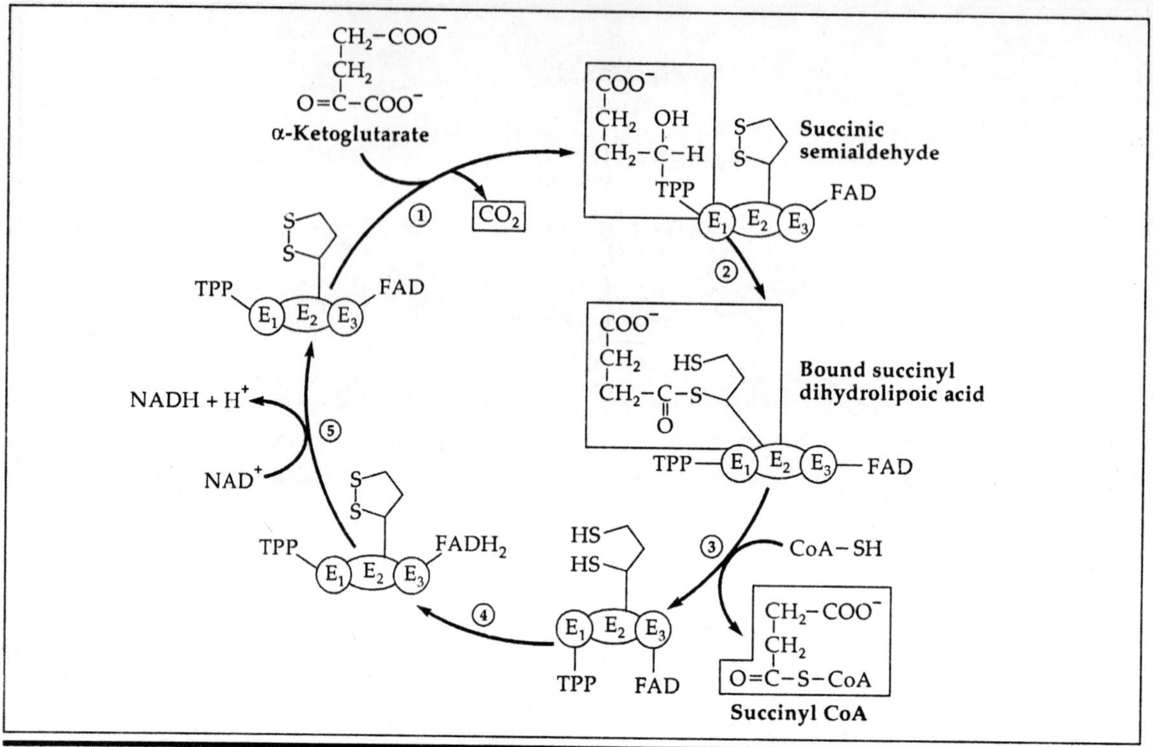

Fig. 3.12a: Oxidative decarboxylation of α-ketoglutarate.

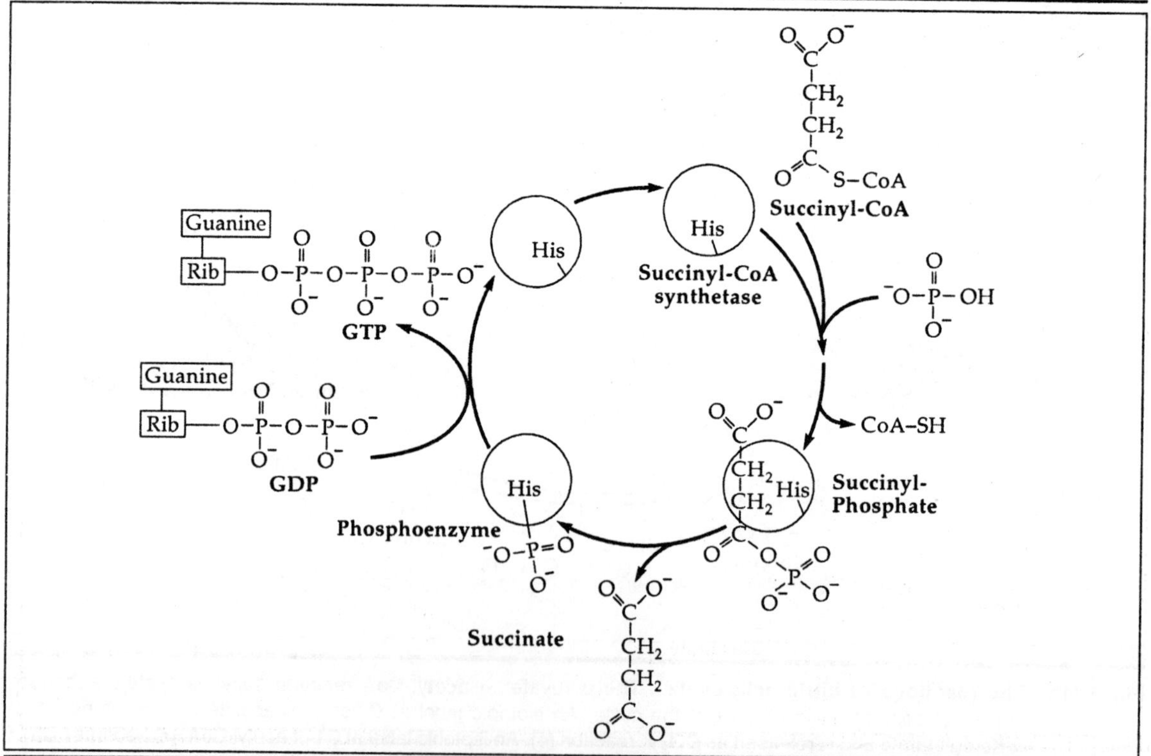

Fig. 3.12b: Intermediates in the succinyl CoA-synthetase reaction.

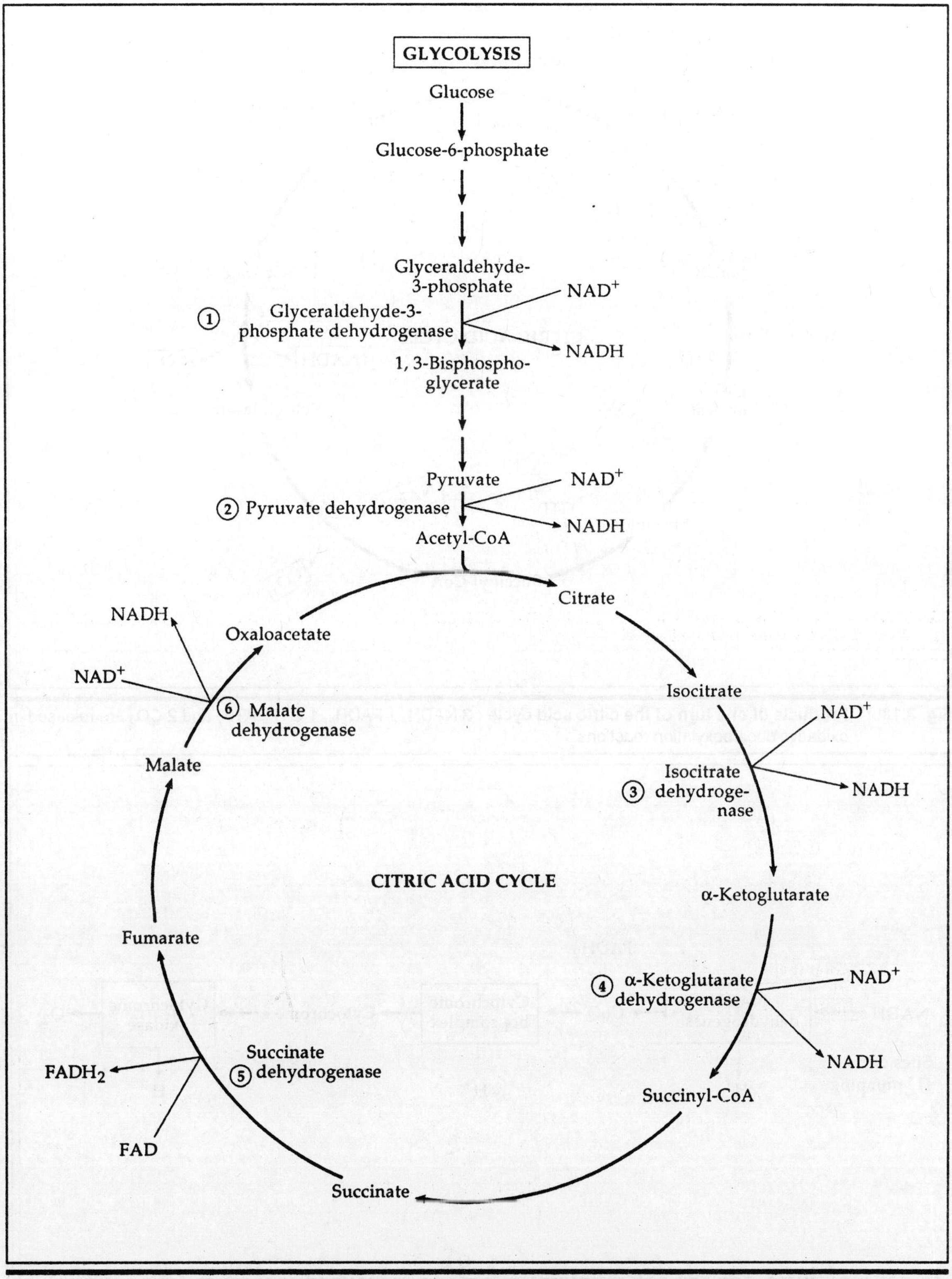

Fig. 3.13a: The sites of electron transfer that form NADH and $FADH_2$ in glycolysis and the citric acid cycle.

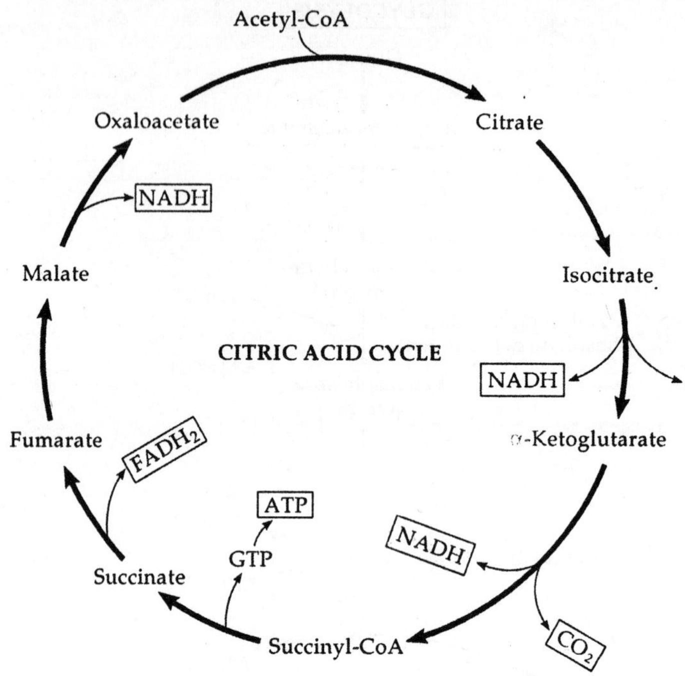

Fig. 3.13b: Products of one turn of the citric acid cycle : 3 NADH, 1 $FADH_2$, 1 GTP (ATP) and 2 CO_2 are released in oxidative decarboxylation reactions.

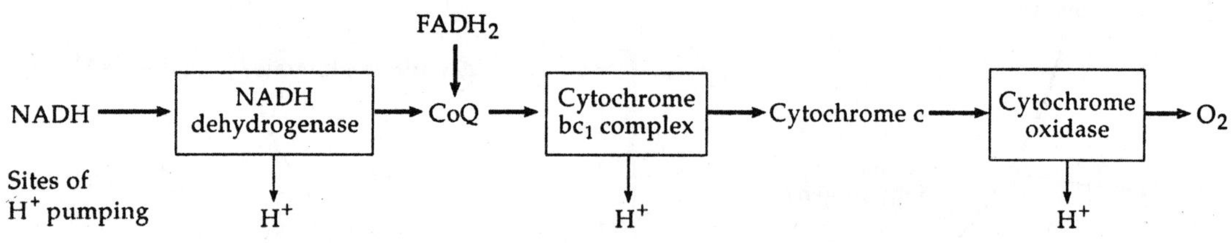

Fig. 3.14a: Respiratory electron transport chain.

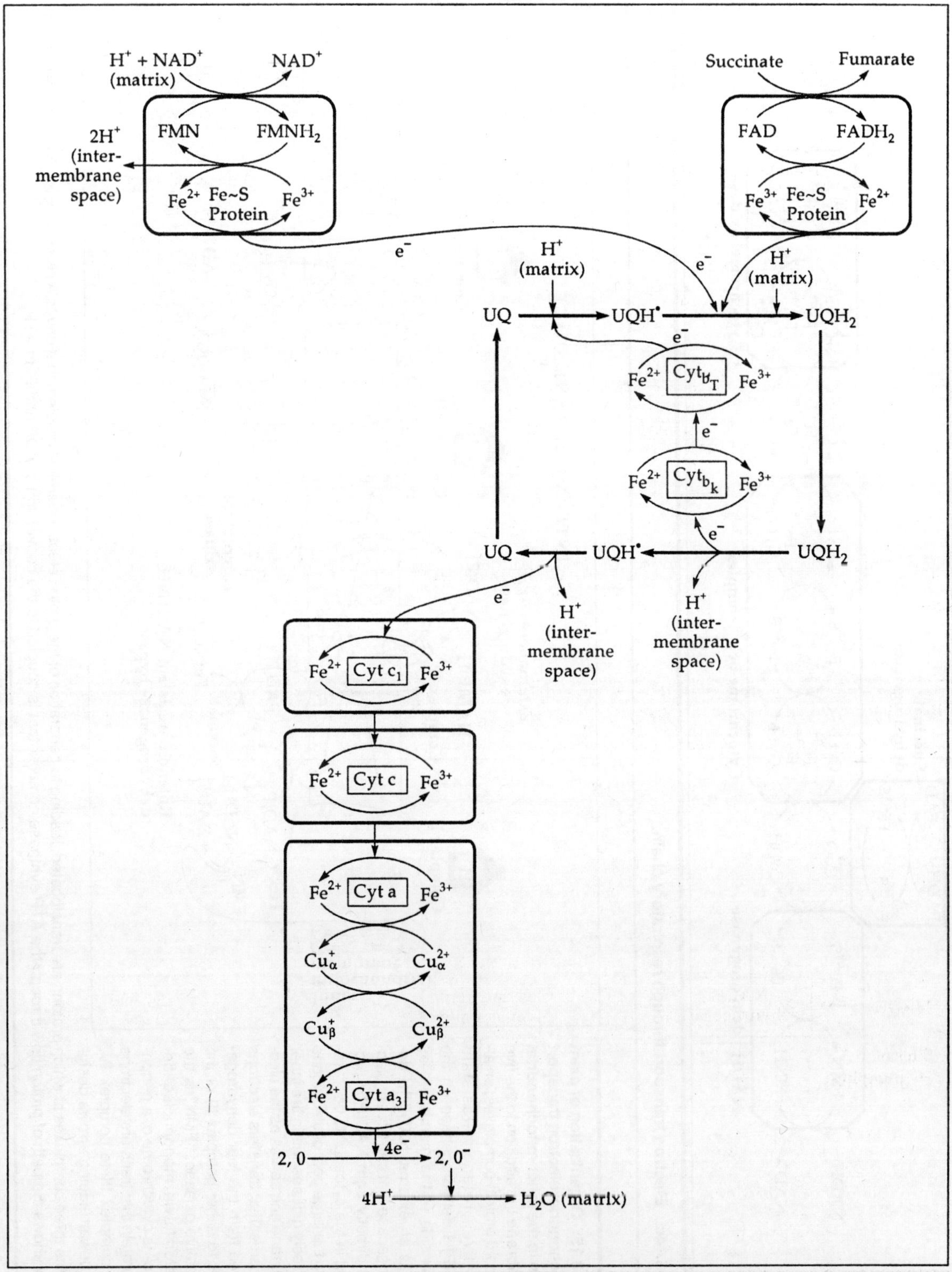

Fig. 3.14b: Schematic representation of electron transport chain in inner mito-membrane.

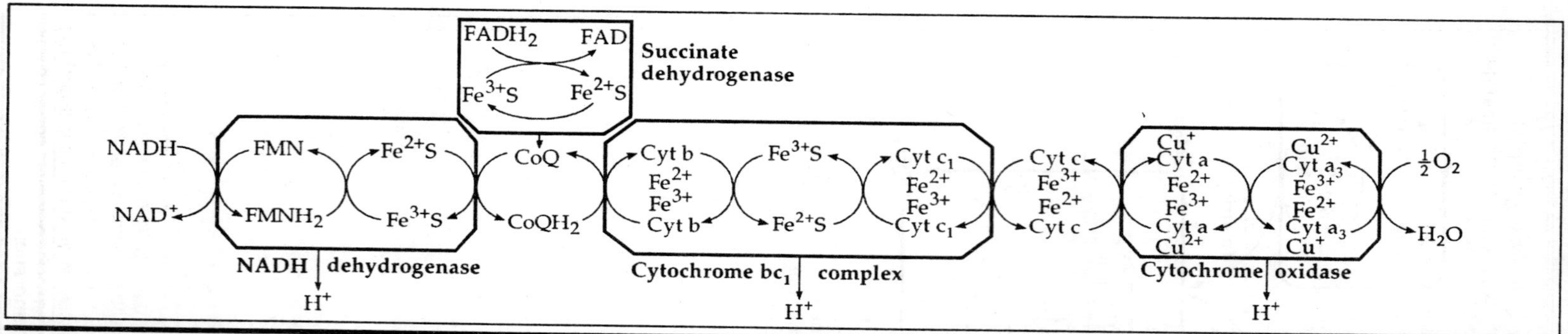

Fig. 3.14c: Electron transport through respiratory chain.

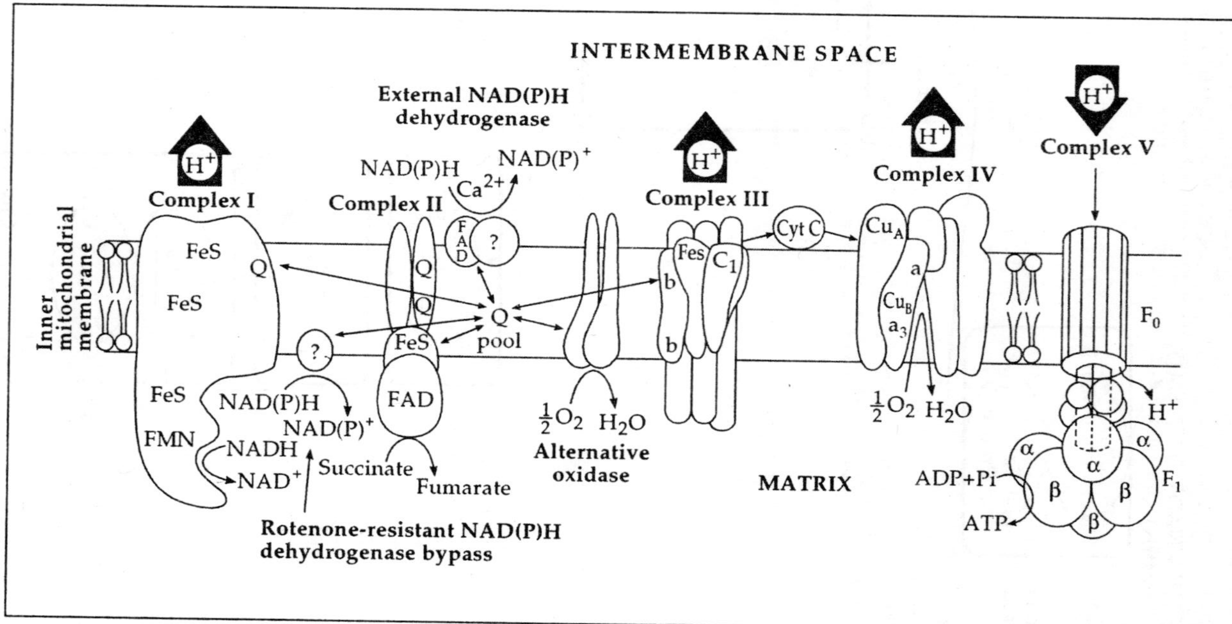

Fig. 3.15: Organization of plant mitochondrial electron transport chain in the inner mitochondrial membrane : In addition to the five standard protein complexes, an alternative rotenone-resistant NAD(P)H dehydrogenase on the matrix side of the membrane as well as an external NAD(P)H dehydrogenase that can accept electrons directly from NAD(P)H produced in the cytosol. "Q pool" refers to the large pool of ubiquinone that freely diffuses within the inner membrane and serves as the electron carrier that transfers electrons derived from the four dehydrogenases to either complex III or the alternative oxidase. "FMN" is the cofactor flavin mononucleotide. "Cyt c" is cytochrome c, a peripheral protein that transfers electrons from complex III to complex IV. Bold upward-facing arrows designate the three complexes at which protons are translocated, leading to generation of the proton electrochemical gradient that drives ATP synthesis. The downward-facing arrow shows the path of proton flow through the ATP synthase complex that is coupled to the conversion of ADP and Pi to ATP.

	Enzyme complex	Number of subunits	Prosthetic group(s)
I.	NADH dehydrogenase	42 (14)	FMN, Fe-S
II.	Succinate dehydrogenase	5	FAD, Fe-S
III.	Ubiquinone : Cytochrome c oxidoreductase	11	Hemes, Fe-S
	Cytochrome c	1	Heme
IV.	Cytochrome oxidase	13(3-4)	Hemes, Cu_A, Cu_B

Fig. 3.16: Protein components of the mitochondrial electron transfer chain.

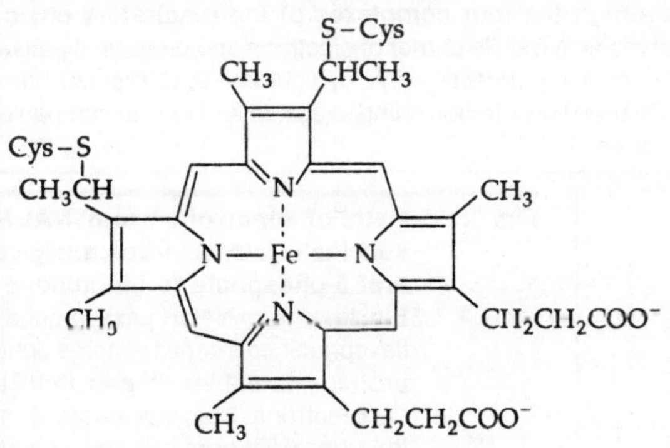

Heme C
[in c-type cytochromes]

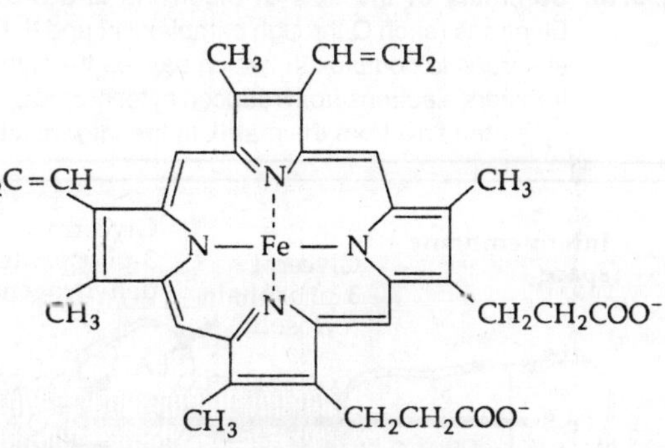

Iron protoporphyrin IX
[in b-type cytochromes]

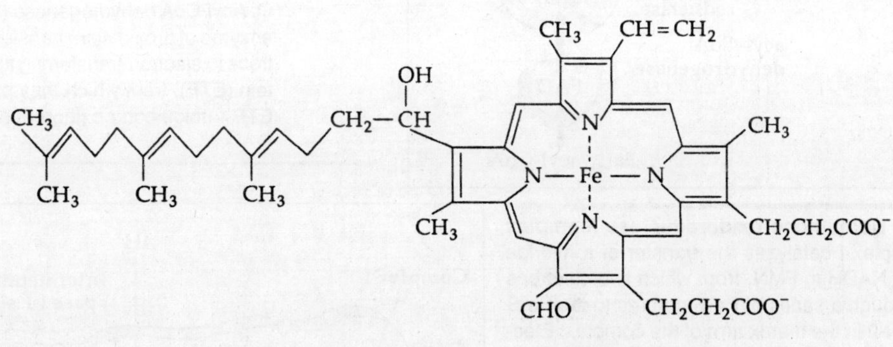

Heme A
[in a-type cytochromes]

Fig. 3.17 : Prosthetic groups of cytochromes : Each consists of four five-membered, nitrogen-containing rings in a cyclic structure called a porphyrin. The four nitrogen atoms are coordinated with a central Fe ion, either Fe^{2+} or Fe^{3+}. Iron protoporphyrin IX is found in b-type cytochromes. Heme C is covalently bound to the protein of cytochrome c through thioether bonds to two Cys residues. Heme A, found in the a-type cytochromes, has a long isoprenoid tail attached to one of the five-membered rings. The conjugated double-bond system of the porphyrin ring accounts for the absorption of visible light by these hemes.

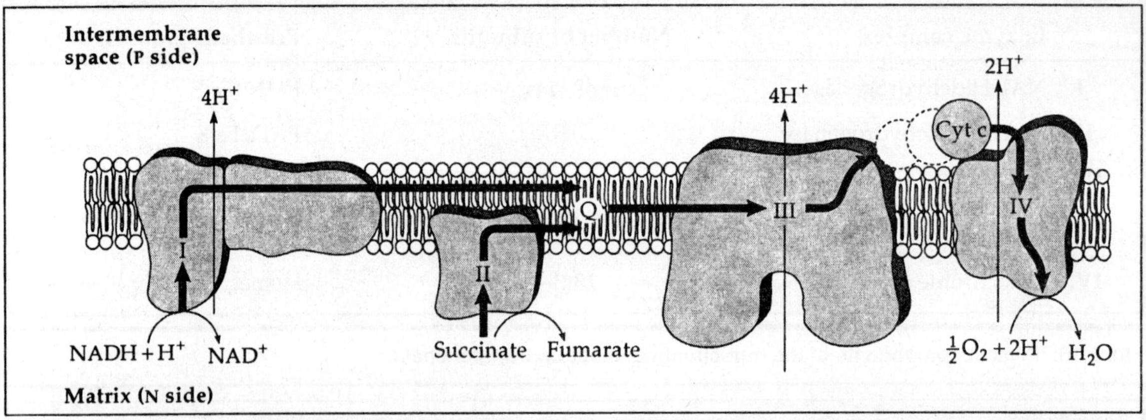

Fig. 3.18: Summary of the flow of electrons and protons through the four complexes of the respiratory chain : Electrons reach Q through complexes I and II. QH_2 serves as a mobile carrier of electrons and protons. It passes electrons to complex III, which passes them to another mobile connecting link, cytochrome c. Complex IV then transfers electrons from reduced cytochrome c to O_2. Electron flow through complexes I, III and IV is accompanied by proton flow from the matrix to the intermembrane space.

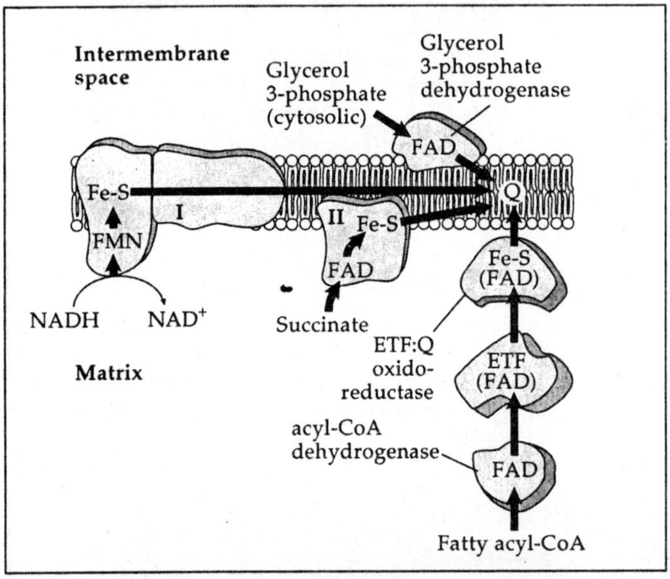

Fig. 3.19: Path of electrons from NADH, succinate, fatty acyl-CoA and glycerol 3-phosphate to ubiquinone : Electrons from NADH pass through a flavoprotein to a series of iron-sulphur proteins (complex I) and then to Q. Electrons from succinate pass through a flavoprotein and several Fe–S centres (complex II) on the way to Q. Glycerol 3-phosphate donates electrons to a flavoprotein (glycerol 3-phosphate dehydrogenase) on the outer face of the inner mitochondrial membrane, from which they pass to Q. Acyl-CoA dehydrogenase (the first enzyme of β-oxidation) transfers electrons to electron-transferring flavoprotein (ETF), from which they pass via ETF : ubiquinone oxidoreductase to Q.

Fig. 3.20: NADH : Ubiquinone oxidoreductase (complex I) : Complex I catalyzes the transfer of a hydride ion from NADH to FMN, from which two electrons pass through a series of Fe-S centres to the Fe-S protein N-2 in the matrix arm of the complex. Electron transfer from N-2 to ubiquinone on the membrane arm forms QH_2 which diffuses into the lipid bilayer. It also drives the expulsion of four protons per pair of electrons from the matrix. This proton flux produces an electrochemical potential across the inner mitochondrial membrane (N side negative, P side positive), which conserves some of the energy released by the electron-transfer reactions. This electrochemical potential drives ATP synthesis.

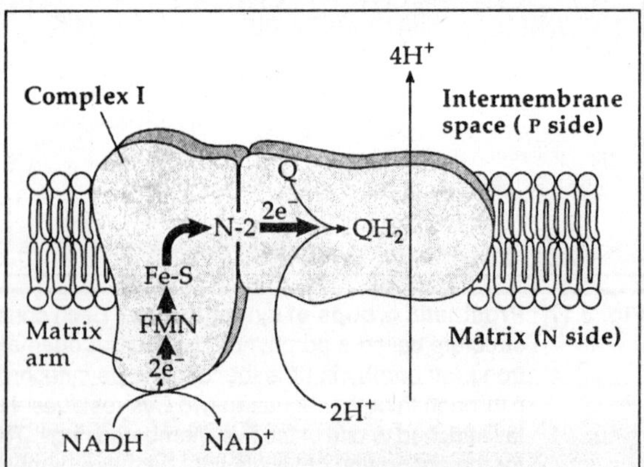

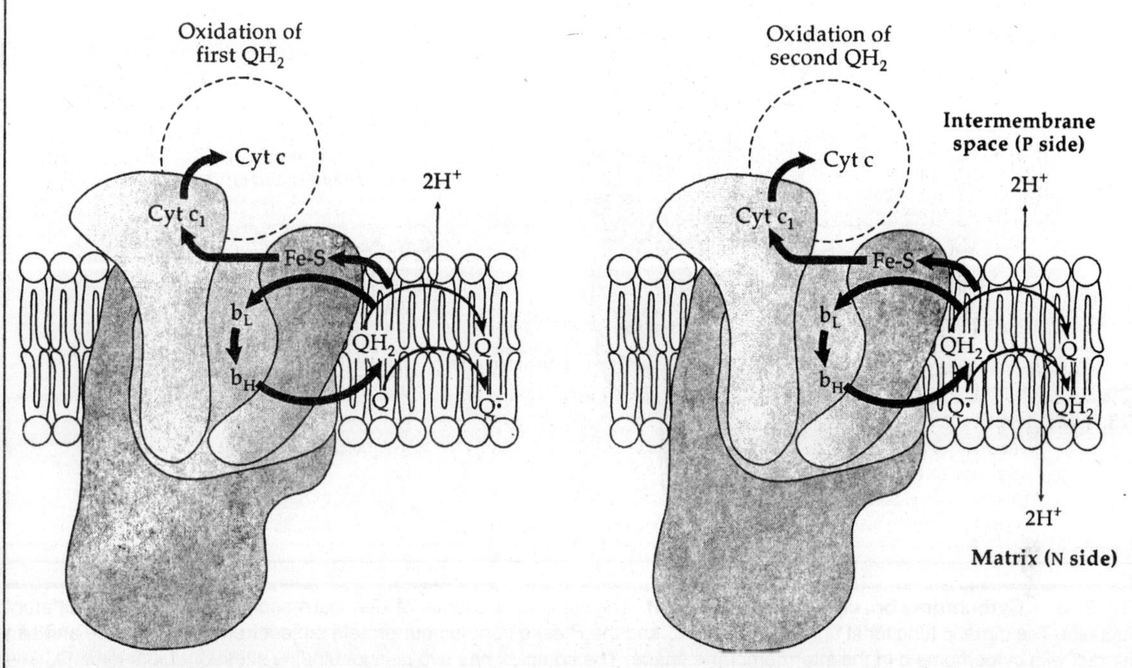

$QH_2 + \text{Cyt } c_1 \text{ (oxidized)} \rightarrow Q^{\bar{\bullet}} + 2H_P^+ + \text{Cyt } c_1 \text{ (reduced)}$

$QH_2 + Q^{\bar{\bullet}} + 2H_N^+ + \text{Cyt } c_1 \text{ (oxidized)} \longrightarrow QH_2 + 2H_P^+ + Q + \text{Cyt } c_1 \text{ (reduced)}$

Net equation : $QH_2 + 2 \text{ Cyt } c_1 \text{ (oxidized)} + 2H_N^+ \longrightarrow Q + 2 \text{ Cyt } c_1 \text{ (reduced)} + 4H_P^+$

Fig. 3.21: The Q cycle : The path of electrons through complex III is shown by arrows. On the P side of the membrane, two molecules of QH_2 are oxidized to Q at the Q_P site, releasing four protons into the intermembrane space. Each QH_2 donates one electron (via the Rieske Fe–S centre) to Cytochrome c_1 and one electron (via Cytochrome b) to a molecule of Q at the Q_N site, reducing it in two steps to QH_2. This reduction also uses two protons taken up from the matrix.

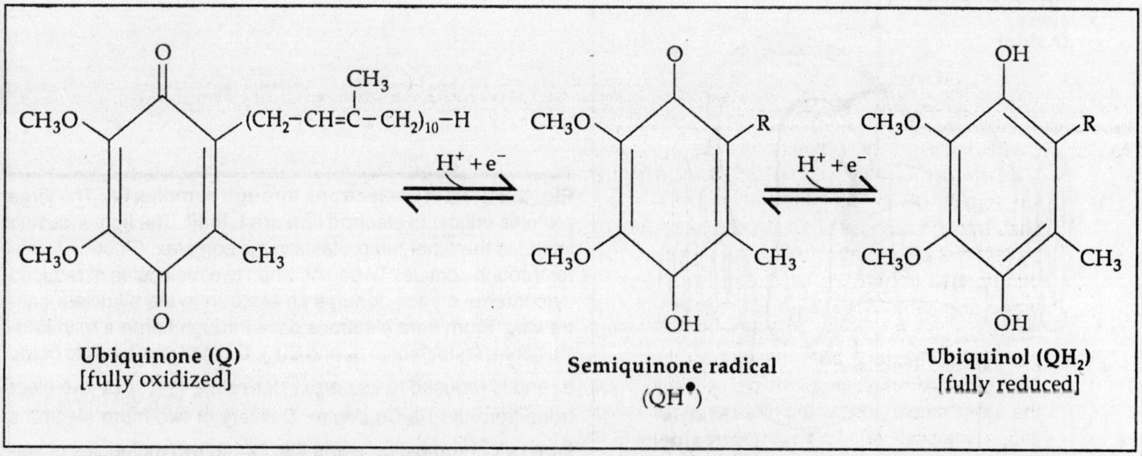

Fig. 3.22: Ubiquinone (Q, or Coenzyme Q) : Complete reduction of ubiquinone requires two electrons and two protons, and occurs in two steps through the semiquinone radical intermediate.

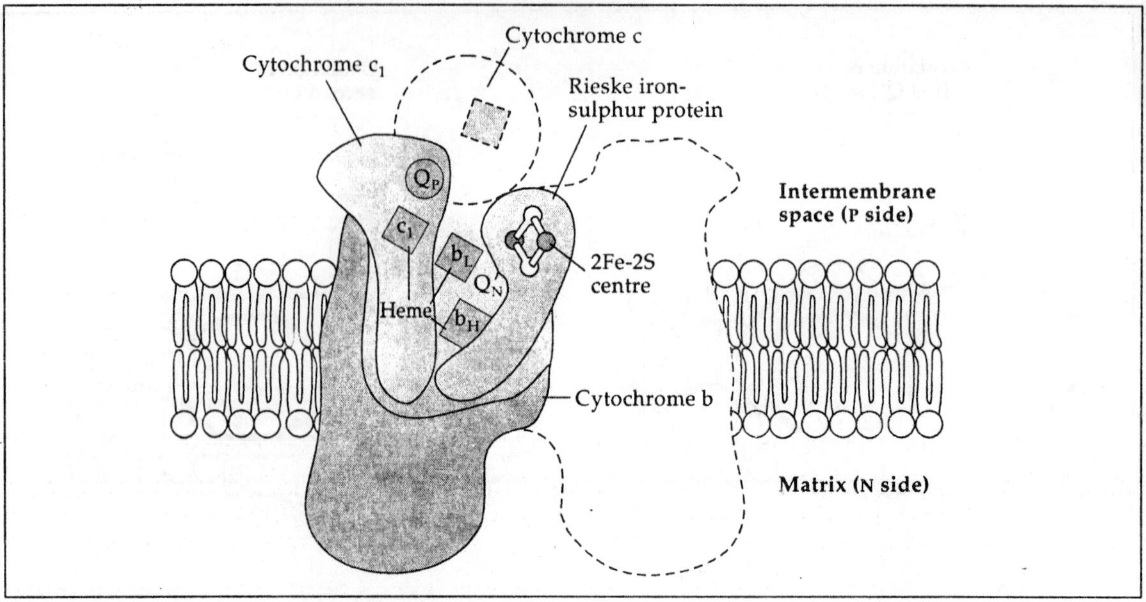

Fig. 3.23: Cytochrome bc$_1$ complex (complex III) : The complex is a dimer of identical monomers, each with 11 different subunits. The dimeric functional unit cytochrome c_1 and the Rieske iron-sulphur protein project from the P surface and can interact with cytochrome c in the intermembrane space. The complex has two distinct binding sites for ubiquinone, Q_N and Q_P, which correspond to the sites of inhibition by two drugs that block oxidative phosphorylation. Antimycin A, which blocks electron flow from heme b_H to Q, binds at Q_N, close to heme b_H on the N (matrix) side of the membrane. Myxothiazol, which prevents electron flow from QH_2 to the Rieske iron-sulphur protein, binds at Q_P, near the 2Fe-2S centre and heme b_L on the P side of the membrane. The dimeric structure is essential to the function of complex III. The interface between monomers forms two pockets, each containing a Q_P site from one monomer and a Q_N site from the other. The movement of ubiquinone intermediates occurs within these sheltered pockets. Complex III crystallizes in two distinct conformations. In one, the Rieske Fe-S centre is close to its electron acceptor, the heme of cytochrome c_1, but relatively distant from cytochrome b and the QH_2-binding site at which it receives electrons. In the other, the Fe-S centre has moved away from cytochrome c_1 and toward cytochrome b. The Rieske protein is thought to oscillate between these two conformations as it is first reduced, then oxidized.

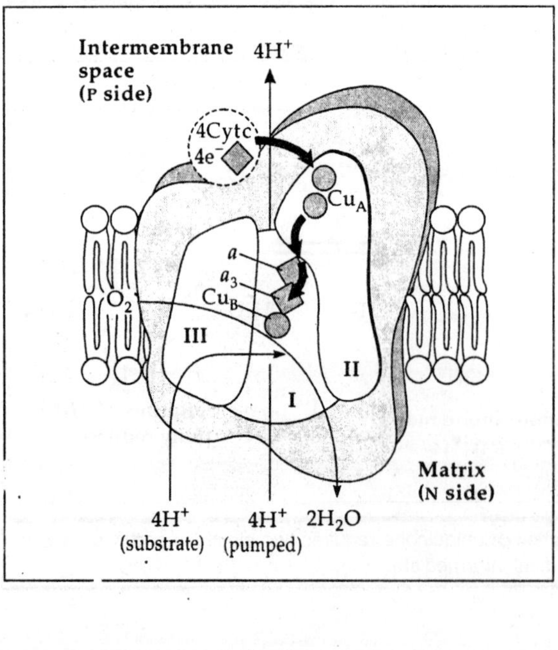

Fig. 3.24: Path of electrons through complex IV : The three proteins critical to electron flow are I, II, III. The lighter outline includes the other ten proteins in the complex. Electron transfer through complex IV begins when two molecules of reduced cytochrome c each donates an electron to the binuclear centre Cu_A. From here electrons pass through heme a to the Fe-Cu centre (cytochrome a_3 and Cu_B). Oxygen now binds to heme a_3 and is reduced to its peroxy derivative (O_2^{2-}) by two electrons from the Fe-Cu centre. Delivery of two more electrons from cytochrome c converts the O_2^{2-} to two molecules of water, consuming four 'substrate' protons from the matrix. At the same time, four more protons are pumped from the matrix.

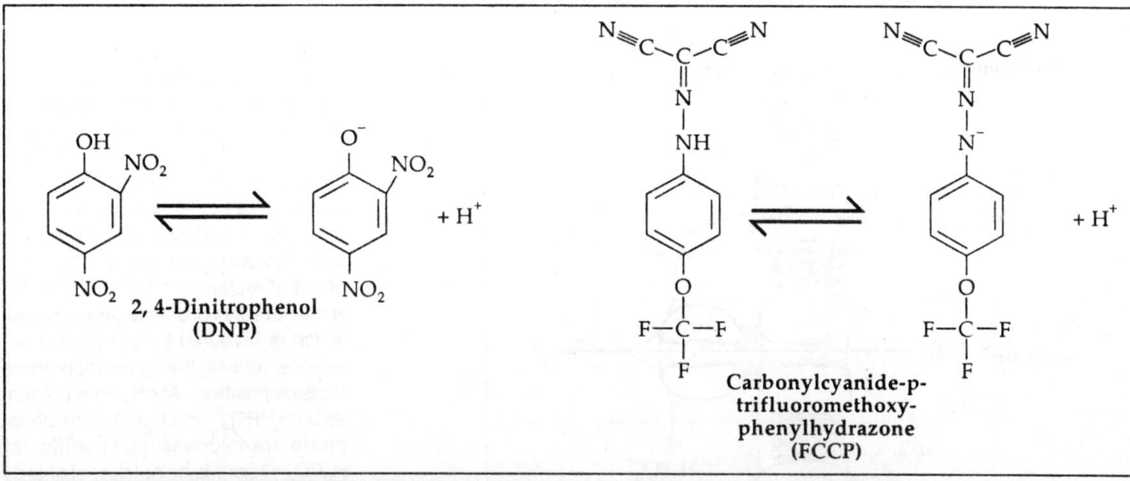

Fig. 3.25: Two chemical uncouplers of oxidative phosphorylation : Both DNP and FCCP have a dissociable proton and are very hydrophobic. They carry protons across the inner mitochondrial membrane, dissipating the proton gradient.

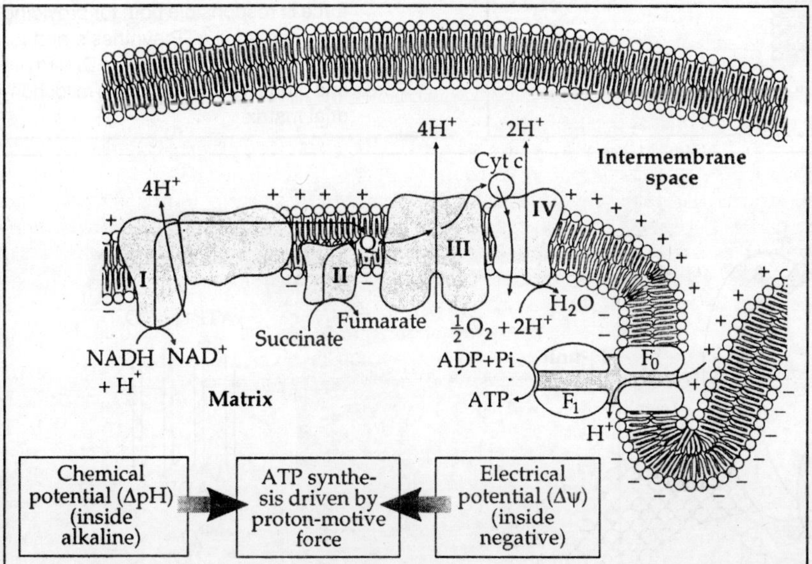

Fig. 3.26: Chemiosmotic model : Electrons from NADH and other oxidizable substrates pass through a chain of carriers arranged asymmetrically in the inner membrane. Electron flow is accompanied by proton transfer across the membrane, producing both a chemical gradient (ΔpH) and an electrical gradient (Δψ). The inner mitochondrial membrane is impermeable to protons; protons can re-enter the matrix only through proton-specific channels (F_0). The proton-motive force that drives proton back into the matrix provides the energy for ATP synthesis, catalyzed by the F_1 complex associated with F_0.

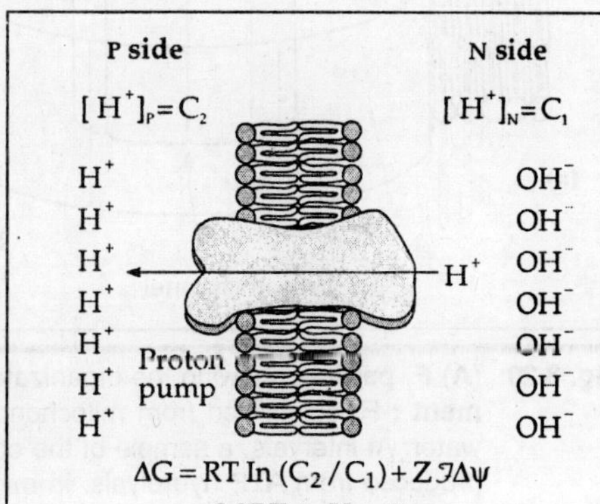

$$\Delta G = RT\ ln\ (C_2/C_1) + Z\mathcal{F}\Delta\psi$$
$$= 2.3RT\ \Delta pH + \mathcal{F}\Delta\psi$$

Fig. 3.27: Proton-motive force : The inner mitochondrial membrane separates two compartments of different [H^+], resulting in differences in chemical concentration (ΔpH) and charge distribution (Δψ), across the membrane. The net effect is the proton-motive force (ΔG), which can be calculated.

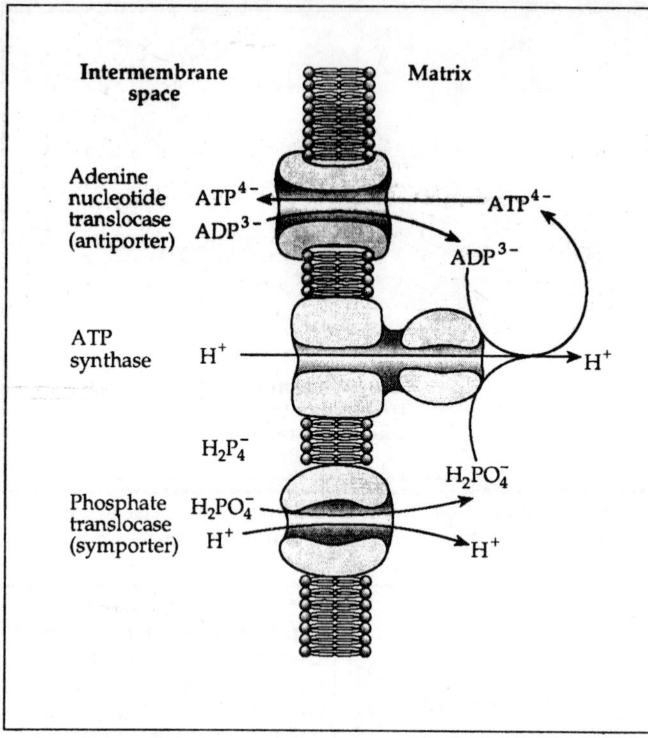

Fig. 3.28: Adenine nucleotide and phosphate translocases : Transport systems of the mitochondrial inner membrane carry ADP+Pi into the matrix and newly synthesized ATP into the cytosol. The adenine nucleotide translocase is an antiporter ; the same protein moves ADP into the matrix and ATP out. The effect of replacing ATP^{4-} with ADP^{3-} is the net efflux of one negative charge, which is favoured by the charge difference across the inner membrane (outside positive). At pH_7, Pi is present as both HPO_4^{2-} and $H_2PO_4^-$; the phosphate translocase is specific for $H_2PO_4^-$. There is no net flow of charge during symport of $H_2PO_4^-$ and H^+, but the relatively low proton concentration in the matrix favours the inward movement of H^+. Thus the proton-motive force is responsible both for providing the energy for ATP synthesis and for transporting substrates (ADP+Pi) in and product (ATP) out of the mitochondrial matrix.

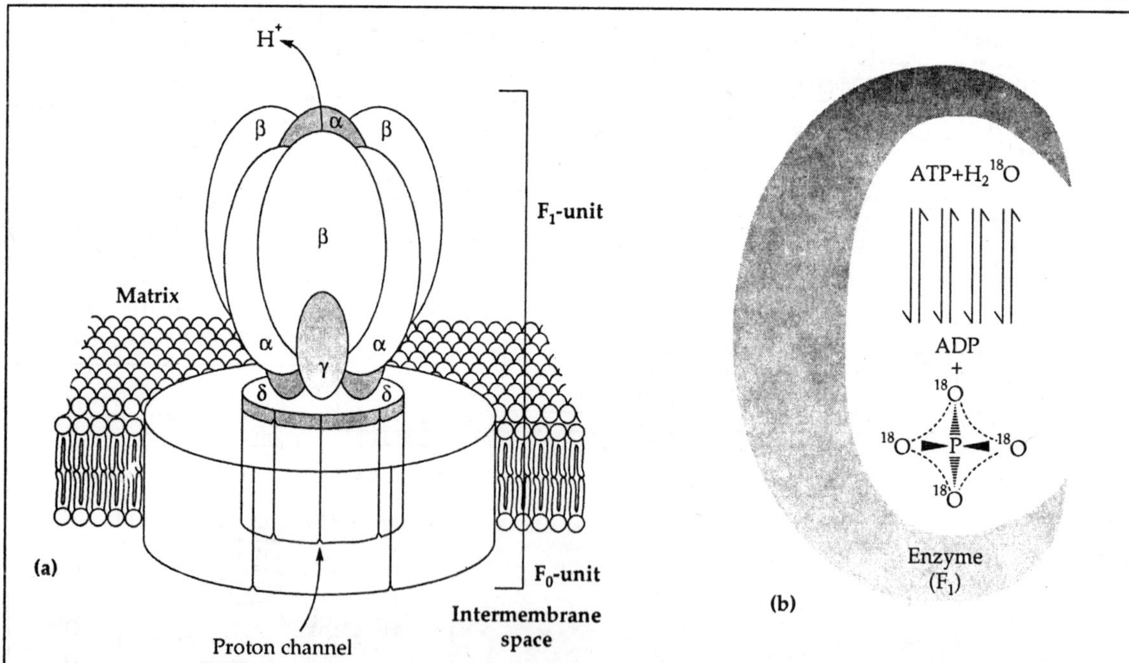

Fig. 3.29: (A) F_1 particle showing the organization of subunits; **(B) Catalytic mechanism of F_1 : ^{18}O exchange experiment :** F_1 solubilized from mitochondrial membranes is incubated with ATP in the presence of ^{18}O-labelled water. At intervals, a sample of the solution is withdrawn and analyzed for the incorporation of ^{18}O into the Pi produced from ATP hydrolysis. In minutes, the Pi contains three or four ^{18}O atoms, indicating that both ATP hydrolysis and ATP synthesis have occurred several times during the incubation.

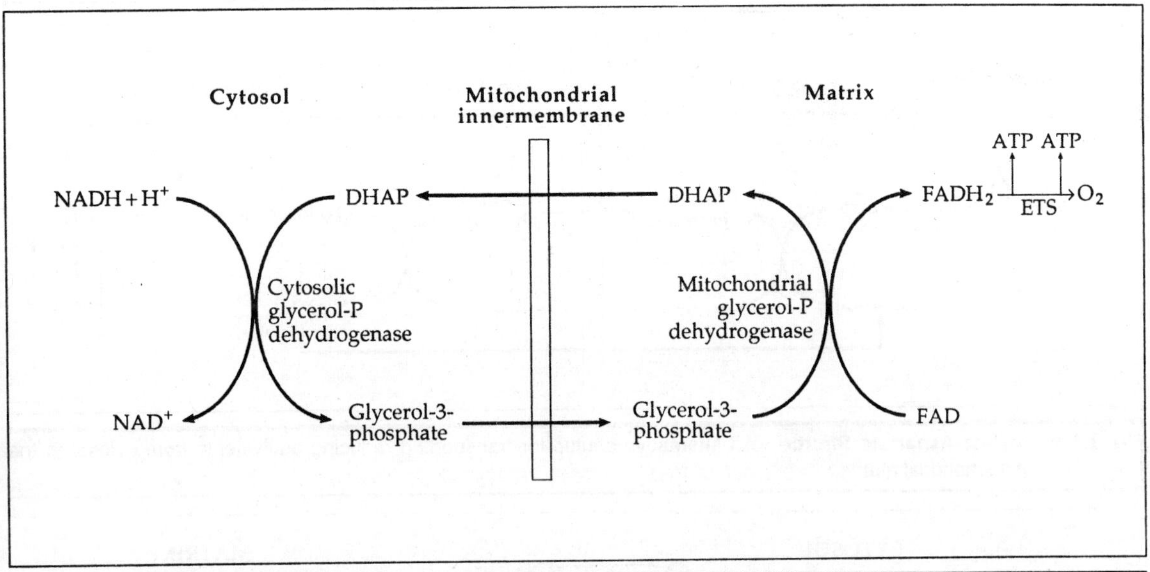

Fig. 3.30a: Glycerol 3-phosphate shultle : This is a shuttle for moving reducing equivalents from the cytosol to the mitochondrial matrix.

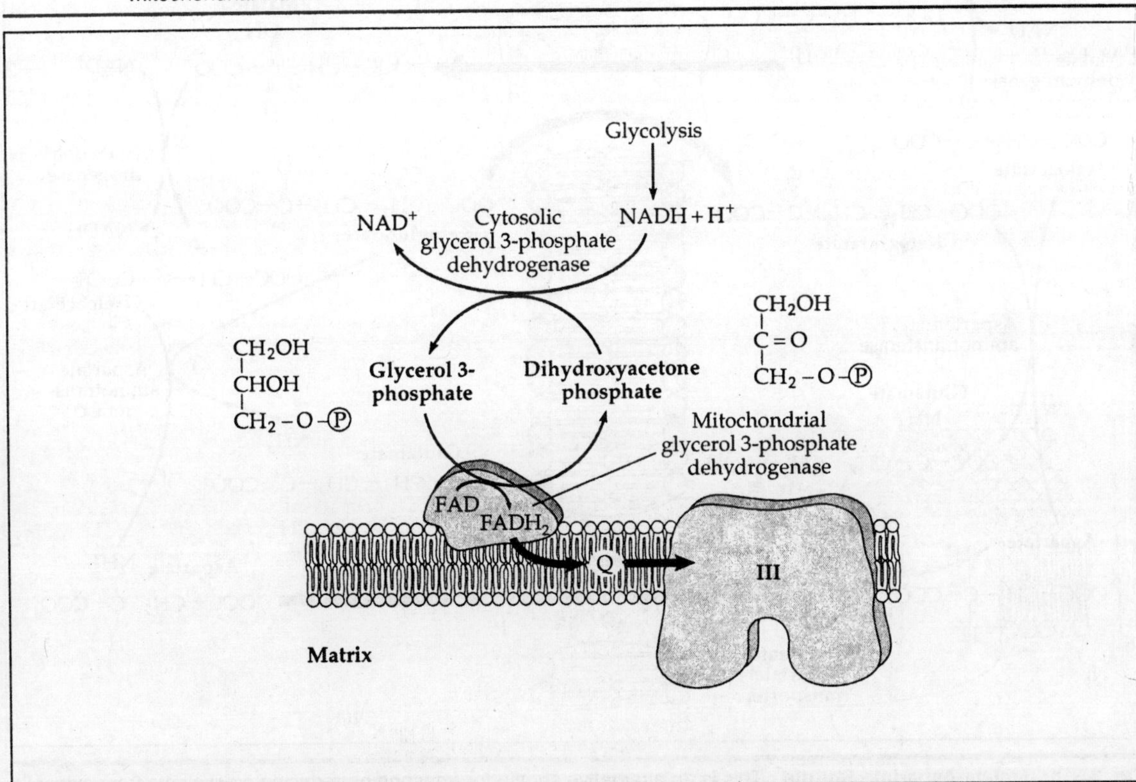

Fig. 3.30b: Glycerol 3-phosphate shultle : In the cytosol, dihydroxyacetone phosphate accepts two reducing equivalents from NADH in a reaction catalyzed by cytosolic glycerol 3-phosphate dehydrogenase. An isozyme of glycerol 3-phosphate dehydrogenase bound to the outer face of the inner membrane then transfers two reducing equivalents from glycerol 3-phosphate in the intermembrane space to ubiquinone. This shuttle does not involve membrane transport system.

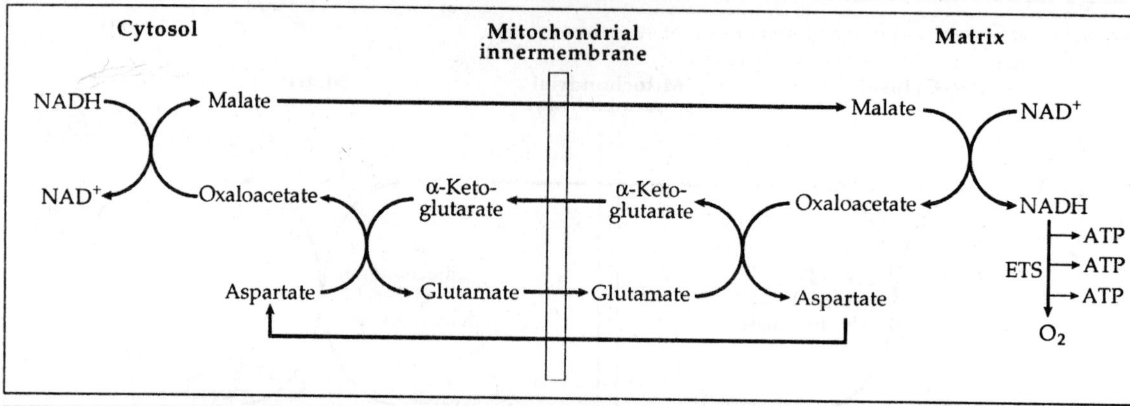

Fig. 3.31a: Malate-Aspartate Shuttle : An alternative shuttle for transporting reducing equivalents from cytosol to the mitochondrial matrix.

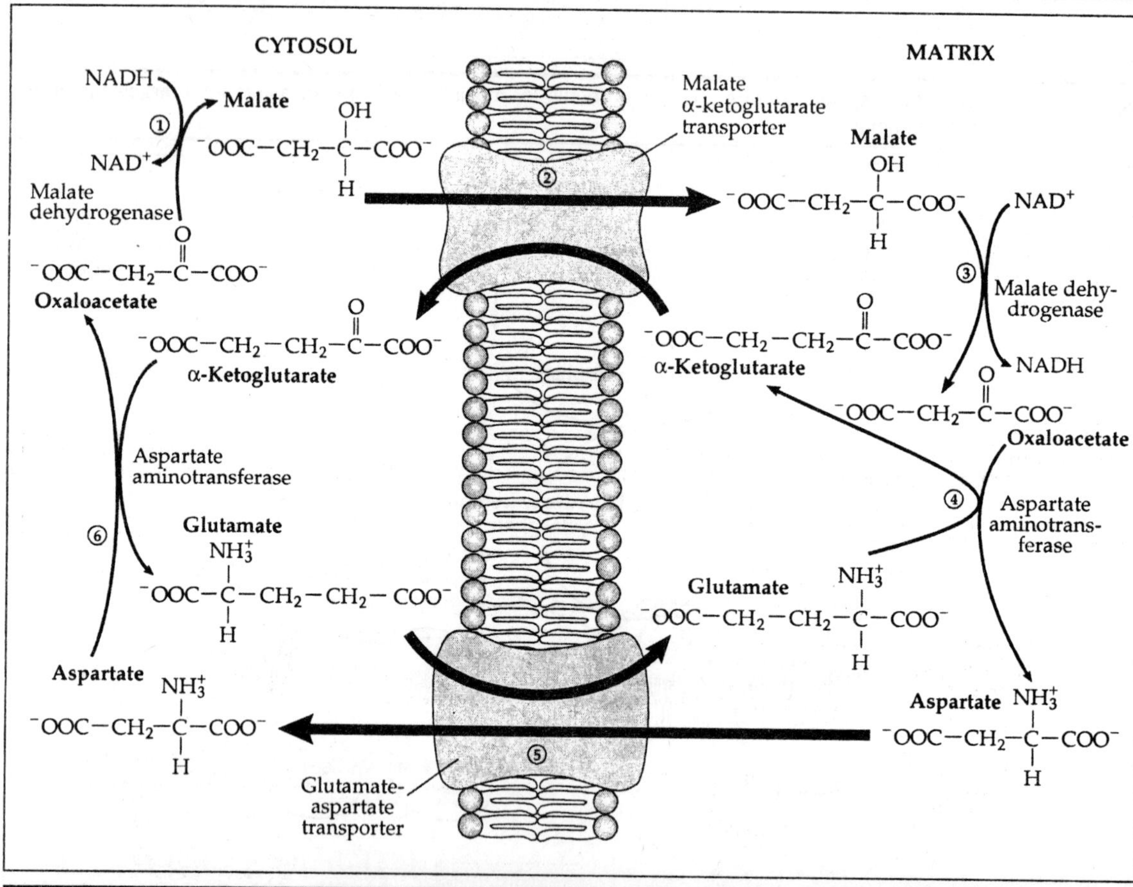

Fig. 3.31b: Malate-Aspartate Shuttle : This is an alternative shuttle for transporting reducing equivalents from cytosolic NADH into the mitochondrial matrix. **(1)** NADH in the cytosol (intermembrane space) passes two reducing equivalents to oxaloacetate, producing malate. **(2)** Malate is transported across the inner membrane by the malate-α-ketoglutarate transporter. **(3)** In the matrix, malate passes two reducing equivalents to NAD^+, and the resulting NADH is oxidized by the respiratory chain. **(4)** The oxaloacetate formed from malate cannot pass directly into the cytosol. Oxaloacetate is transaminated to aspartate. **(5)** Aspartate leaves via the glutamate-aspartate transporter. **(6)** Oxaloacetate is regenerated in the cytosol completing the cycle.

Fig. 3.32: Bacterial respiratory chain : Respiratory carriers of the inner membrane of *E. coli* contain minimal form of complex I, containing all the prosthetic groups normally associated with the mitochondrial complex, but only 14 polypeptides. This plasma membrane complex carries NADH to ubiquinone or menaquinone, the bacterial equivalent of ubiquinone, while pumping protons outward and creating an electrochemical potential that drives ATP

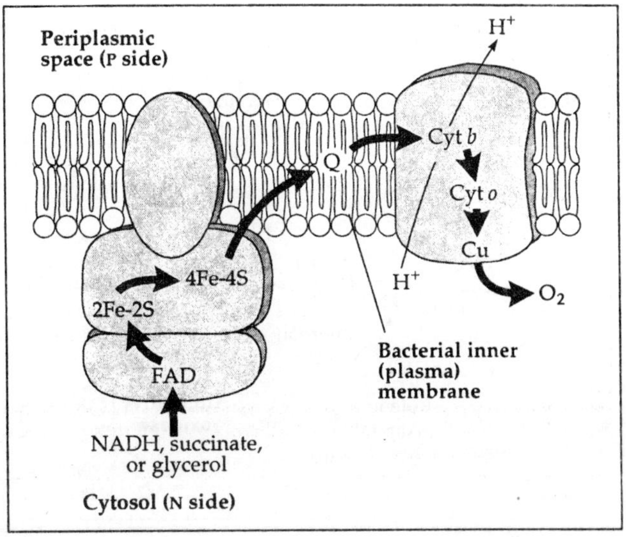

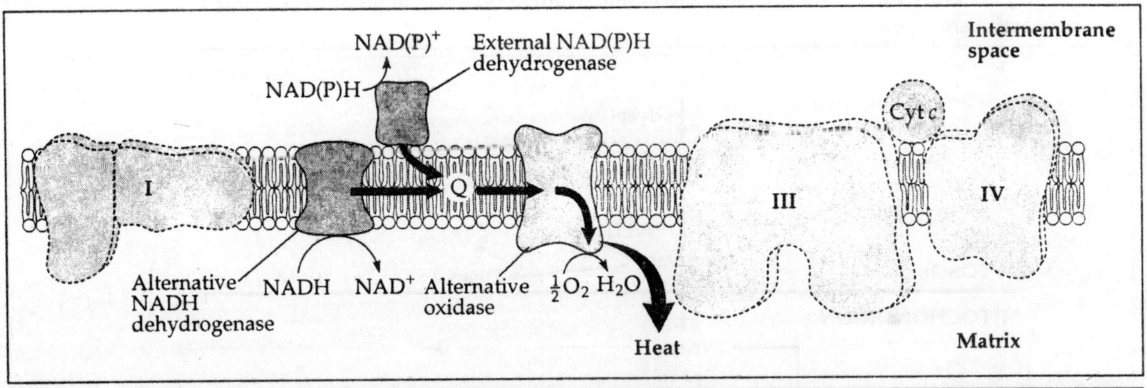

Fig. 3.33: Electron carriers of the inner membrane of plant mitochondria : Electrons can flow through complexes I, III, IV or through alternative carriers by the paths shown with arrows.

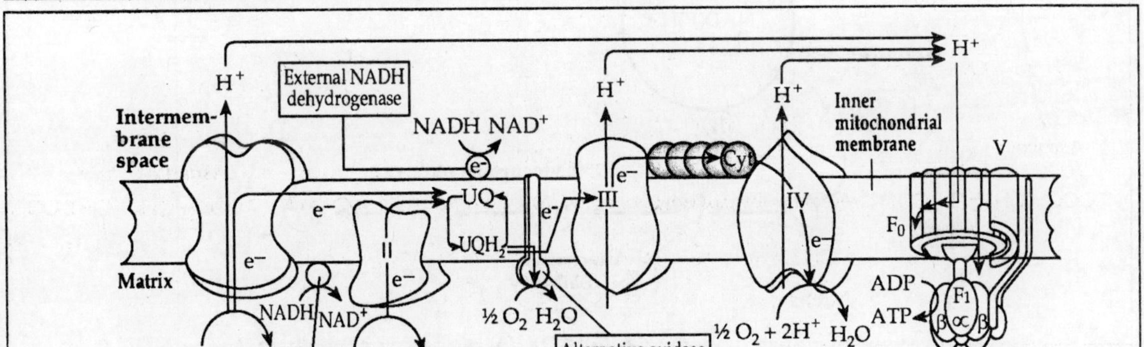

Fig. 3.34: Organization of the plant mitochondrial electron transfer chain including additional rotenone-resistant NADH dehydrogenases and the alternative oxidase.

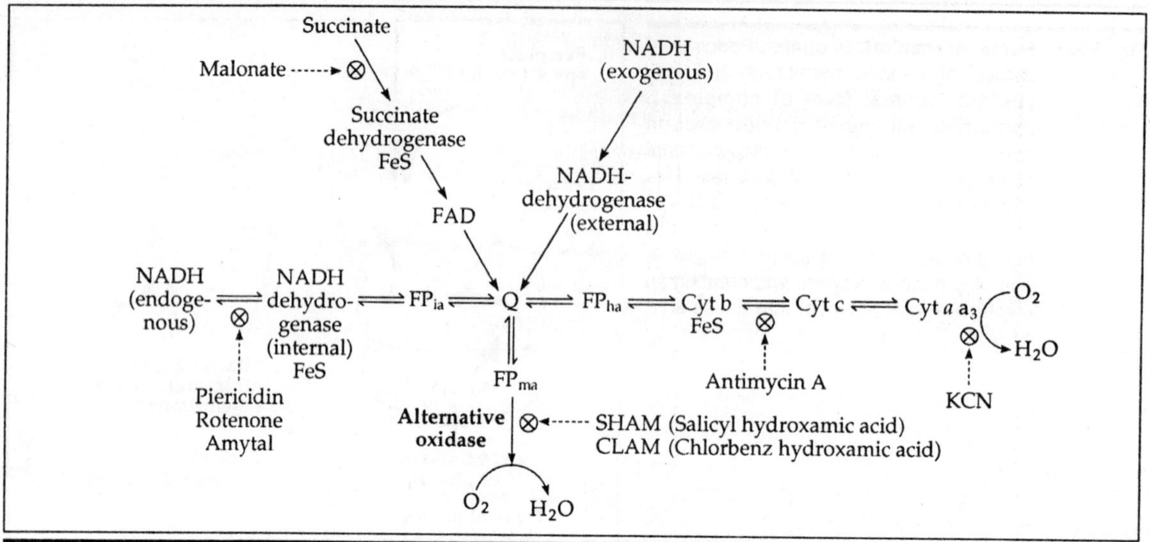

Fig. 3.35: Cyanide-sensitive cytochrome pathway and cyanide-resistant alternative pathway : The sites of inhibitor action are shown.

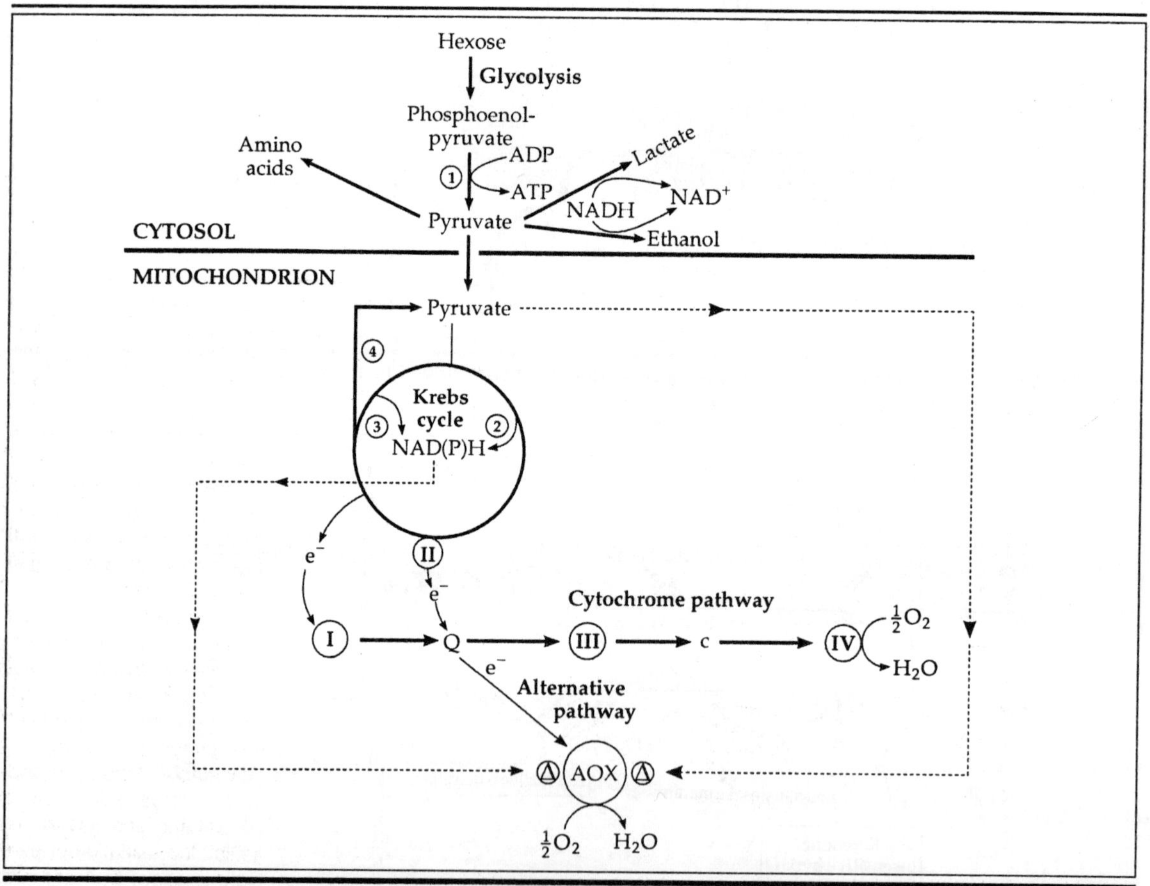

Fig. 3.36: Regulation of alternative oxidase : Accumulation of pyruvate, intramitochondrial NAD(P)H or reduced ubiquinone have the potential to increase electron transport to alternative oxidase. (1) Pyruvate kinase, (2) Isocitrate dehydrogenase, (3) Malate dehydrogenase, (4) Malic enzyme are involved in the process.

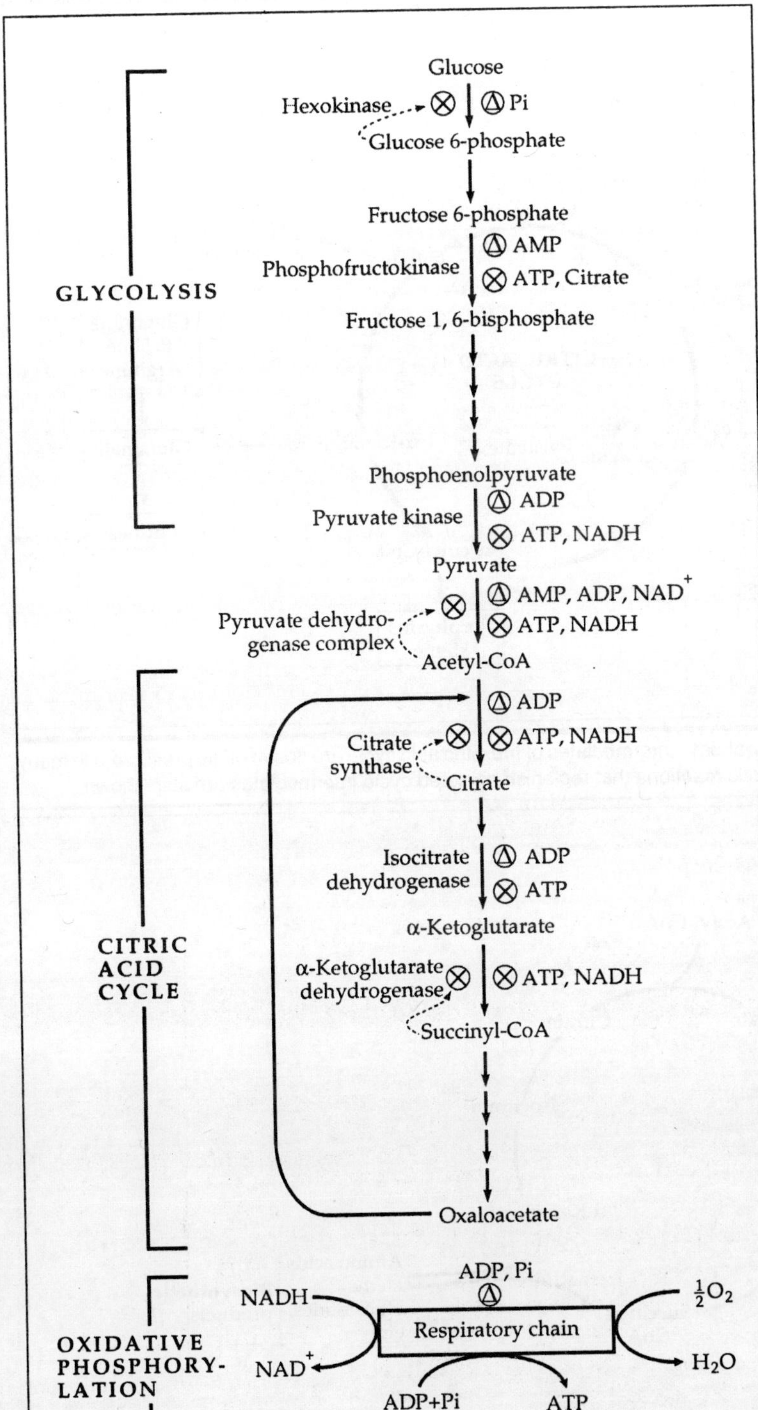

Fig. 3.37: Regulation of ATP-producing pathways–glycolysis, pyruvate oxidation, citric acid cycle, oxidative phosphorylation : High ATP or low ADP/AMP produces low rates of glycolysis, pyruvate oxidation, acetate oxidation via the citric acid cycle and oxidative phosphorylation. All four pathways are accelerated when the use of ATP and the formation of ADP, AMP and Pi increase. Interlocking of glycolysis and the citric acid cycle by citrate, which inhibits glycolysis, supplements the action of the adenine nucleotide system. In addition, increased levels of NADH and acetyl-CoA also inhibit the oxidation of pyruvate to acetyl-CoA and high $NADH/NAD^+$ ratios inhibit the dehydrogenase reactions of the citric acid cycle.

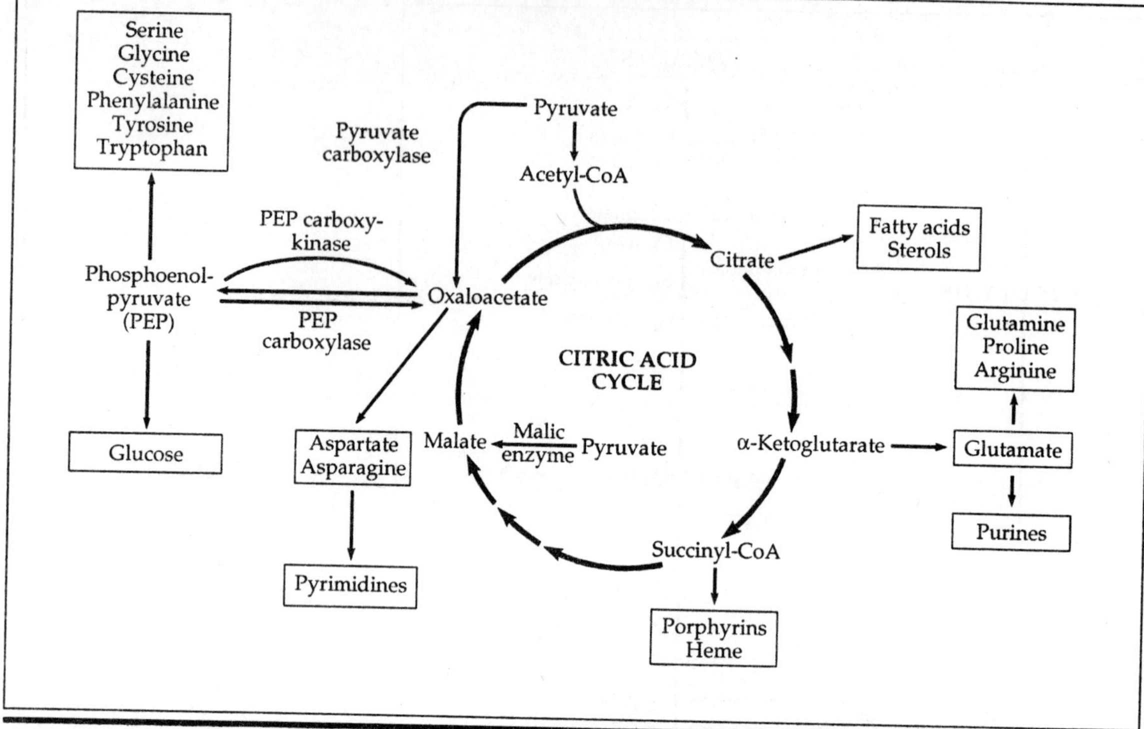

Fig. 3.38: Role of citric acid cycle in anabolism : Intermediates of the citric acid cycle are drawn off as precursors in many biosynthetic pathways. Anaplerotic reactions that replenish depleted cycle intermediates are also shown.

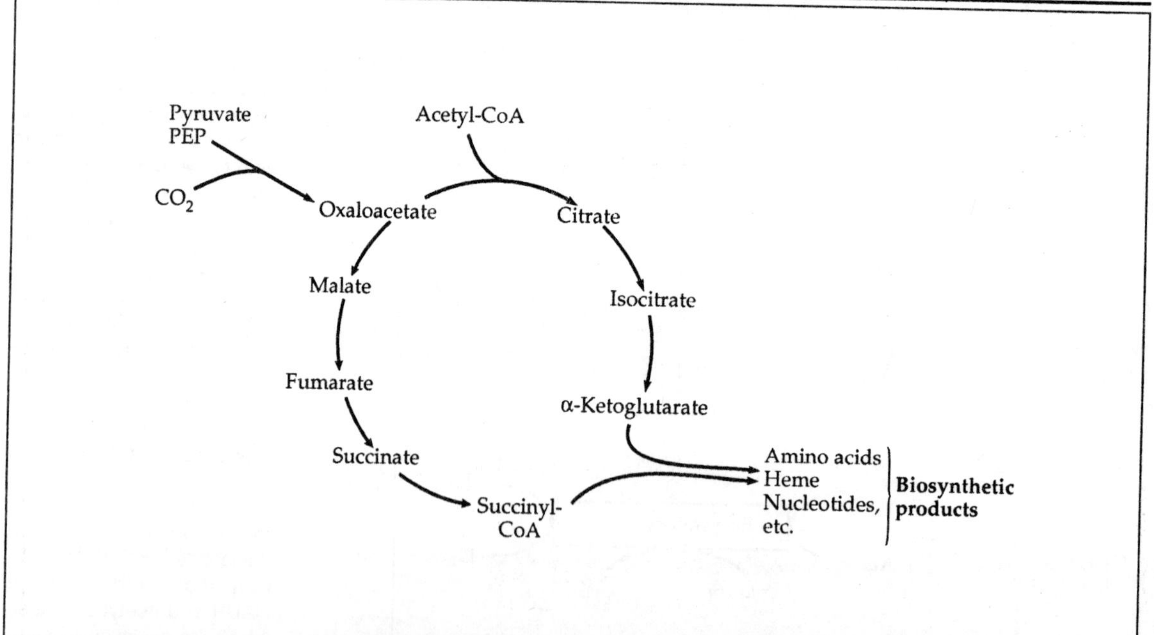

Fig. 3.39: Biosynthetic precursors produced by an incomplete citric acid cycle in anaerobic bacteria : These anaerobes lack α-ketoglutarate dehydrogenase and therefore cannot carry out the complete citric acid cycle. α-Ketoglutarate and succinyl-CoA serve as precursors in a variety of biosynthetic pathways.

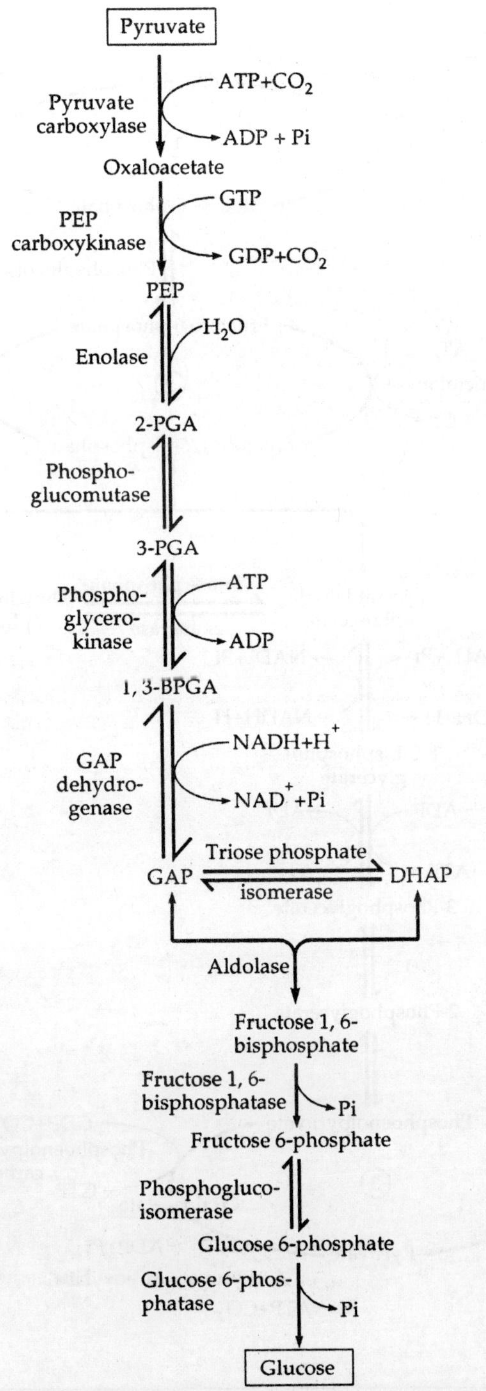

Fig. 3.40: Steps of Gluconeogenesis : Synthesis of glucose from pyruvate.

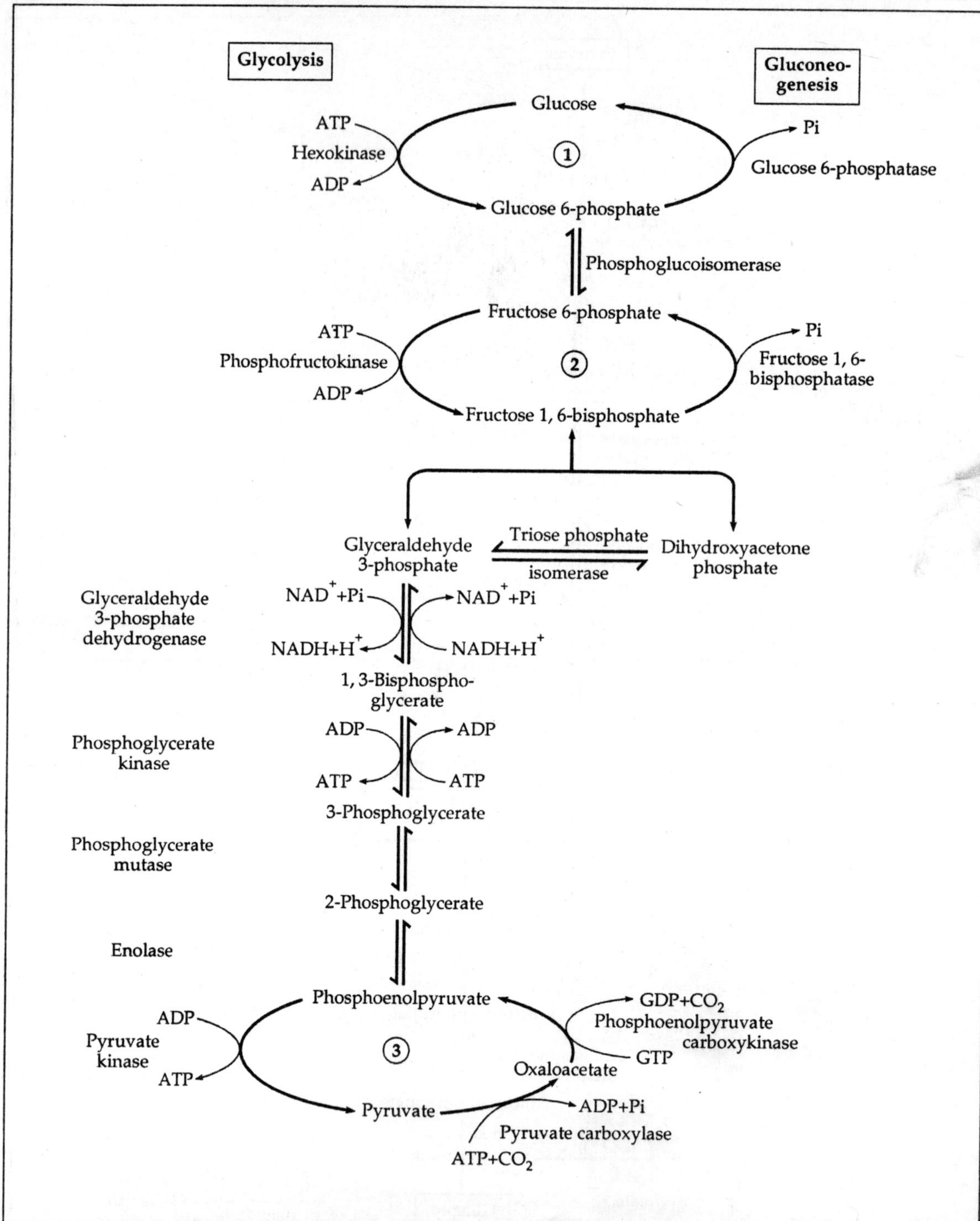

Fig. 3.41: Comparison of gluconeogenesis and glycolysis :

(1) Hexokinase activity in glycolysis is reversed by glucose 6-phosphatase in gluconeogenesis.

(2) Phosphofructokinase in glycolysis is reversed by fructose 1, 6-bisphosphatase in gluconeogenesis.

(3) Pyruvate kinase in glycolysis is reversed by two sequential reactions in gluconeogenesis catalyzed by pyruvate carboxylase and PEP carboxykinase.

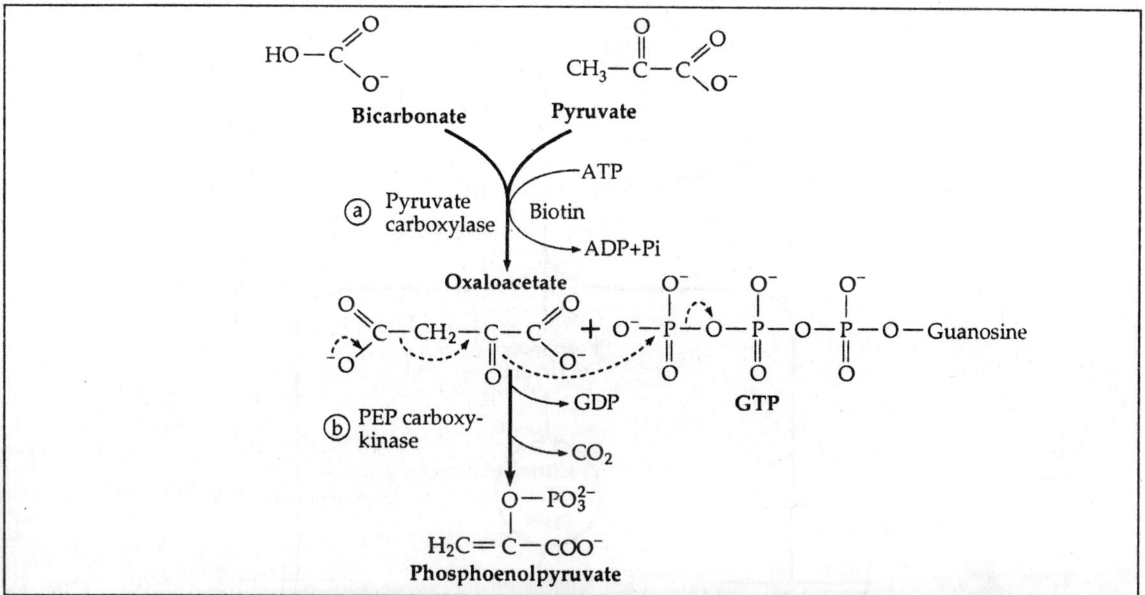

Fig. 3.42: Synthesis of phosphoenolpyruvate from pyruvate : (a) In mitochondria, pyruvate is converted to oxaloacetate in a biotin-requiring reaction catalysed by pyruvate carboxylase. **(b)** In the cytosol, oxaloacetate is converted to phosphoenolpyruvate by PEP carboxykinase. The CO_2 incorporated in the pyruvate carboxylase reaction is lost here as CO_2. The decarboxylation leads to a rearrangement of electrons that facilitates attack of the carbonyl oxygen of the pyruvate moiety on the γ-phosphate of GTP.

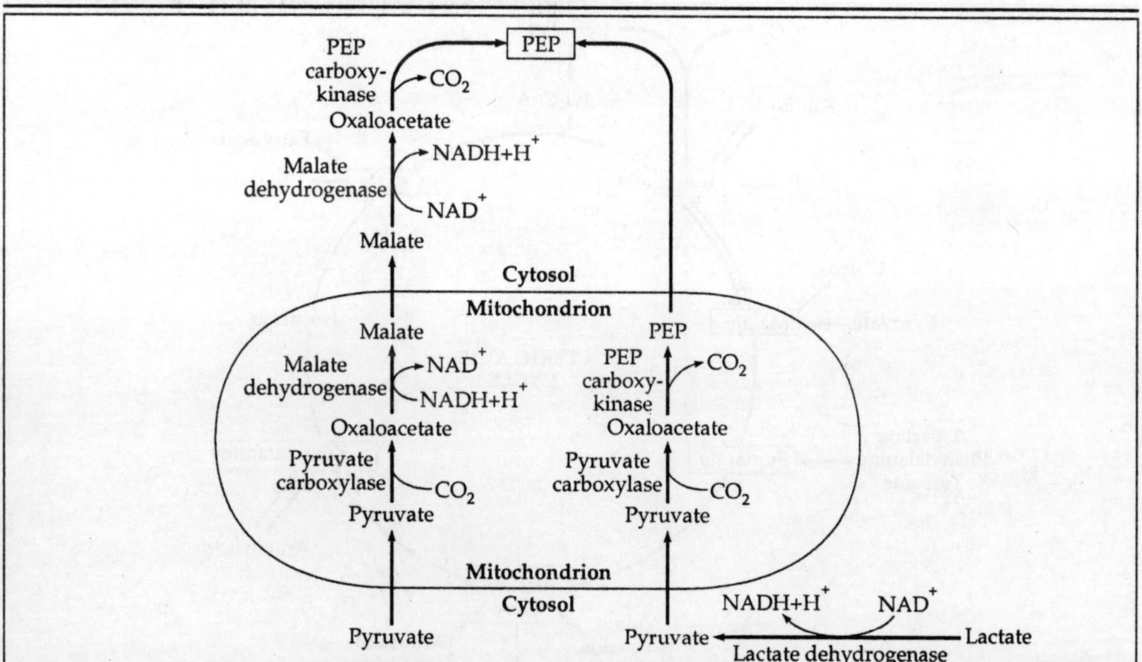

Fig. 3.43: Alternative paths from pyruvate to phosphoenolpyruvate : The path on the right predominates when lactate is the precursor, because cytosolic NADH is generated in the lactate dehydrogenase reaction and does not have to be shuttled out of the mitochondrion. The path on the left uses pyruvate as glucogenic precursor. The relative importance of two pathways depends on the availability of lactate and the cytosolic requirements for NADH in gluconeogenesis.

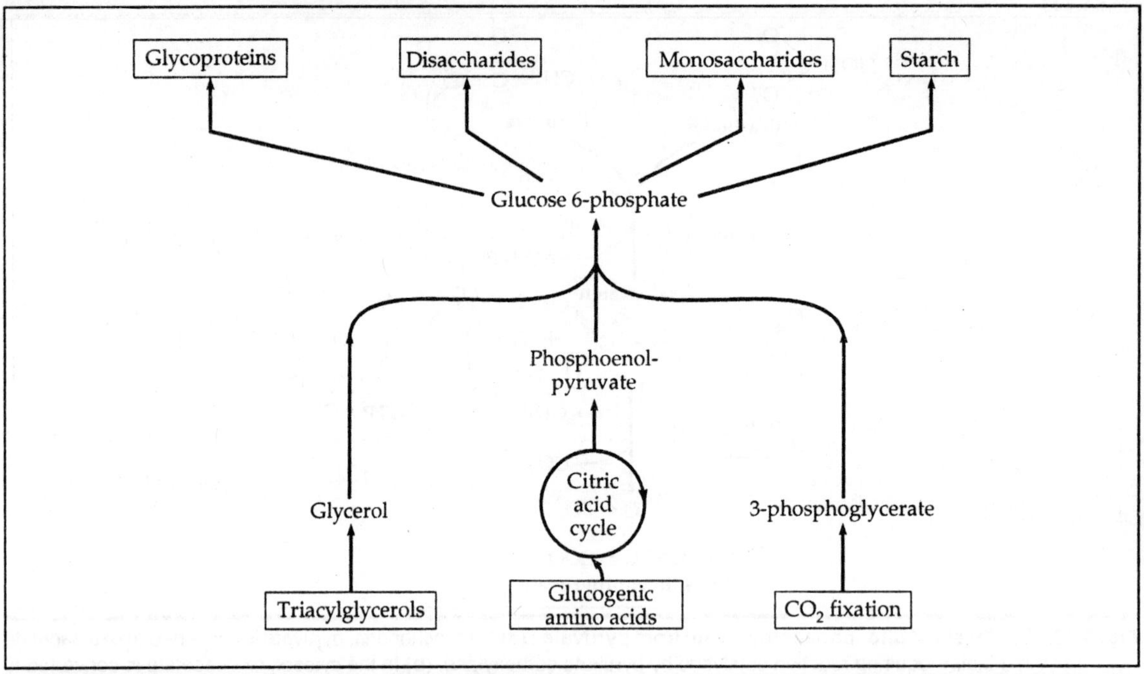

Fig. 3.44: Carbohydrate synthesis from simple precursors : Glucose 6-phosphate is common to the biosynthetic conversion of many different precursors to carbohydrates.

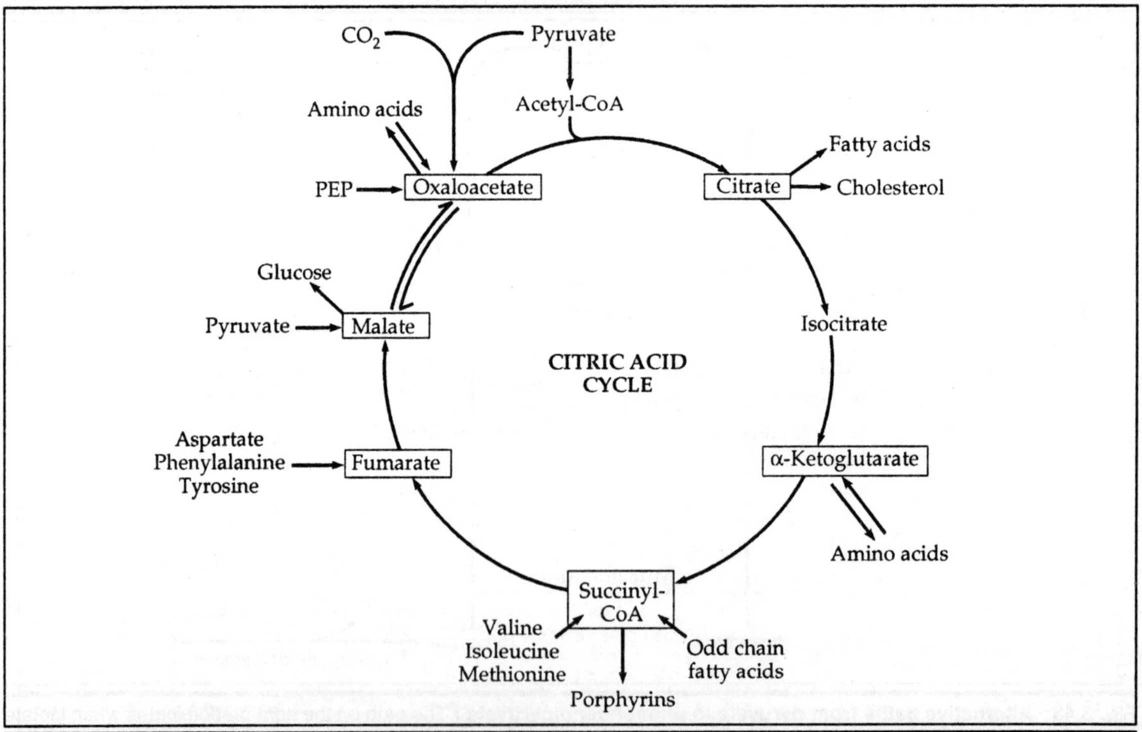

Fig. 3.45: Amphibolic functions of the citric acid cycle : The diagram indicates the positions at which intermediates are drawn off by cataplerotic reactions for use in anabolic pathways and the points where anaplerotic reactions replenish cycle intermediates.

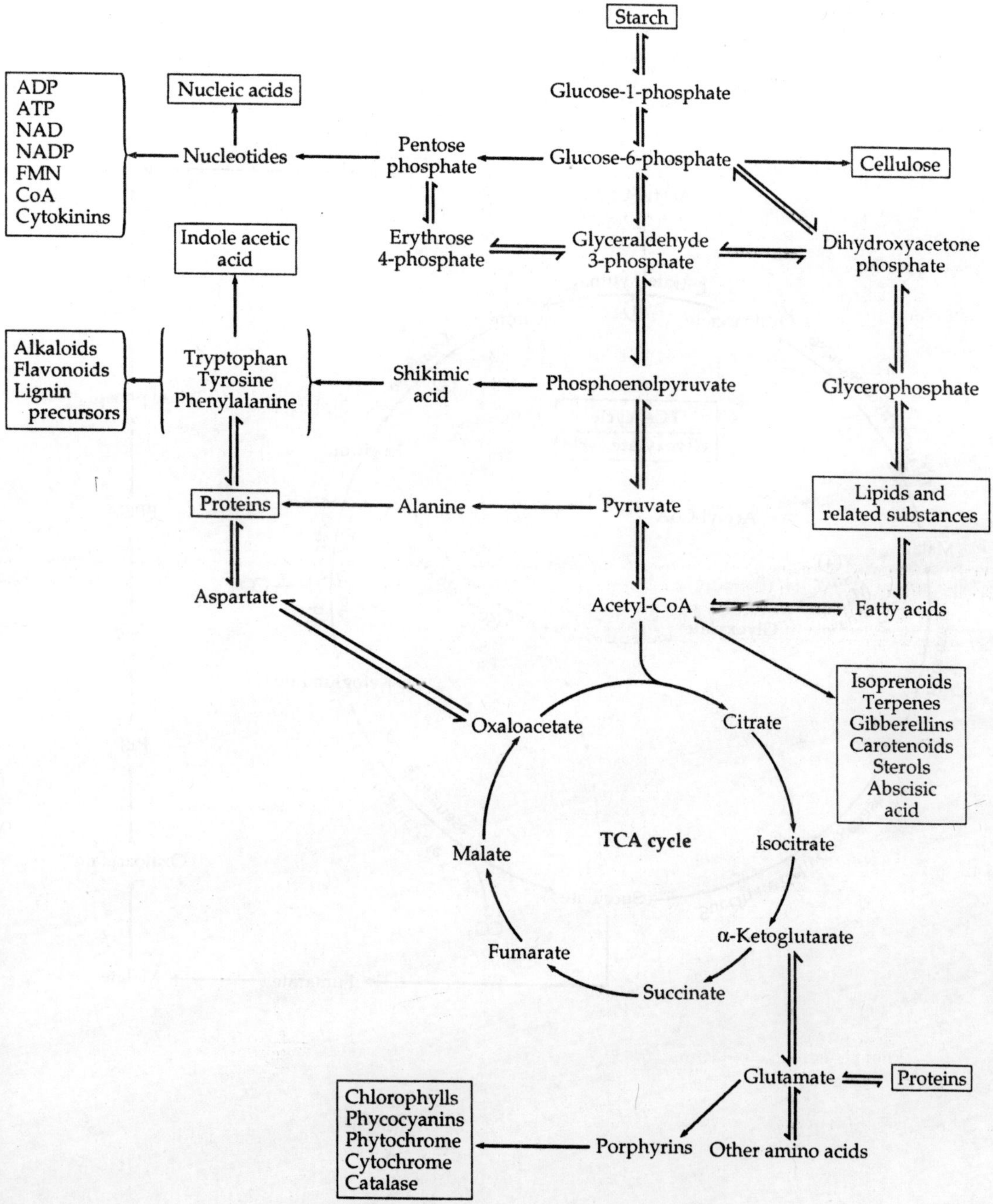

Fig. 3.46: Glycolysis and TCA cycle contribute intermediates to a variety of other metabolic pathways : Not all the carbon that enters the glycolytic pathway is oxidised to CO_2, rather some carbon flux to other pathways to synthesize different metabolites.

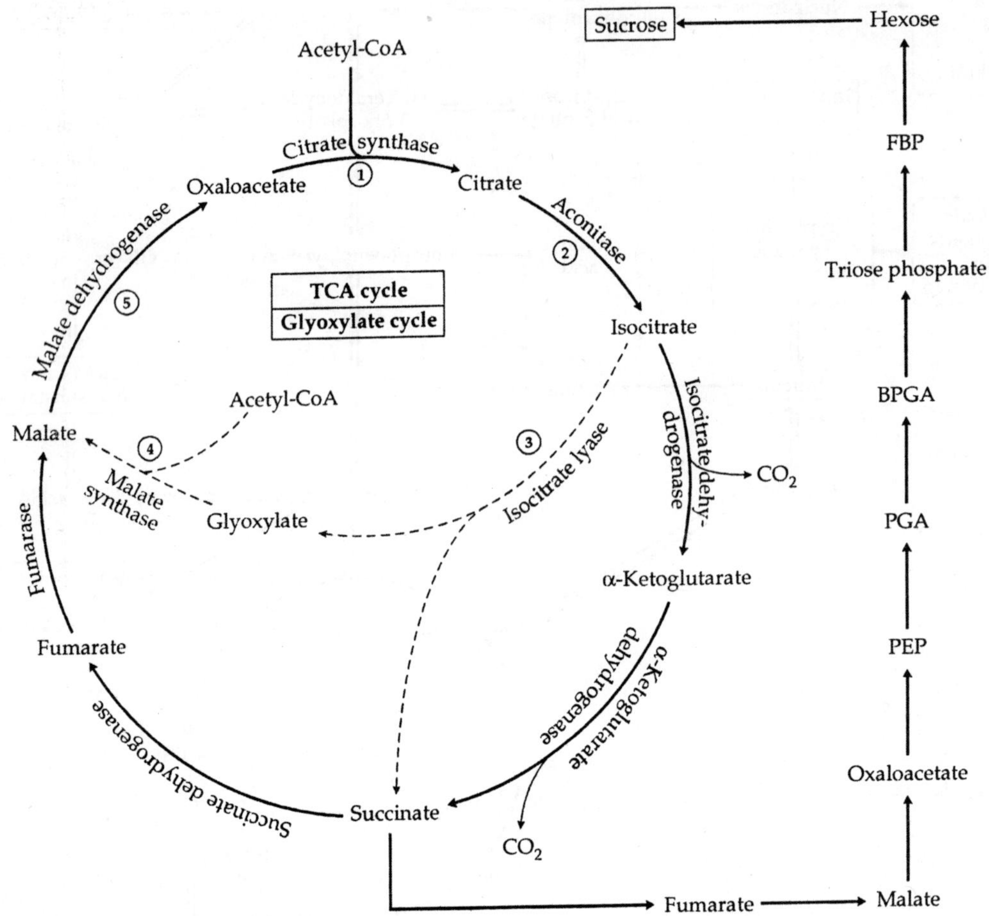

Fig. 3.47: Glyoxylate cycle : This cycle is a bypass of the TCA cycle. In glyoxysome isocitrate instead of being converted to α-ketoglutarate by isocitrate dehydrogenase, may be converted to glyoxylate and succinate with the help of isocitrate lyase. Glyoxylate with the help of another acetyl-CoA molecule will be converted to malate. Succinate, the product of glyoxylate cycle, is converted to sucrose by gluconeogenesis involving both mitochondrion and cytosol.

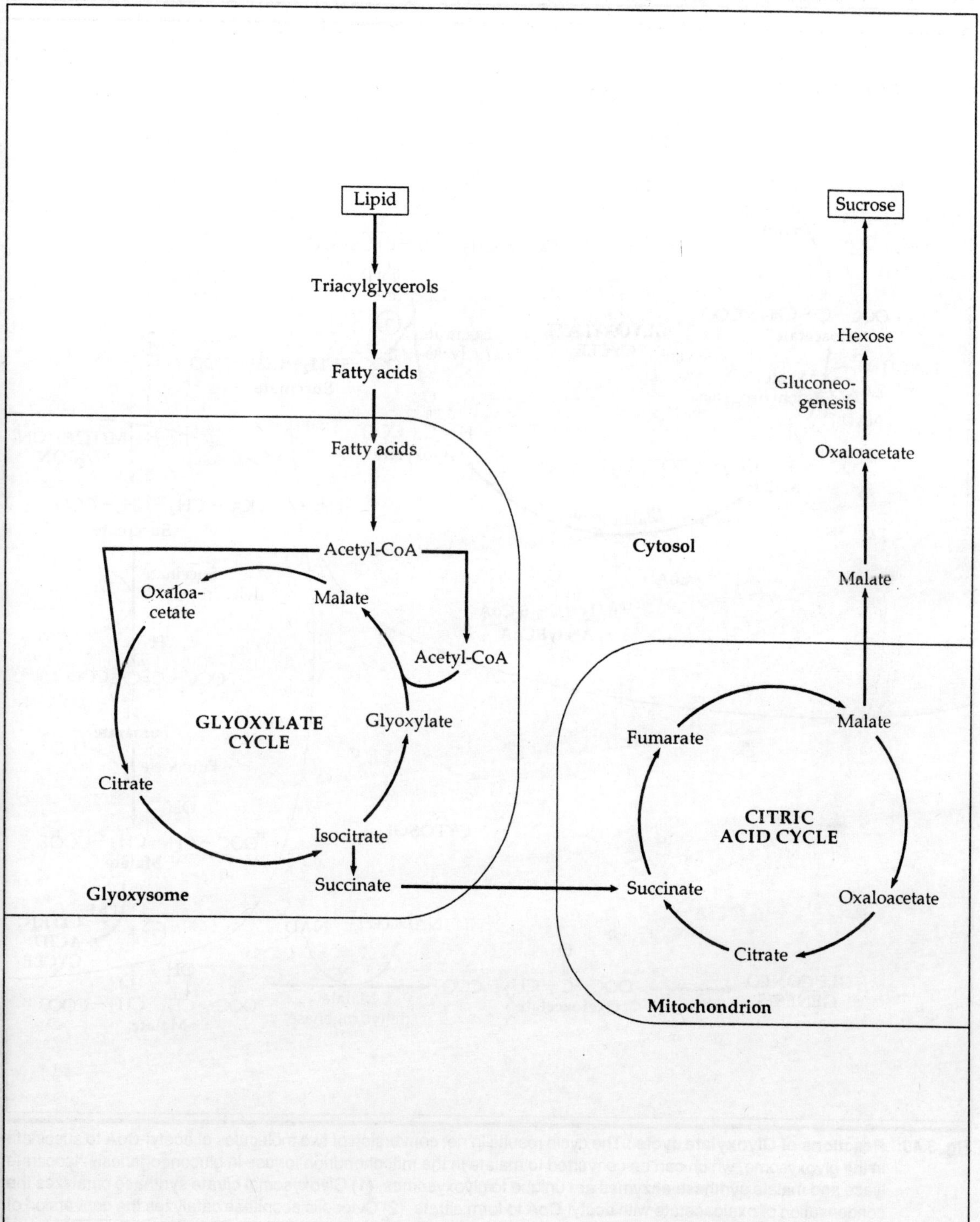

Fig. 3.48: Relationship between the glyoxylate and citric acid cycle.

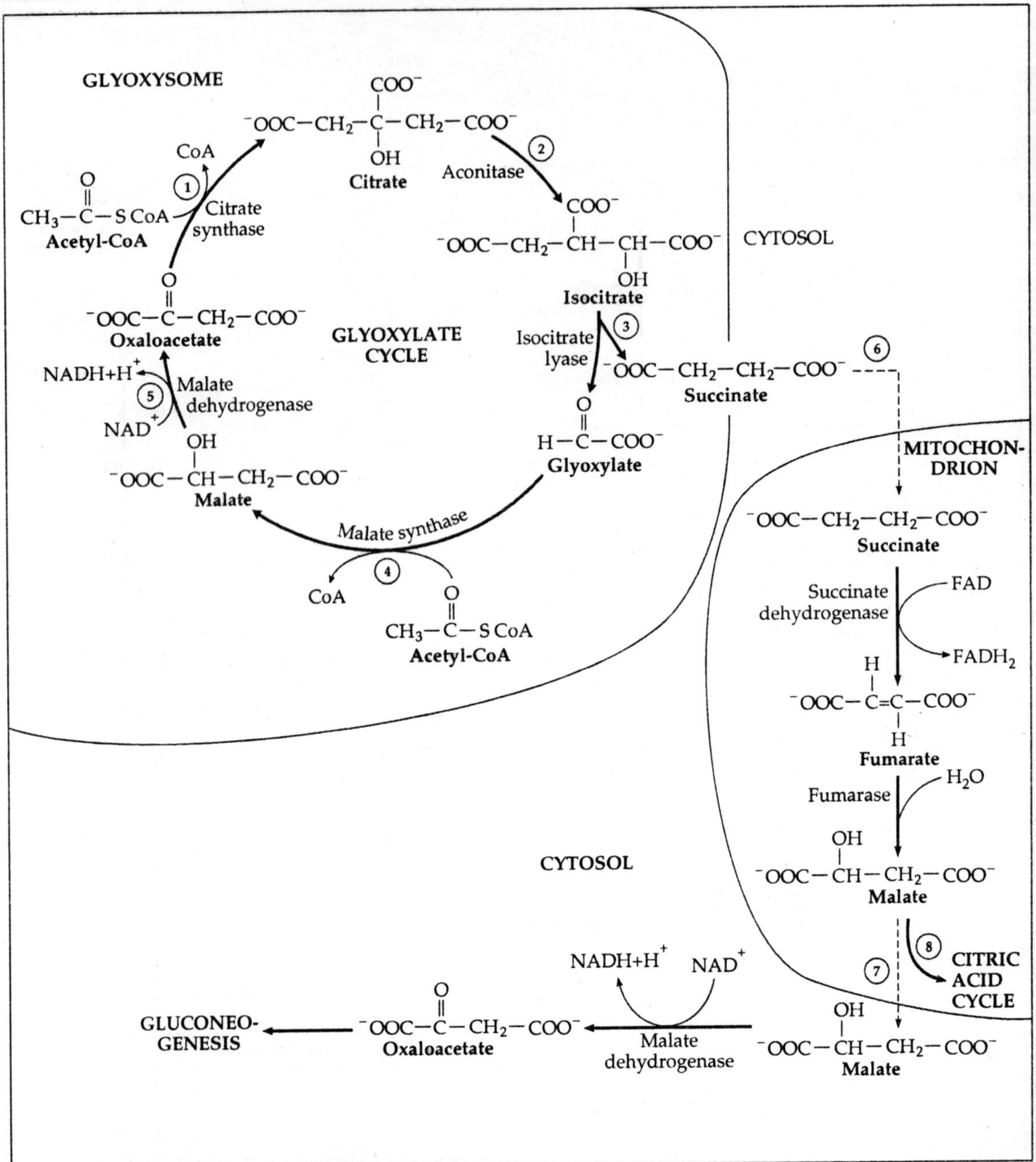

Fig. 3.49: Reactions of Glyoxylate cycle : The cycle results in net conversion of two molecules of acetyl-CoA to succinate in the glyoxysome, which can be converted to malate in the mitochondrion for use in gluconeogenesis. Isocitrate lyase and malate synthase enzymes are unique to glyoxysomes. **(1)** Glyoxysomal citrate synthase catalyzes the condensation of oxaloacetate with acetyl-CoA to form citrate. **(2)** Cytosolic aconitase catalyzes the conversion of citrate to isocitrate. **(3)** Isocitrate lyase catalyzes the cleavage of isocitrate to succinate and glyoxylate. **(4)** Malate synthase catalyzes the condensation of glyoxylate with acetyl-CoA to form malate. **(5)** Glyoxysomal malate dehydrogenase catalyzes the oxidation of malate to oxaloacetate completing the cycle. **(6)** Succinate is transported to mitochondrion, where it is converted to malate via the citric acid cycle. **(7)** Malate is transported to cytosol and oxidised to oxaloacetate and can be used in gluconeogenesis. **(8)** Malate can continue citric acid cycle, making the glyoxylate cycle anaplerotic.

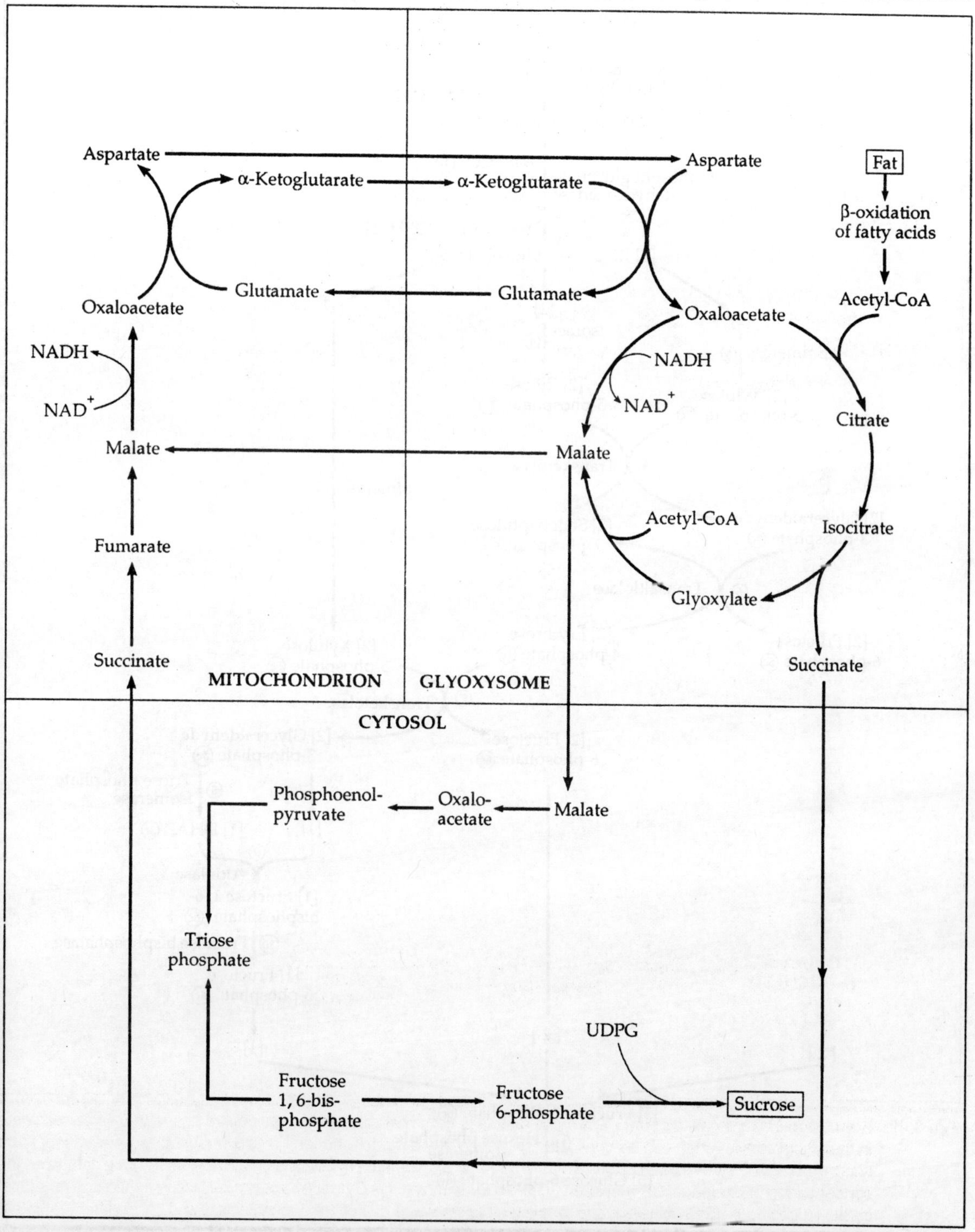

Fig. 3.50: Compartmentation of the principal steps during conversion of fat to sucrose in germinating fatty seeds : The product of glyoxylate cycle, the malate is converted to sucrose in cytosol and succinate is used to replenish oxaloacetate in citrate synthesis. A malate-aspartate shuttle is responsible for transfer of reducing equivalents between glyoxysome and mitochondrion for oxidation of NADH produced by glyoxysomal β-oxidation.

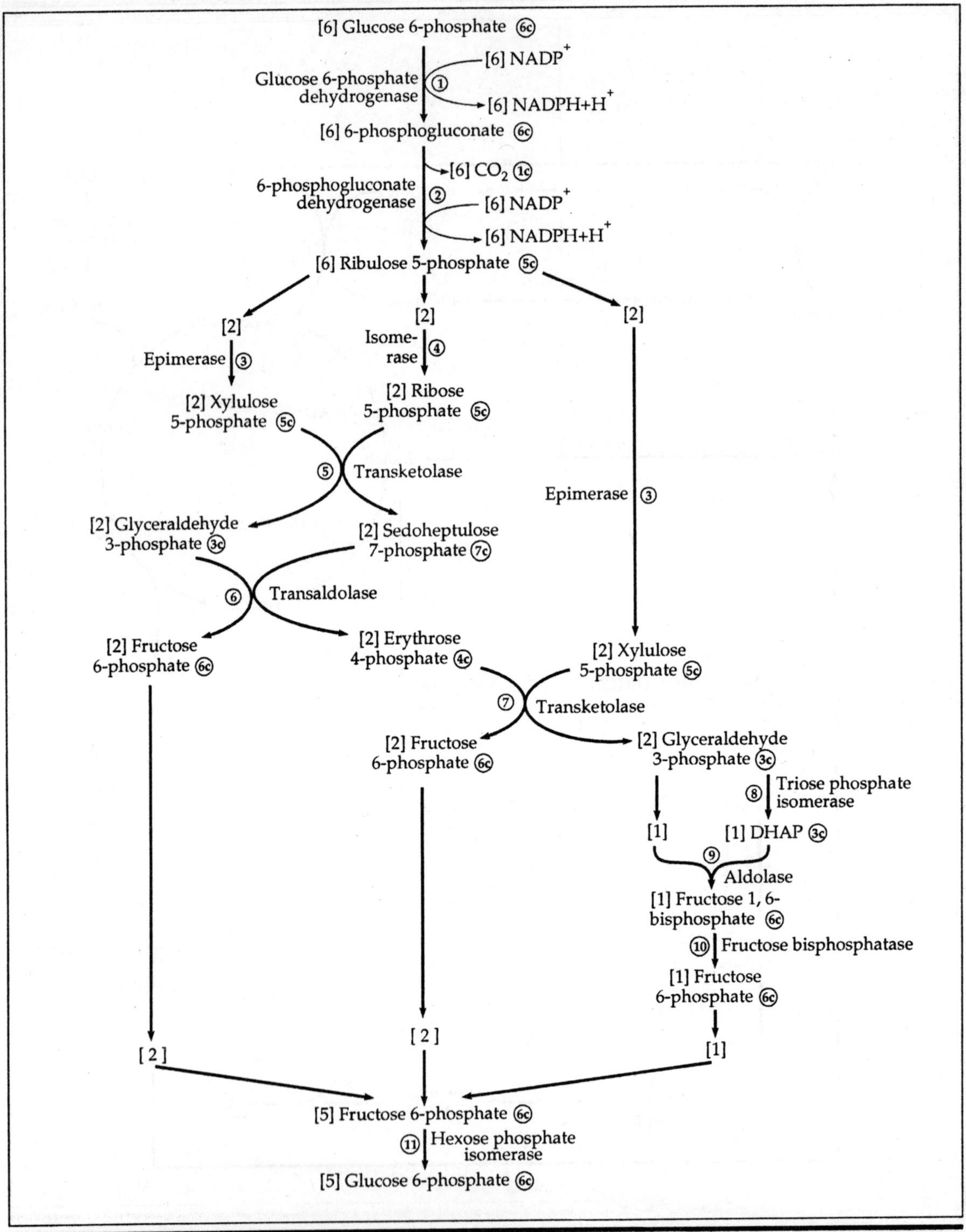

Fig. 3.51: Pentose phosphate pathway : The pathway starts with oxidative reactions of C_6 sugars to yield C_5 sugars. Six C_5 sugars are converted to four C_6 and two C_3 sugars, the later condense to one C_6 sugar and thus total five C_6 sugars regenerated. Thus one molecule C_6 sugar oxidizes to liberate six molecules CO_2 and twelve molecules NADPH+H+.

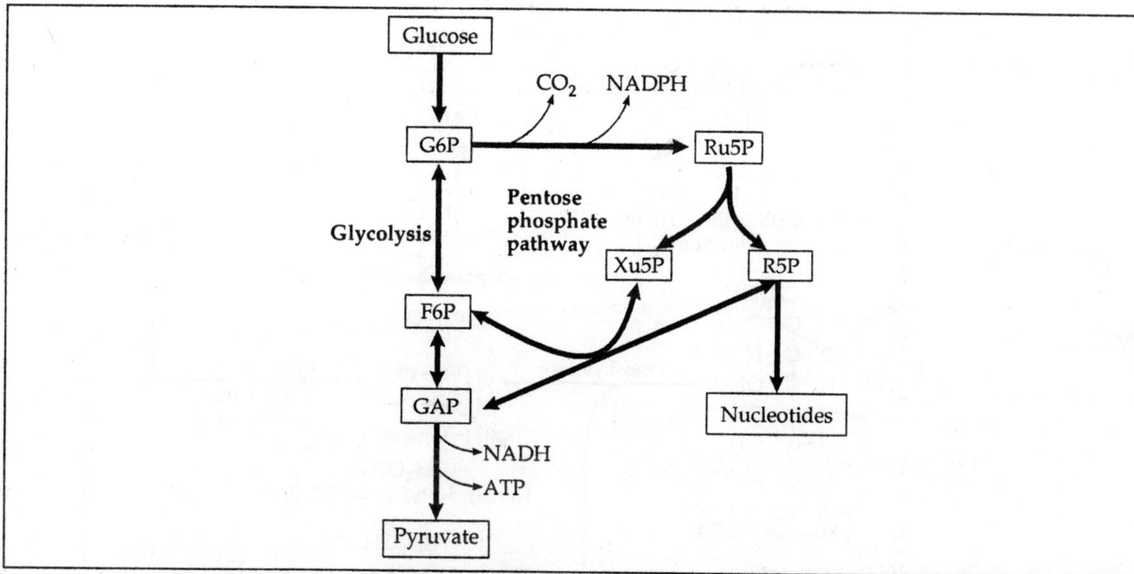

Fig. 3.52: Relationship between glycolysis and pentose phosphate pathway : The pentose phosphate pathway, which begins with G6P produced in second step of glycolysis, generates NADPH for use in reductive reactions and R5P for nucleotide synthesis. Excess R5P is converted to glycolytic intermediates by a sequence of reactions that can operate in reverse to generate additional R5P, if required.

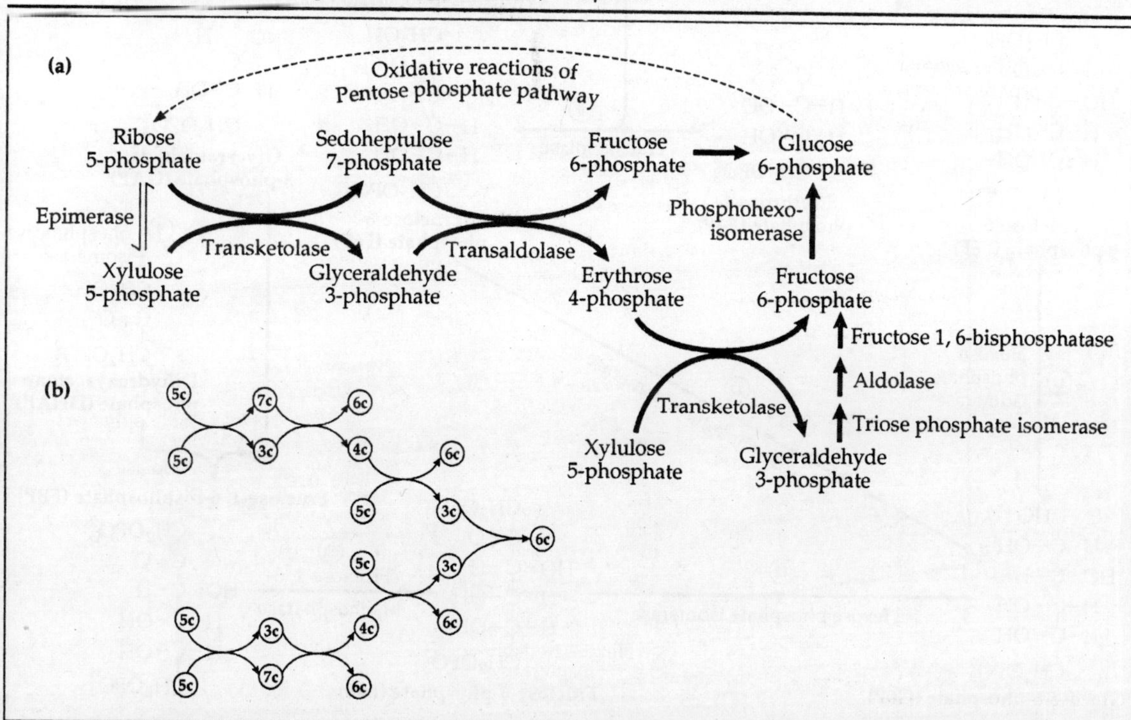

Fig. 3.53: Non-oxidative reactions of the pentose phosphate pathway : (a) These reactions convert pentose phosphates back to hexose phosphates, allowing the oxidative reactions to continue. The enzymes transketolase and transaldolase are specific to this pathway; the other enzymes also serve in the glycolytic or gluconeogenic pathways. **(b)** A schematic diagram showing the pathway from six pentoses (5C) to five hexoses (6C). Every reaction of this pathway is reversible; in dark reactions of photosynthesis, the direction of these reactions is reversed.

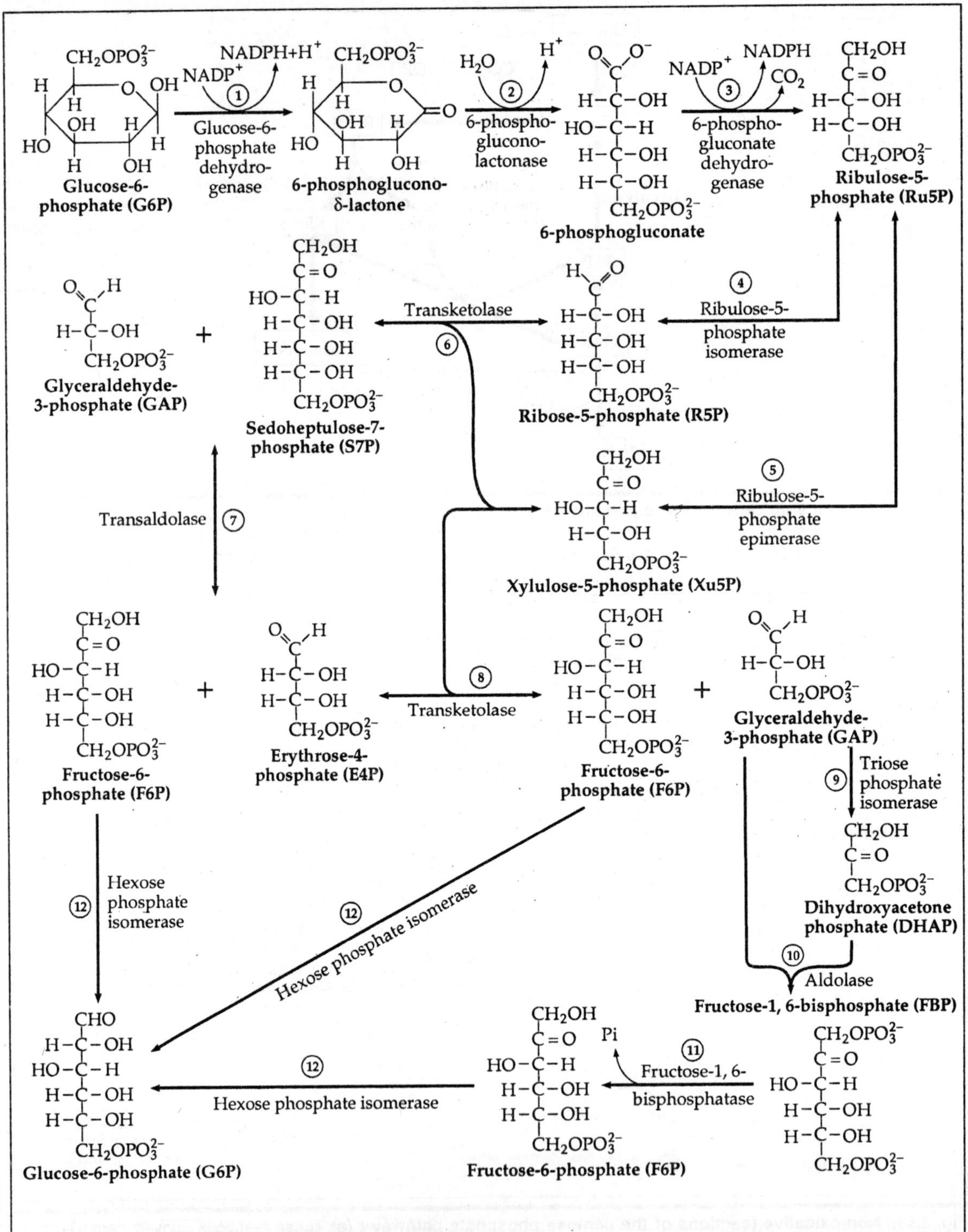

Fig. 3.54: The reactions of pentose phosphate pathway : The pathway starts with oxidative reactions [(1)–(3)] of C_6 sugar (G6P) yielding C_5 sugar (Ru5P). A series of carbon-carbon bond formations and cleavages regenerate C_6 sugar from C_5 sugar [(4)–(12)].

$$2\,[C_5 + C_5 \overset{⑥}{\rightleftharpoons} C_7 + C_3\,]$$
$$2\,[C_7 + C_3 \overset{⑦}{\rightleftharpoons} C_6 + C_4\,]$$
$$2\,[C_5 + C_4 \overset{⑧}{\rightleftharpoons} C_6 + C_3\,]$$
$$C_3 + C_3 \overset{⑩}{\rightleftharpoons} C_6$$
$$\text{(Sum) } 6\,C_5 \rightleftharpoons 5\,C_6$$

Fig. 3.55: Summary of the carbon skeleton rearrangements in the pentose phosphate pathway : The number in each reaction is keyed to the corresponding reaction in Fig. 3.54.

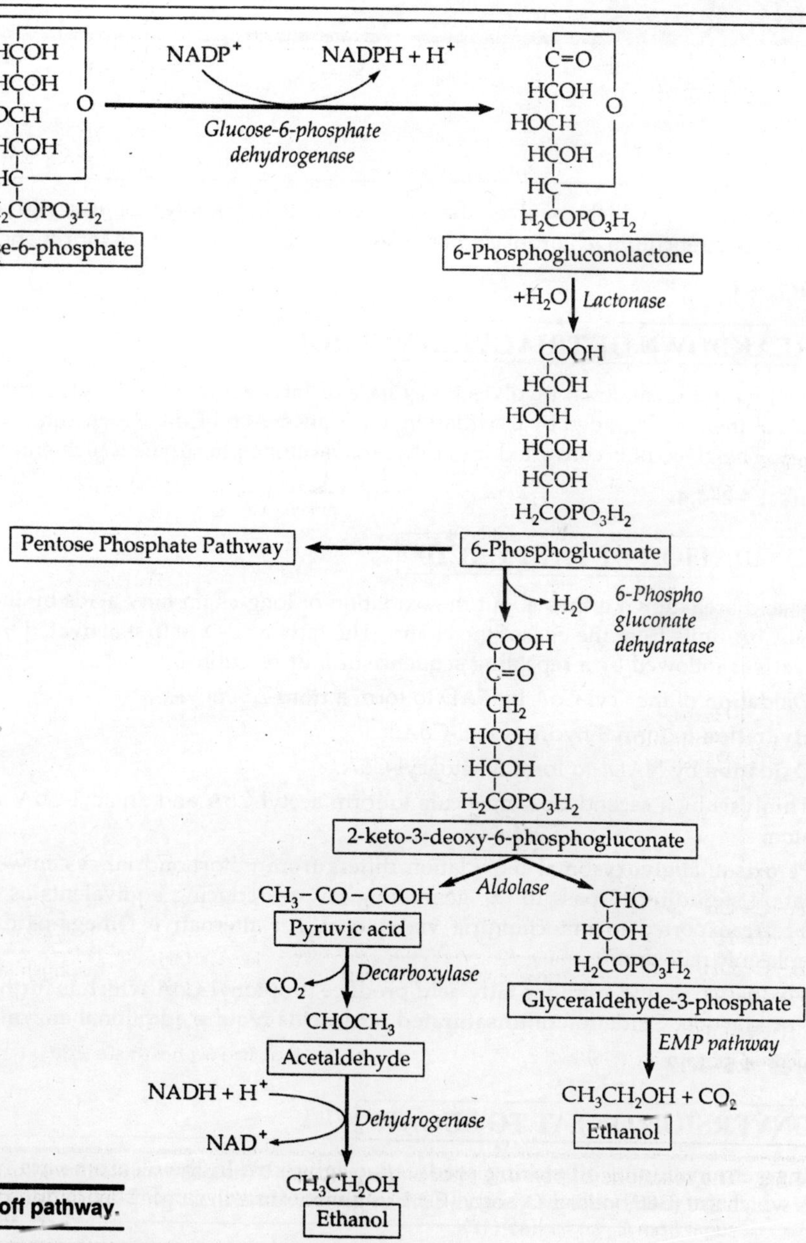

Fig. 3.56: Entner-Doudoroff pathway.

CHAPTER

4

LIPID METABOLISM

Lipid metabolism which involves the synthesis and breakdown of **triacylglycerols,** occurs mainly in endoplasmic reticulum, chloroplast, peroxisome/glyoxysome, mitochondria and cytosol.

Figure: 4.1

BREAKDOWN OF TRIACYLGLYCEROL

Triacylglycerol breakdown involves the release of **fatty acids** from the **glycerol** backbone. The free fatty acids can then be degraded by β-oxidation to produce acetyl-CoA which enters in Krebs cycle to produce energy. The glycerol is converted into dihydroxyacetone phosphate which enters in glycolysis.

Figures: 4.2–4.4

β-OXIDATION OF FATTY ACID

Fatty acid breakdown brings about the oxidation of long-chain fatty acids by the removal of two carbon acetyl-CoA units from the end of the chains. The fatty acids are first activated to their **acyl-Coenzyme A derivatives** followed by a repeating sequence of four reactions:

1. **Oxidation** of the acyl-CoA by FAD to form a trans-Δ^2-enoyl-CoA.
2. **Hydration** to form 3-hydroxyacyl-CoA.
3. **Oxidation** by NAD^+ to form 3-ketoacyl-CoA.
4. **Thiolysis** by a second CoA molecule to form acetyl-CoA and an acyl-CoA shortened by two carbon atoms.

Peroxisomal/glyoxysomal β-oxidation differs from mitochondrial system—the electron during first oxidation step directly pass to O_2, generating H_2O_2 ; reducing equivalents as well as acetyl-CoA produced are exported to mitochondria via cytosol. An alternative **Omega-oxidation** pathway exists in endoplasmic reticulum.

β-oxidation of **odd-number fatty acid** produce propionyl-CoA which is further oxidised via succinyl-CoA production. Oxidation of **unsaturated fatty acids** require additional enzymes.

Figures: 4.5–4.12

CONVERSION OF FAT TO SUGAR

During germination of **oil-storing seeds,** triacylglycerols in the oleosome are hydrolysed to yield fatty acids which are metabolised to acetyl-CoA in glyoxysome through β-oxidation pathway. Acetyl-CoA via

glyoxylate and **citric acid cycles** yields oxaloacetate which is converted to phosphoenol pyruvate. The resulting PEP is then metabolised to produce glucose via the **gluconeogenic pathway.**

Figures: 4.13–4.16

SYNTHESIS OF FATTY ACID

Fatty acid synthesis involves the condensation of two carbon units, **acetyl-CoA**, to form long hydrocarbon chains in a series of reactions in cytosol. Acetyl-CoA produced from pyruvate passes out of mitochondria as citrate and in cytosol it is cleaved to regenerate the acetyl-CoA.

The first committed step is the **carboxylation** of acetyl-CoA to form malonyl-CoA, followed by the formation of **ACP-derivatives.**

The **elongation cycle** of fatty acid synthesis has four stages:

1. **Condensation** of acetyl-ACP and malonyl-ACP to form acetoacetyl-ACP.
2. **Reduction** of acetoacetyl-ACP to form hydroxybutyryl-ACP.
3. **Dehydration** of hydroxybutyryl-ACP to crotonyl-ACP.
4. **Reduction** of crotonyl-ACP to yield 4-carbon butyryl-ACP.

Further rounds of elongation add more two carbon units from malonyl-ACP on to the growing hydrocarbon chain until the 16-carbon palmitate is formed. Further **elongation** and/or **desaturation** takes place on the cytosolic surface of the SER resulting in the formation of longer chain saturated/unsaturated fatty acids.

Figures: 4.17–4.31

TRIACYLGLYCEROL SYNTHESIS

Triacylglycerols are synthesized from **glycerol 3-phosphate,** derived from glycolytic intermediate DHAP and **fatty acyl-CoAs. Phosphatidic acid,** an intermediate, is produced by transfer of fatty acyl-CoA molecules to ester linkage with glycerol 3-phosphate. The phosphate group of phosphatidic acid is removed to form **diacylglycerol** and further acylated to **triacylglycerol.**

Figures: 4.32–4.35

SYNTHESIS OF PHOSPHOLIPIDS

Glycerophospholipids are synthesized both from diacylglycerol and phosphatidic acid. **CDP** supplies the phosphate group of the phosphodiester bond of phospholipid. Attachment of different types of **head group** to diacylglycerol results in the synthesis of different types of glycerophospholipids.

Sphingolipids are based on sphingosine to which a single fatty acid chain and a phosphorylated head group are attached.

Figures: 4.36–4.42

SYNTHESIS OF STEROLS

The first stage in the synthesis of sterol is the formation of **mevalonate** from acetyl-CoA. This is followed by conversion into **isoprene units;** six of these units combine to form **squalene.** Ring closure converts linear squalene into the cyclized **steroid** nucleus—stigmasterol, ergosterol, lanosterol and cholesterol.

Figures: 4.43–4.48

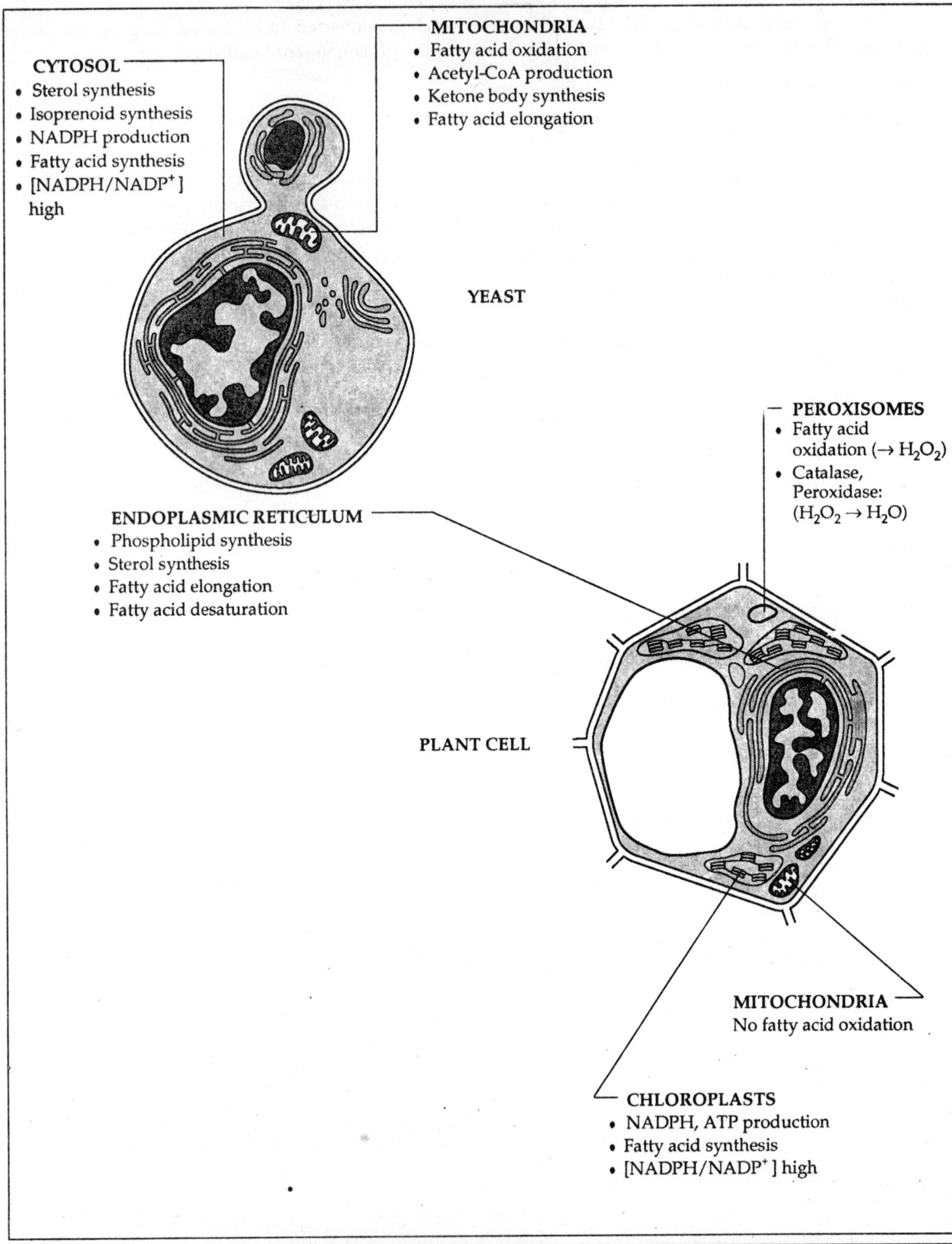

Fig. 4.1: Subcellular localization of lipid metabolism: Yeast differs from higher plant cells in the compartmentation of lipid metabolism. Fatty acid synthesis takes place in the compartment in which NADPH is available for reductive synthesis (i.e., where the NADPH/NADP$^+$ ratio is high).

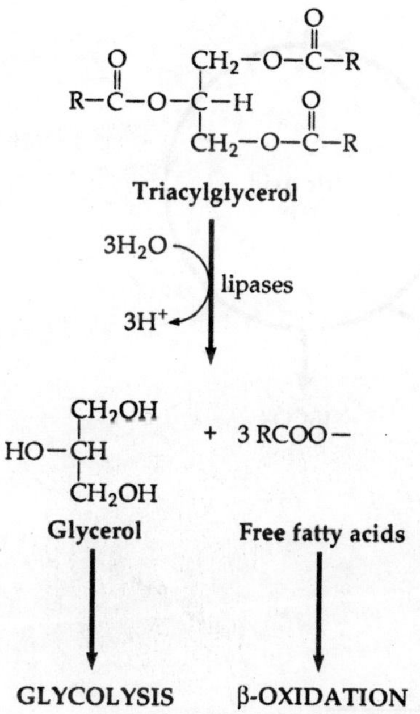

Fig. 4.2: Breakdown of triacylglycerols.

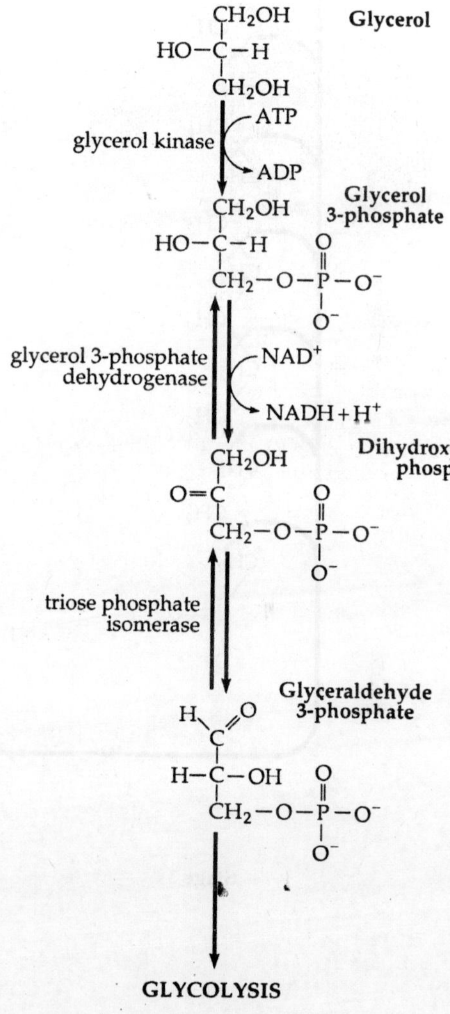

Fig. 4.3: Entry of glycerol into the glycolytic pathway.

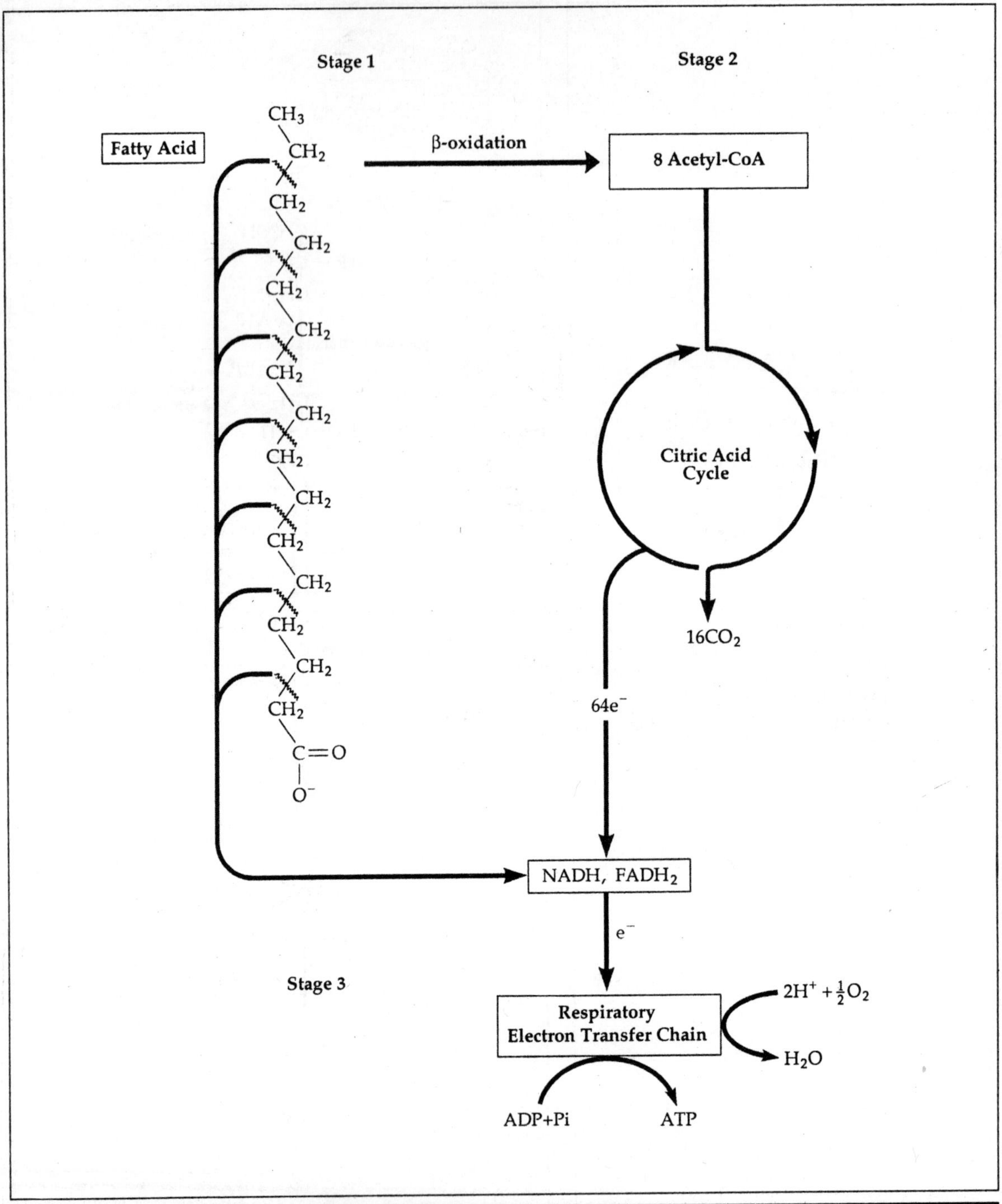

Fig. 4.4: Stages of fatty acid oxidation.

Stage 1 : A long-chain fatty acid is oxidised to yield acetyl residues in the form of Acetyl-CoA. This process is called β-oxidation.

Stage 2 : The acetyl groups are oxidised to CO_2 via the Citric acid cycle.

Stage 3 : Electrons derived from the oxidation of stage 1 and 2 pass to O_2 via the mitochondrial respiratory chain, providing the energy for ATP synthesis by oxidative phosphorylation.

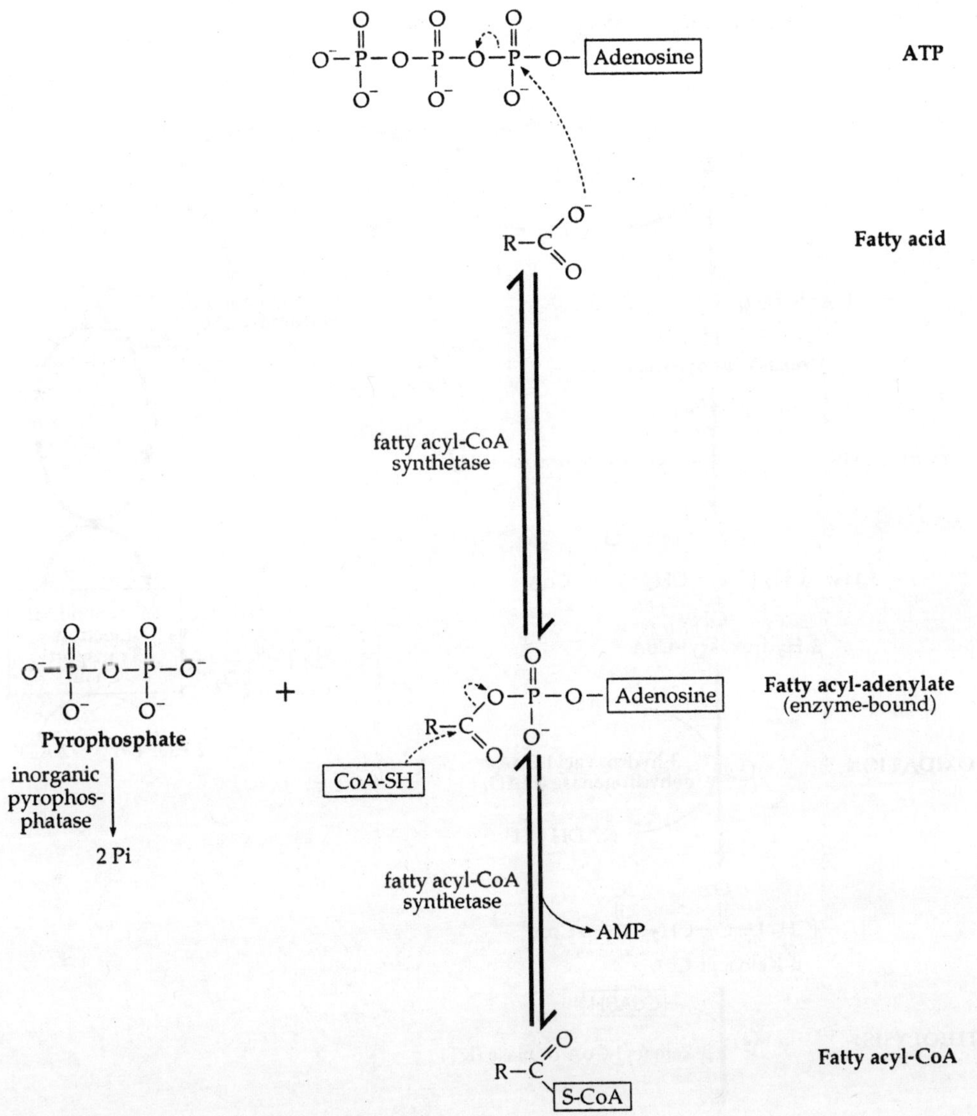

Fig. 4.5: Conversion of a fatty acid to fatty acyl-CoA: The conversion is catalysed by fatty acyl-CoA synthetase and inorganic pyrophosphatase. Fatty acid activation by formation of the fatty acyl-CoA derivative occurs in two steps. First, the carboxylate ion displaces the outer two (β and γ) phosphates of ATP to form a fatty acyl-adenylate, the mixed anhydride of a carboxylic acid and a phosphoric acid. The other product is PPi, an excellent leaving group that is immediately hydrolysed to two Pi, pulling the reaction in the forward direction. The thiol group of coenzyme A carries out nucleophilic attack on the enzyme-bound mixed anhydride, displacing AMP and forming the thioester fatty acyl-CoA.

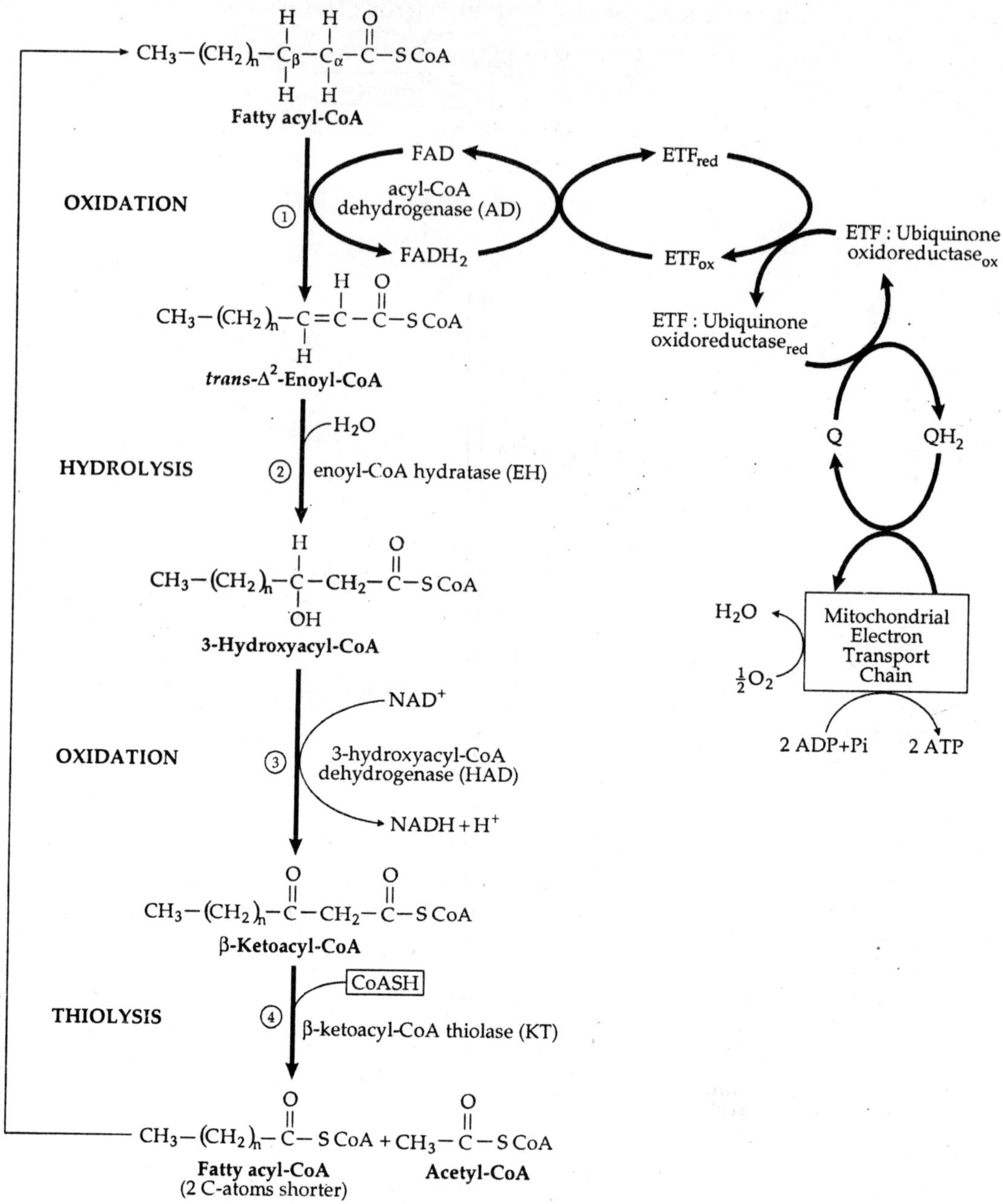

Fig. 4.6: The β-oxidation pathway of fatty acyl-CoA.

(1) Formation of a trans-α, β double bond through dehydrogenation by the flavoenzyme acyl-CoA dehydrogenase (AD).

(2) Hydration of the double bond by enoyl-CoA hydratase (EH) to form a 3-hydroxyacyl-CoA.

(3) NAD^+ dependent dehydrogenation of this β-hydroxyacyl-CoA by 3-hydroxyacyl-CoA dehydrogenase (HAD) to form the corresponding β-ketoacyl-CoA.

(4) C_α-C_β cleavage in a thiolysis reaction with CoA as catalysed by β-ketoacyl-CoA thiolase (KT ; also called just thiolase) to form acetyl-CoA and a new acyl-CoA containing two fewer C-atoms than the original one.

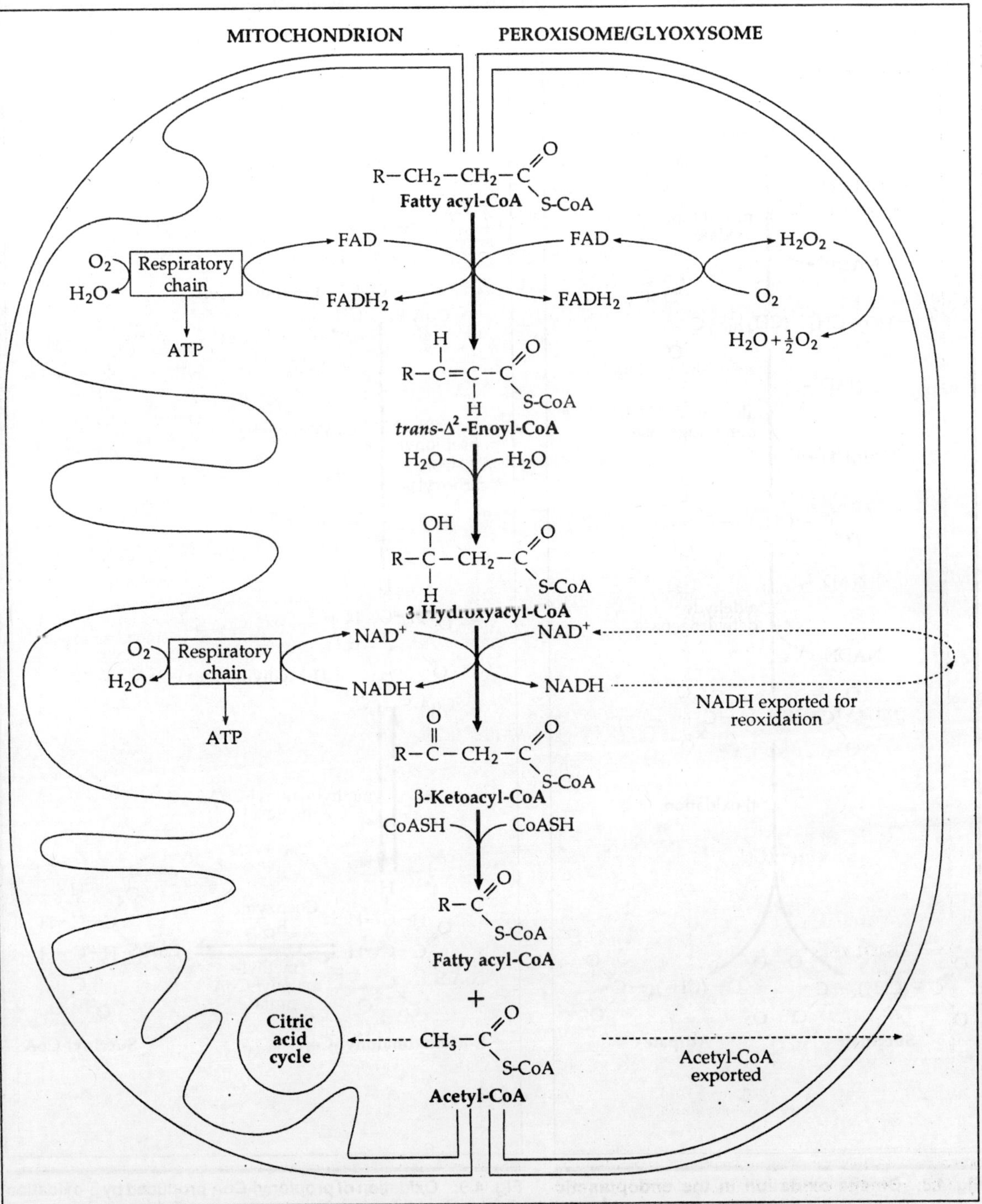

Fig. 4.7: Comparison of β-oxidation in mitochondria and in peroxisomes and glyoxysomes : The peroxisomal/glyoxysomal system differs from the mitochondrial system in two respects. (1) In the first oxidative step electrons pass directly to O_2, generating H_2O_2 and (2) the NADH formed in the second oxidative step cannot be reoxidised, so reducing equivalents are exported from the peroxisome and glyoxysome to cytosol, eventually entering mitochondria. Acetyl-CoA produced by peroxisomes/glyoxysomes is also exported, whereas acetyl-CoA from mitochondria is further oxidised in the citric acid cycle.

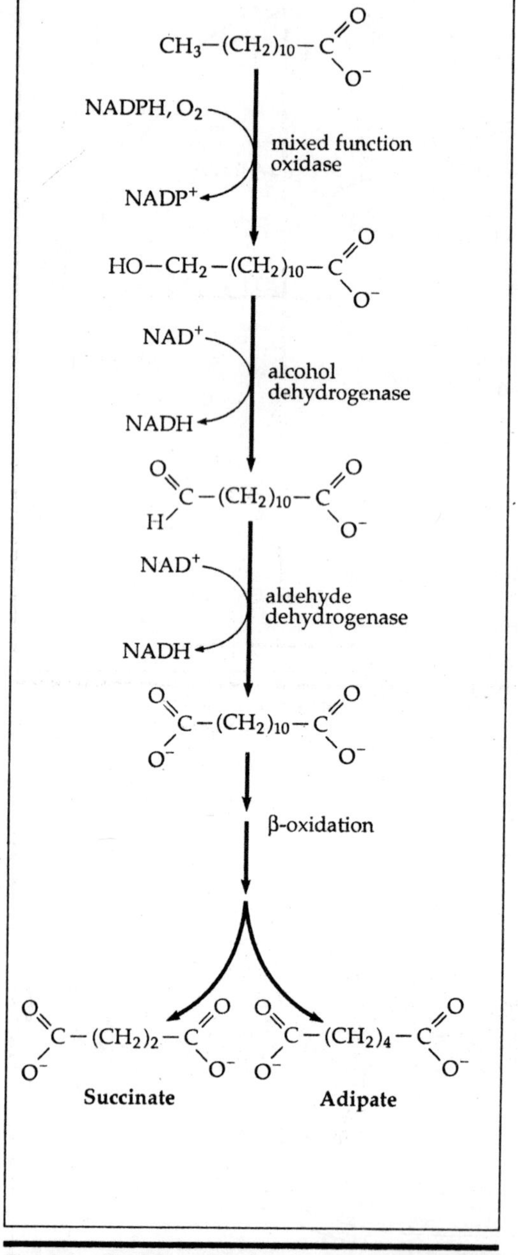

Fig. 4.8: Omega oxidation in the endoplasmic reticulum : This alternative to β-oxidation begins with oxidation of the carbon most distant from the α-carbon–the omega (ω) carbon. The substrate is usually a medium-chain fatty acid. This pathway is generally not the major route for oxidative catabolism of fatty acids.

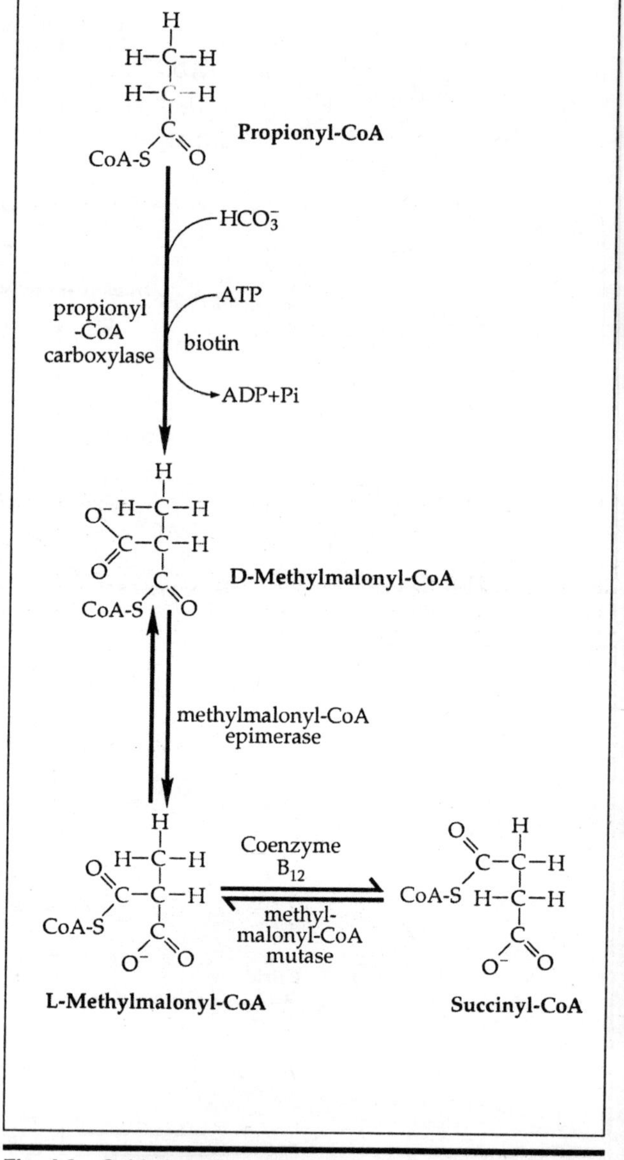

Fig. 4.9: Oxidation of propionyl-CoA produced by β-oxidation of odd-number fatty acids : The sequence involves the carboxylation of propionyl-CoA to methylmalonyl-CoA and then conversion to succinyl-CoA. The conversion requires epimerisation of methylmalonyl-CoA, followed by a remarkable reaction in which substituents on adjacent carbon atoms exchange positions.

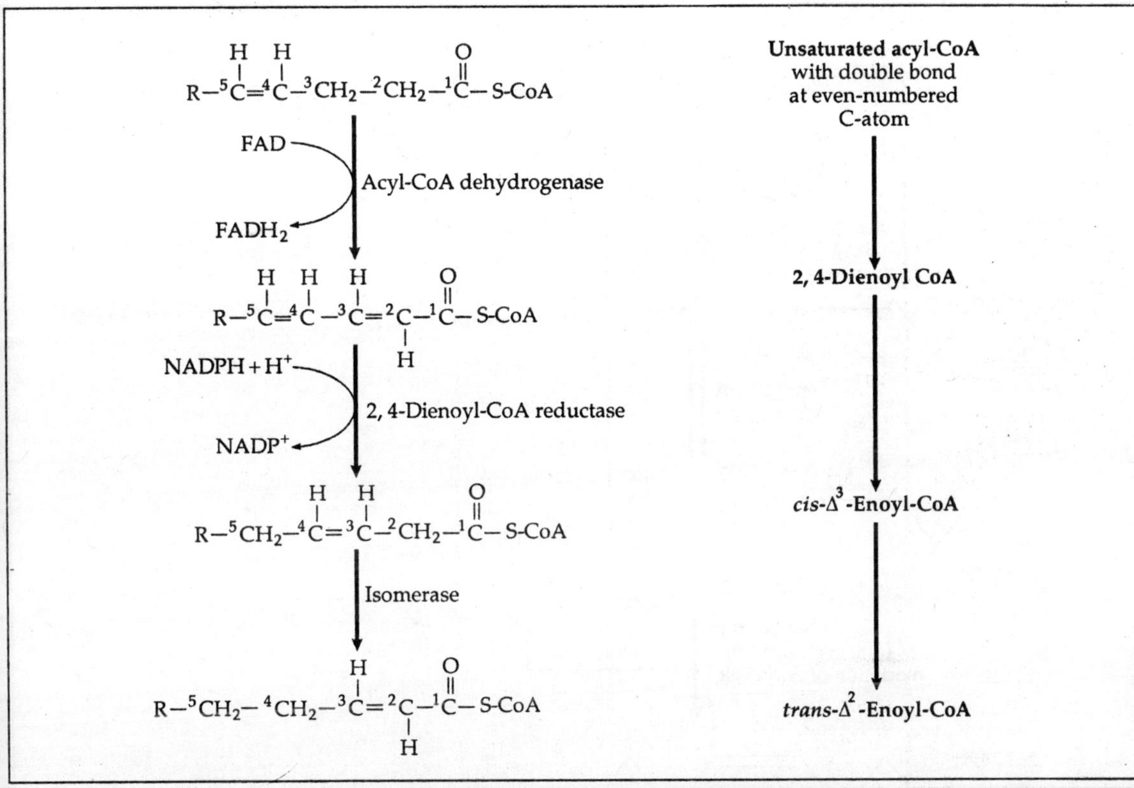

Fig. 4.10: Accessory enzymes required for the metabolism of unsaturated fatty acids.

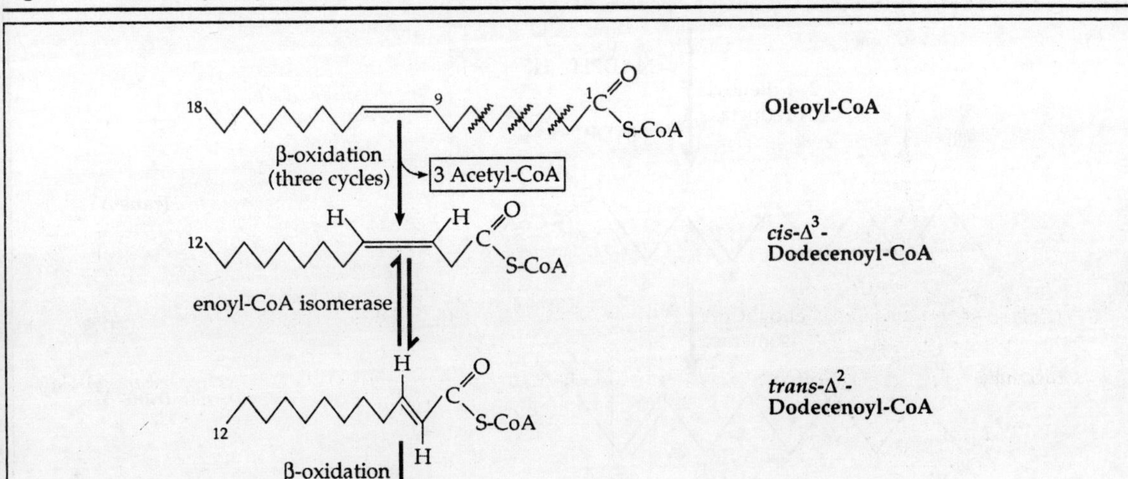

Fig. 4.11: Oxidation of a monounsaturated fatty acid : Oleic acid, as oleoyl-CoA (Δ^9), is the example used here. Oxidation requires an additional enzyme, enoyl-CoA isomerase, to reposition the double bond, converting the 'cis', isomer to a 'trans' isomer, a normal intermediate in β-oxidation.

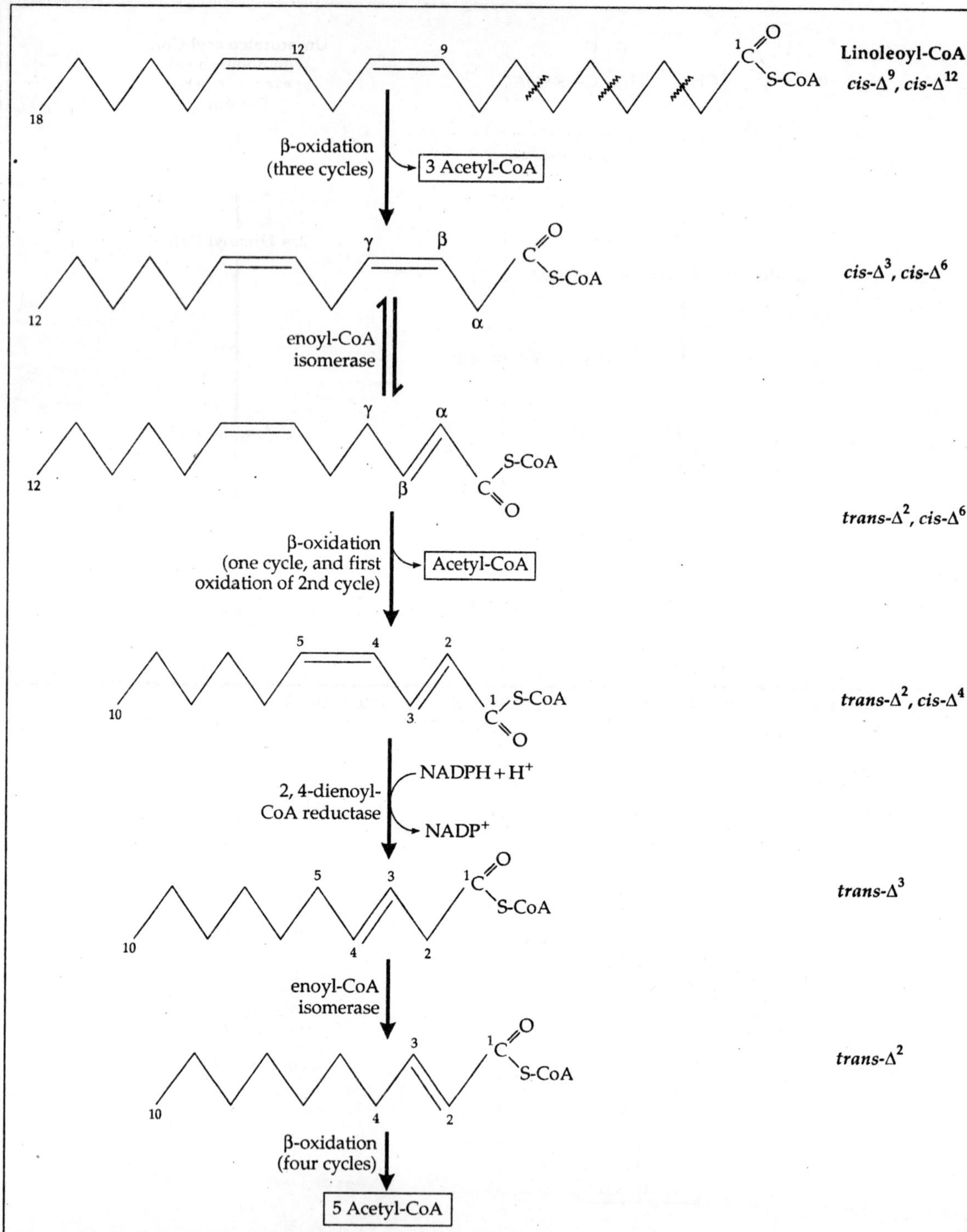

Fig. 4.12: Oxidation of polyunsaturated fatty acid : The example here is Linoleic acid, as Linoleyol-CoA ($\Delta^{9,12}$). Oxidation requires a second auxiliary enzyme in addition to Enoyl-CoA isomerase: NADPH-dependent 2, 4-Dienoyl-CoA reductase. The combined action of these two enzymes converts a *trans*-Δ^2, *cis*-Δ^4-dienoyl-CoA intermediate into the *trans*-Δ^2-enoyl-CoA substrate necessary for β-oxidation.

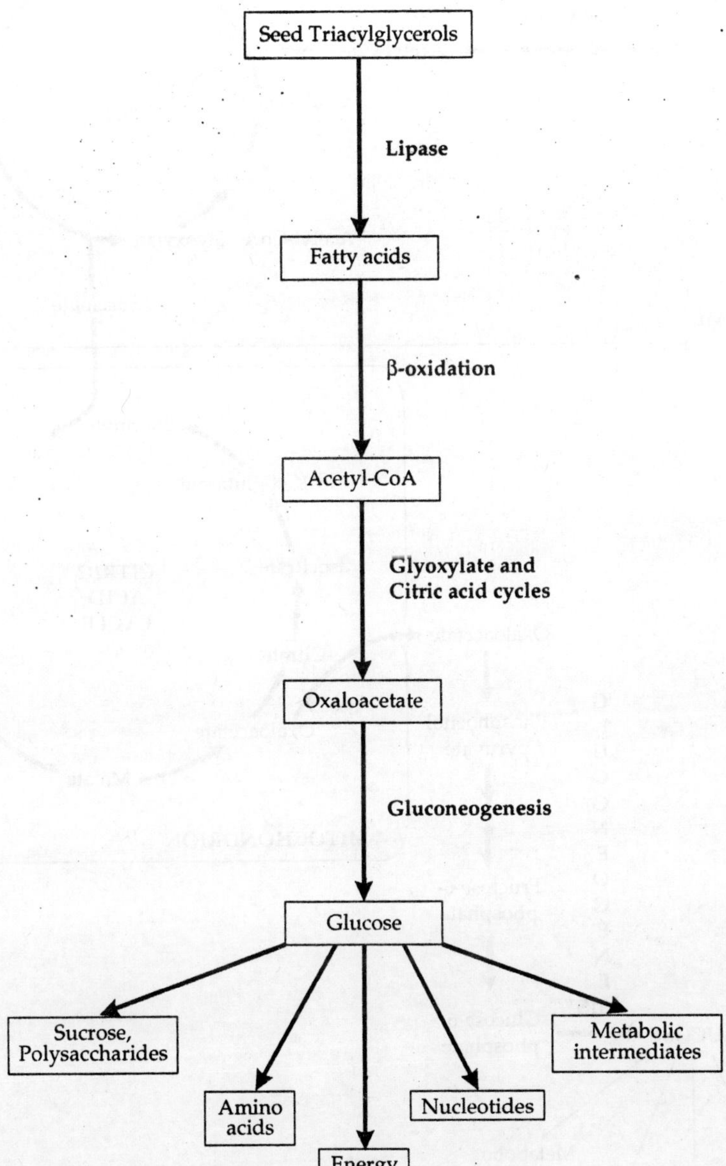

Fig. 4.13: Role of β-oxidation in the conversion of stored triacylglycerols to glucose in germinating seeds.

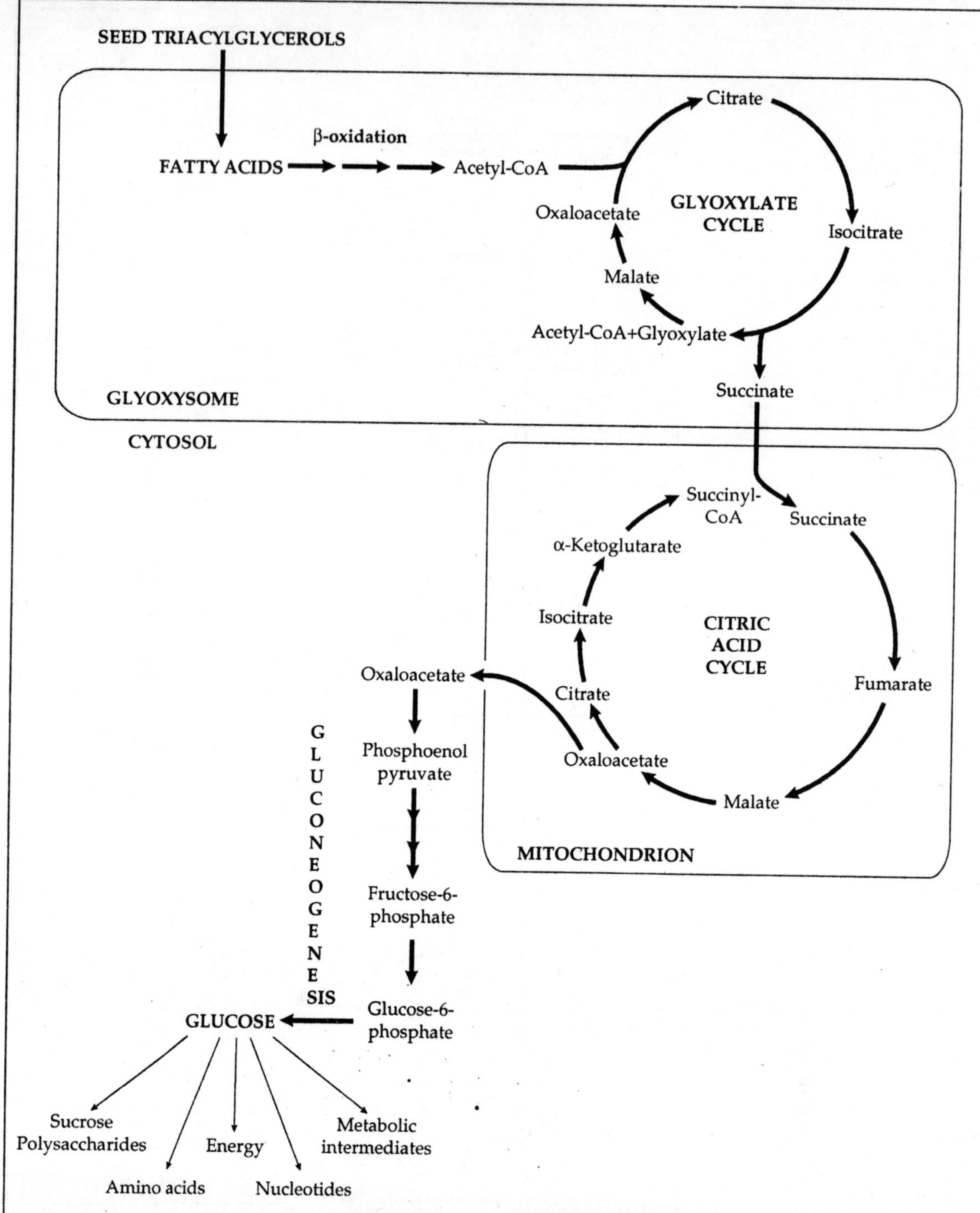

Fig. 4.14: Conversion of stored Triacylglycerol to Glucose in germinating fatty seeds : This pathway begins in glyoxysomes. Succinate is produced and exported to mitochondria, where it is converted to Oxaloacetate by enzymes of the citric acid cycle. Oxaloacetate enters the cytosol and serves as the starting material for gluconeogenesis. The Glucose is then utilised for the synthesis of sucrose, the transport form of carbon in plants, different amino acids, nucleotides, as a source of energy and other metabolic intermediates.

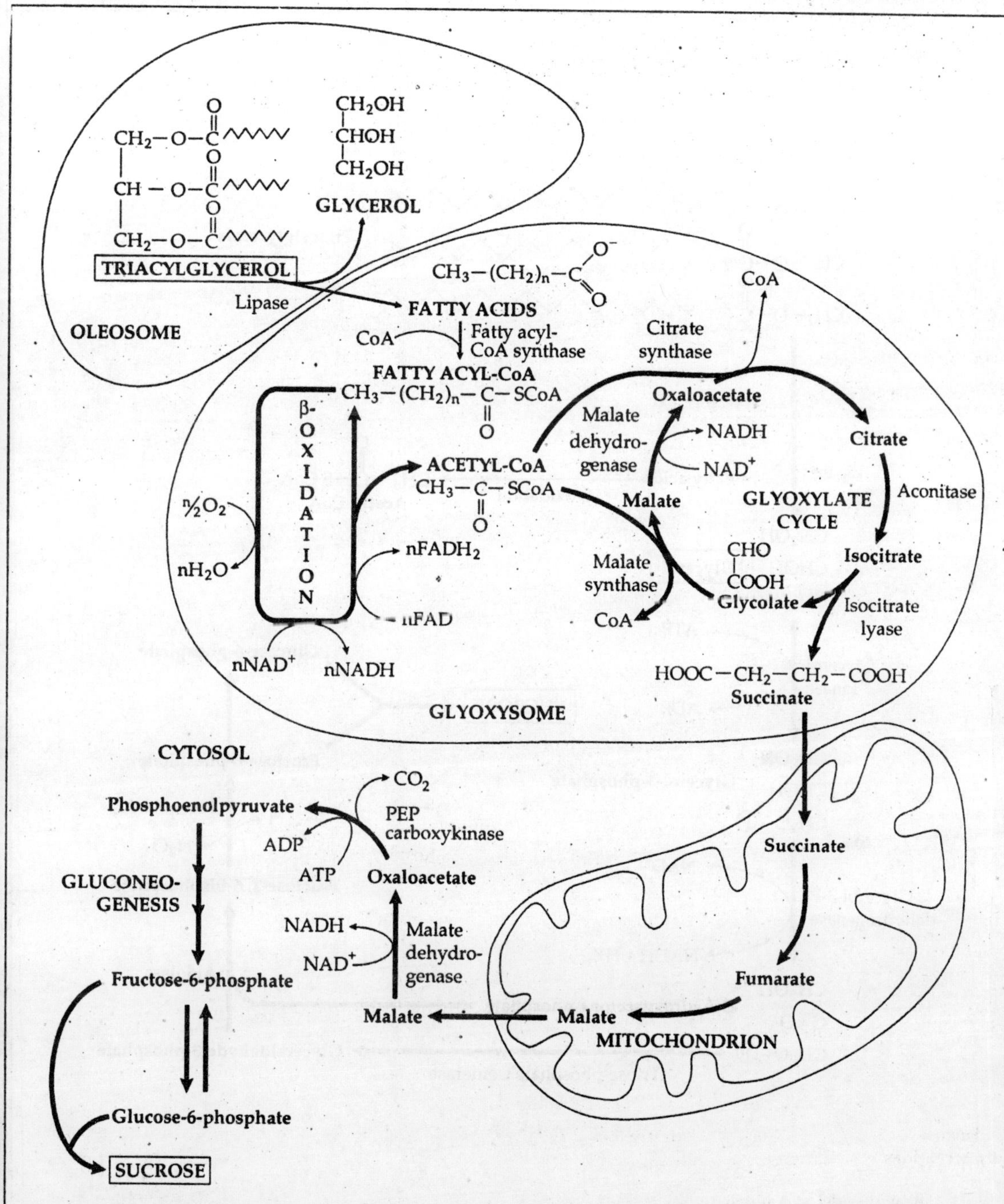

Fig. 4.15: The conversion of fats to sugars during germination in oil-storing seeds : Carbon flow during fatty acid breakdown and gluconeogenesis. Triacylglycerols in the oleosome are hydrolysed to yield fatty acids, which are metabolised to Acetyl-CoA in the glyoxysome through β-oxidation pathway. In this pathway produced Acetyl-CoA then enters to glyoxylate cycle and generates Succinate, which is then transported to mitochondrion and converted ultimately to Malate. The resulting Malate is then transported into the cytosol and oxidised to Oxaloacetate, and converted into Phosphoenolpyruvate which through the process of gluconeogenesis is ultimately converted into F-6-P and G-6-P which make together sucrose.

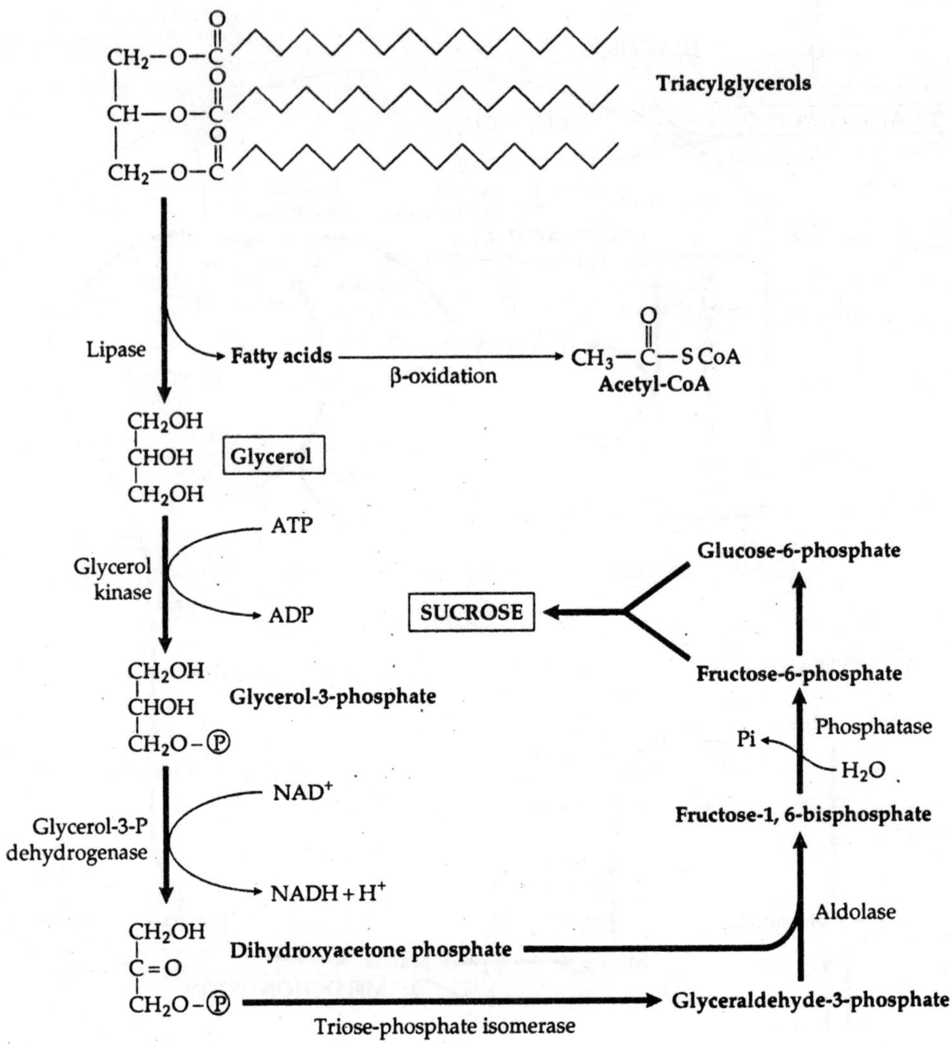

Fig. 4.16: Conversion of the glycerol moiety of Triacylglyceerols to Sucrose in germinating seeds : The Glycerol of Triacylglycerols is oxidised to Dihydroxyacetone phosphate, which enters the gluconeogenic pathway at the triose phosphate isomerase reaction. Through this pathway ultimately the transportable form of Sucrose is formed.

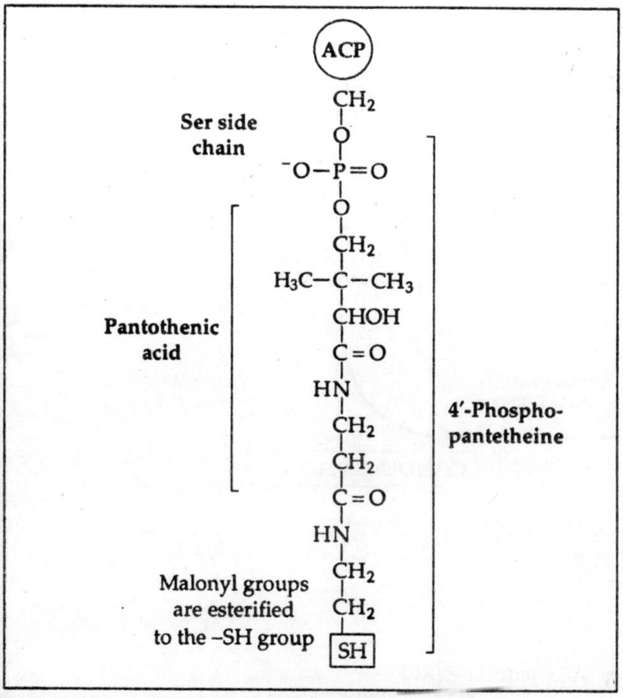

Fig. 4.17: Acyl Carrier Protein (ACP) : The prosthetic group is 4-carbon phosphopantetheine, which is covalently attached to the hydroxyl group of a Ser residue in ACP. Phosphopantetheine contains the B vitamin pantothenate, also found in the coenzyme A molecule. Its –SH group is the site of entry of malonyl groups during fatty acid synthesis.

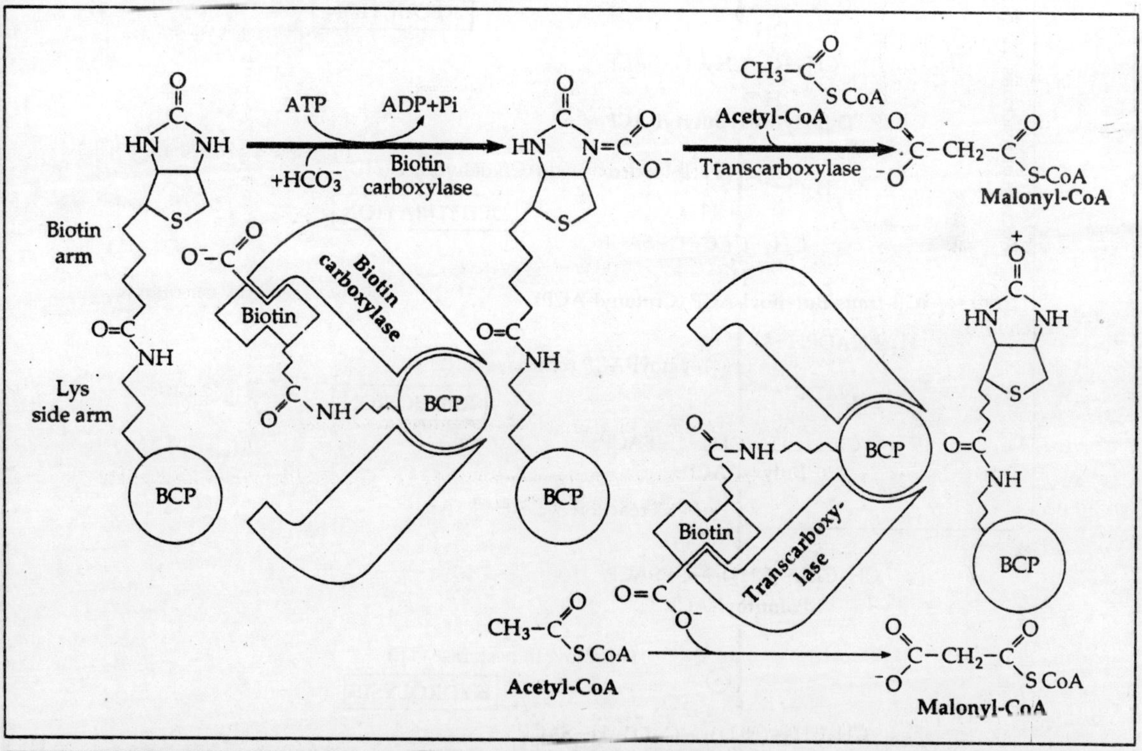

Fig. 4.18: The Acetyl-CoA Carboxylase reaction : Acetyl-CoA carboxylase has three functional regions : biotin carrier protein (BCP); biotin carboxylase, which activates CO_2 by attaching it to a nitrogen in the biotin ring in an ATP-dependent reaction; and transcarboxylase transfers activated CO_2 from biotin to Acetyl-CoA, producing Malonyl-CoA.

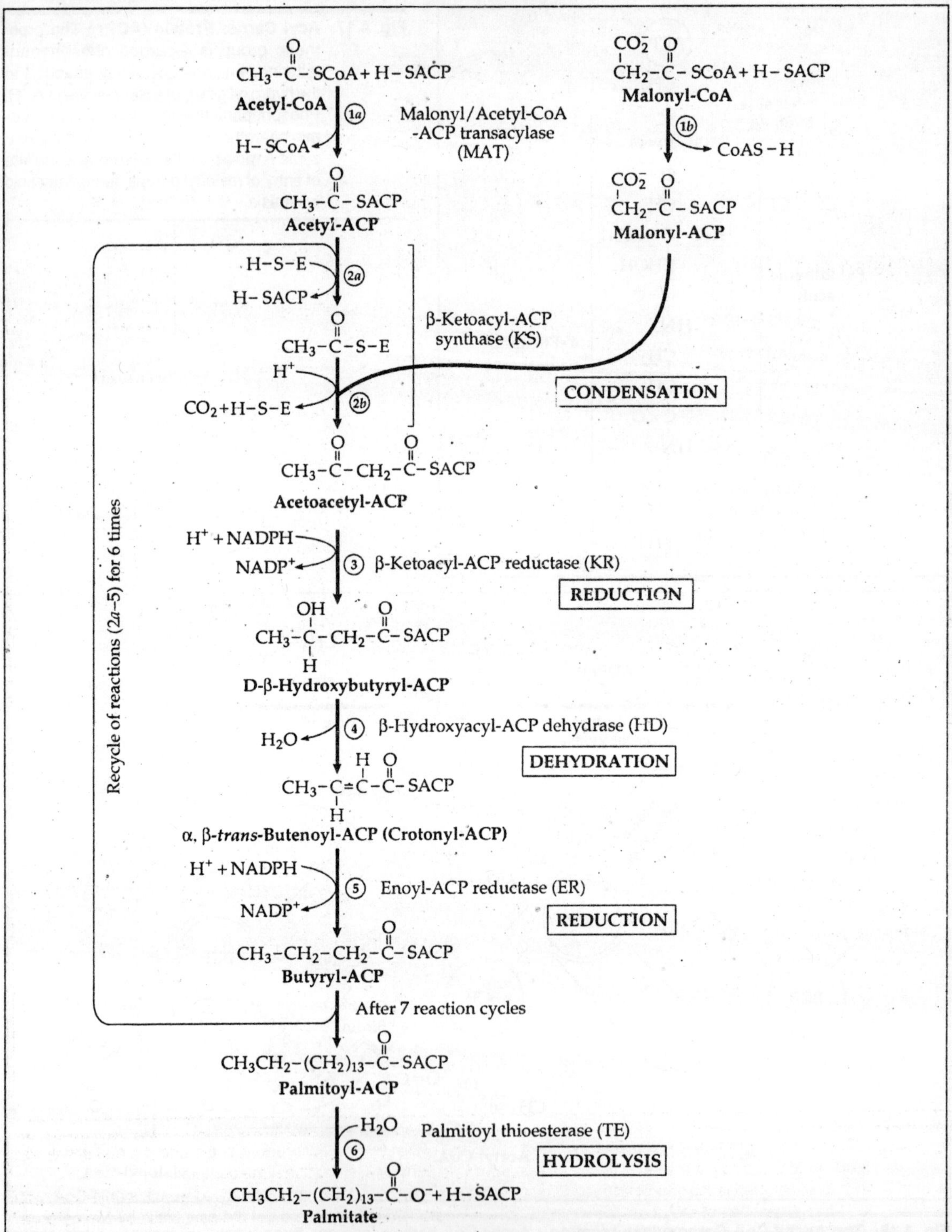

Fig. 4.19: Reaction sequence for the biosynthesis of fatty acids : In forming Palmitate, the pathway is repeated for seven cycles of C_2 elongation followed by a final hydrolysis step.

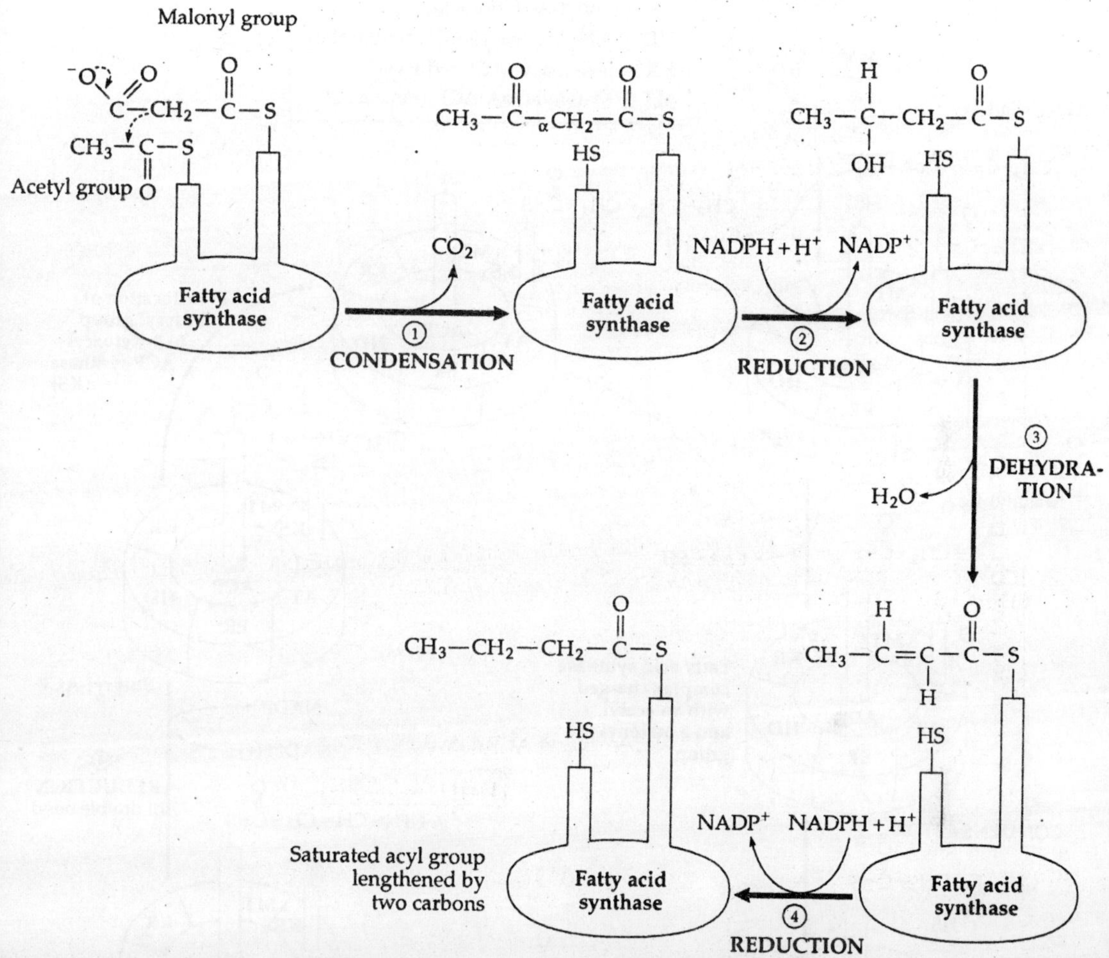

Fig. 4.20: A four-step sequence lengthens a growing fatty acyl chain by two carbons : Each malonyl group and acetyl group is activated by a thioester that links it to fatty acid synthase, a multienzyme complex.

(1) The first step is the condensation of an activated acyl group and two carbons derived from Malonyl-CoA, with the elimination of CO_2 from the malonyl group; the net effect is extension of the acyl chain by two carbons.

(2) The β-Keto group is reduced to an alcohol.

(3) The elimination of H_2O creates a double bond.

(4) The double bond is reduced to form the corresponding saturated fatty acyl group.

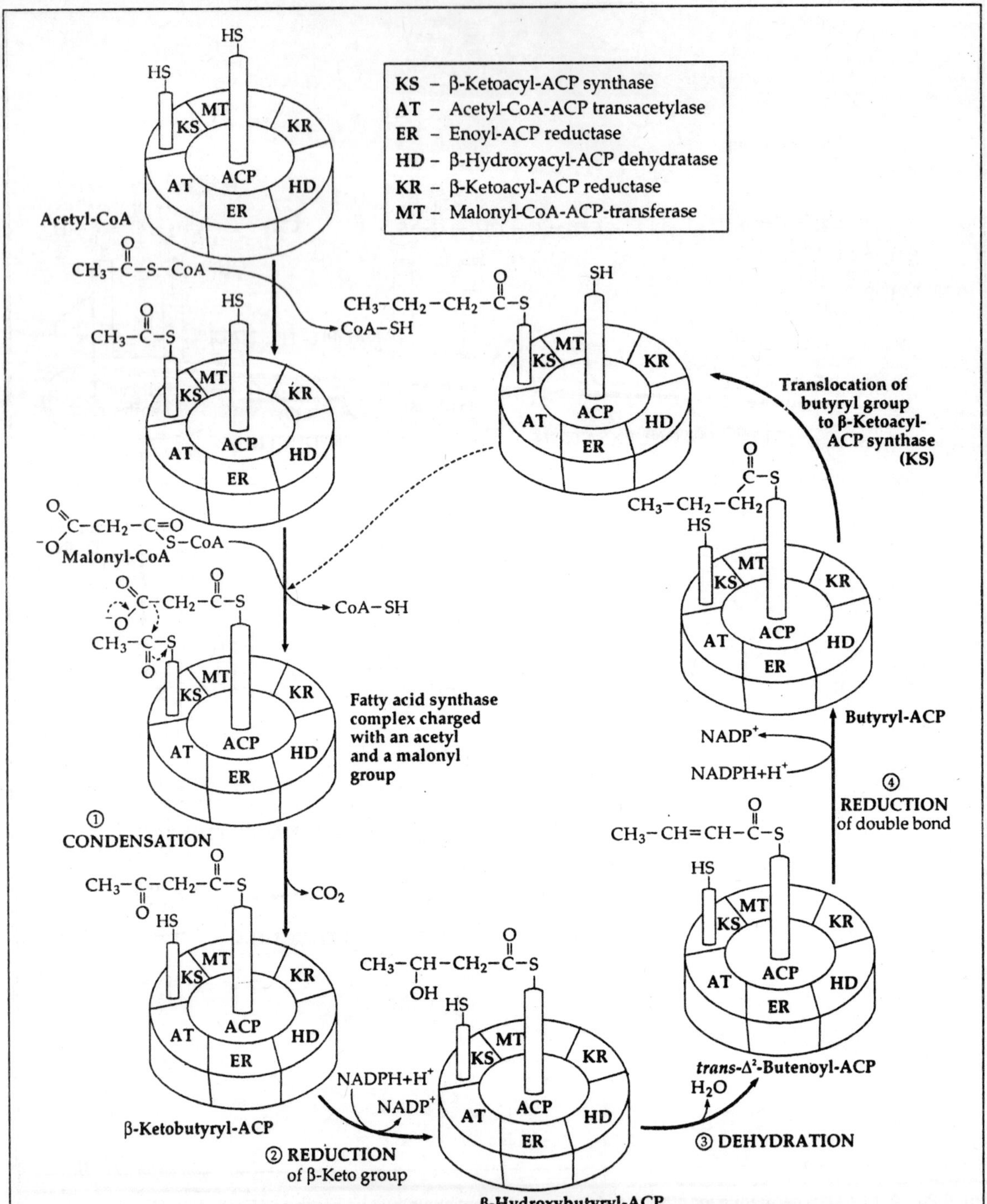

Fig. 4.21: Sequence of events during Fatty acid synthesis : The fatty acid synthase complex is shown schematically. Each segment of the disk represents one of the six enzymatic activities of the complex. At the centre is ACP with its phosphopantetheine arm.

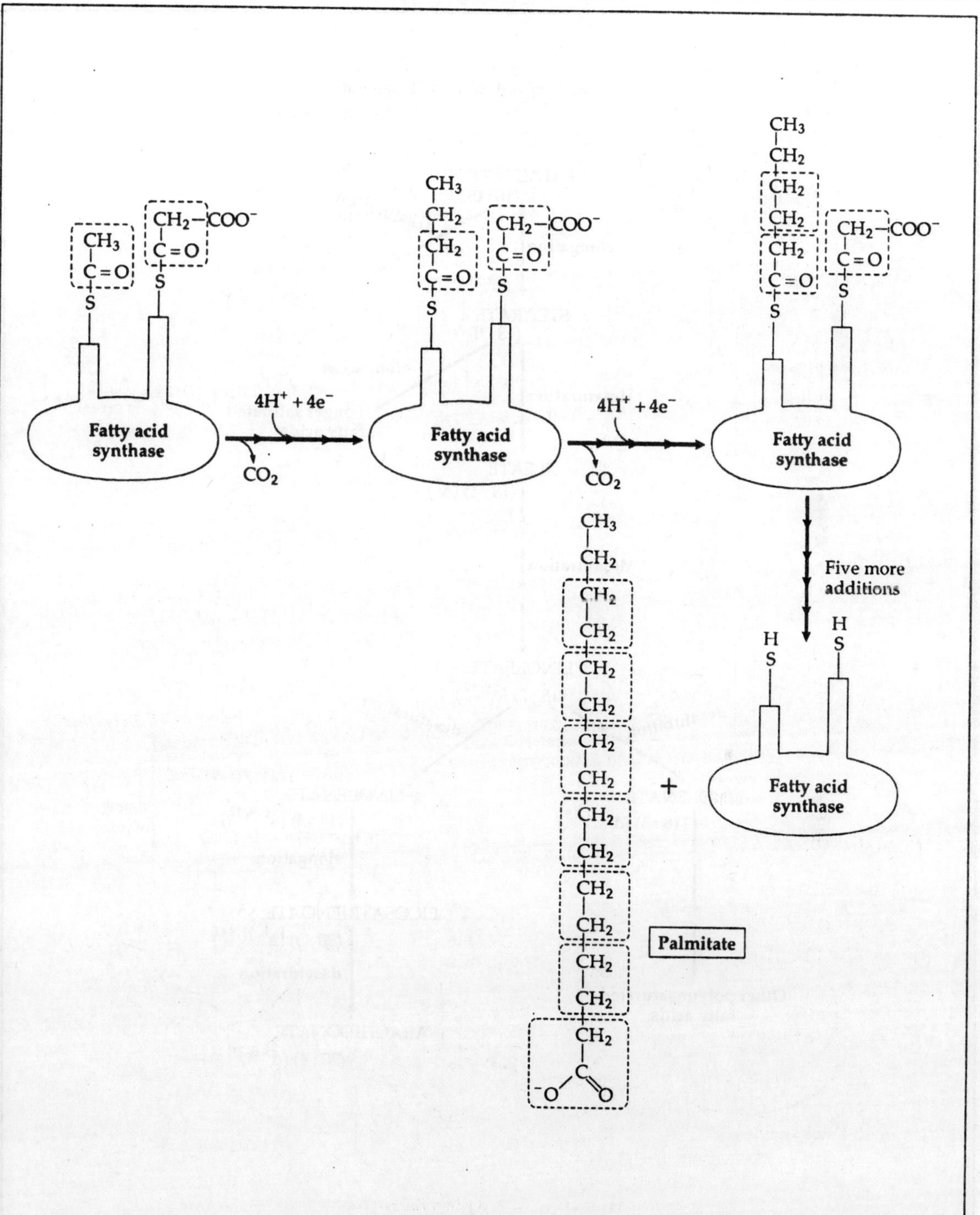

Fig. 4.22: The overall process of Palmitate synthesis : The fatty acyl chain grows by two-carbon units donated by activated malonate, with loss of CO_2 at each step. After each two-carbon addition, reductions convert the growing chain to a saturated fatty acid of four, then six, then eight carbons, and so on. Total 7 cycles are required to get the final product Palmitate (16 : 0).

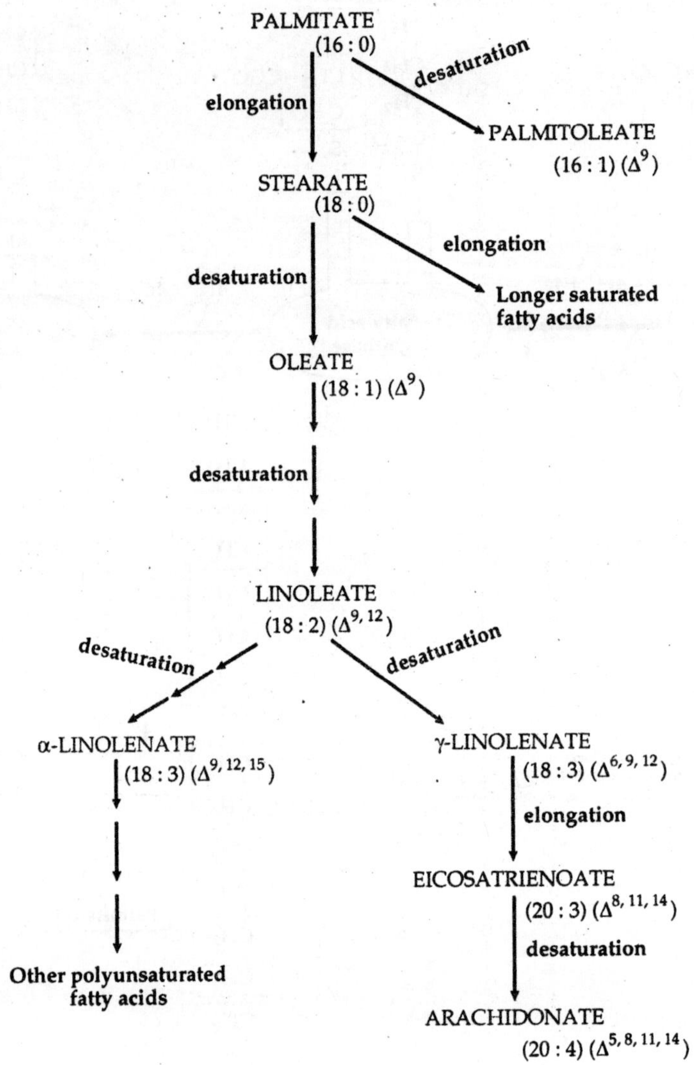

Fig. 4.23: Routes of Synthesis of other fatty acids : Palmitate is the precursor of stearate and longer-chain saturated fatty acids, as well as the monounsaturated acids, Palmitoleate and Oleate. Conversion of Linoleate into other polyunsaturated fatty acids and Eicosanoids is shown here. Unsaturated fatty acids are symbolized by indicating the number of carbons and the number and position of double bonds.

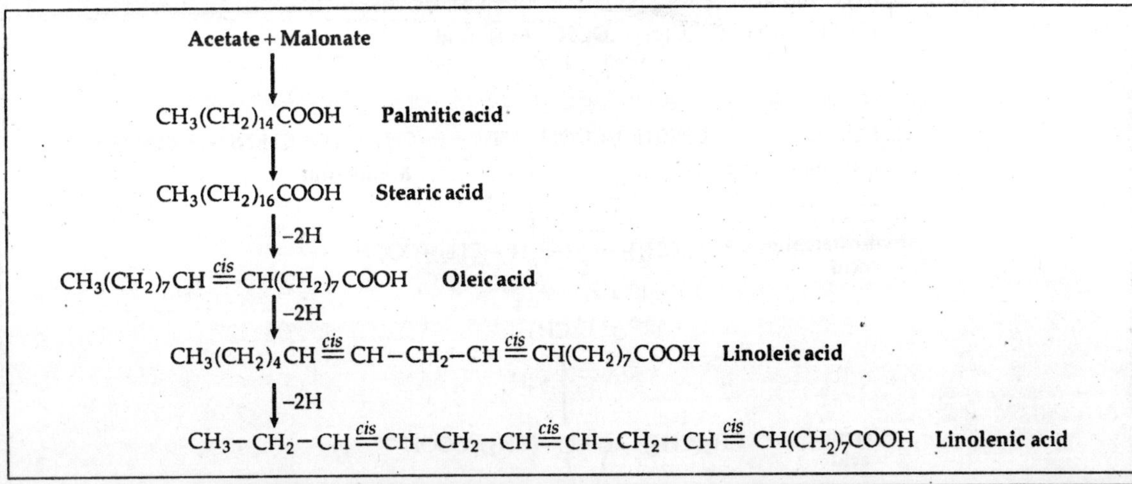

Fig. 4.24: Sequence of formation of olefinic fatty acids.

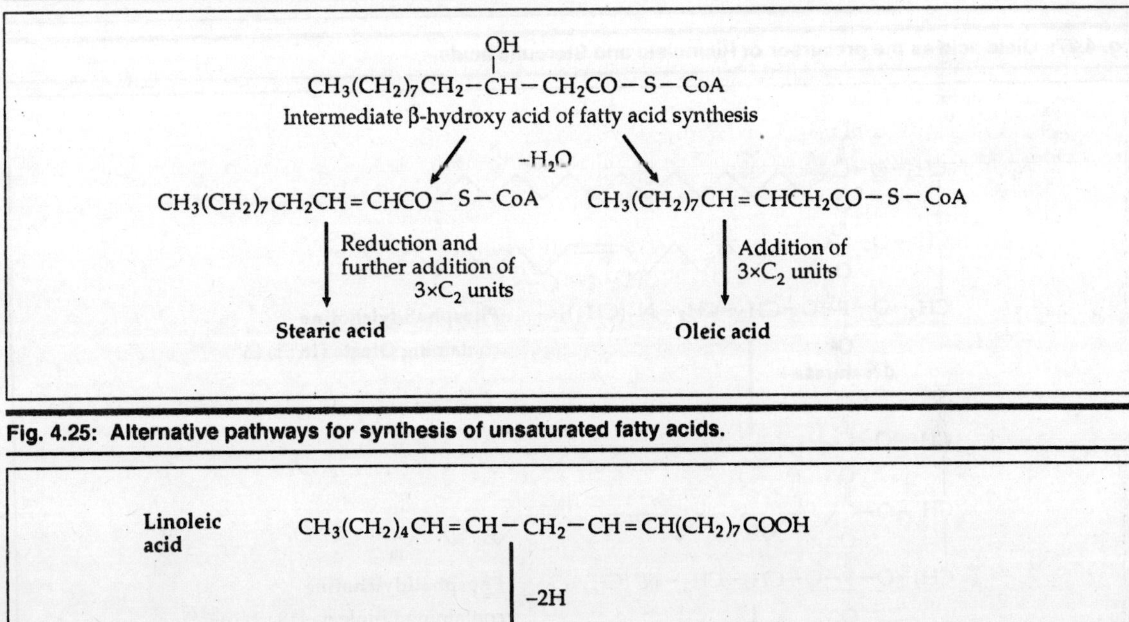

Fig. 4.25: Alternative pathways for synthesis of unsaturated fatty acids.

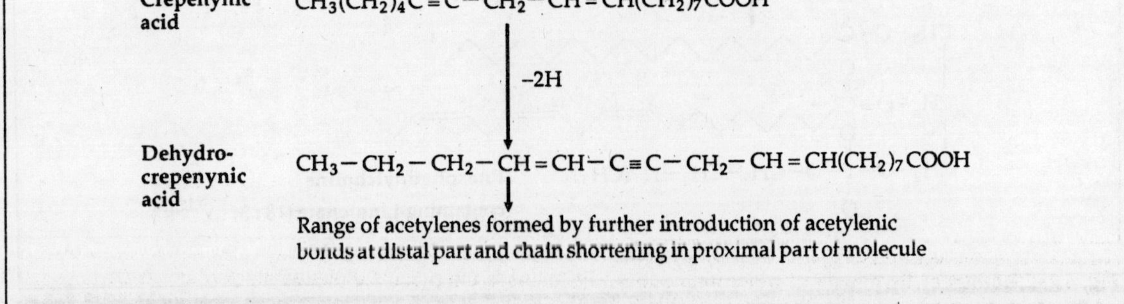

Fig. 4.26: Formation of acetylenic fatty acids.

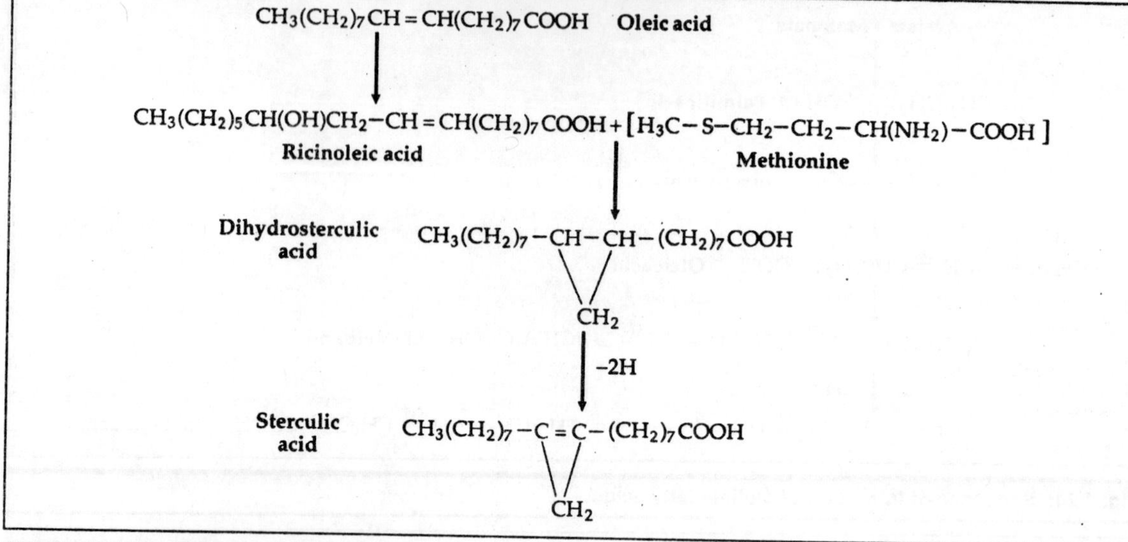

Fig. 4.27: Oleic acid as the precursor of Ricinoleic and Sterculic acids.

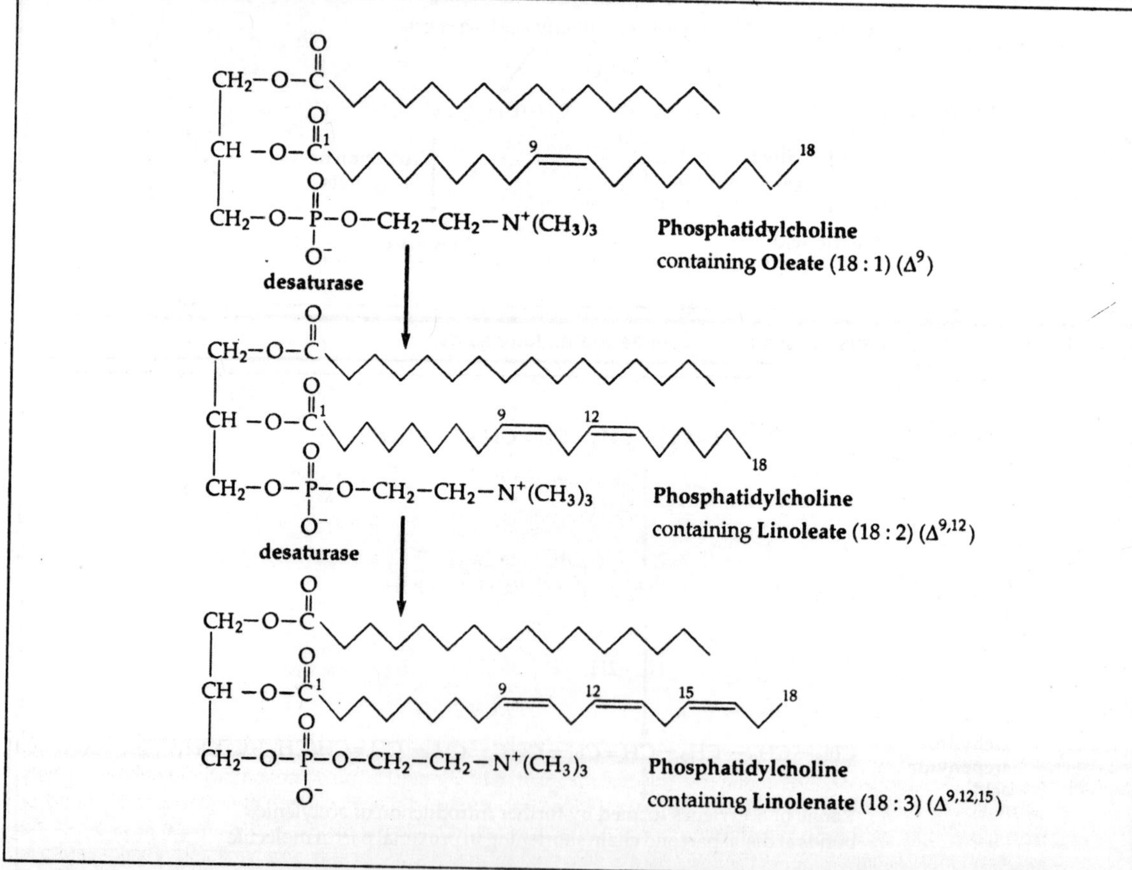

Fig. 4.28: Action of desaturases : Desaturases oxidise phosphatidylcholine-bound Oleate, producing polyunsaturated fatty acids. Some of the products are released from phosphatidylcholine by hydrolysis.

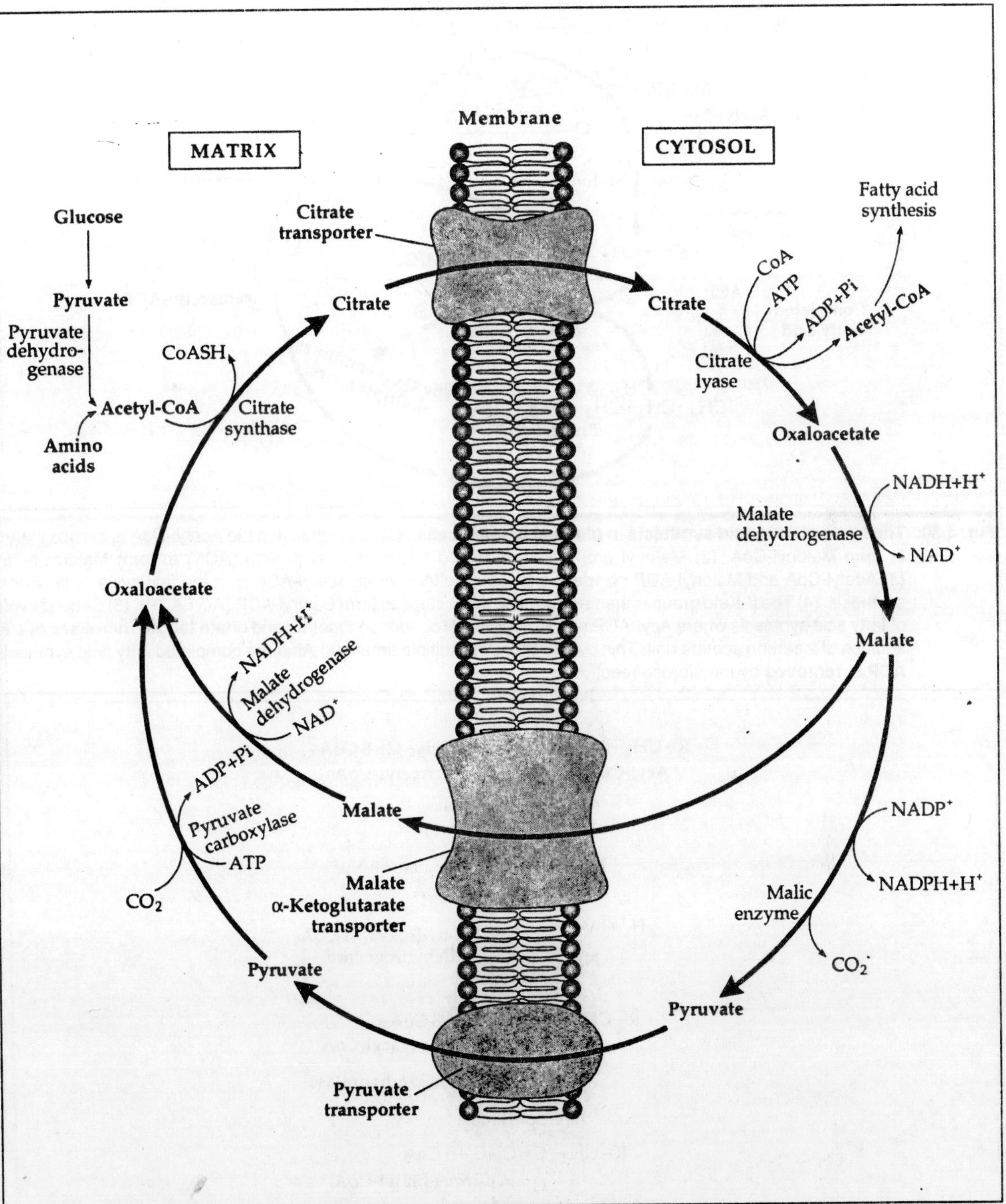

Fig. 4.29: Shuttle for transfer of acetyl groups from mitochondria to Cytosol : The mitochondrial membrane is freely permeable to all of these compounds. Pyruvate derived from amino acid catabolism in the mitochondrial matrix or from glucose by glycolysis in the cytosol, is converted to acetyl-CoA in the matrix. Acetyl groups pass out of the mitochondrion as citrate; in the cytosol they are delivered as acetyl-CoA for fatty acid synthesis. Oxaloacetate is reduced to malate, which returns to the mitochondrial matrix and is converted to oxaloacetate. An alternative fate for cytosolic malate is oxidation by malic enzyme to generate cytosolic NADPH; the pyruvate produced returns to the mitochondrial matrix.

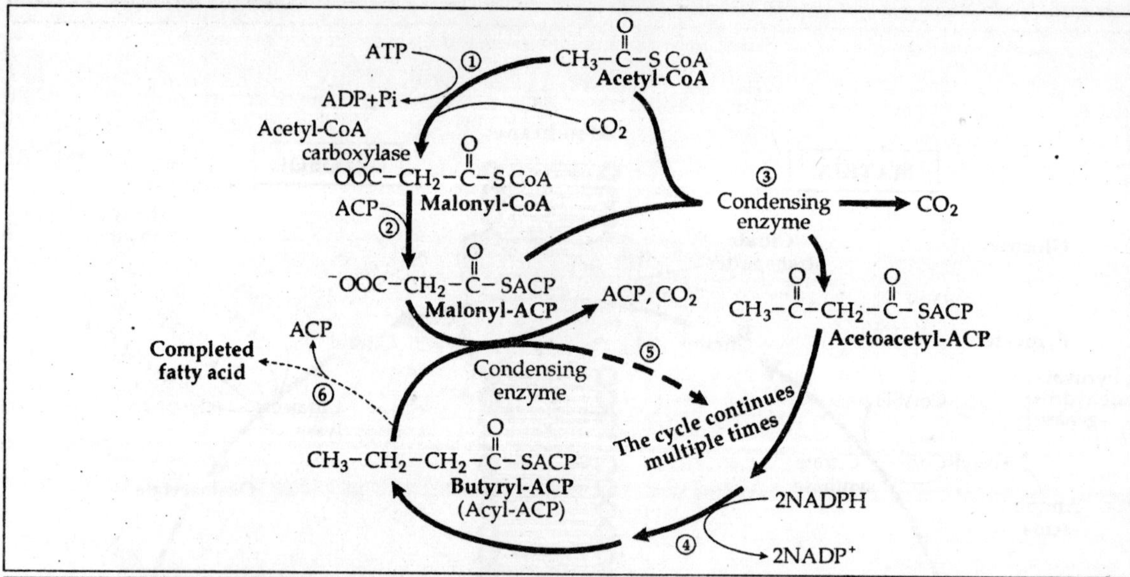

Fig. 4.30: The cycle of fatty acid synthesis in plastids of plant cells : (1) In the first step the Acetyl-CoA is **carboxylated** to form Malonyl-CoA. (2) Malonyl group is **transferred** to acyl carrier protein (ACP) to form Malonyl-ACP. (3) Acetyl-CoA and Malonyl-ACP **condense** together to form Acetoacetyl-ACP is in the first cycle of fatty acid synthesis. (4) The β-Keto group is then **removed** in three steps to form Butyryl-ACP (Acyl-ACP). (5) Second cycle of fatty acid synthesis where Acyl-ACP and Malonyl-ACP condense together and **chain length increases** due to addition of 2 carbon acetate unit. This cycle continues multiple times. (6) After the completed fatty acid synthesis **ACP is removed** by transferase reaction.

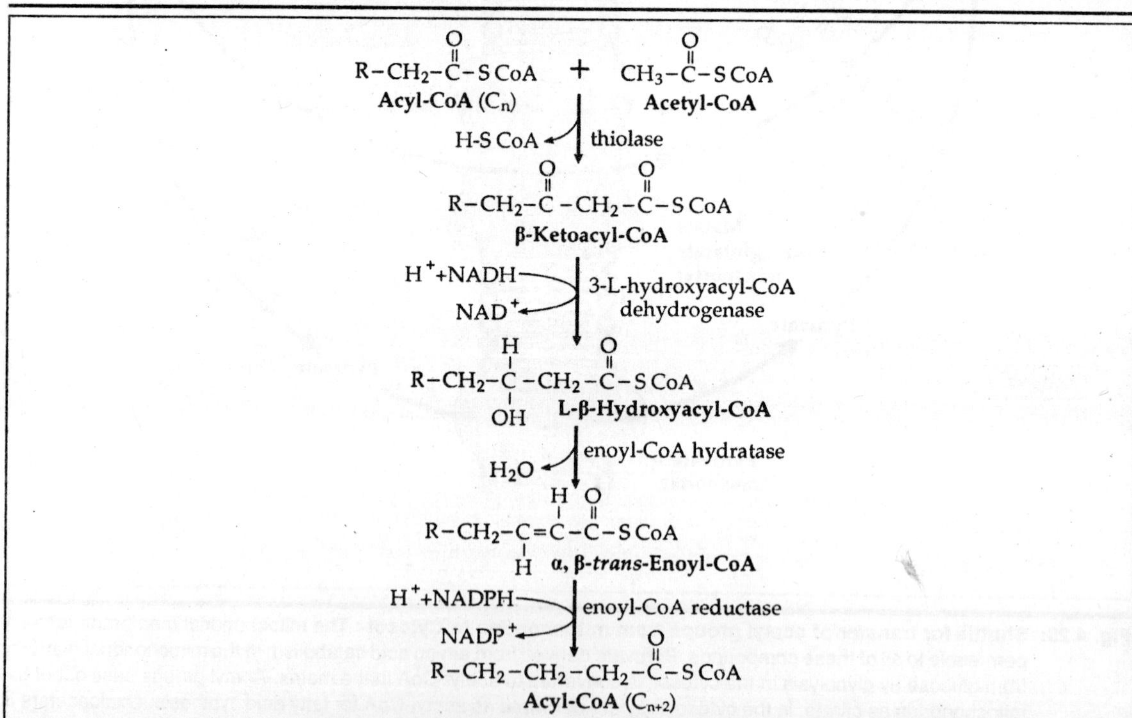

Fig. 4.31: The steps for Mitochondrial fatty acid elongation : This process is reverse of fatty acid oxidation except that the final reaction employs NADPH, not $FADH_2$, as its redox coenzyme.

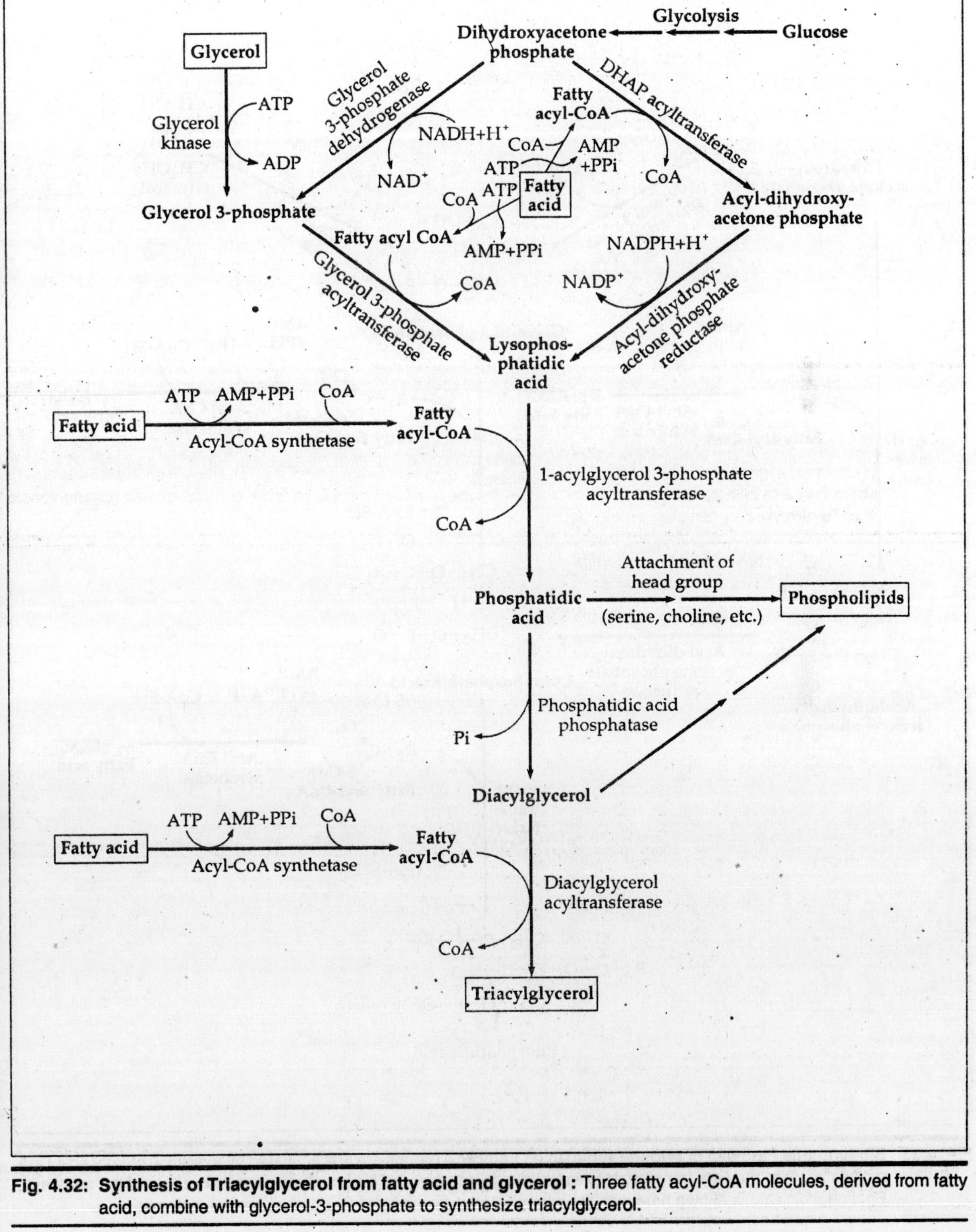

Fig. 4.32: Synthesis of Triacylglycerol from fatty acid and glycerol : Three fatty acyl-CoA molecules, derived from fatty acid, combine with glycerol-3-phosphate to synthesize triacylglycerol.

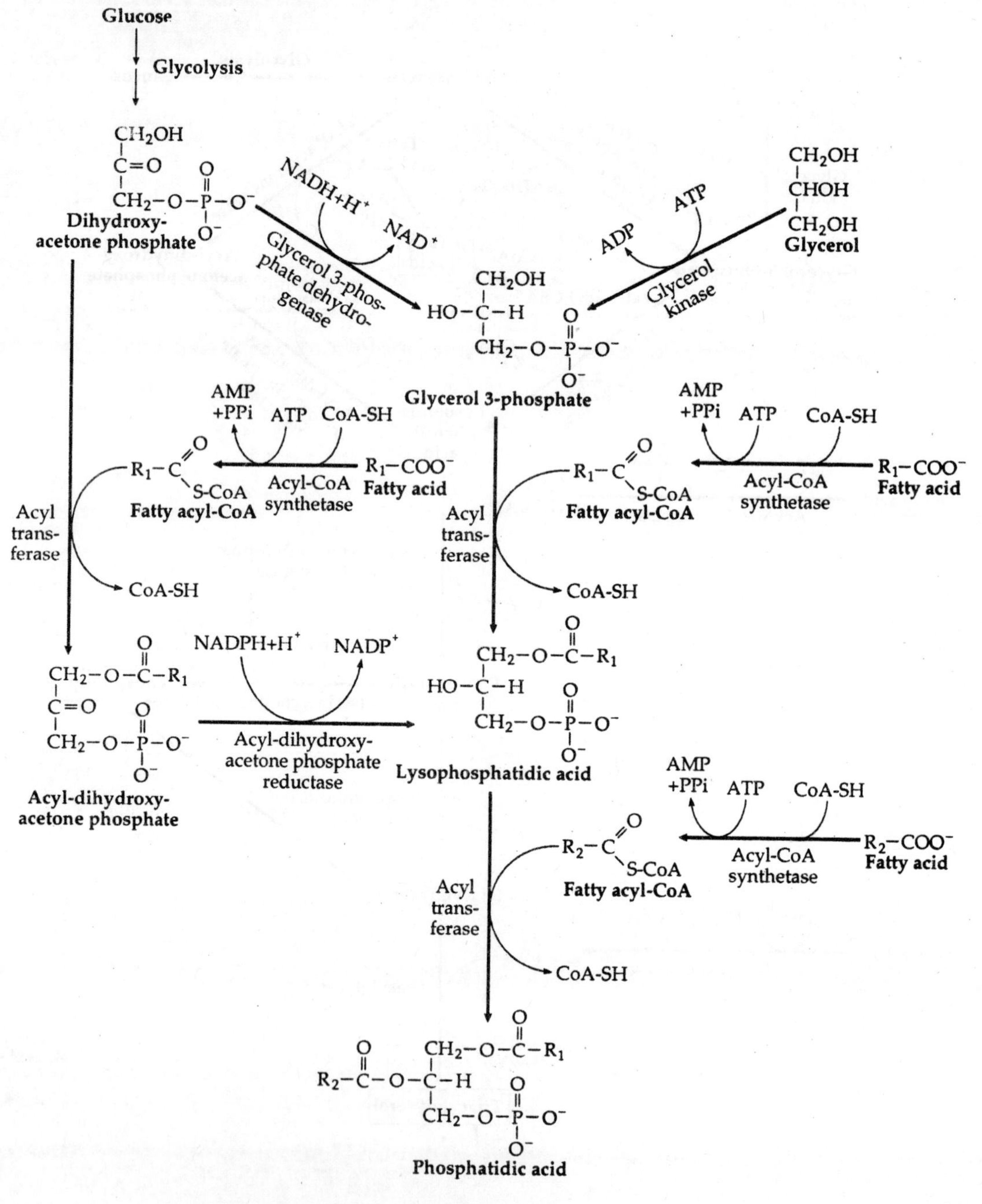

Fig. 4.33: Biosynthetic Pathway of phosphatidic acid : Fatty acyl groups are first activated by formation of fatty acyl-CoA molecules, then they are transferred to ester linkage with glycerol 3-phosphate, formed in either of the two ways. Phosphatidic acid is shown here with the correct stereochemistry at C-2 of the glycerol molecule.

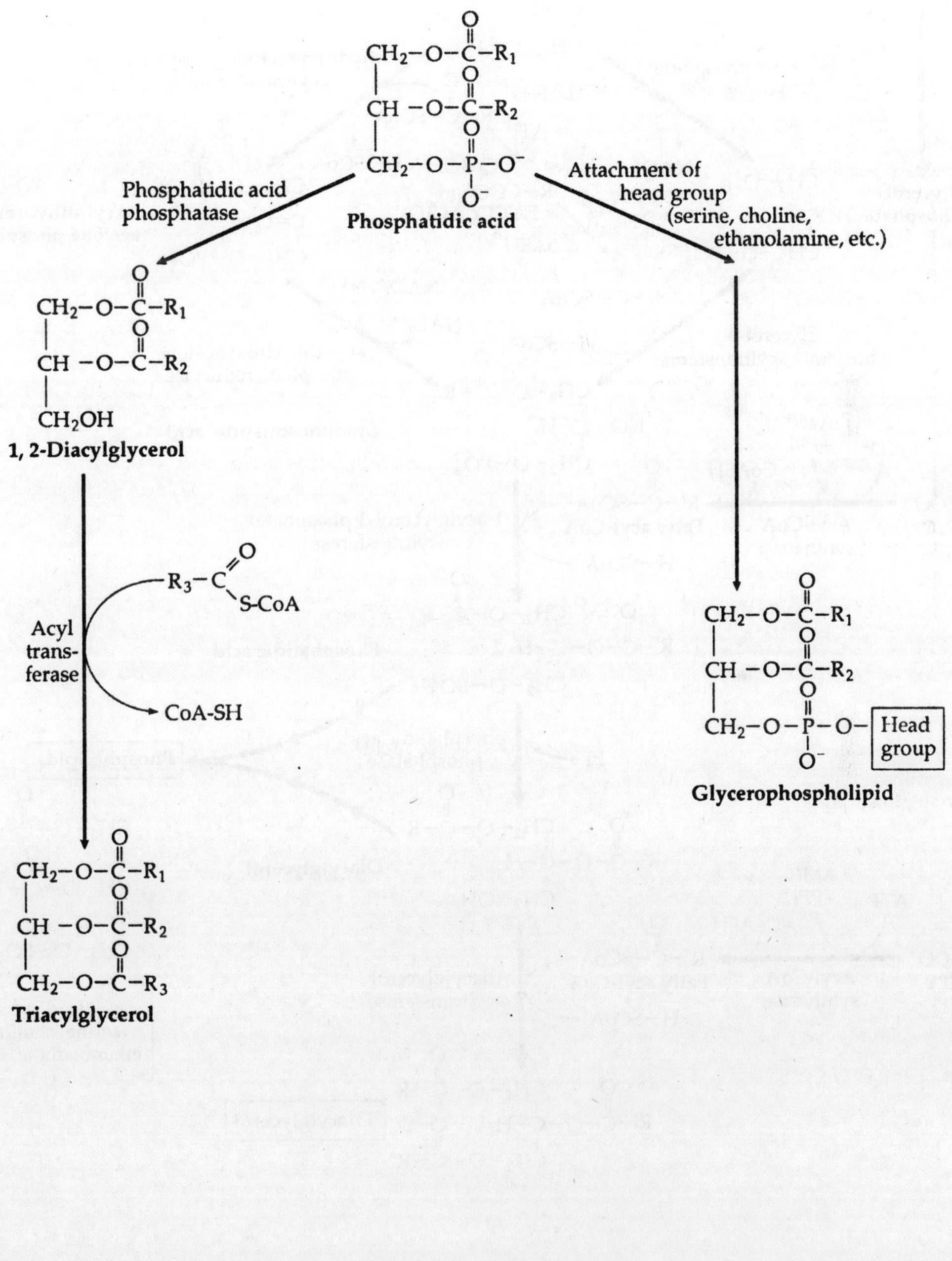

Fig. 4.34: Phosphatidic acid in lipid biosynthesis : Phosphatidic acid is the precursor of both triacylglycerols and glycerophospholipids.

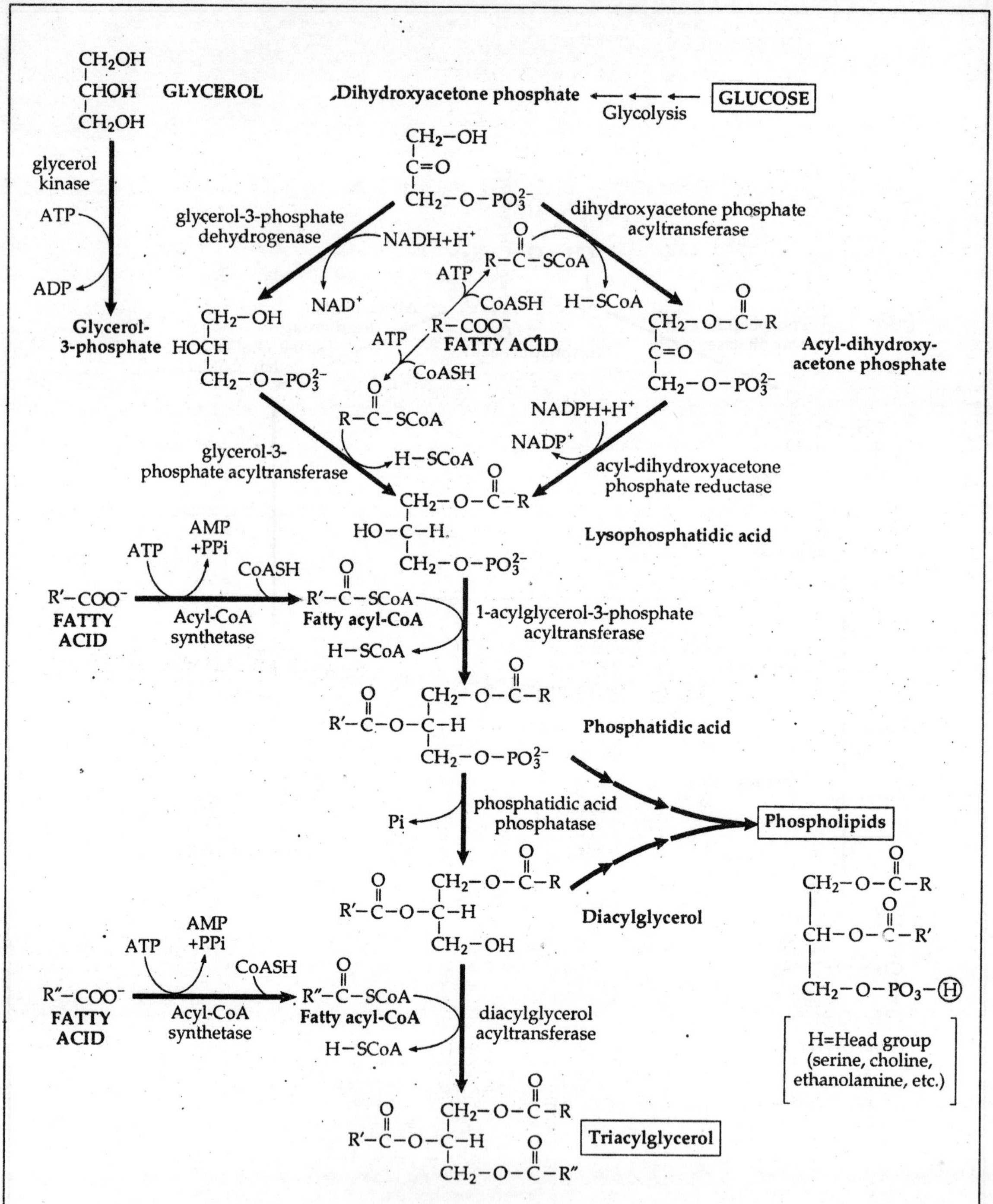

Fig. 4.35: The reactions of Triacylglycerol biosynthesis : Fatty acyl groups are first activated by formation of fatty acyl-CoA molecules, then they are transferred to ester linkage with Glycerol-3-phosphate, formed in either of the two ways shown. Phosphatidic acid is the precursor of both triacylglycerols and phospholipids.

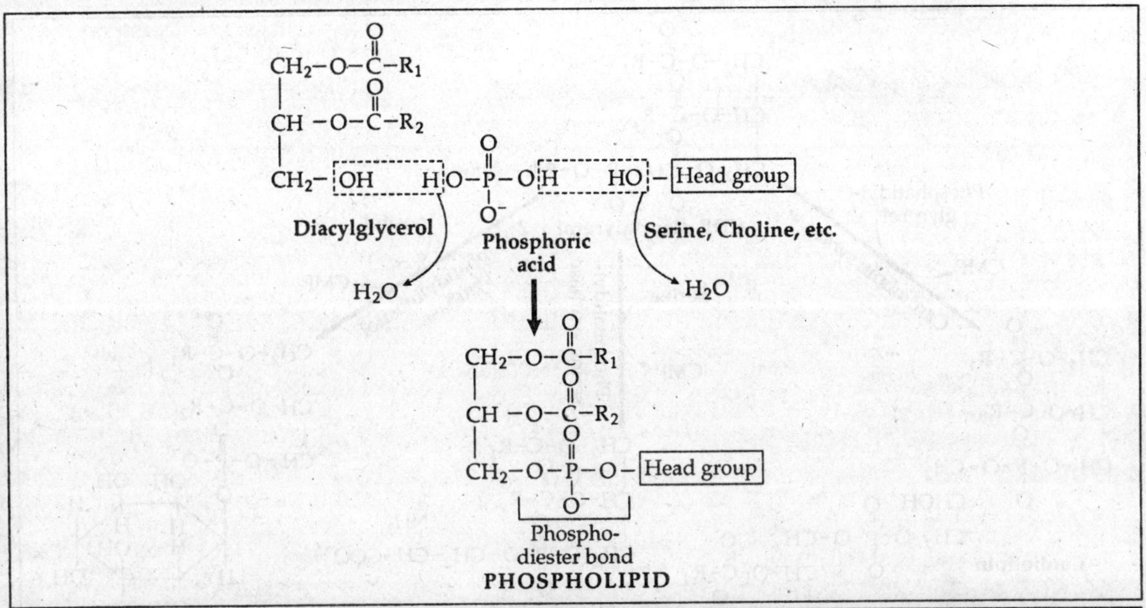

Fig. 4.36: Formation of phosphodiester bond in phospholipid : The phospholipid head group is attached to a diacylglycerol by a phosphodiester bond, formed when phosphoric acid reacts with –OH groups eliminating two molecules of H_2O.

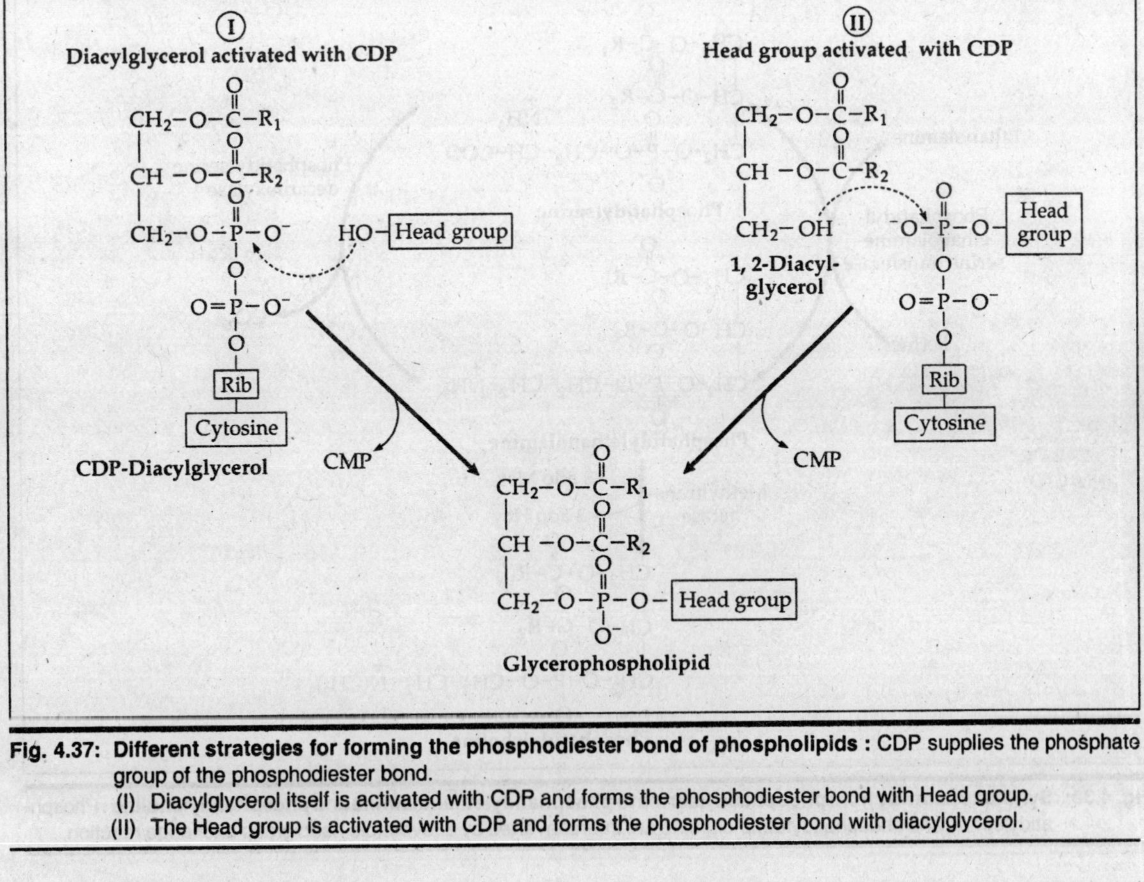

Fig. 4.37: Different strategies for forming the phosphodiester bond of phospholipids : CDP supplies the phosphate group of the phosphodiester bond.

(I) Diacylglycerol itself is activated with CDP and forms the phosphodiester bond with Head group.

(II) The Head group is activated with CDP and forms the phosphodiester bond with diacylglycerol.

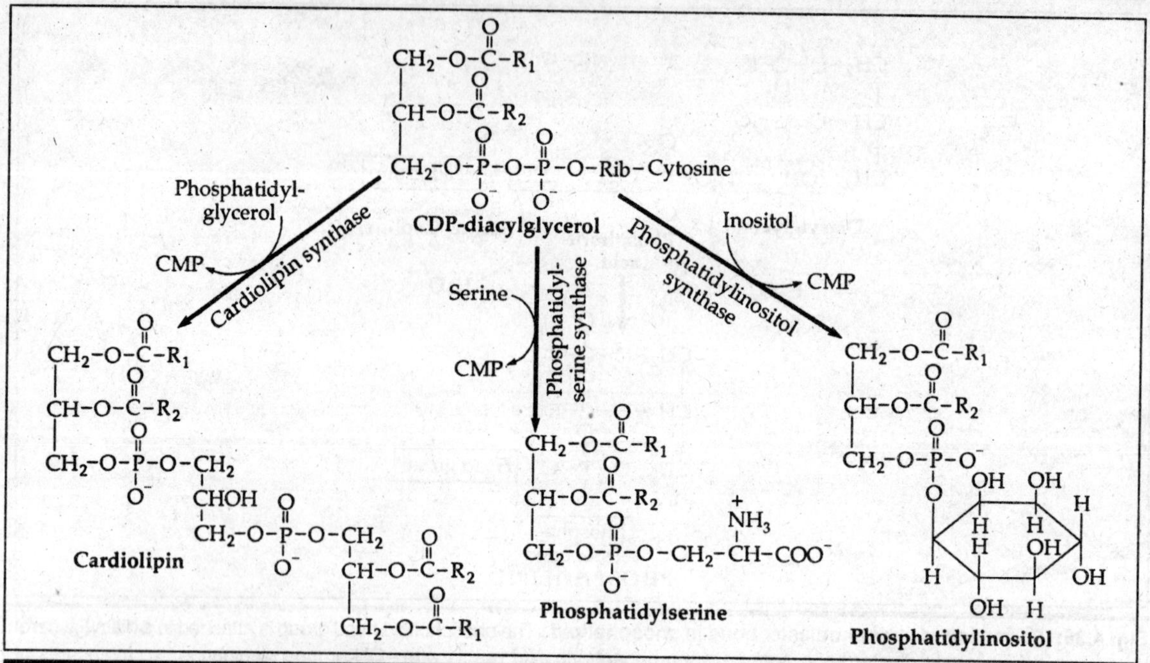

Fig. 4.38: Synthesis of cardiolipin, phosphatidylserine, phosphatidylinositol from CDP-diacylglycerol.

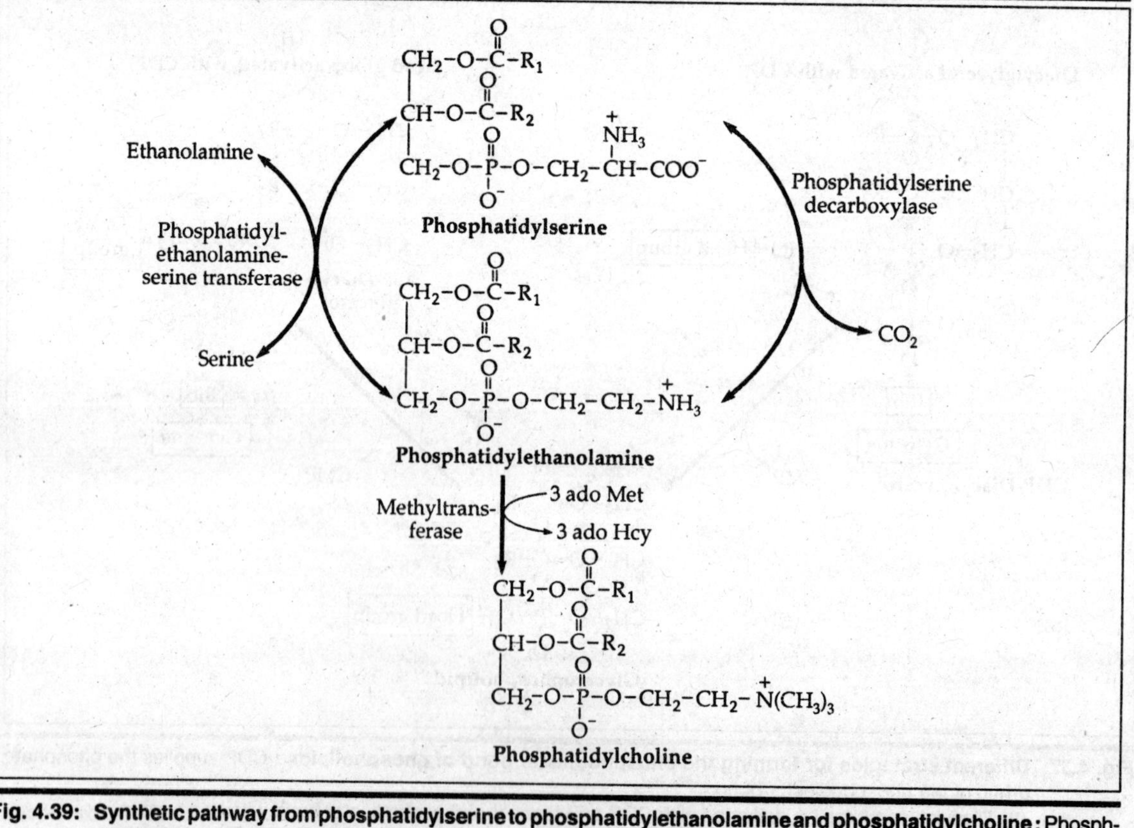

Fig. 4.39: Synthetic pathway from phosphatidylserine to phosphatidylethanolamine and phosphatidylcholine: Phosphatidylserine and phosphatidylethanolamine are interconverted by a reversible head group exchange reaction.

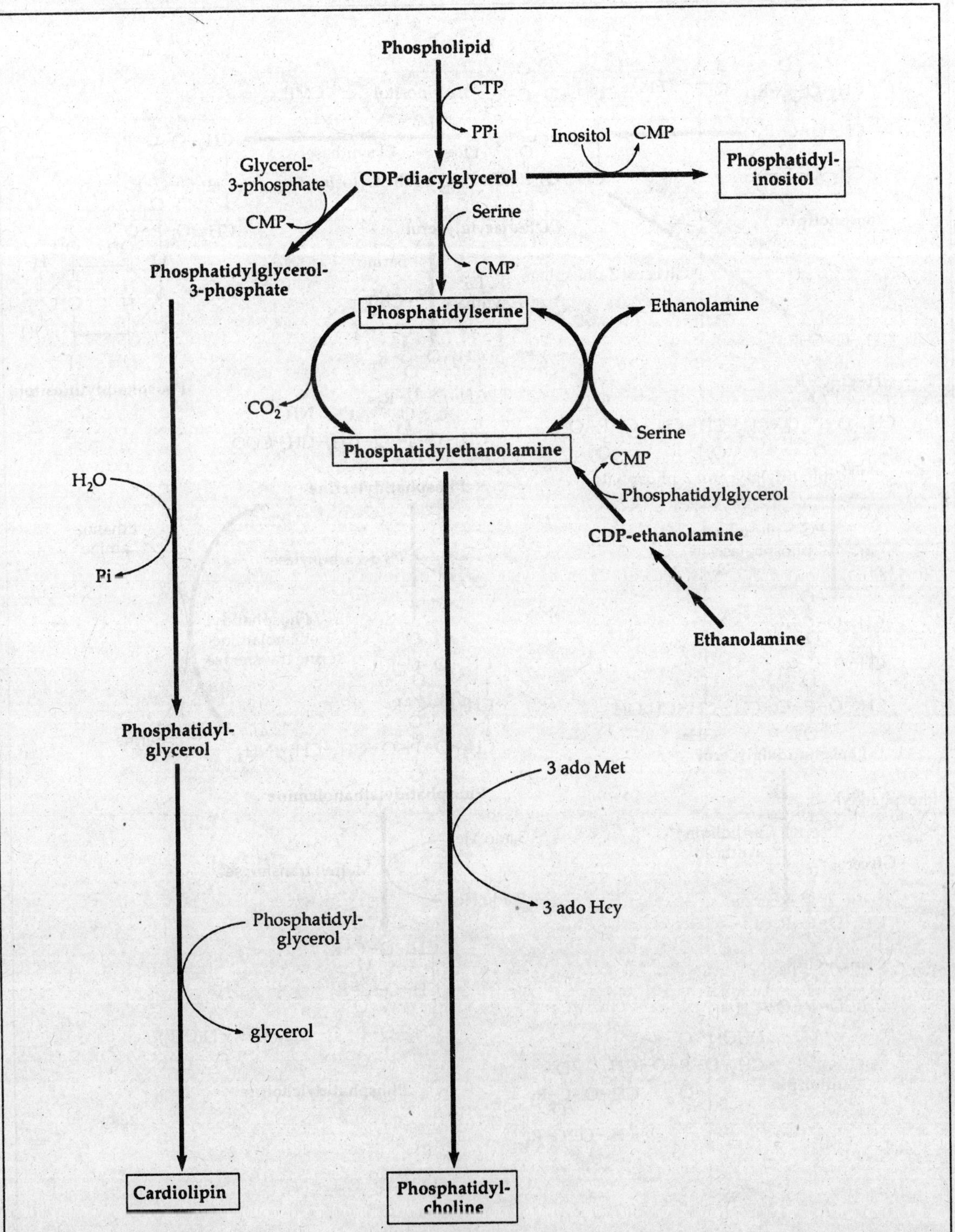

Fig. 4.40: Summary of the pathways leading to the synthesis of cardiolipin, phosphatidylserine, phosphatidylinositol, phosphatidylcholine, phosphatidylethanolamine.

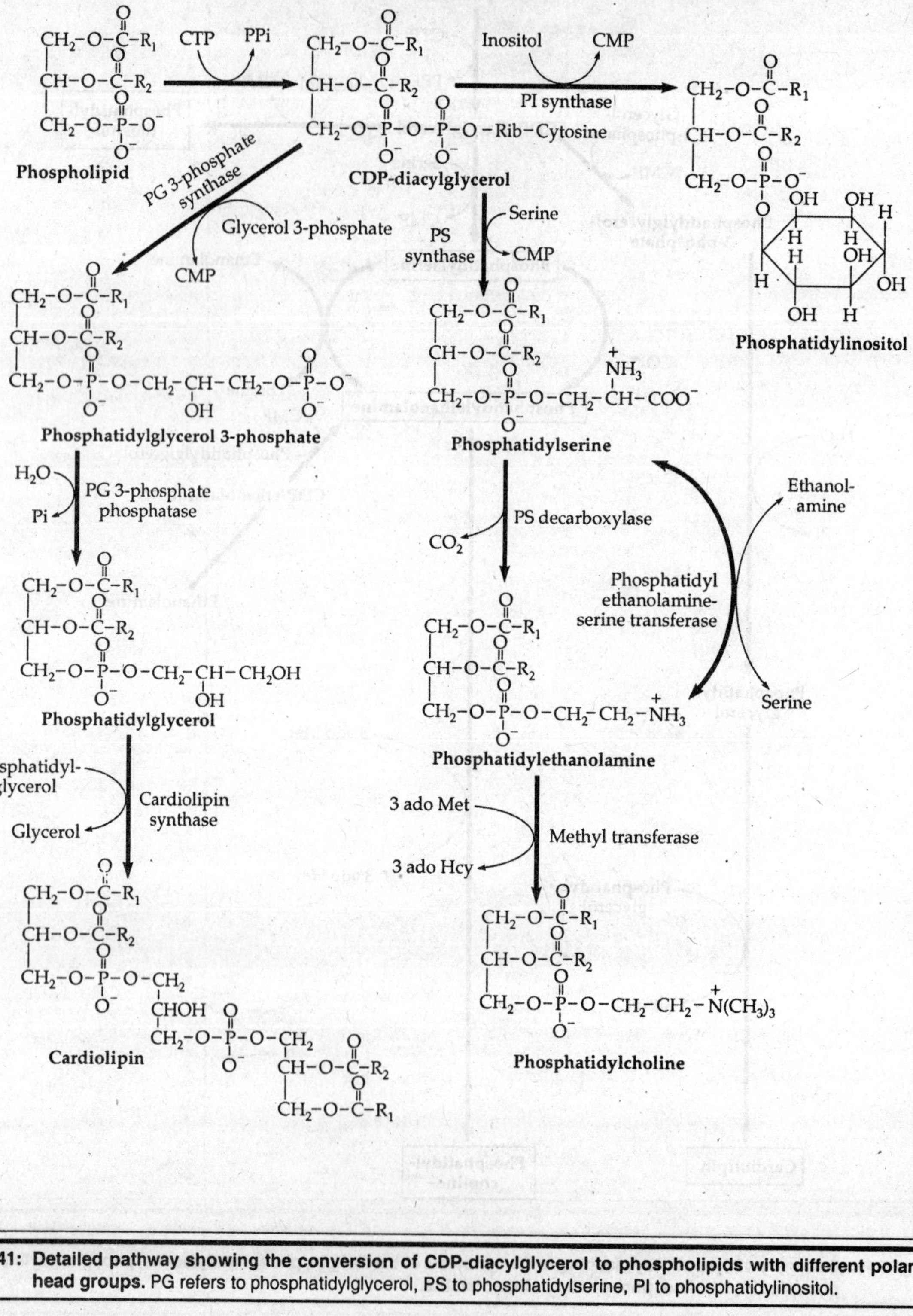

Fig. 4.41: Detailed pathway showing the conversion of CDP-diacylglycerol to phospholipids with different polar head groups. PG refers to phosphatidylglycerol, PS to phosphatidylserine, PI to phosphatidylinositol.

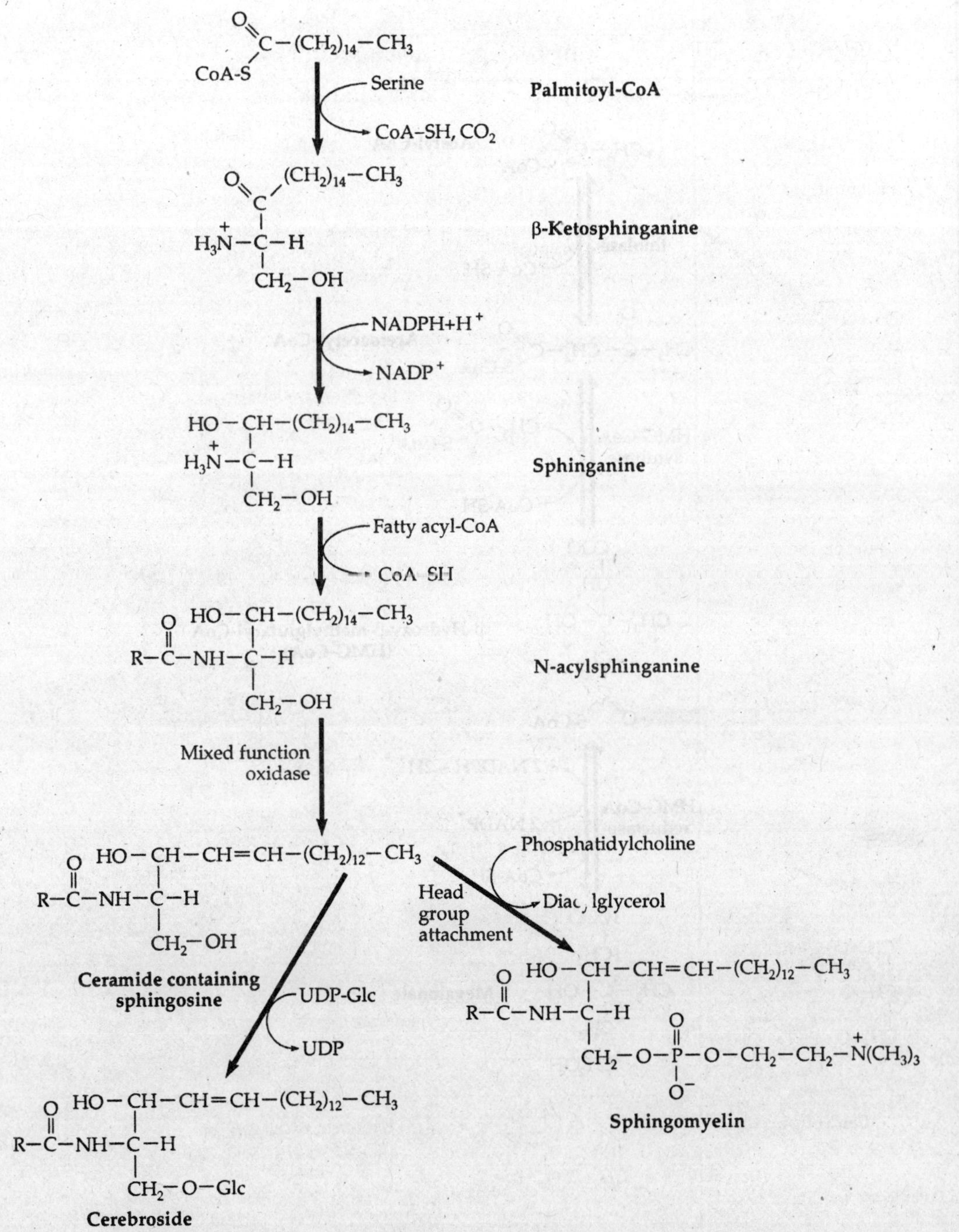

Fig. 4.42: Biosynthesis of sphingolipids : Condensation of Palmitoyl-CoA and Serine followed by reduction with NADPH yields Sphinganine, which is then acylated to form N-acylsphinganine (Ceramide). A double bond is formed by oxidase, before the final addition of a head group–phosphatidylcholine, to form Sphingomyelin or glucose to form a Cerebroside.

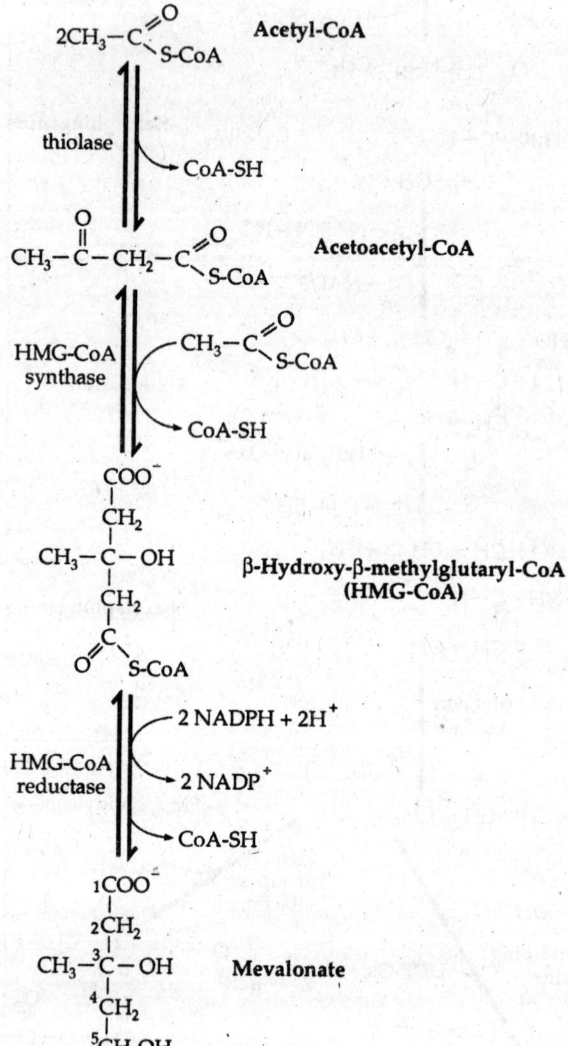

Fig. 4.43: Formation of Mevalonate from Acetyl-CoA. The origin of C1 and C2 of Mevalonate from Acetyl-CoA is shown.

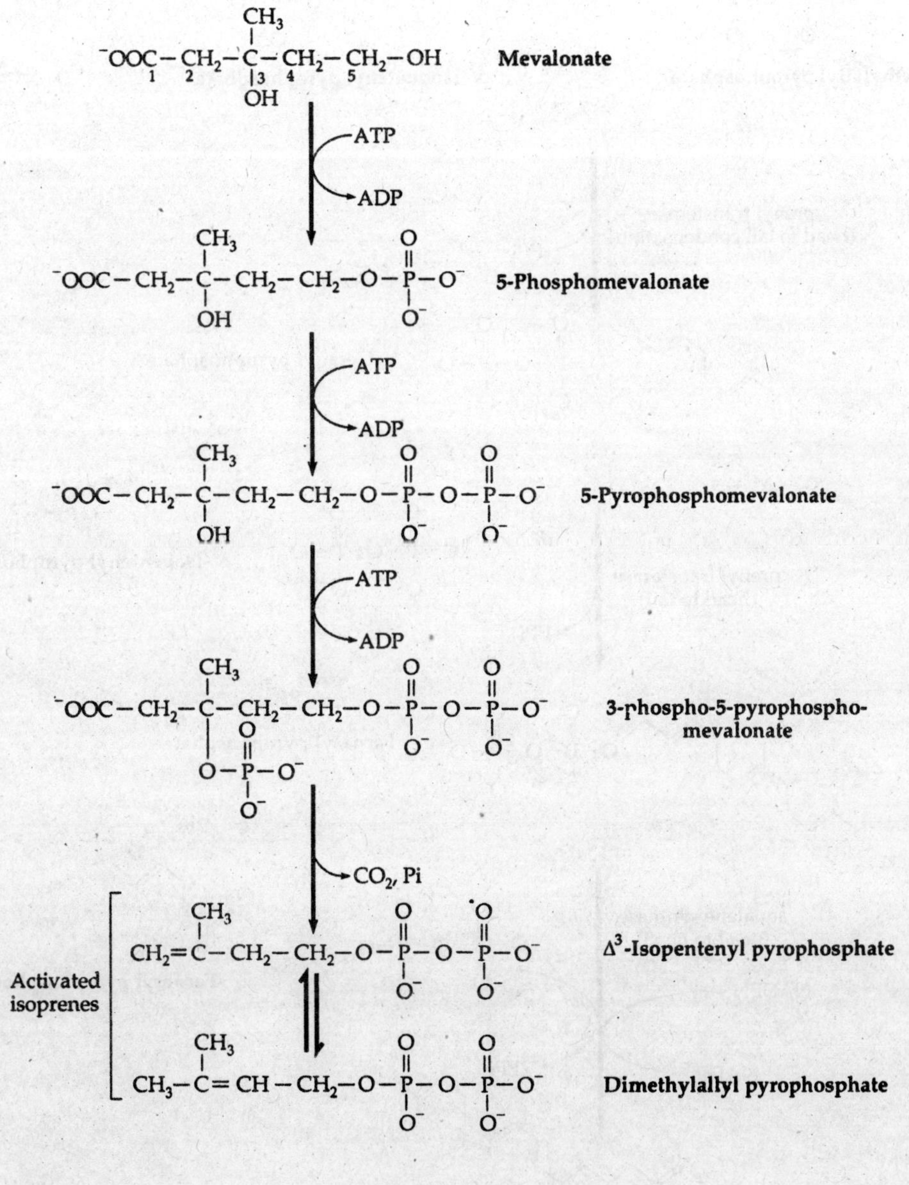

Fig. 4.44: Conversion of Mevalonate into activated Isoprene.

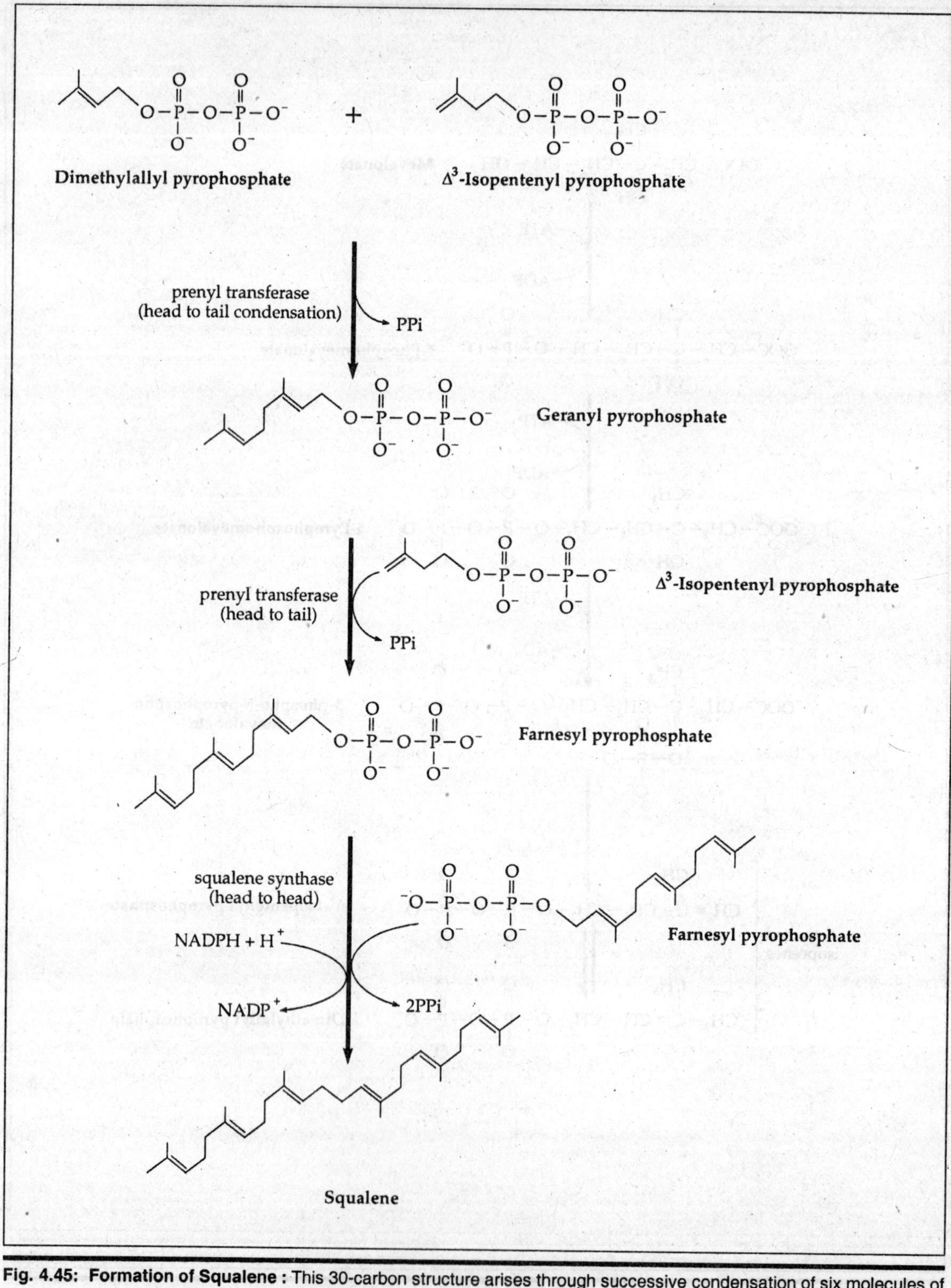

Fig. 4.45: Formation of Squalene : This 30-carbon structure arises through successive condensation of six molecules of activated isoprene (five-carbon) units.

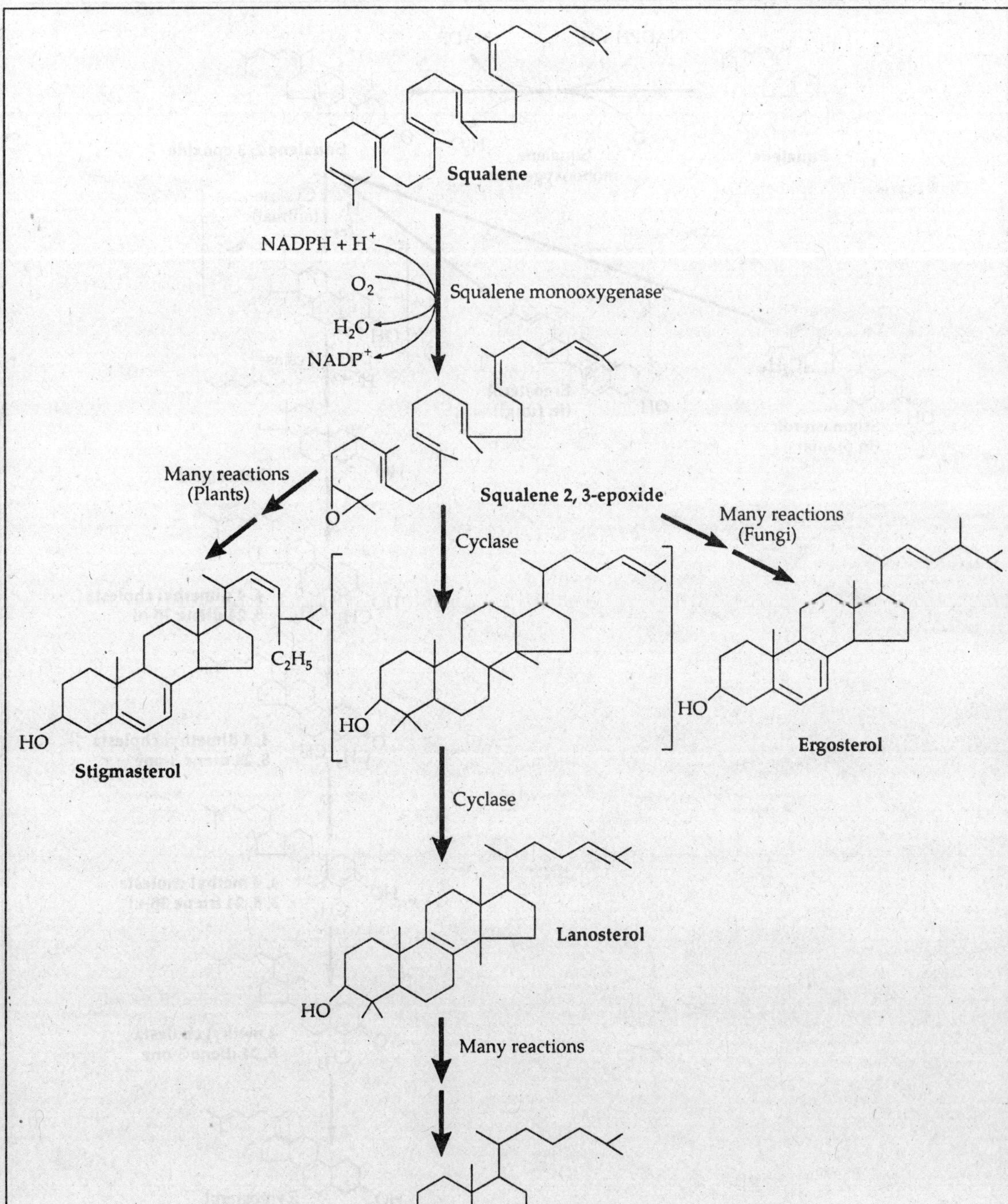

Fig. 4.46a: Ring closure converts linear Squalene into the condensed steroid structure nucleus : The first step in this sequence is catalysed by a mixed function oxidase (a monooxygenase), for which the cosubstrate is NADPH. The product is an epoxide, which in the next step is cyclized to the steroid nucleus.

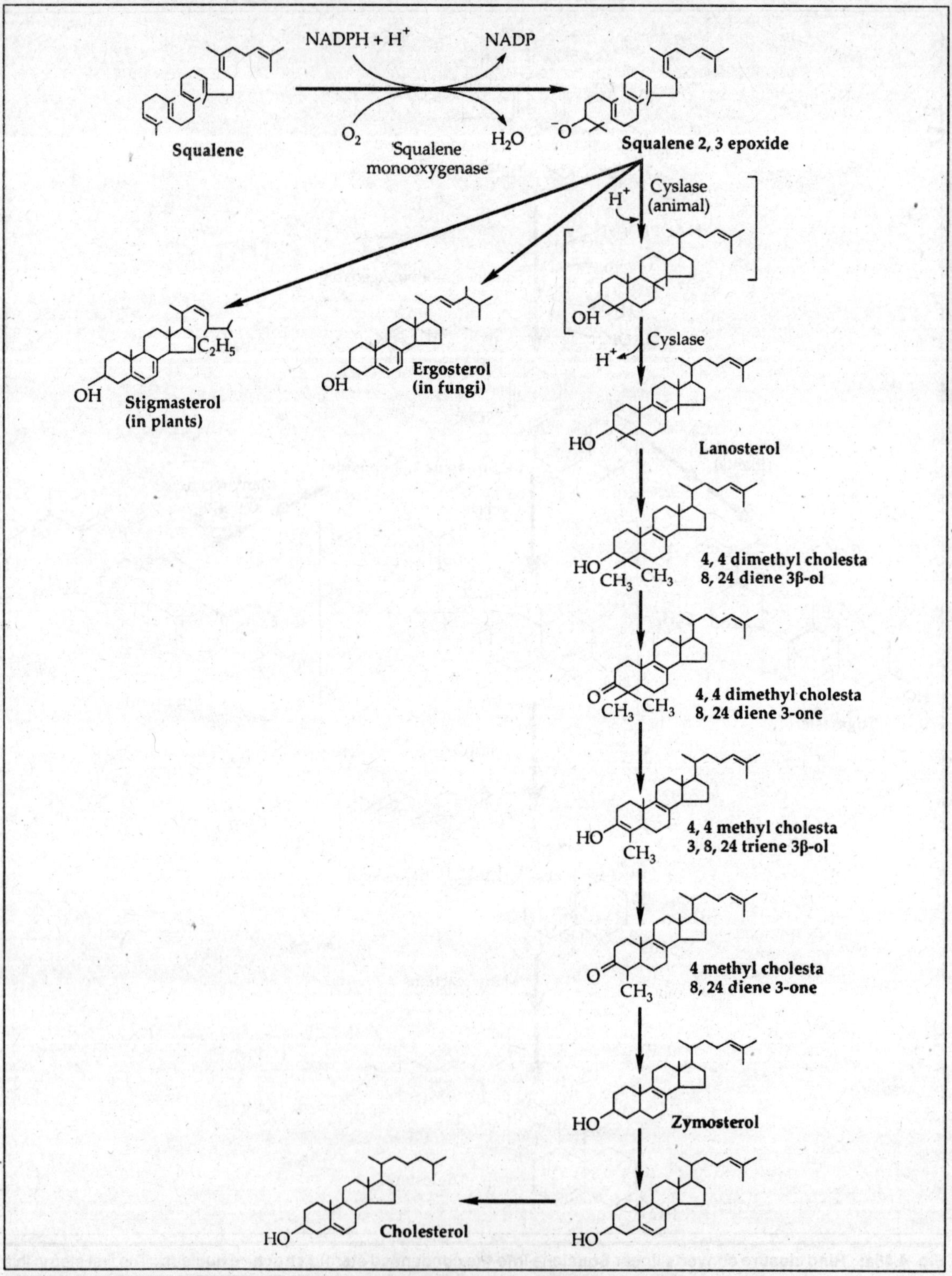

Fig. 4.46b: Conversion of squalene to cholesterol (Detail reactions from lanosterol).

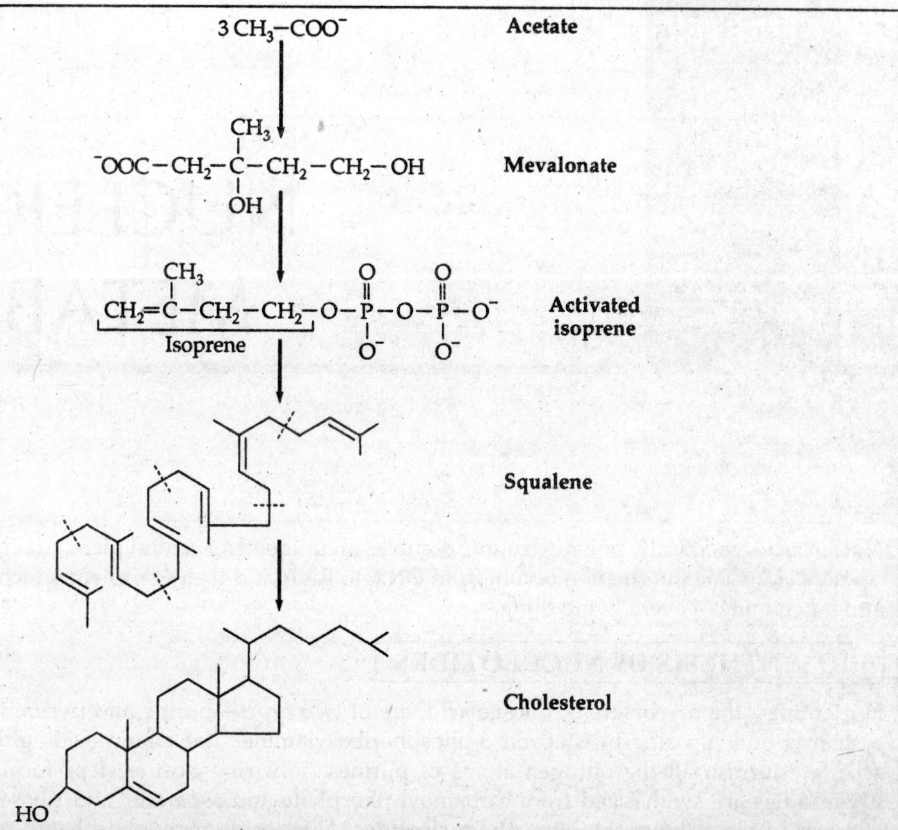

Fig. 4.47: Summary of cholesterol biosynthesis.

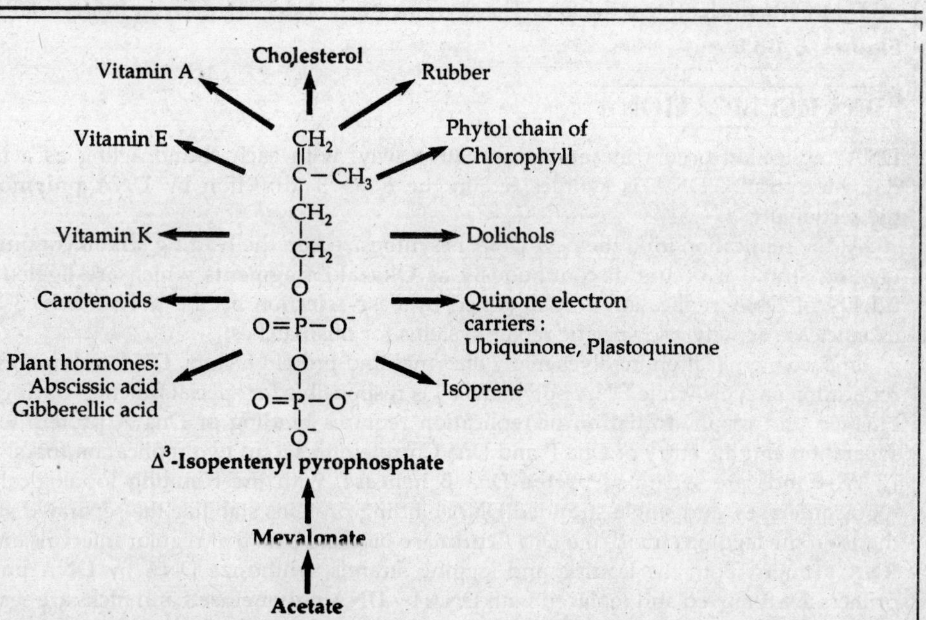

Fig. 4.48: An overview of Isoprenoid Biosynthesis.

CHAPTER 5

NUCLEIC ACID METABOLISM

Nucleic acid, chemically polynucleotide, occupies an unique and central place among biomolecules as the passage of genetic information occurs from DNA to RNA and then to protein which guides the structure and functioning of every living thing.

BIOSYNTHESIS OF NUCLEOTIDES

Nucleotides, the precursors of nucleic acid, are of two types—purine and pyrimidine. The **purine** ring system is built up step by step on 5-phosphoribosylamine. The amino acids glutamine, glycine and aspartate furnish all the nitrogen atoms of purines. Two ring-closure steps form the purine nucleus. **Pyrimidines** are synthesized from carbamoyl phosphate and aspartate, and ribose-5-phosphate is then attached to yield the pyrimidine **ribonucleotides**. Nucleoside monophosphates are converted to their triphosphates by enzymatic phosphorylation reactions. Ribonucleotides are converted to **deoxyribonucleotides** by ribonucleotide reductase. The thymine nucleotides are derived from dCDP and dUMP.

Figures: 5.1-5.8

DNA REPLICATION

DNA replication occurs in **semiconservative** way, with each strand acting as a template for a new daughter strand. DNA is synthesized in the 5′ → 3′ direction by **DNA polymerase** and proceeds bidirectionally.

At the replication fork, the new DNA is synthesized on the **leading strand** continuously while on the **lagging strand** it occurs discontinuously as **Okazaki fragments** which are ligated subsequently. The fidelity of DNA replication is maintained by base selection by the polymerase; 3′ → 5′ proofreading exonuclease activity and specific repair systems for mismatches.

In *E. coli*, replication involves many enzymes and protein factors. **DNA polymerase** III is the primary replication enzyme while DNA polymerase I is responsible for special function during replication, recombination and repair. **Initiation** of replication requires binding of Dna A protein to the origin, strand separation and the entry of Dna B and Dna C protein to set up two replication forks. During **elongation,** DNA strands are separated by the Dna B **helicase,** with the resulting topological strain relieved by **topoisomerases,** and single-stranded DNA-binding proteins stabilize the separated strands. During synthesis on the lagging strand, the Dna G **primase** binds to Dna B at regular intervals and synthesizes short **RNA primers**. Both the leading and lagging strands synthesize DNA by **DNA polymerase III**. RNA primers are removed and replaced with DNA by **DNA polymerase I** and nicks are sealed by DNA **ligase.**

In contrast to single replication origin in *E. coli,* **eukaryotic** chromosomes have multiple replication origin. Sets of 20-80 replicons act as replication units that are activated in sequence. Five eukaryotic **DNA**

polymerases (α, β, γ, δ **and** ε) have been identified—DNA polymerase α synthesizes DNA on lagging strand, DNA polymerase δ synthesizes DNA on leading strand, RNA primers are synthesized by DNA polymerase α, DNA polymerases β and ε repair DNA, and DNA polymerase γ replicates mitochondrial DNA; **telomerase** is used to replicate DNA at telomeres.

Figures: 5.9–5.16

DNA REPAIR

Cells have multiple systems of DNA repair. **Mismatch repair** corrects error of DNA replication by identifying mismatched nucleotides and replacing the incorrectly inserted base. This system determines which of the mismatched bases is correct by identifying the parental DNA strand which is methylated.

In **excision repair**, single-strand nicks are created in the DNA adjacent to a damaged nucleotide by a repair enzyme. A nuclease removes the damaged base and adjacent bases. The gap is then filled with new DNA by a DNA polymerase and closed by DNA ligase.

Direct repair involves the reversal of structural alterations that occur in nucleotides. Photoreactivation is an important example which involves repair of thymine dimers by enzymes called DNA photolyases which break the links formed on dimerization.

Figures: 5.17–5.23

DNA RECOMBINATION

The rearrangement of genetic information within and among DNA molecules fall into at least three classes. **Homologous genetic recombination** involves genetic exchanges between any two DNA molecules of nearby identical sequence and it creates genetic diversity. A Holliday intermediate, a crossover between the strands of two homologous DNAs, is formed during the process. The Rec A protein promotes both formation of Holliday intermediates and branch migration to extend heteroduplex DNA. **Site specific recombination** involves the exchanges only at a particular DNA sequence. Recombinases cleave the DNA at specific points and ligate the strands to new partners. **DNA transposition** involves a short segment of DNA, called transposon, with capacity to move from one location to another within or between chromosomes.

Figures: 5.24–5.29

RNA TRANSCRIPTION

Transcription is catalysed by DNA-dependent RNA polymerase synthesizing RNA complementary to the template strand of duplex DNA. The first committed step in transcription is **binding of RNA polymerase** to DNA site called promoter and **initiation of transcript** synthesis which are carefully regulated. Eukaryotic cells have three different types of RNA polymerases. Binding of RNA polymerase II to its promoters requires the activity of **transcription factors**. The next step is **elongation** of RNA transcript followed by **termination** of transcription at terminators which may be rho (ρ) dependent or independent.

Figures: 5.30–5.32

RNA PROCESSING

Eukaryotic **messenger RNAs** are made from longer primary RNA transcripts containing noncoding regions called **introns**, which are removed by splicing. **Ribosomal RNAs** and **transfer RNAs** are also being derived from longer precursor RNAs that are trimmed by nucleases. Some bases are modified enzymatically during the maturation process. The splicing of group I introns found in rRNAs requires a guanosine factor while in group II introns adenosine acts as nucleophile. The splicing of group III introns in nuclear mRNA precursors occurs with the aid of RNA-protein complexes called SnRNPs, assembled into elaborate structures called **spliceosomes**. mRNAs are also modified by addition of a 7-methyl guanosine residue at the 5′ end called **cap** followed by cleavage and polyadenylation at the 3′ end to form a long **Poly (A) tail**.

Figures: 5.33-5.46

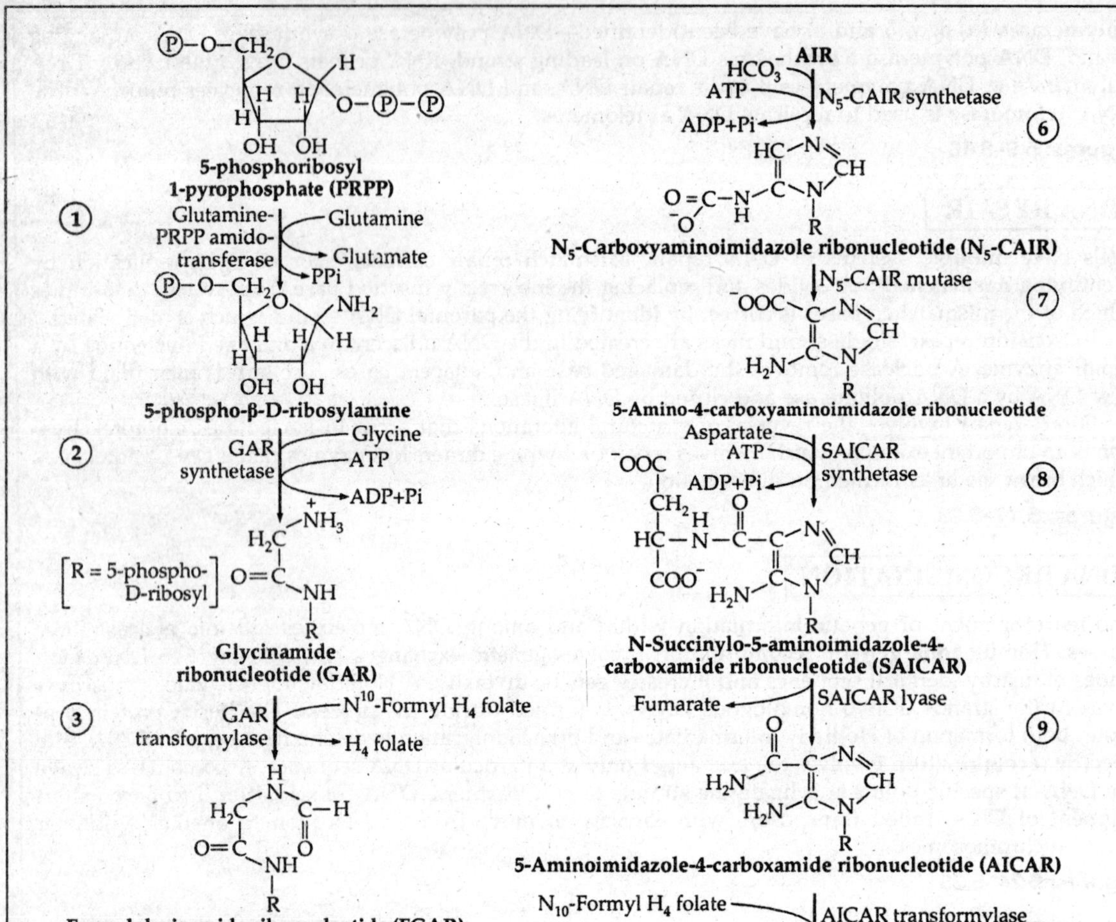

Fig. 5.1: Synthesis of purine nucleotides: Construction of purine ring of inosinate (IMP).

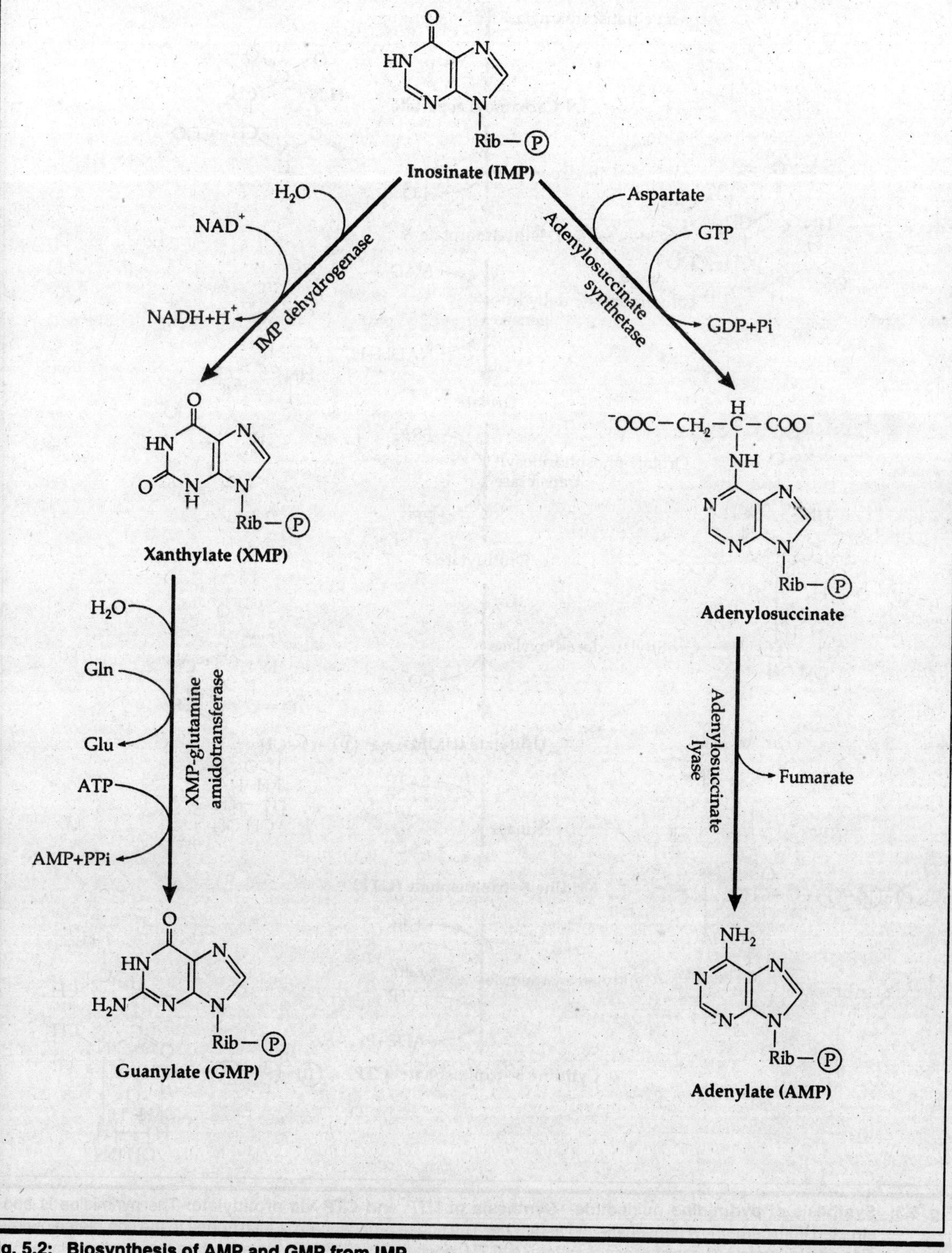

Fig. 5.2: Biosynthesis of AMP and GMP from IMP.

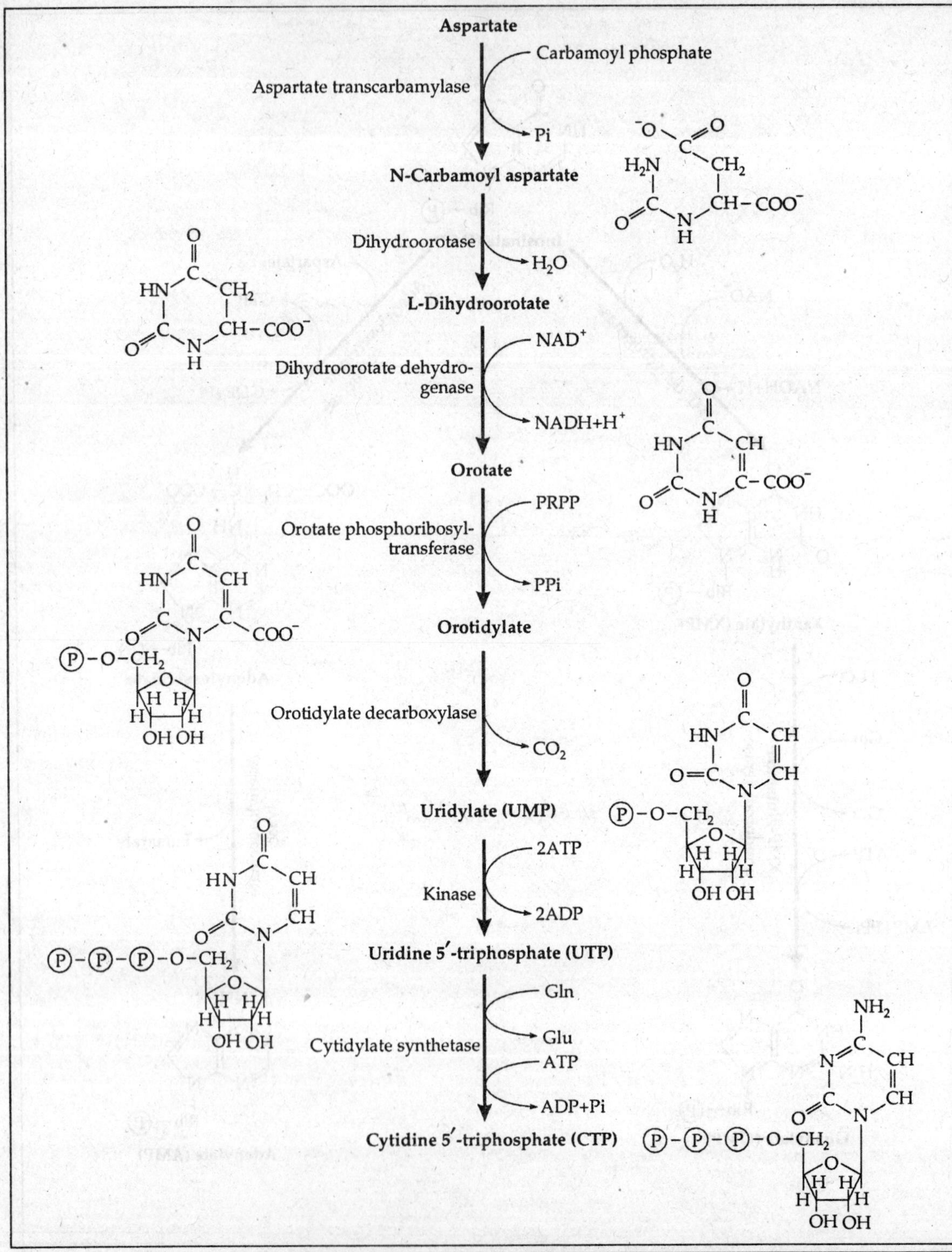

Fig. 5.3: Synthesis of pyrimidine nucleotide—Synthesis of UTP and CTP via orotidylate: The pyrimidine is constructed from carbamoyl phosphate and aspartate. The ribose 5-phosphate is then added to the completed pyrimidine ring by orotate phosphoribosyltransferase. CTP regulates the first enzyme aspartate transcarbamylase.

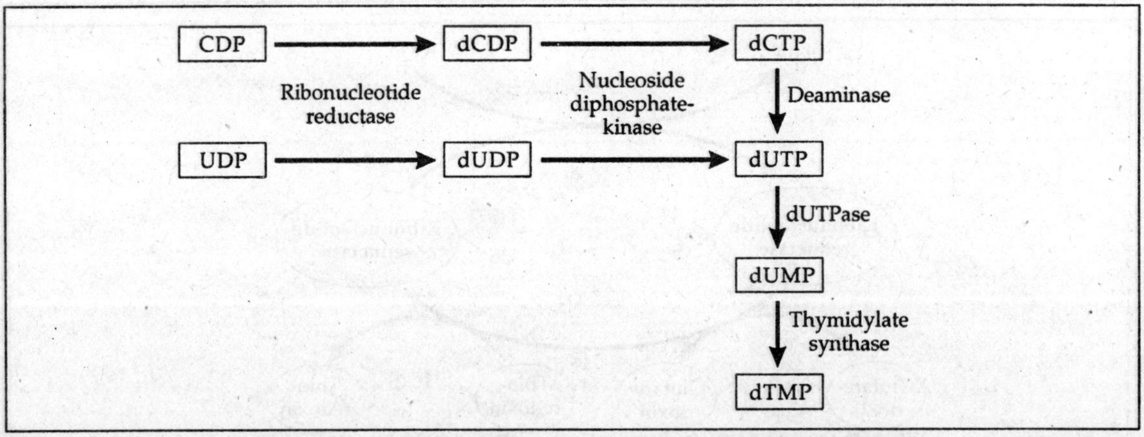

Fig. 5.4: Synthesis of thymidylate (dTMP): The reaction is catalysed by ribonucleotide reductase.

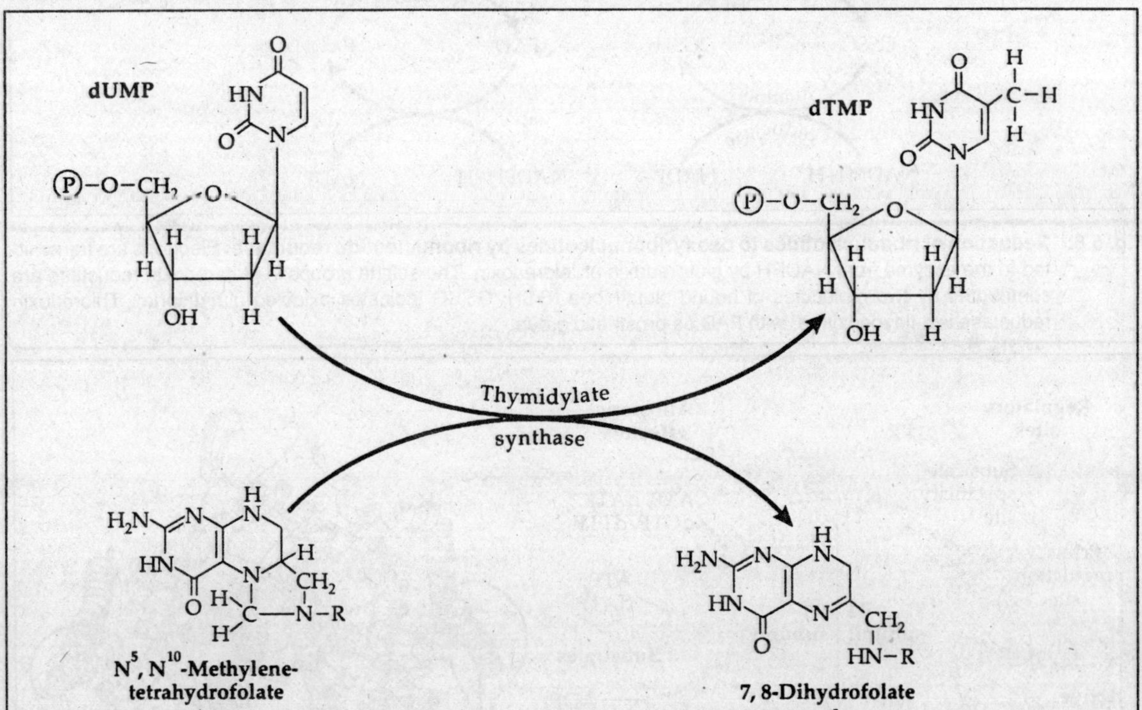

Fig. 5.5: Conversion of dUMP to dTMP by thymidylate synthase and dihydrofolate reductase: Serine hydroxymethyltransferase is required for regeneration of the N^5, N^{10}-methylene form of tetrahydrofolate. In the synthesis of dTMP, all three hydrogens of the added methyl group are derived from N^5, N^{10}-methylene-tetrahydrofolate.

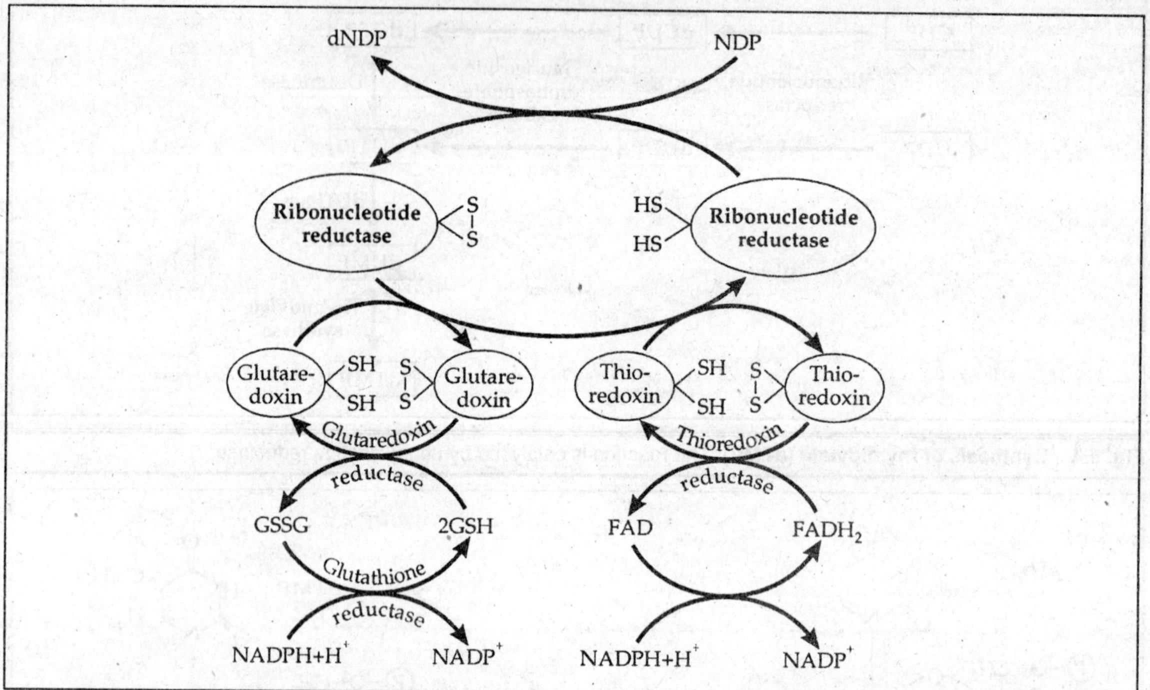

Fig. 5.6: Reduction of ribonucleotides to deoxyribonucleotides by ribonucleotide reductase: Electrons are transmitted to the enzyme from NADPH by glutaredoxin or thioredoxin. The sulfide groups in glutaredoxin reductase are contributed by two molecules of bound glutathione (GSH; GSSG indicates oxidized glutathione). Thioredoxin reductase is a flavoenzyme, with FAD as prosthetic group.

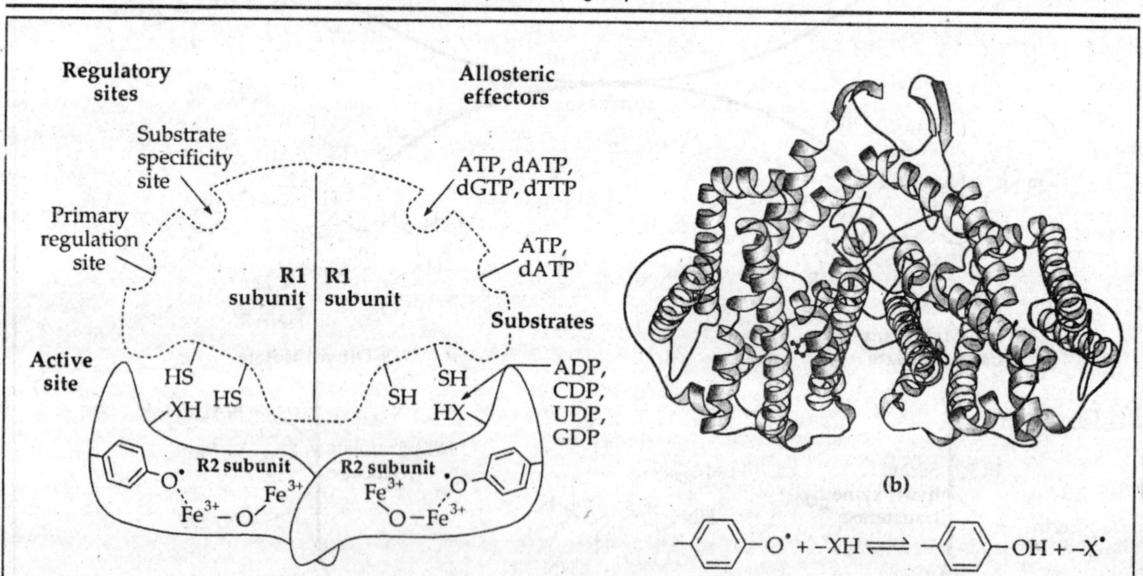

Fig. 5.7: Ribonucleotide reductase: (a) Subunit structure—the functions of two regulatory sites are explained. The active site contains two thiols and a(–XH) group that can be converted to an active-site radical (–X•). **(b)** The R2 subunits of *E. coli* ribonucleotide reductase. **(c)** The tyrosyl radical functions to generate the active-site radical (–X•).

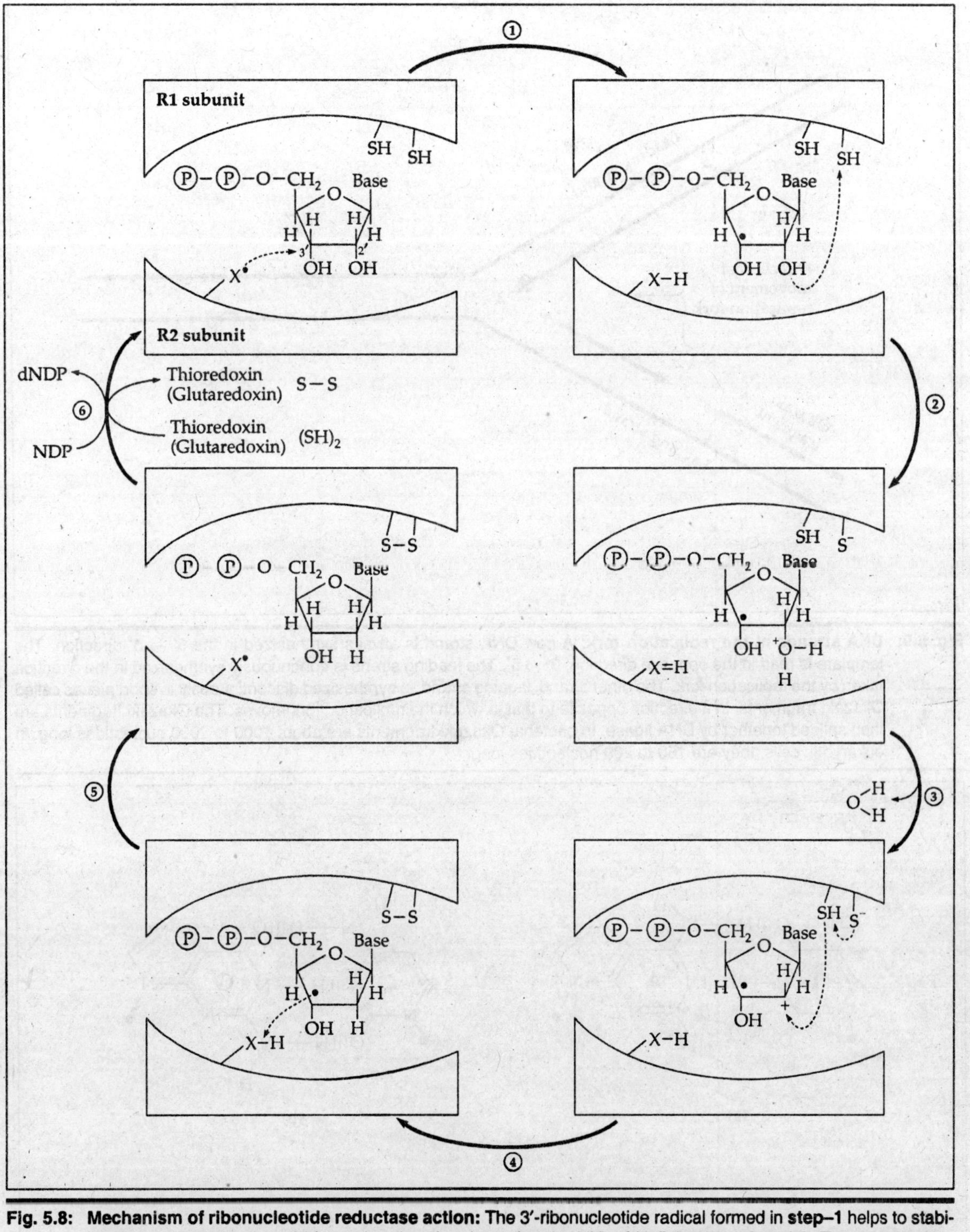

Fig. 5.8: Mechanism of ribonucleotide reductase action: The 3′-ribonucleotide radical formed in **step–1** helps to stabilize the cation subsequently formed at the 2′ carbon after the loss of H_2O in **steps 2 and 3**. Two one-electron transfers accompanied by oxidation of the dithiol reduce the radical cation and regenerate the 3′-ribonucleotide radical in **step-4**. **Step-5** is the reverse of step-1, regenerating the tyrosyl radical and forming the deoxy product. The oxidized dithiol is reduced to complete the cycle in **step-6**.

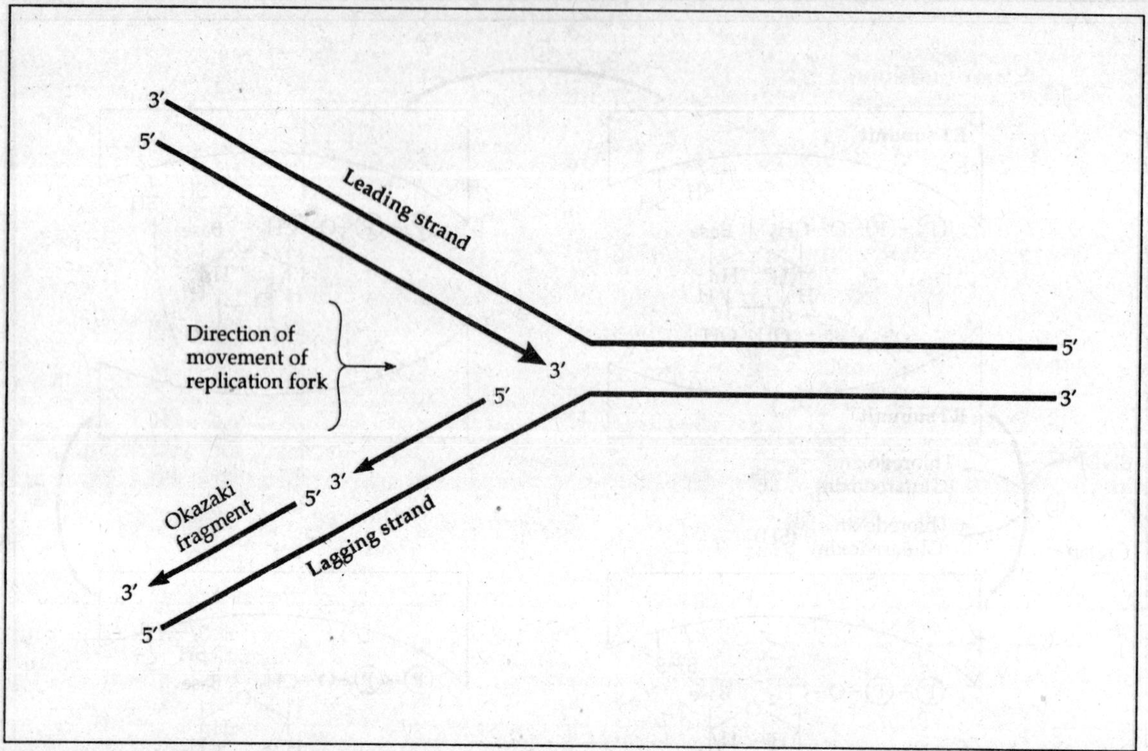

Fig. 5.9: DNA strands at the replication fork: A new DNA strand is always synthesized in the 5′ → 3′ direction. The template is read in the opposite direction, 3′ → 5′. The leading strand is continuously synthesized in the direction taken by the replication fork. The other strand, lagging strand, is synthesized discontinuously in short pieces called Okazaki fragments, in a direction opposite to that in which the replication fork moves. The Okazaki fragments are then spliced together by DNA ligase. In bacteria, Okazaki fragments are about 1000 to 2000 nucleotides long. In eukaryotic cells, they are 150 to 200 nucleotides long.

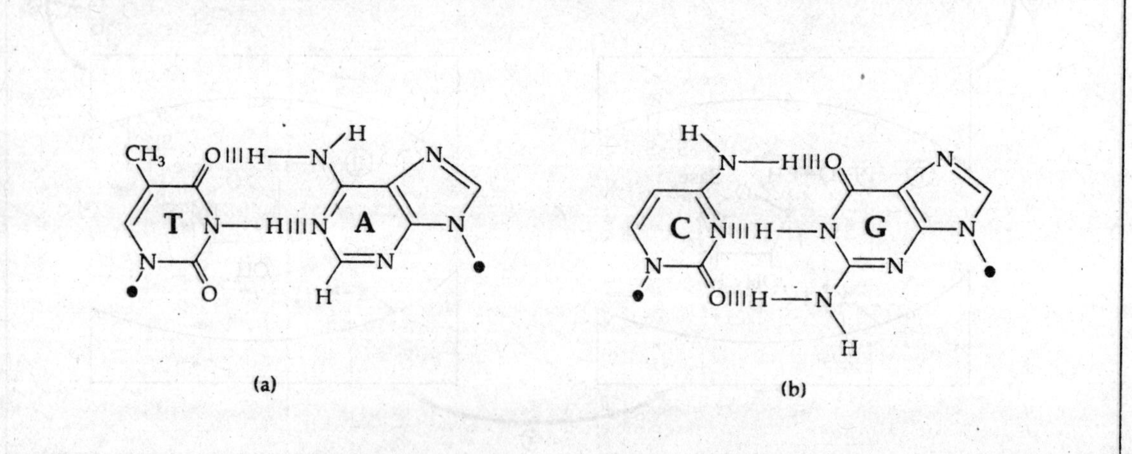

Fig. 5.10: Contribution of base pair geometry to the fidelity of DNA replication: The standard A = T and G ≡ C base pairs have very similar geometries and an active site sized to fit one will generally accommodate the other.

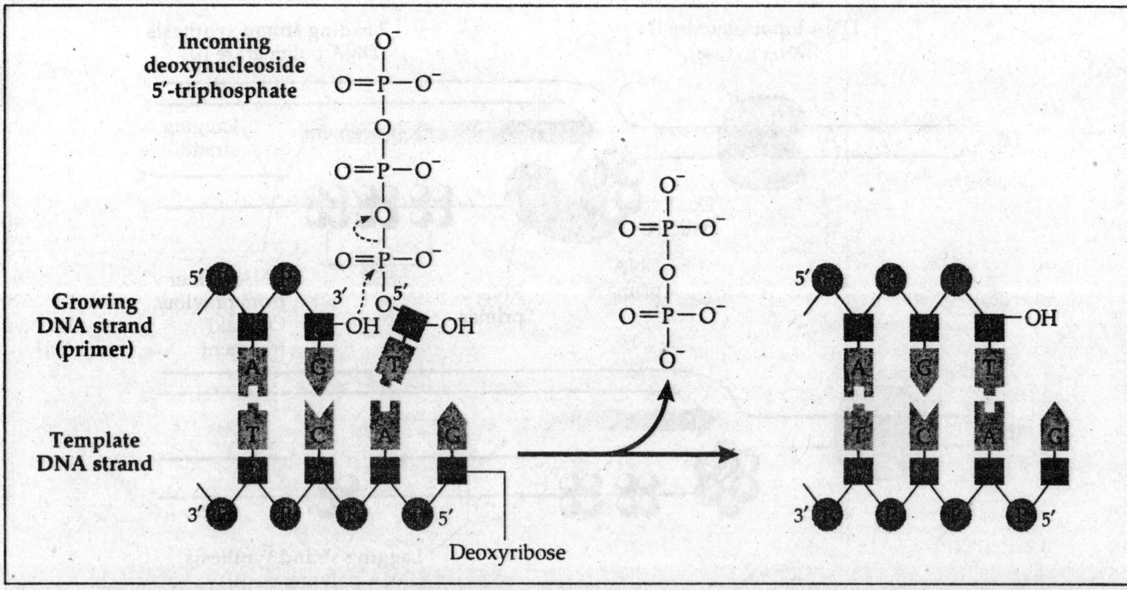

Fig. 5.11: Elongation of a DNA chain: A single unpaired strand will act as template and a primer strand will provide a free hydroxyl group at the 3′ end to which a new nucleotide unit is added. Incoming nucleotide is selected by base pairing to the appropriate nucleotide in the template strand. The reaction product has a new free 3′ OH, allowing the addition of another nucleotide.

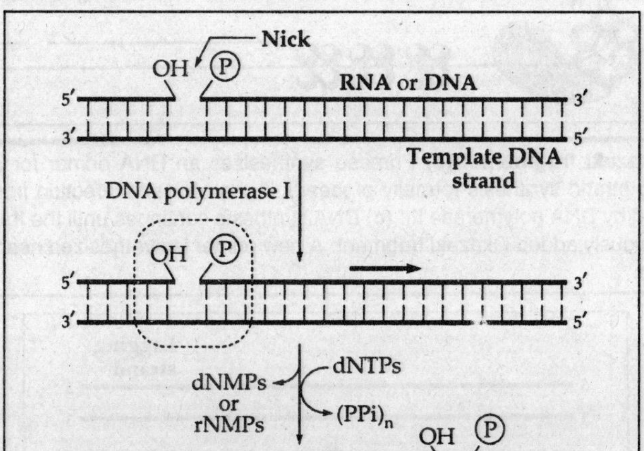

Fig. 5.12: Nick translation: An RNA or DNA strand paired to a DNA template is simultaneously degraded by the 5′ → 3′ exonuclease activity of DNA polymerase I and replaced by the polymerase activity of the same enzyme. These activities have a role in both DNA repair and the removal of RNA primers during replication. A nick (broken phosphodiester bond, leaving a free 3′ OH and a free 5′ phosphate) occurs where DNA synthesis is to start. Polymerase I extends the nontemplate DNA strand and moves the nick along the DNA, the process is called nick translation. A nick remains where DNA polymerase I dissociates, until it is sealed by another enzyme.

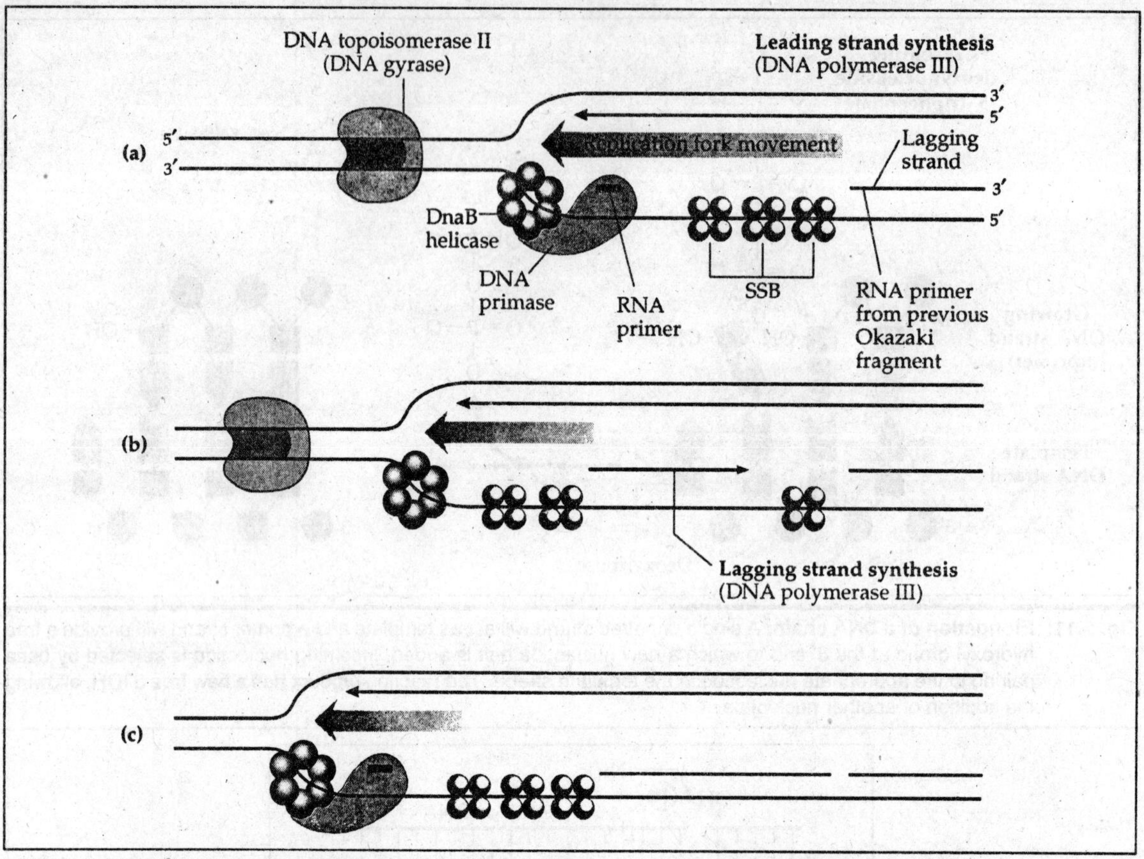

Fig. 5.13: Synthesis of Okazaki fragments: (a) Primase synthesizes an RNA primer for a new Okazaki fragment at intervals. Lagging strand synthesis formally proceeds in the opposite direction from fork movement. **(b)** Each primer is extended by DNA polymerase III. **(c)** DNA synthesis continues until the fragment extends as far as the primer of the previously added Okazaki fragment. A new primer is synthesized near the replication fork to begin the process again.

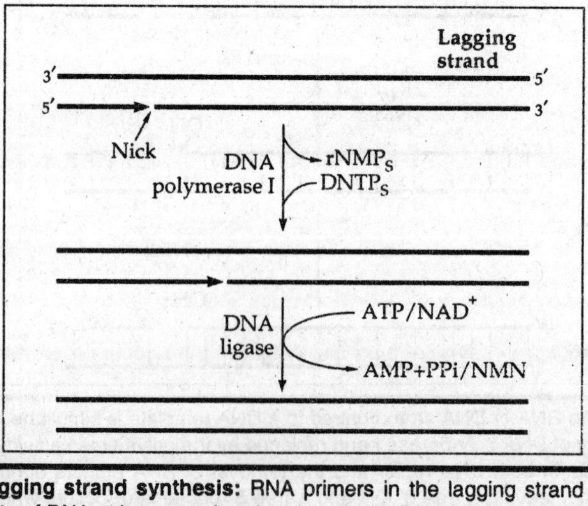

Fig. 5.14: Completion of lagging strand synthesis: RNA primers in the lagging strand are removed by the 5′ → 3′ exonuclease activity of DNA polymerase I and replaced with DNA by the same enzyme. The remaining nick is sealed by DNA ligase.

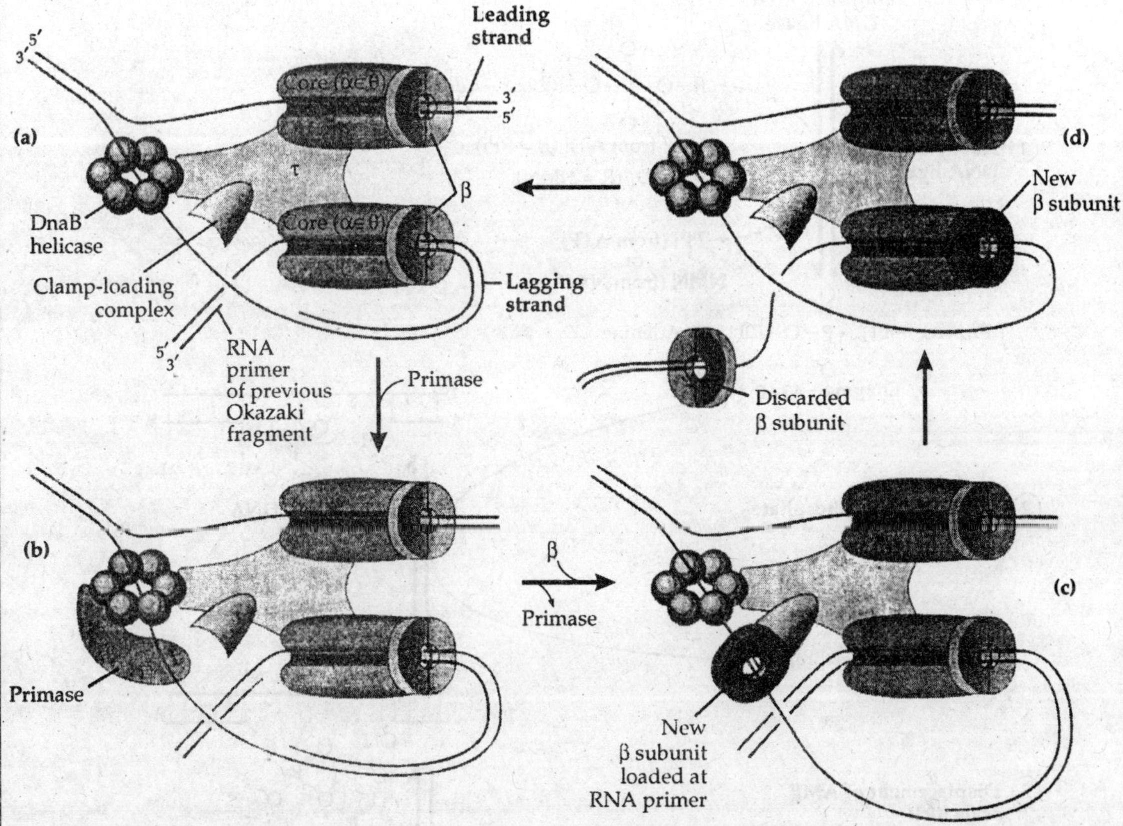

Fig. 5.15: DNA synthesis on the leading and lagging strands: Events at the replication fork are coordinated by a single DNA polymerase III dimer, in an integrated complex with DnaB helicase. The lagging strand is looped so that DNA synthesis proceeds steadily on both the leading and lagging strand templates at the same time, catalyzed by two sets of DNA polymerase III core subunits. Arrows indicate the 3′ end of two new strands and the direction of DNA synthesis. An Okazaki fragment is being synthesized on the lagging strand. **(a)** Continuous synthesis on the leading strand proceeds as DNA is unwound by the DnaB helicase. **(b)** DNA primase binds to DnaB, synthesizes a new primer, then dissociates. **(c)** The DNA polymerase III clamp-loading complex catalyzes the loading of a new β sliding clamp at the new RNA primer. Meanwhile the Okazaki fragment that was being synthesized is completed. **(d)** The core subunits of DNA polymerase III that synthesized the lagging strand release both the completed Okazaki fragment and the β sliding clamp used in its synthesis. The clamp-loading complex probably facilitates the recycling of β sliding clamps. The same core subunits then bind to the new β sliding clamp and begin synthesis of another Okazaki fragment at its primer.

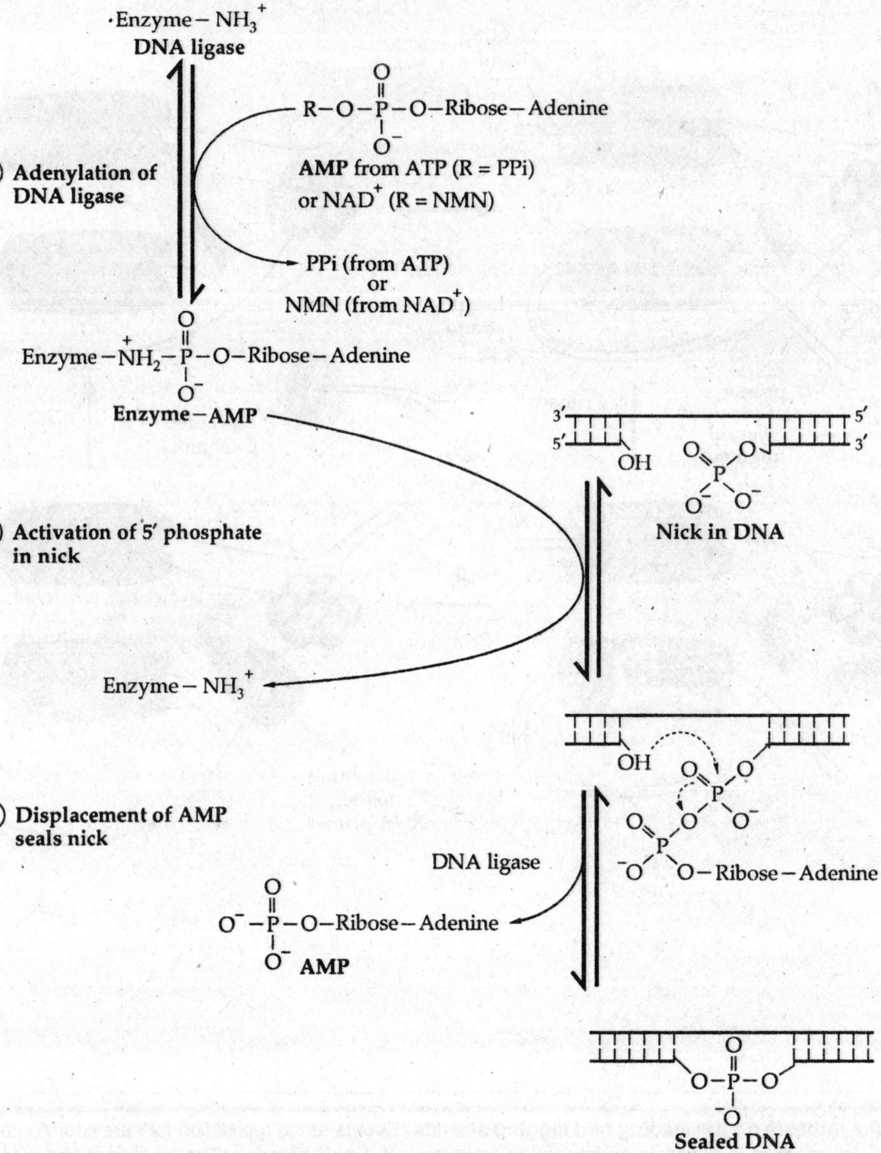

Fig. 5.16: Mechanism of DNA ligase reaction: In each step one phosphodiester bond is formed at the expense of another. **Step 1 and 2** lead to activation of the 5′ phosphate in the nick. An AMP group is transferred first to a Lys residue on the enzyme and then to the 5′ phosphate in the nick. In **step 3**, the 3′-OH group attacks this phosphate and displaces AMP, producing a phosphodiester bond to seal the nick. In the *E. coli* DNA ligase reaction, AMP is derived from NAD^+. The DNA ligases isolated from a number of viral and eukaryotic sources use ATP rather than NAD^+, and they release pyrophosphate (PPi) rather than nicotinamide mononucleotide (NMN) in step 1.

ENZYMES / PROTEINS	TYPE OF DAMAGE
I. Mismatch repair	
Dam methylase	Mismatches
MutH, MutL, MutS proteins	
DNA helicase II	
SSB	
DNA polymerase III	
Exonuclease I	
Exonuclease VII	
RecJ nuclease	
DNA ligase	
Exonuclease X	
II. Base-excision repair	
DNA glycosylases	Abnormal bases (uracil, xanthine, hypoxanthine), alkylated bases, pyrimidine dimers
AP endonucleases	
DNA polymerase I	
DNA ligase	
III. Nucleotide-excision repair	
ABC excinuclease	DNA lesions that cause large structural changes (e.g., pyrimidine dimers)
DNA polymerase I	
DNA ligase	
IV. Direct repair	
DNA photolyases	Pyrimidine dimers
O^6-Methylguanine-DNA methyltransferase	O^6-Methylguanine

Fig. 5.17: Types of DNA repair systems in *E. coli*.

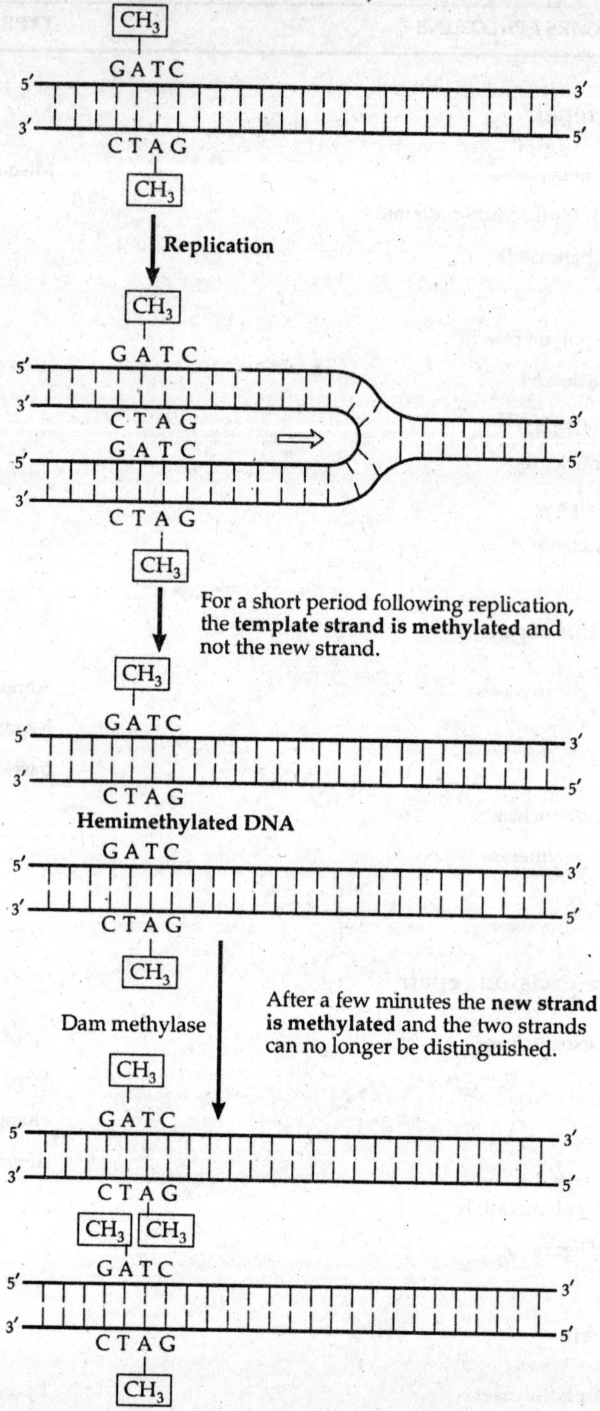

Fig. 5.18: Methylation and mismatch repair: Methylation of DNA strands can serve to distinguish parental template strands from newly synthesized strands in *E. coli* DNA, a function that is critical to mismatch repair. The methylation occurs at the N^6 of adenines in (5′) GATC sequences. This sequence is a palindrome and thus represent in opposite orientations on the two strands.

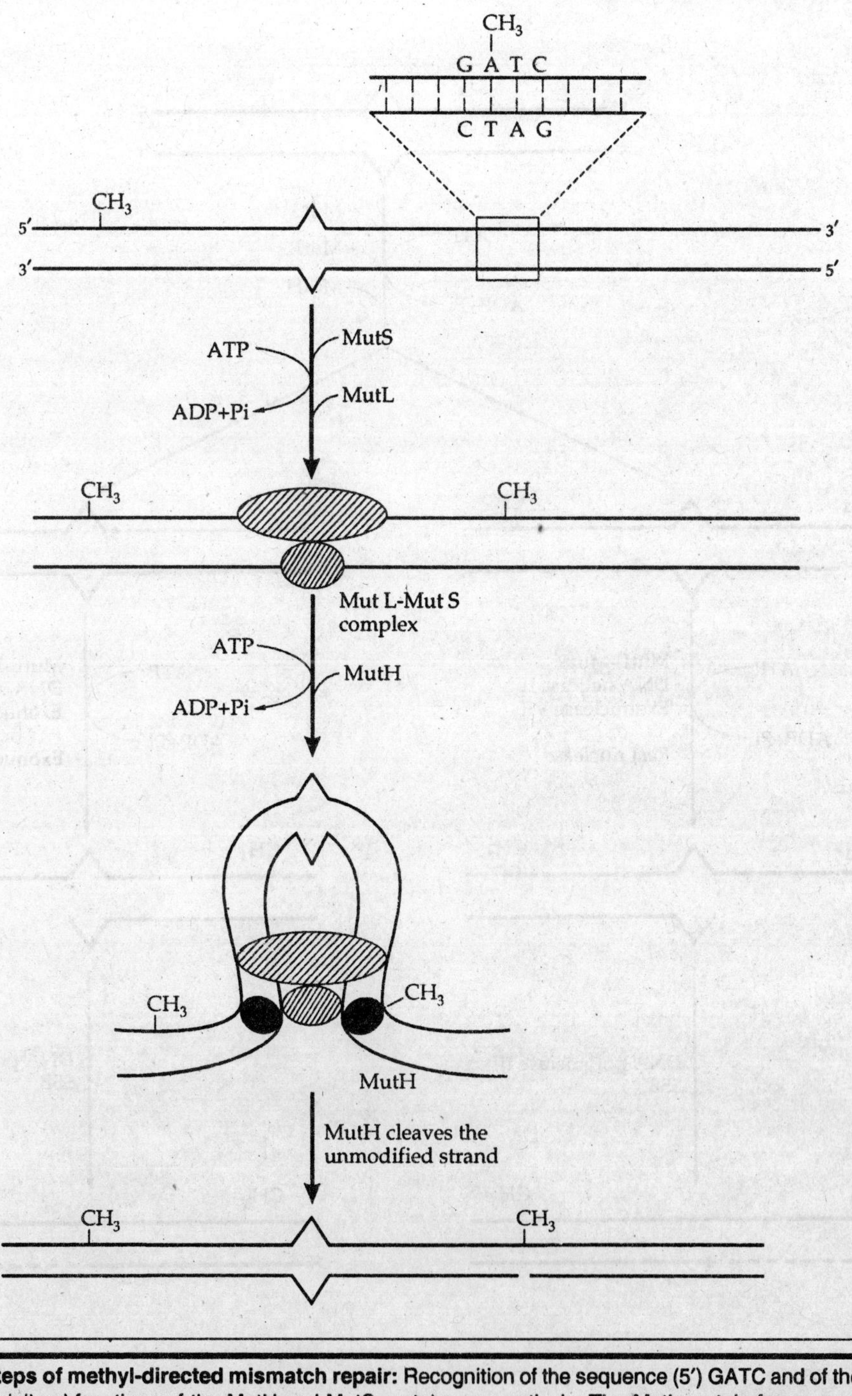

Fig. 5.19: Early steps of methyl-directed mismatch repair: Recognition of the sequence (5′) GATC and of the mismatch are specialised functions of the MutH and MutS proteins respectively. The MutL protein forms a complex with MutS at the mismatch. DNA is threaded through this complex such that the complex moves simultaneously in both directions along the DNA until it encounters a MutH protein bound at a hemimethylated GATC sequence. MutH cleaves the unmethylated strand on the 5′ side of the G in the GATC sequence. A complex consisting of DNA helicase II and one of several exonucleases then degrades the unmethylated DNA strand from that point toward the mismatch.

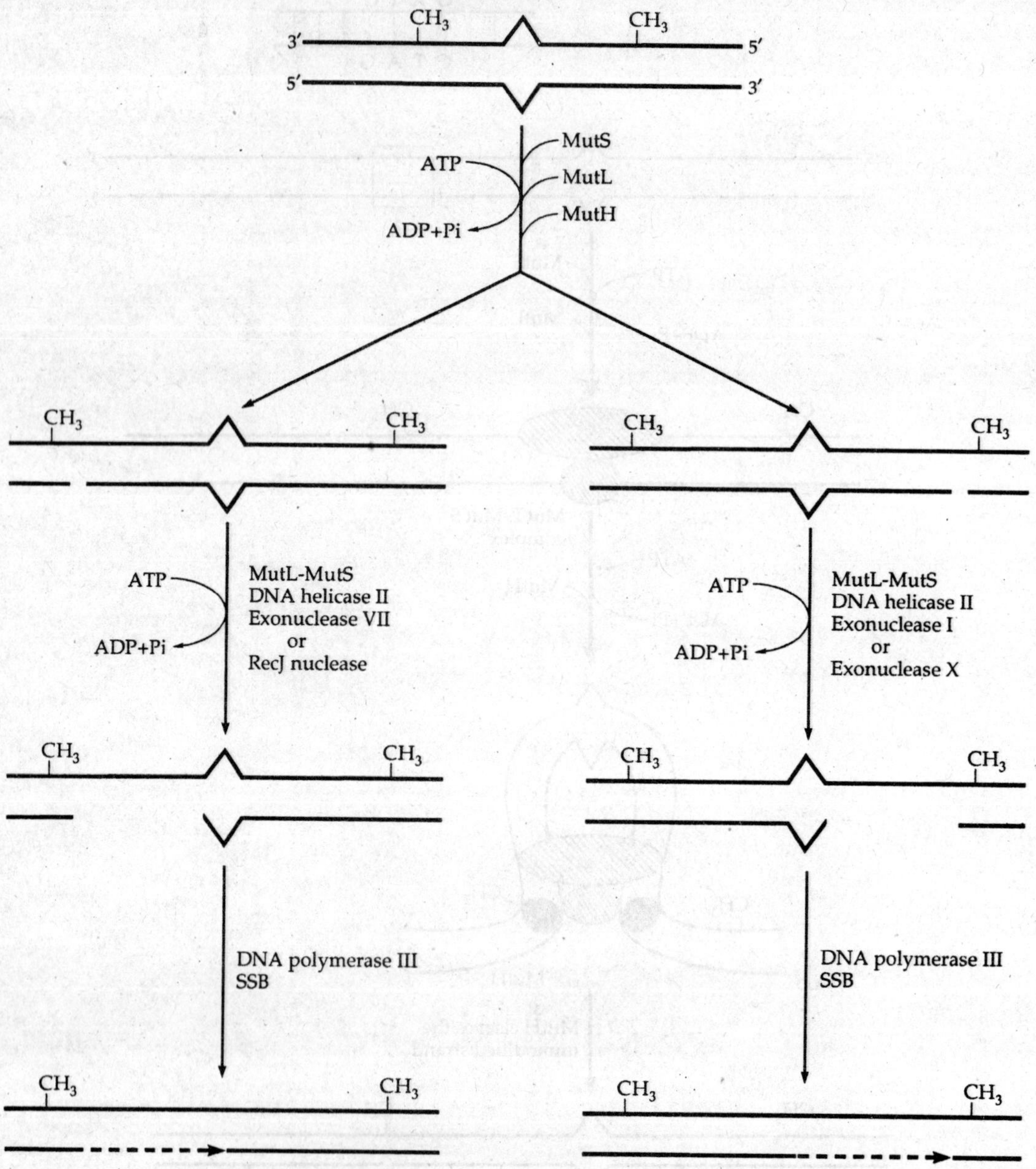

Fig. 5.20: Completing methyl-directed mismatch repair: The combined action of DNA helicase II, SSB, and one of four different exonucleases removes a segment of the new strand between the MutH cleavage site and a point just beyond the mismatch. The exonuclease used depends on the location of the cleavage site relative to the mismatch. The resulting gap is filled in by DNA polymerase III, and the nick is sealed by DNA ligase.

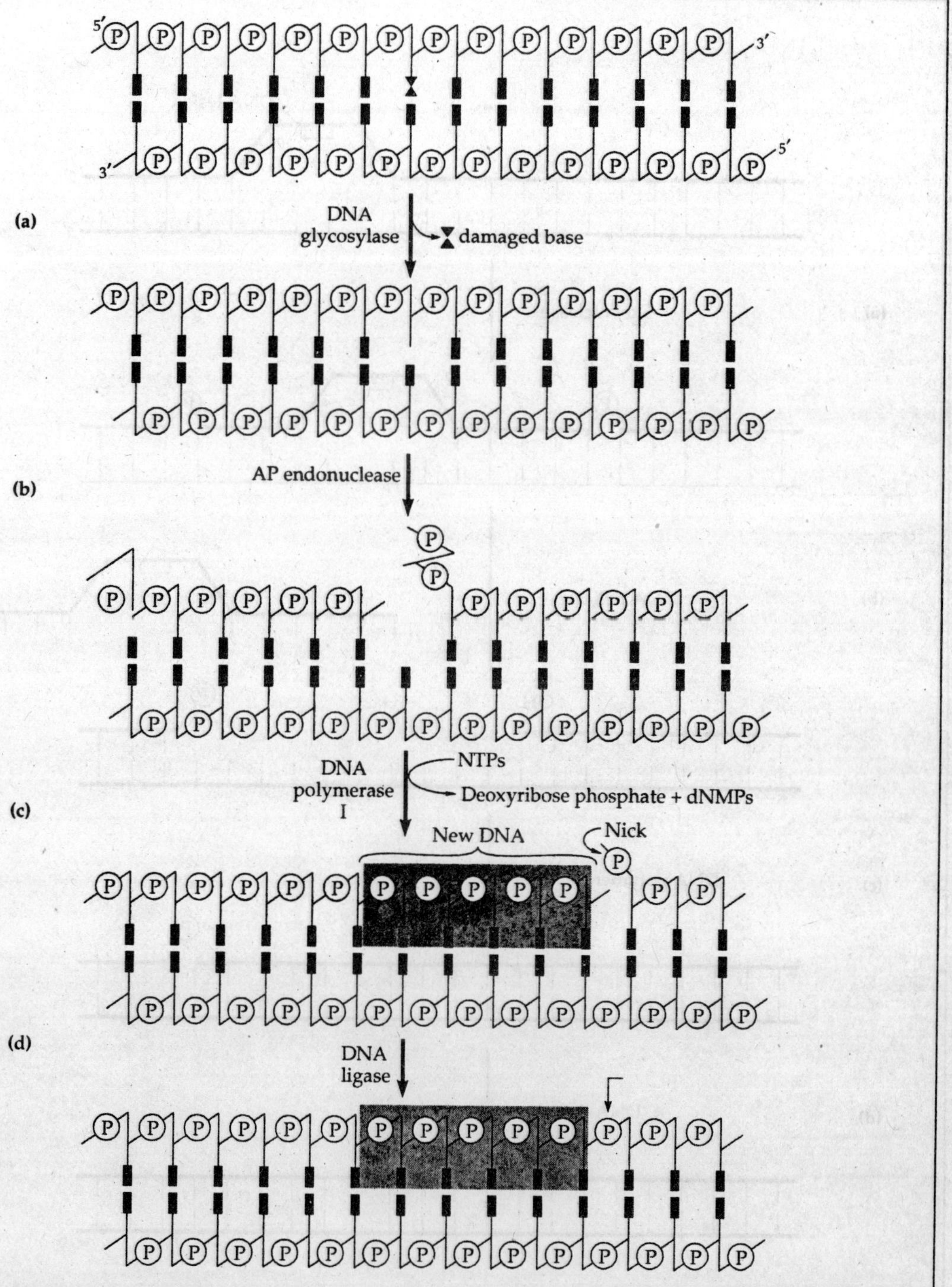

Fig. 5.21: DNA repair by the base-excision repair pathway: (a) DNA glycosylase recognizes a damaged base and cleaves between the base and deoxyribose in the backbone. **(b)** AP endonuclease cleaves the phosphodiester backbone near the AP site. **(c)** DNA polymerase I initiates repair synthesis from the free 3′ OH at the nick, removing a portion of the damaged strand (with its 5′ → 3′ exonuclease activity) and replacing it with undamaged DNA. **(d)** The nick remaining after DNA polymerase I has dissociated is sealed by DNA ligase.

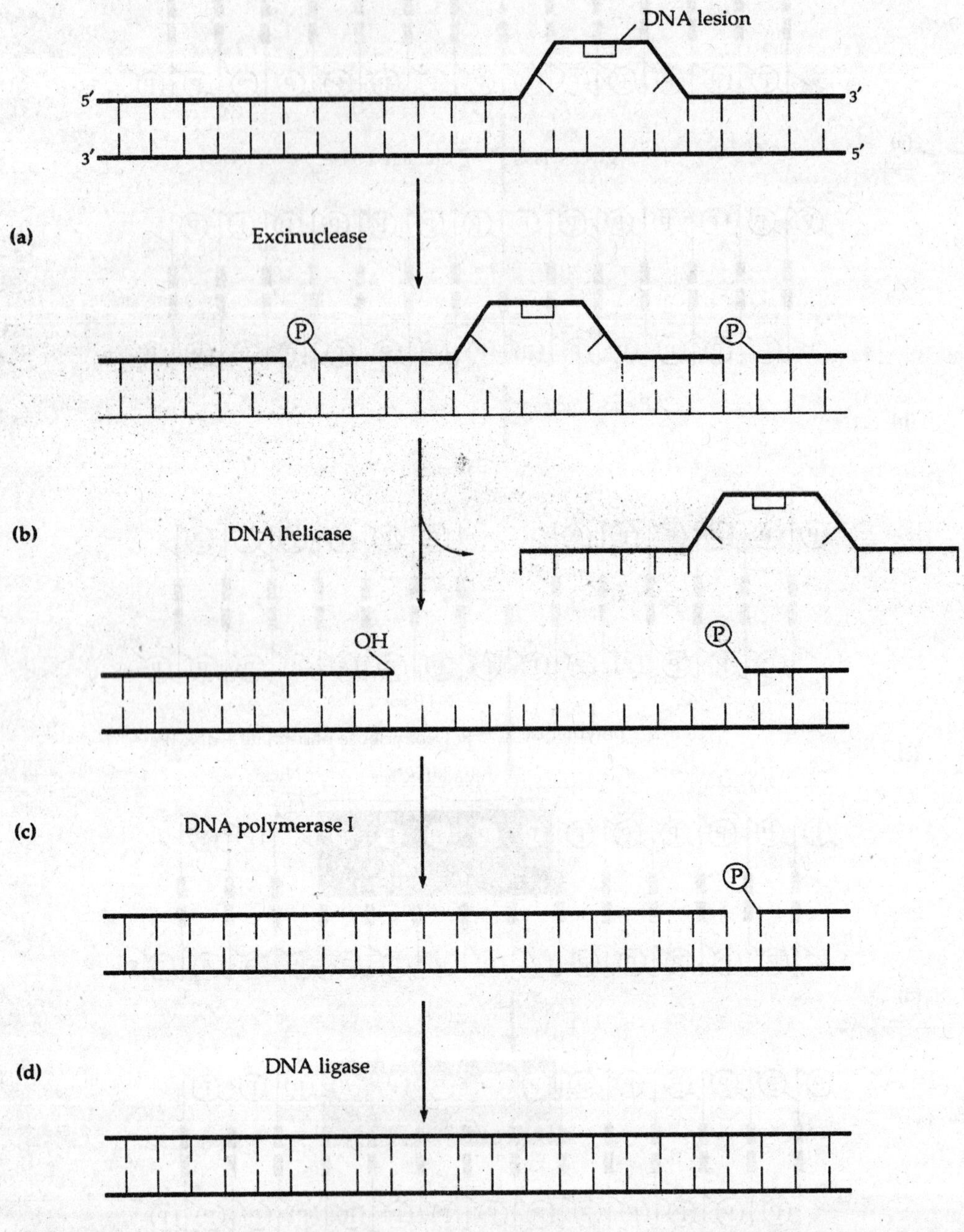

Fig. 5.22: Nucleotide-excision repair in *E. coli*: (a) Excinuclease binds to DNA at the site of a bulky lesion. **(b) Excinuclease cleaves the damaged DNA strand on either side of the lesion and the DNA segment is removed with the help of helicase. (c) The gap is filled in by DNA polymerase. (d) The remaining nick is sealed with DNA ligase.**

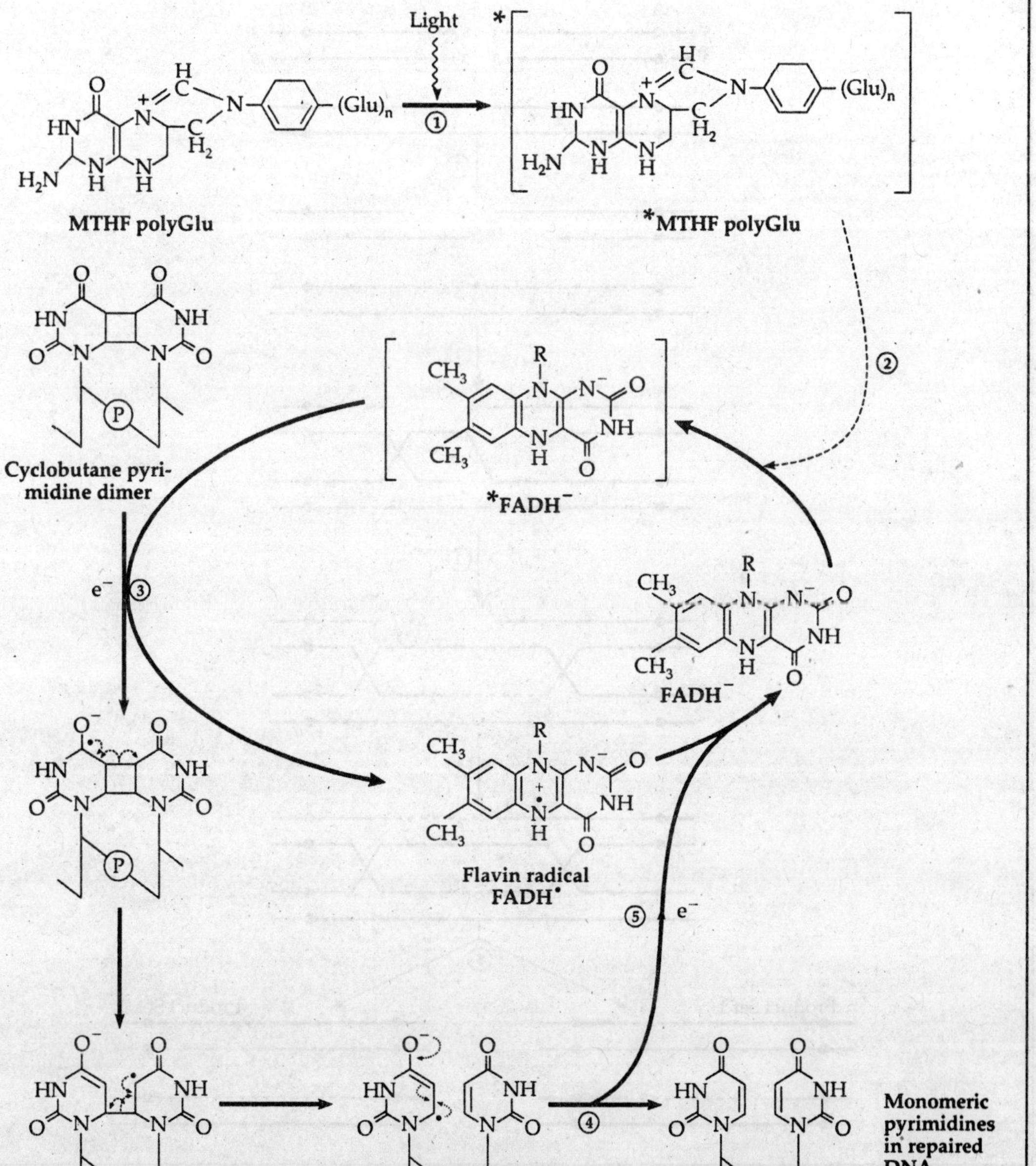

Fig. 5.23: Repair of pyrimidine dimers with photolyase: Energy derived from absorbed light is used to reverse the photoreaction that caused the lesion. The two chromophores in photolyase, N^5-N^{10} methenyltetrahydrofolyl-polyglutamate (MTHF polyGlu) and $FADH^-$, perform complementary functions. On binding of photolyase to a pyrimidine dimer, repair proceeds as follows: **(1)** A blue light photon is absorbed by MTHF polyGlu which functions as photoantenna. **(2)** The excitation energy is transferred to $FADH^-$ in the active site of the enzyme. **(3)** The excited flavin (*$FADH^-$) donates an electron to the pyrimidine dimer to generate an unstable dimer radical. **(4)** Electronic rearrangement restores the monomeric pyrimidines. **(5)** The electron is transferred back to the flavin radical to regenerate $FADH^-$.

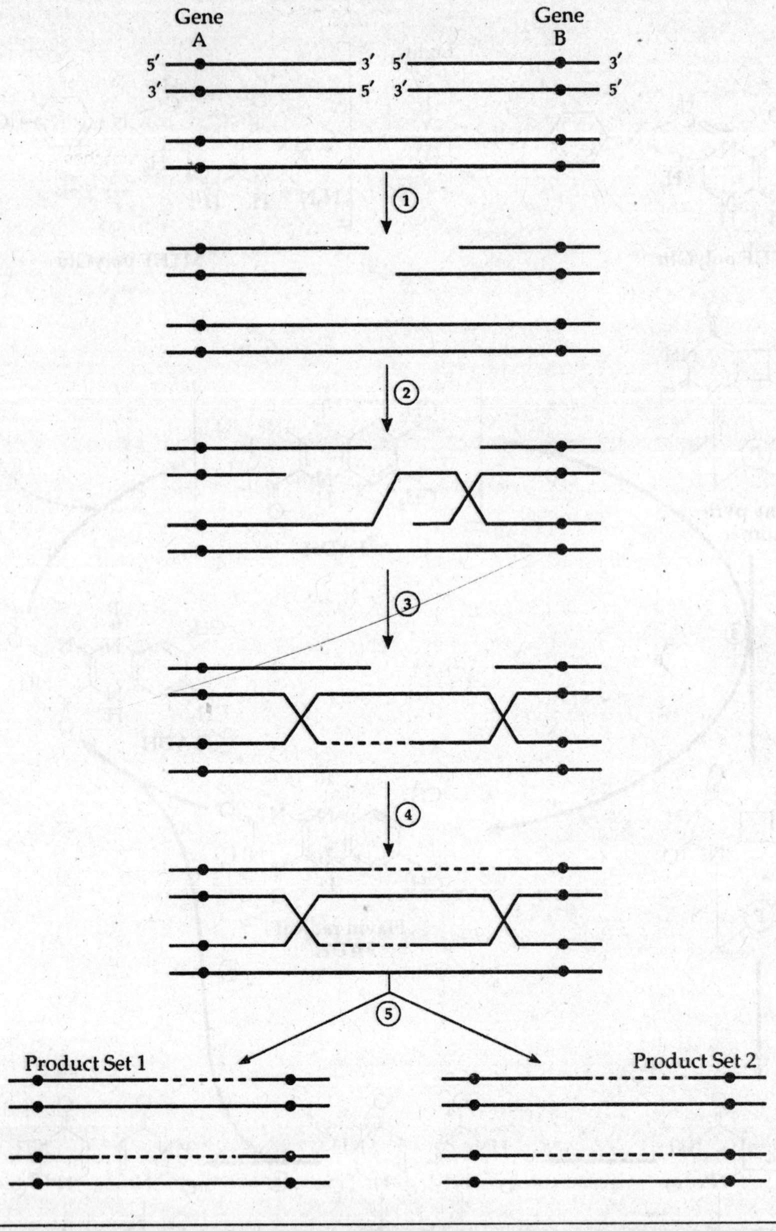

Fig. 5.24: Recombination during meiosis: Double-strand break repair model for homologous genetic recombination. The two homologous chromosomes involved in this recombination event have similar sequences. Each of the two genes shown has different alleles on the two chromosomes.

(1) A double-strand break in one of two homologs is converted to a double-strand gap by the action of exonucleases. Strands with 3′ ends are degraded less than those with 5′ ends, producing 3′ single-strand extensions.

(2) An exposed 3′ end pairs with its complement in the intact homolog. The other strand of the duplex is displaced.

(3) The invading 3′ end is extended by DNA polymerase plus branch migration, eventually generating a DNA molecule with two cross-overs called Holliday intermediates.

(4) Further DNA replication replaces the DNA missing from the site of the original double-strand break.

(5) Cleavage of the Holliday intermediates by specialised nucleases generates either of the two recombination products. In product set 2, the DNA on either side of the region undergoing repair is recombined.

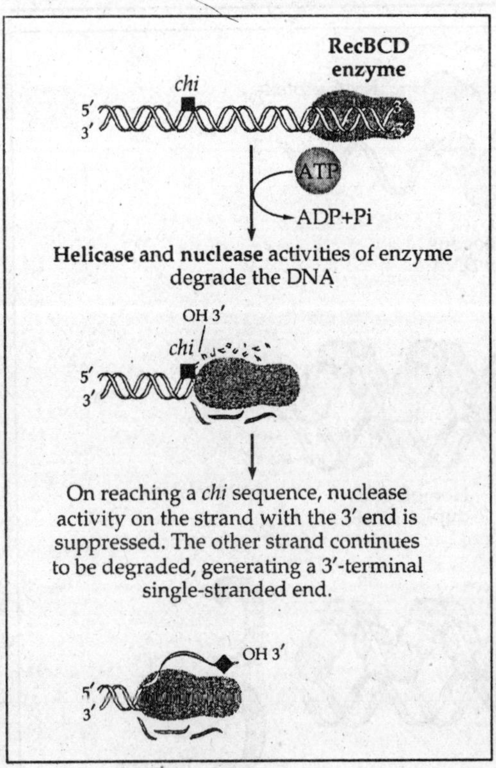

Fig. 5.25: Helicase and nuclease activities of the RecBCD enzyme: Entering at a double-stranded end, RecBCD unwinds and degrades the DNA until it encounters a *chi* sequence. The interaction with *chi* alters the activity of RecBCD so that it generates a single-stranded DNA with a 3′ end, suitable for subsequent steps in recombination. Movement of the enzyme requires ATP hydrolysis. This enzyme is believed to help initiate homologous genetic recombination. It is also involved in the repair of double-strand breaks at collapsed replication forks.

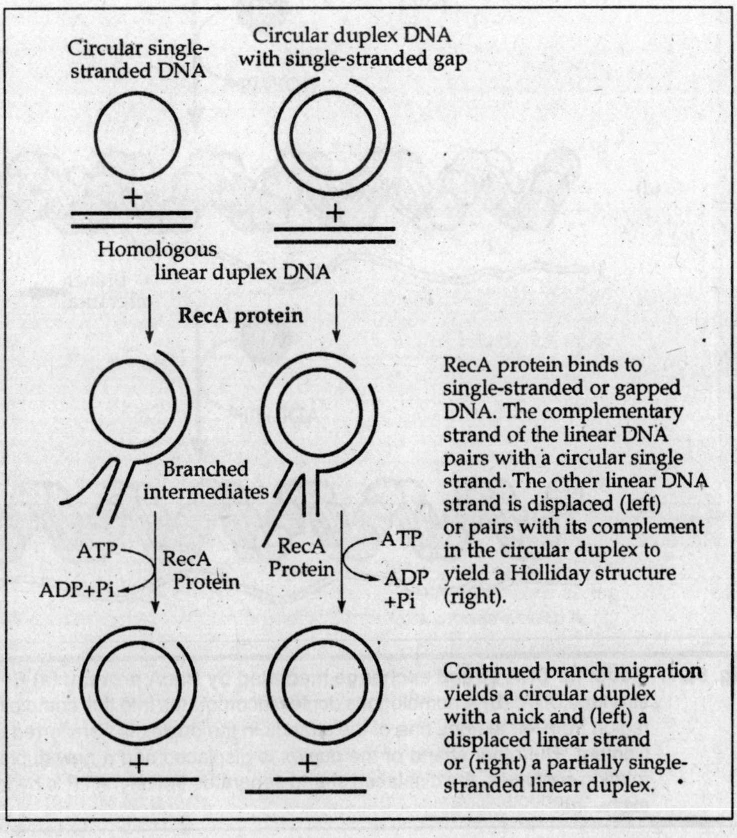

Fig. 5.26: DNA strand-exchange reactions promoted by RecA protein *in vitro*: Strand exchange involves the separation of one strand of a duplex DNA from its complement and transfer of the strand to an alternative complementary strand to form a heteroduplex DNA. The transfer forms a branched intermediate. Formation of the final product depends on branch migration, which is facilitated by RecA. The reaction can involve three strands (left) or a reciprocal exchange between two homologous duplexes—four strands in all (right). When four strands are involved, the branched intermediate that results is a Holliday structure. RecA protein promotes the branch migration phases of these reactions using energy derived from ATP hydrolysis.

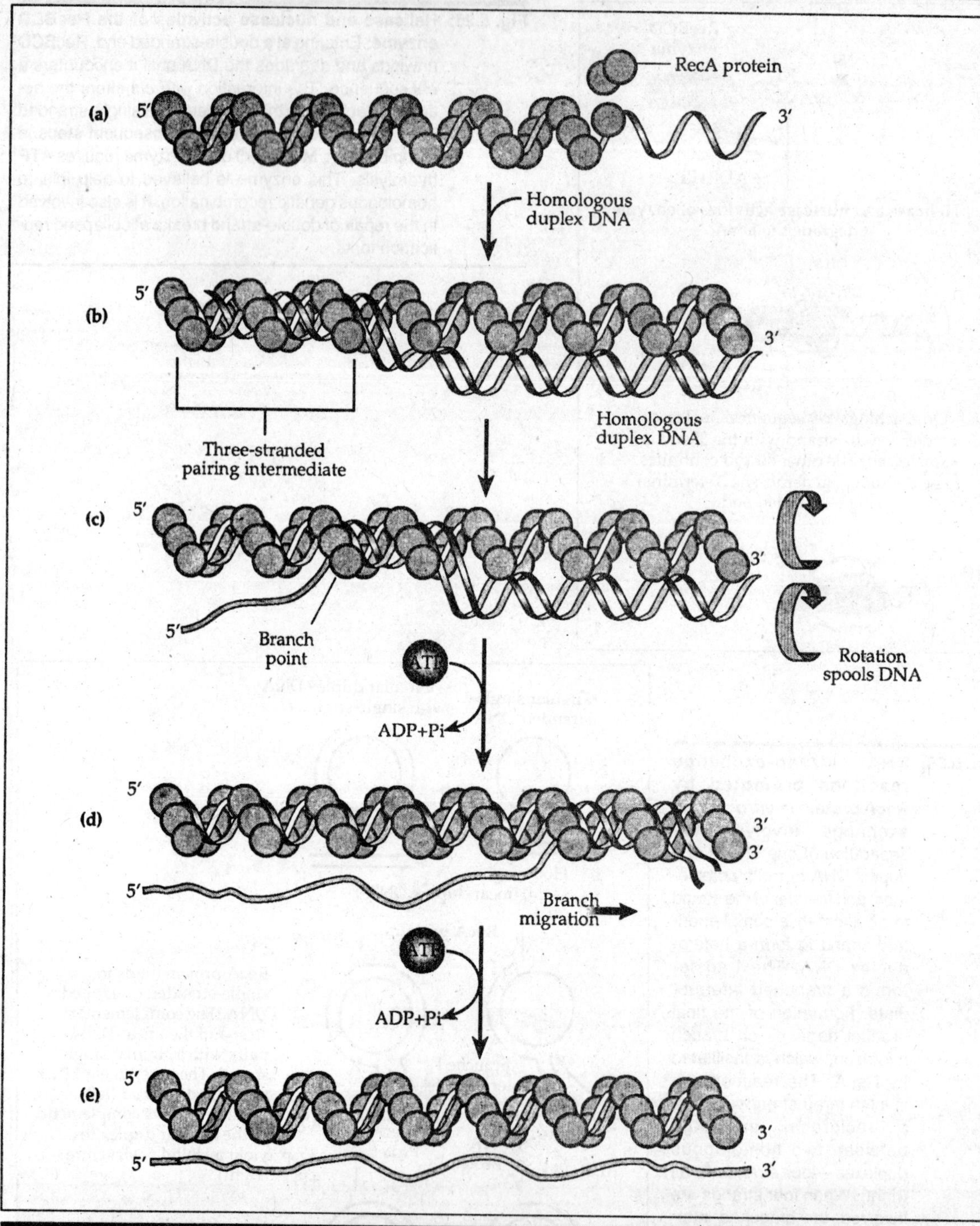

Fig. 5.27: Model for DNA strand exchange mediated by RecA protein: **(a)** RecA protein forms a filament on the single-stranded DNA. **(b)** A homologous duplex incorporates into this complex. **(c)** As spooling shifts the three-stranded region from left to right, one of the strands in the duplex is transferred to the single strand originally bound in the filament. The other strand of the duplex is displaced and a new duplex is formed within the filament. **(d, e)** As rotation continues, the displaced strand separates entirely. ATP is hydrolyzed by RecA protein during this strand exchange.

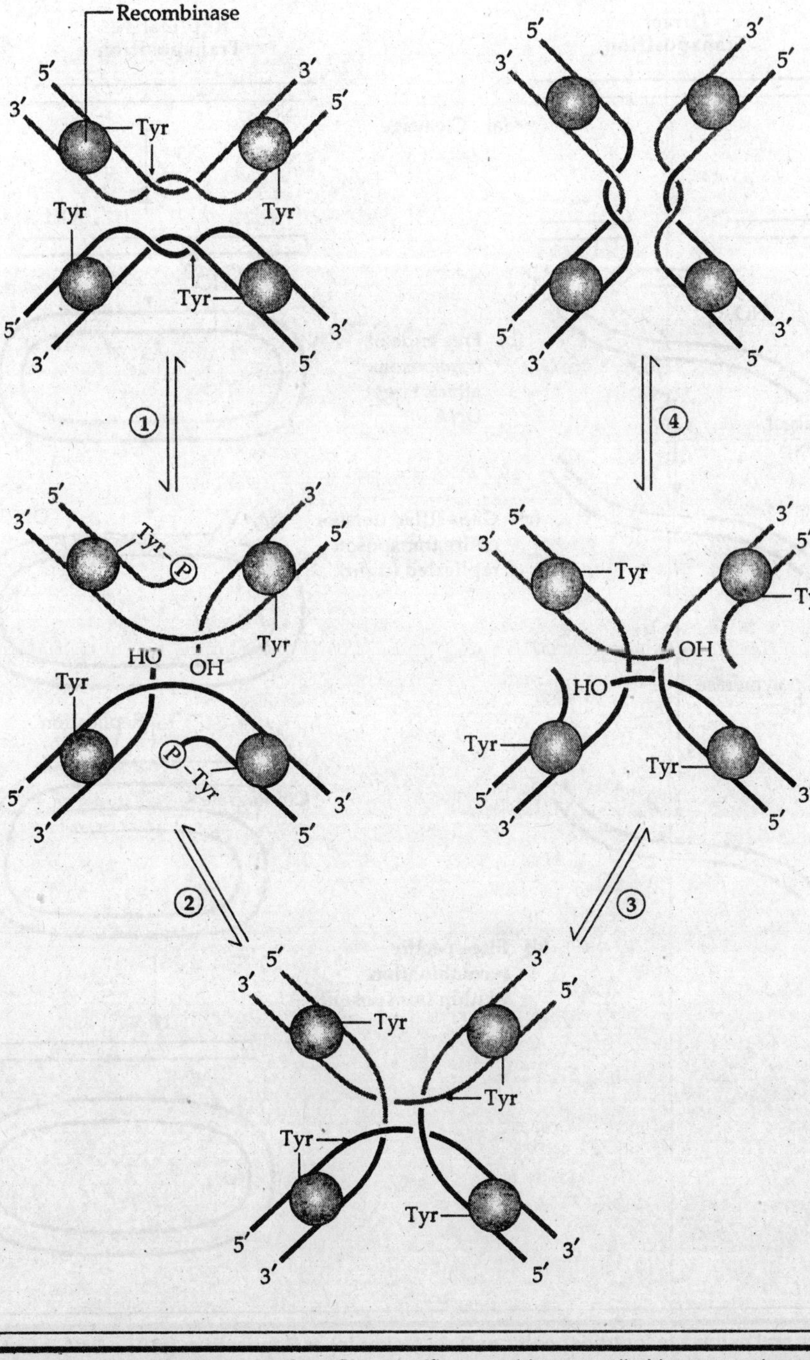

Fig. 5.28: A site-specific recombination reaction: Site-specific recombinases called integrase-class recombinases involved in the reaction carried out within a tetramer of identical subunits. Recombinase subunits bind to specific sequence called recombination site. **(1)** One strand in each DNA is cleaved at particular points within the sequence. The nucleophile is the OH group of an active-site Tyr residue and the product is a covalent phosphotyrosine link between protein and DNA. **(2)** The cleaved strands join to new partners, producing a Holliday intermediate. **(3)** and **(4)** Completion of the reaction by a process similar to the first two steps. The original sequence of the recombination site is regenerated after recombining the DNA flanking the site.

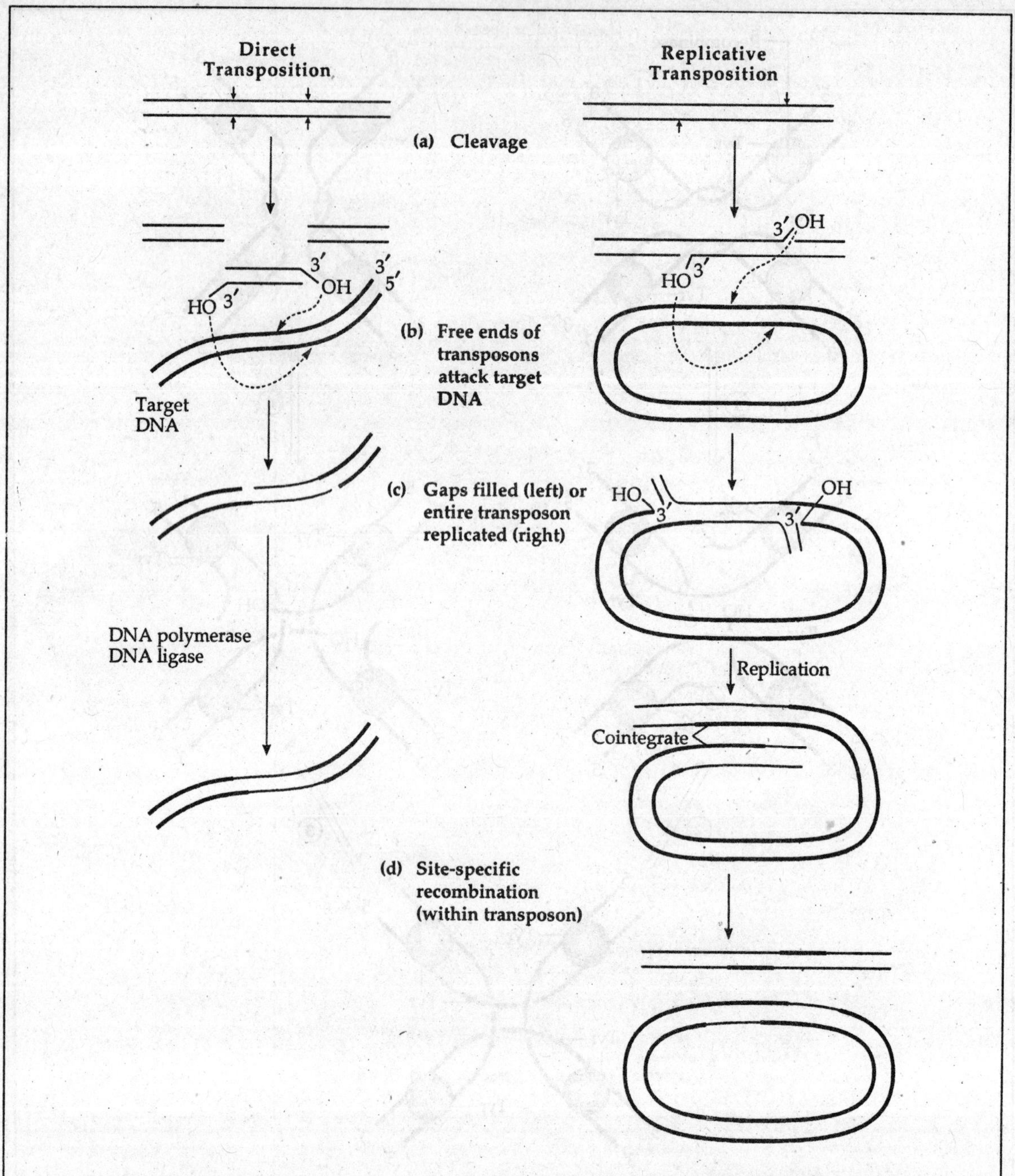

Fig. 5.29: Two general pathways for transposition: Direct (simple) or Replicative: (a) The DNA is first cleaved on each side of the transposon, at the sites indicated by arrows. **(b)** The liberated 3′-hydroxyl groups at the ends of the transposon act as nucleophiles in a direct attack on phosphodiester bonds in the target DNA. The target phosphodiester bonds are staggered in the two DNA strands. **(c)** The transposon is then linked to the target DNA. In direct transposition, replication fills in gaps at each end. In replicative transposition the entire transposon is replicated, creating a cointegrate intermediate. **(d)** The cointegrate is often resolved later with the aid of a separate site-specific recombination system. The cleaved host DNA left behind after direct transposition is either repaired by DNA end-joining or degraded.

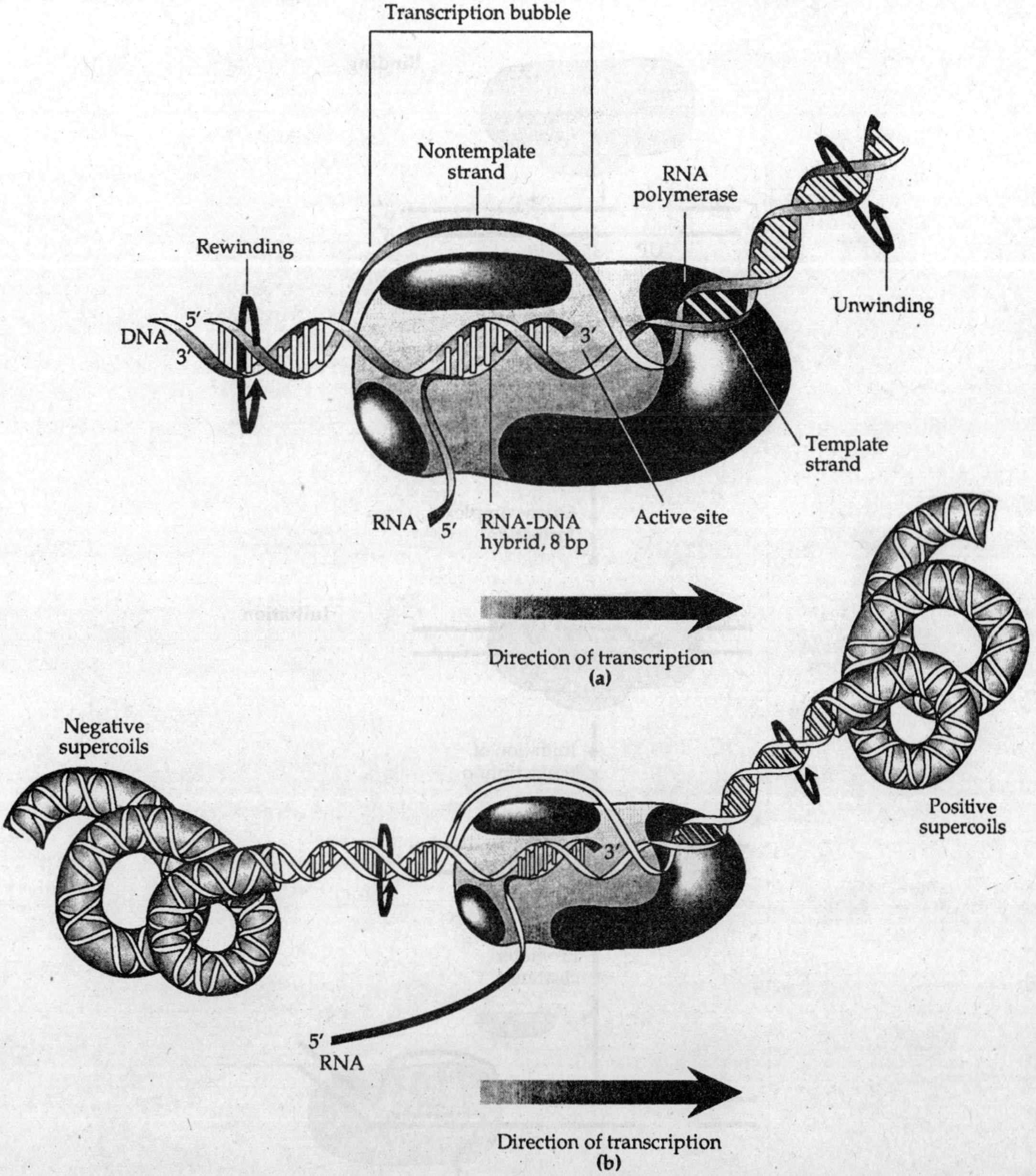

Fig. 5.30: Transcription by RNA polymerase in *E.coli*: To synthesize an RNA strand complementary to one of two DNA strands in a double helix, the DNA is transiently unwound.

(a) About 17 base pairs are unwound at any given time. The transcription bubble moves from left to right keeping pace with RNA synthesis. The DNA is unwound ahead and rewound behind as RNA is transcribed. Arrows show the direction in which the DNA and the short RNA-DNA hybrid must rotate to permit this process. As the DNA is rewound, the RNA-DNA hybrid is displaced and the RNA strand is extruded. The RNA polymerase is in close contact with the DNA ahead of the transcription bubble, as well as with the separated DNA strands and the RNA within and immediately behind the bubble. The polymerase footprint encompasses about 35 base pairs of DNA during elongation.

(b) Supercoiling of DNA brought about by transcription. Positive supercoils form ahead of the transcription bubble and negative supercoils form behind.

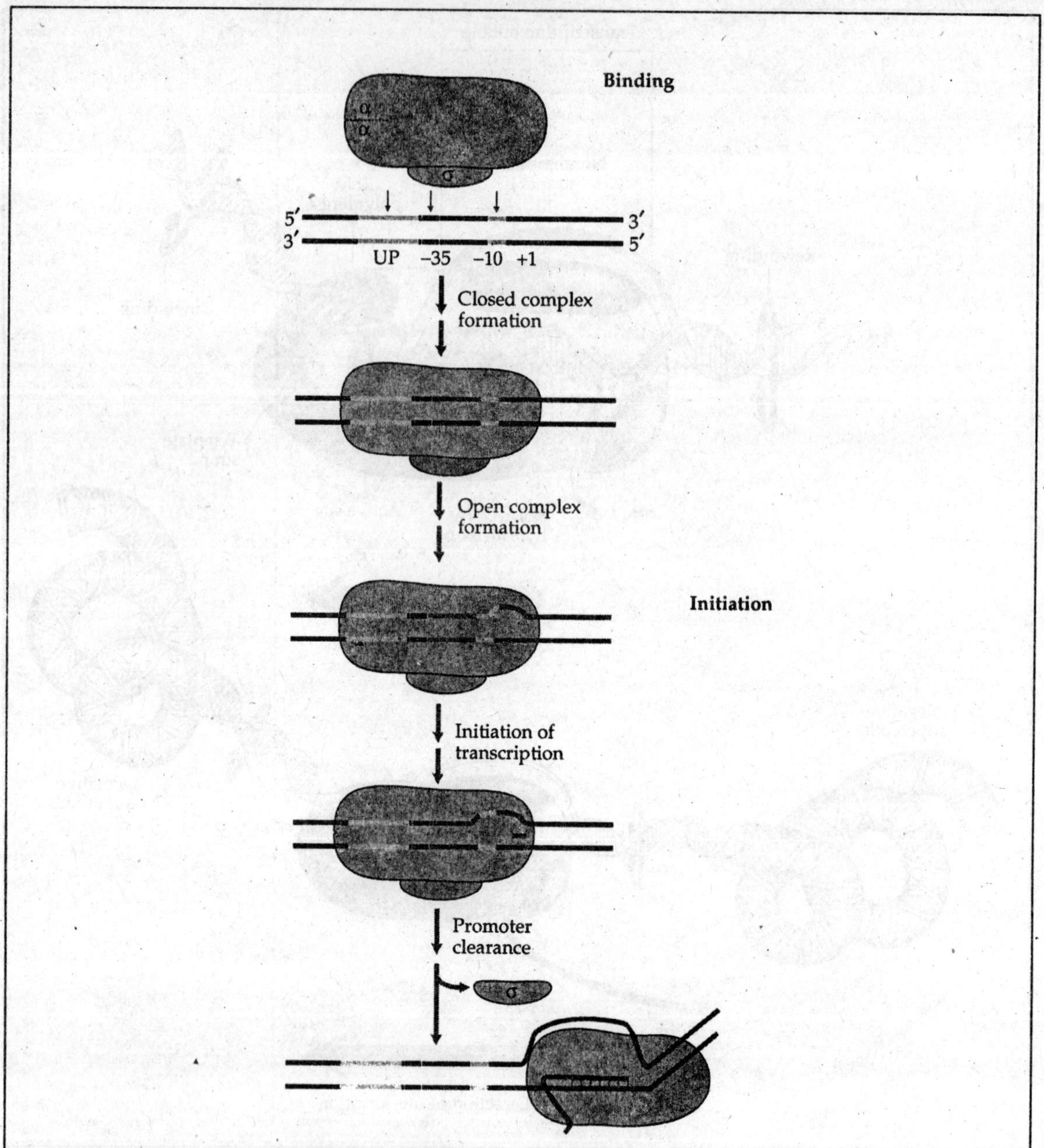

Fig. 5.31: Steps in the initiation of transcription by *E. coli* RNA polymerase: Initiation occurs in two phases—the binding and initiation phases. In the binding phase, the initial interaction of the RNA polymerase with the promoter leads to the formation of a closed complex, in which the promoter DNA is stably bound but not unwound. A 12 to 15 base pair region of DNA—from within the −10 region to position +2 or +3—is then unwound to form an open complex. Additional intermediates have been detected in the pathways leading to the closed and open complexes, along with several changes in protein conformation. The initiation phase encompasses transcription initiation and promoter clearance. Once the first eight or nine nucleotides of a new RNA are synthesized, the σ subunit is released and the polymerase leaves the promoter and becomes committed to productive elongation of the RNA.

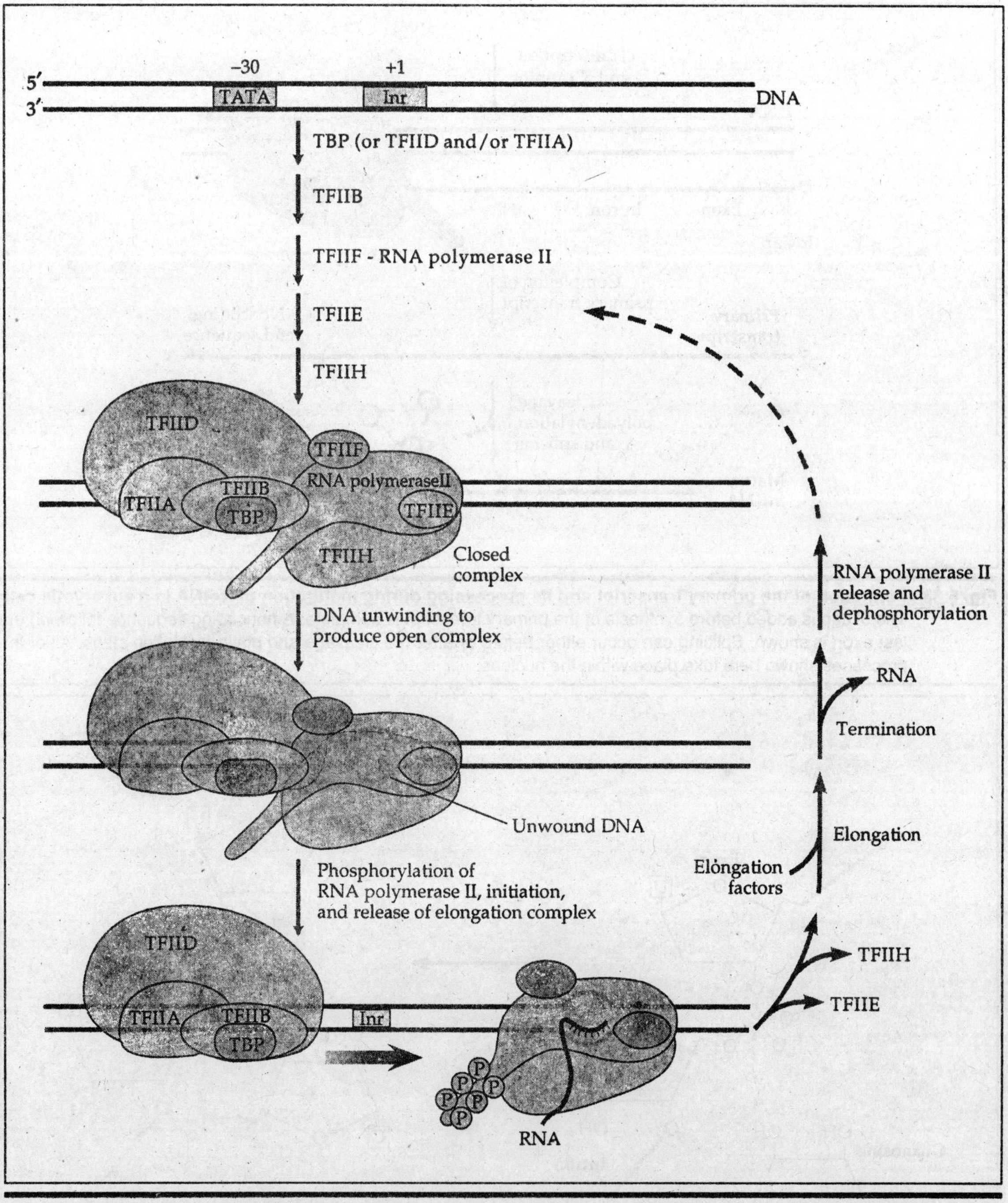

Fig. 5.32: Transcription at RNA polymerase II promoters: The sequential assembly of a complex consisting of TBP (often with TFIIA), TFIIB plus RNA polymerase II, TFIIE, TFIIF and TFIIH results in a closed complex. TBP often binds as a part of a larger complex called TFIID. Some of the TFIID subunits play a role in transcription regulation. The DNA is unwound at the Inr region by the helicase activities of TFIIH and TFIIE, creating an open complex. The carboxyl-terminal domain of the largest RNA polymerase II subunit is phosphorylated by TFIIH and the polymerase then escapes the promoter and begins transcription. Elongation is accompanied by the release of many transcription factors and is also enhanced by elongation factors. After termination, the RNA polymerase is released, dephosphorylated and recycled.

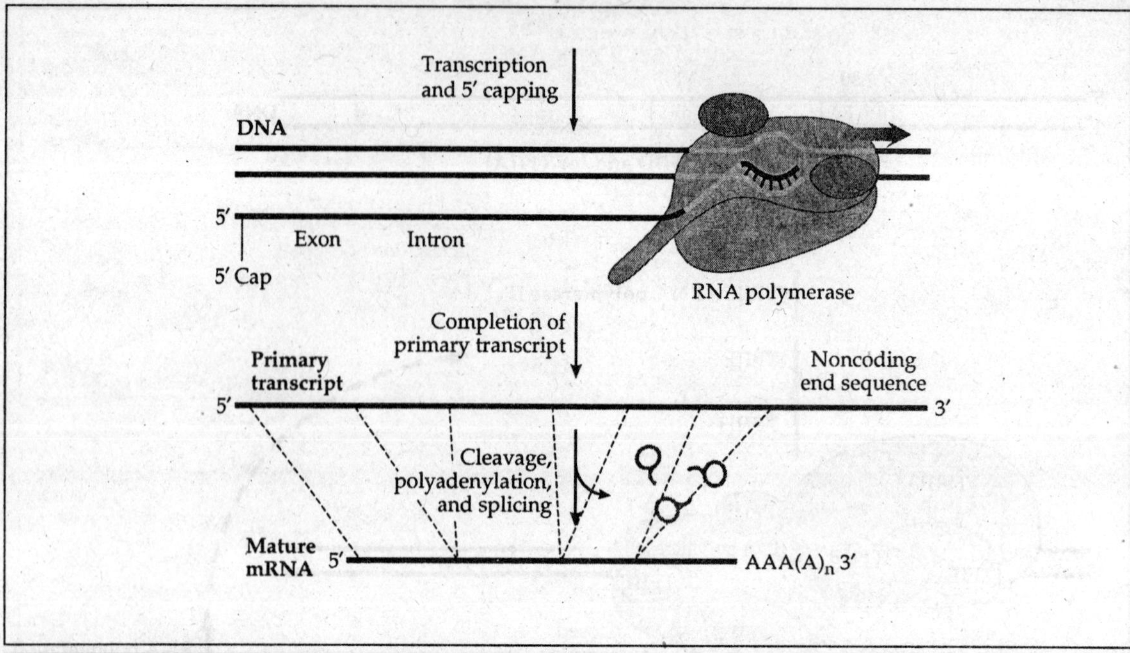

Fig. 5.33: Formation of the primary transcript and its processing during maturation of mRNA in a eukaryotic cell: The 5′ cap is added before synthesis of the primary transcript is complete. A noncoding sequence following the last exon is shown. Splicing can occur either before or after the cleavage and polyadenylation steps. All of the processes shown here take place within the nucleus.

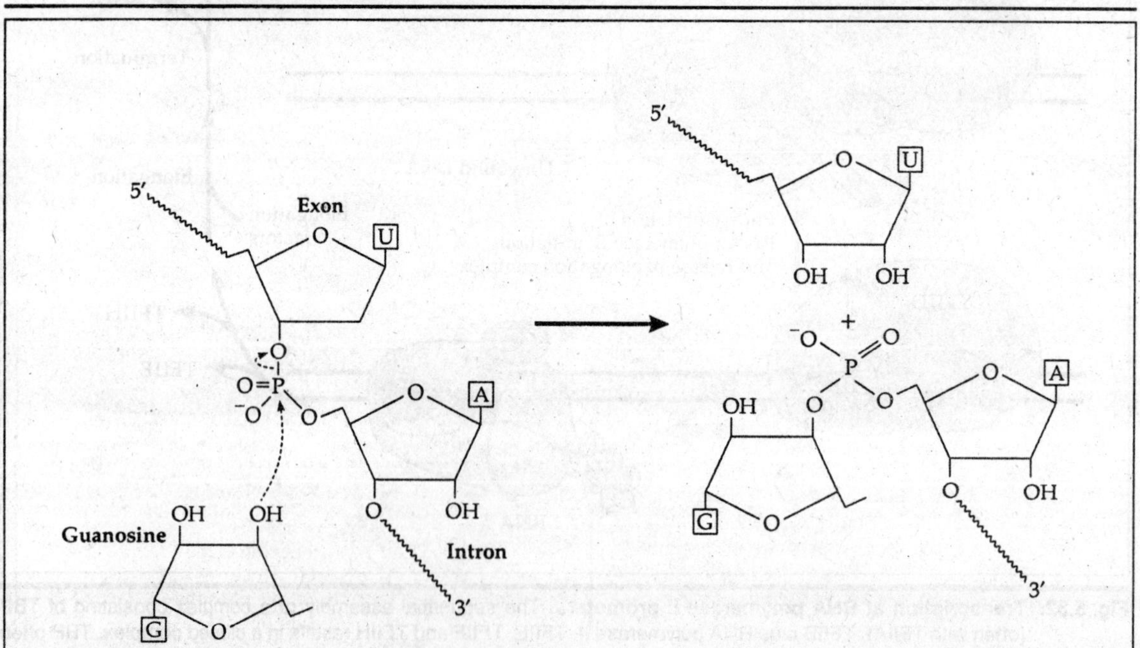

Fig. 5.34: Transesterification reaction: This is the first step in the splicing of group I introns. The 3′ OH of guanosine molecule acts as nucleophile.

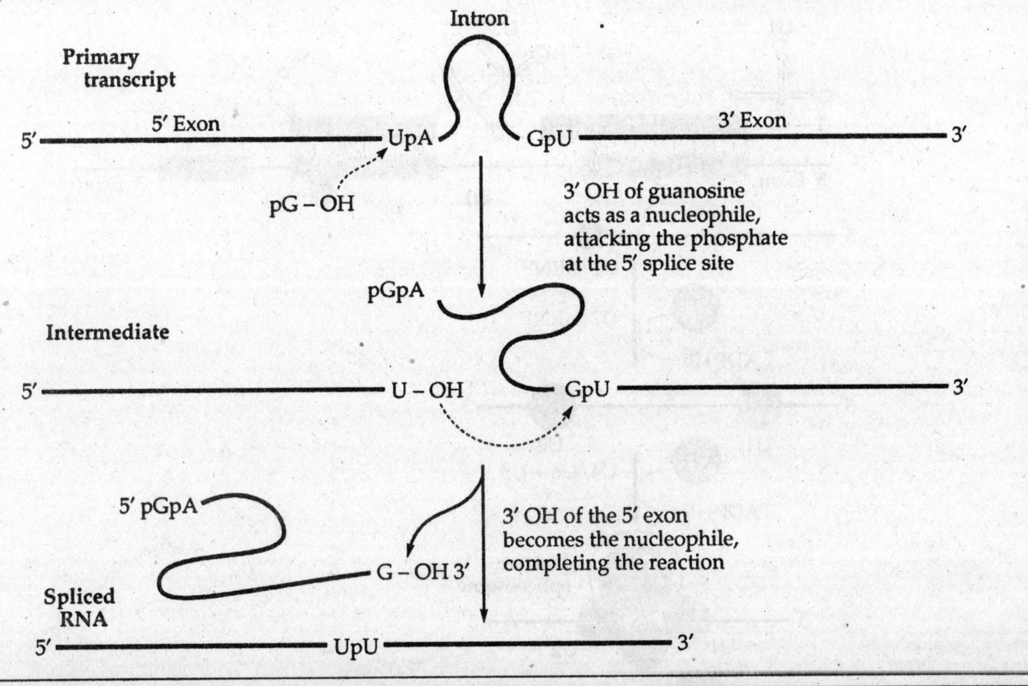

Fig. 5.35: Splicing mechanism of group I introns: The nucleophile in the first step may be guanosine, GMP, GDP or GTP. The spliced intron is eventually degraded.

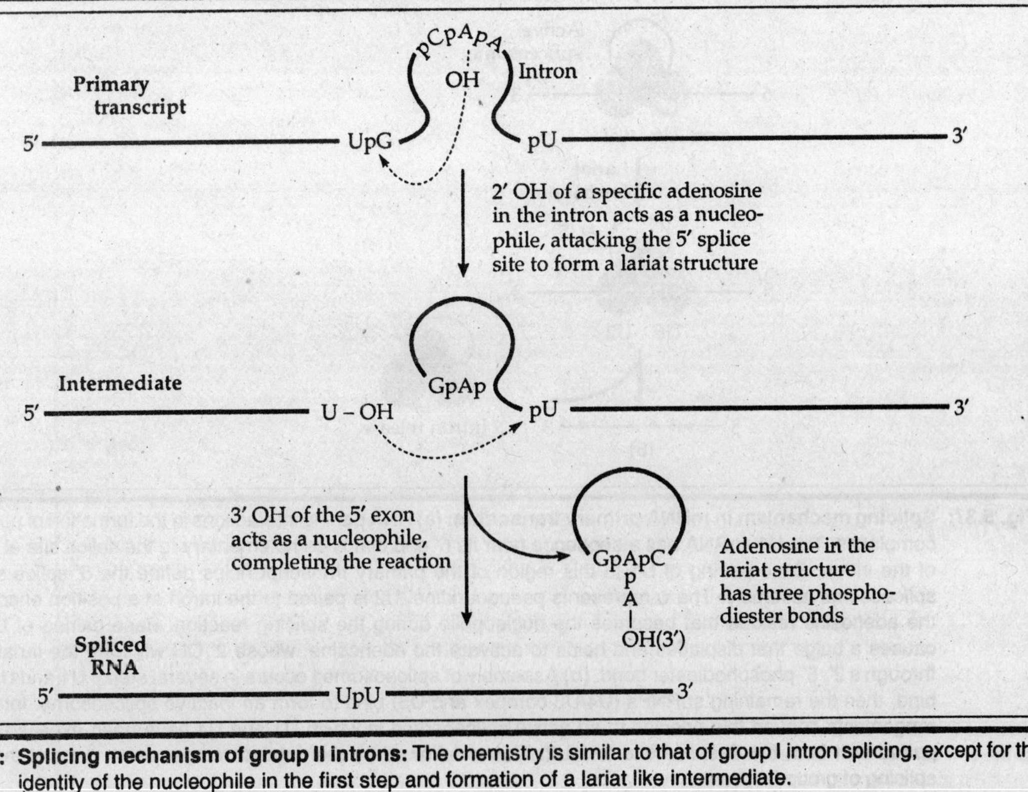

Fig. 5.36: Splicing mechanism of group II introns: The chemistry is similar to that of group I intron splicing, except for the identity of the nucleophile in the first step and formation of a lariat like intermediate.

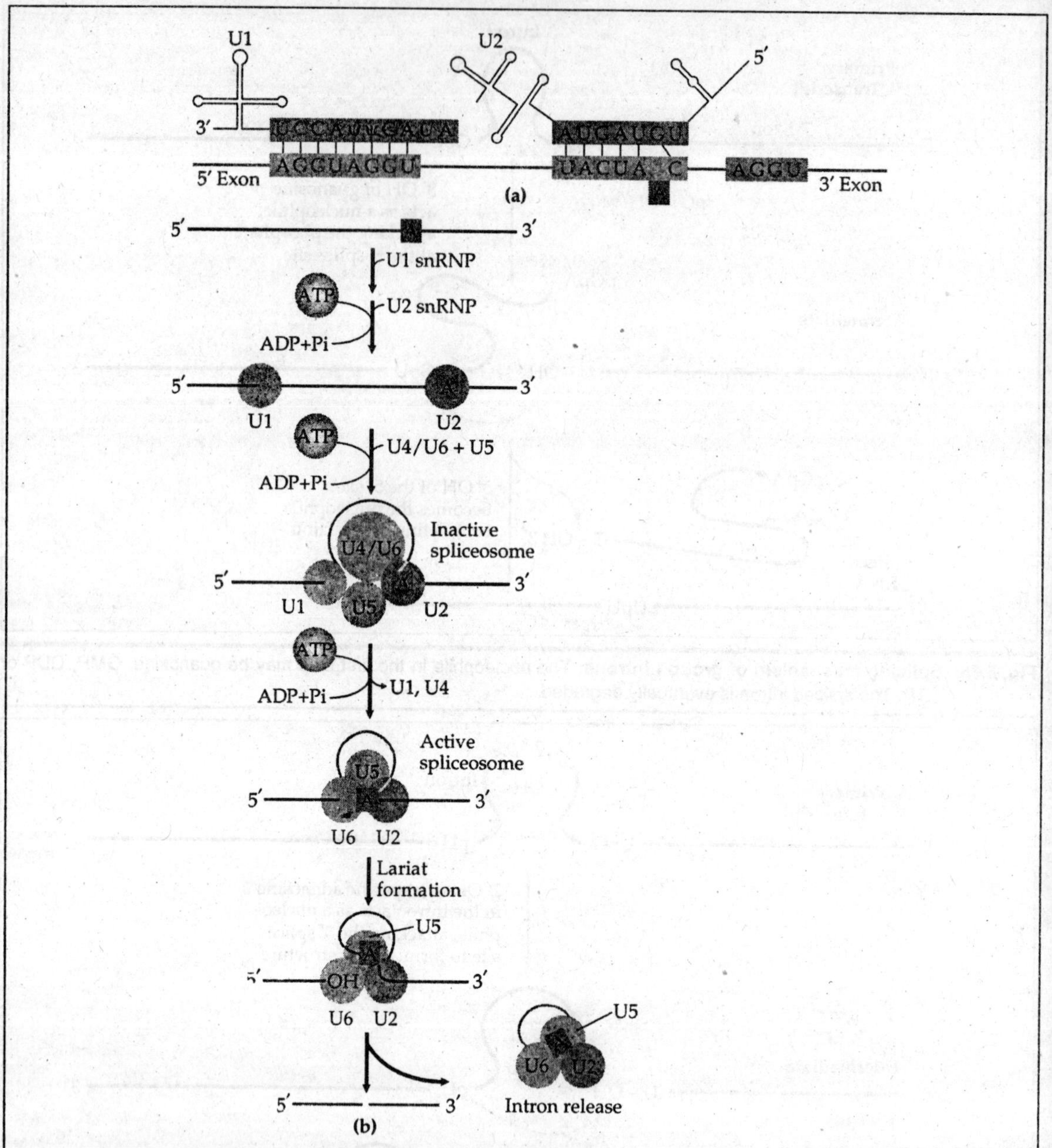

Fig. 5.37: Splicing mechanism in mRNA primary transcripts: (a) RNA pairing interactions in the formation of spliceosome complexes. The U1 snRNA has a sequence near its 5′ end that is complementary to the splice site at the 5′ end of the intron. Base pairing of U1 to this region of the primary transcript helps define the 5′ splice site during spliceosome assembly. The ψ represents pseudouridine. U2 is paired to the intron at a position encompassing the adenosine residue that becomes the nucleophile during the splicing reaction. Base pairing of U2 snRNA causes a bulge that displaces and helps to activate the adenosine, whose 2′ OH will form the lariat structure through a 2′, 5′-phosphodiester bond. **(b)** Assembly of spliceosomes occurs in several steps. U1 and U2 snRNPs bind, then the remaining snRNPs (U4/U6 compex and U5) bind to form an inactive spliceosome. Internal rearrangements convert this species to an active spliceosome in which U1 and U4 have been expelled and U6 is paired with both the 5′ splice site and U2. This is followed by the catalytic steps, which parallel those of the splicing of group II introns.

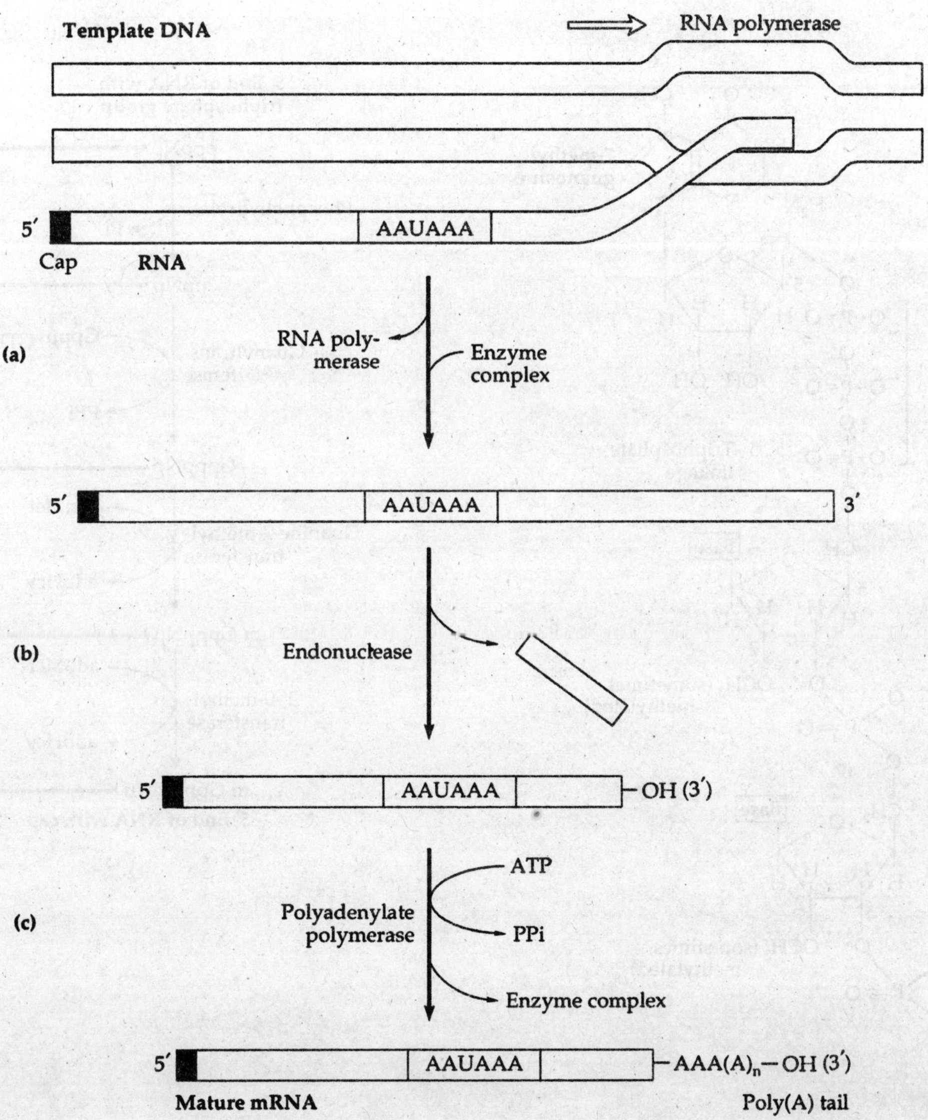

Fig. 5.38: Addition of the poly(A) tail to the primary RNA transcript of eukaryotes: RNA polymerase synthesizes RNA beyond the segment of the transcript containing the cleavage signal sequences, including the highly conserved upstream sequence (5′) AAUAAA.

(a) The cleavage signal sequence is bound by an enzyme complex that includes an endonuclease, a polyadenylate polymerase and several other multisubunit proteins involved in sequence recognition, stimulation of cleavage and regulation of the length of the poly (A) tail.

(b) The RNA is cleaved by the endonuclease at a point 10-30 nucleotides 3′ to (downstream of) the sequence AAUAAA.

(c) The polyadenylate polymerase synthesizes a poly(A) tail 80-250 nucleotides in length, beginning at the cleavage site.

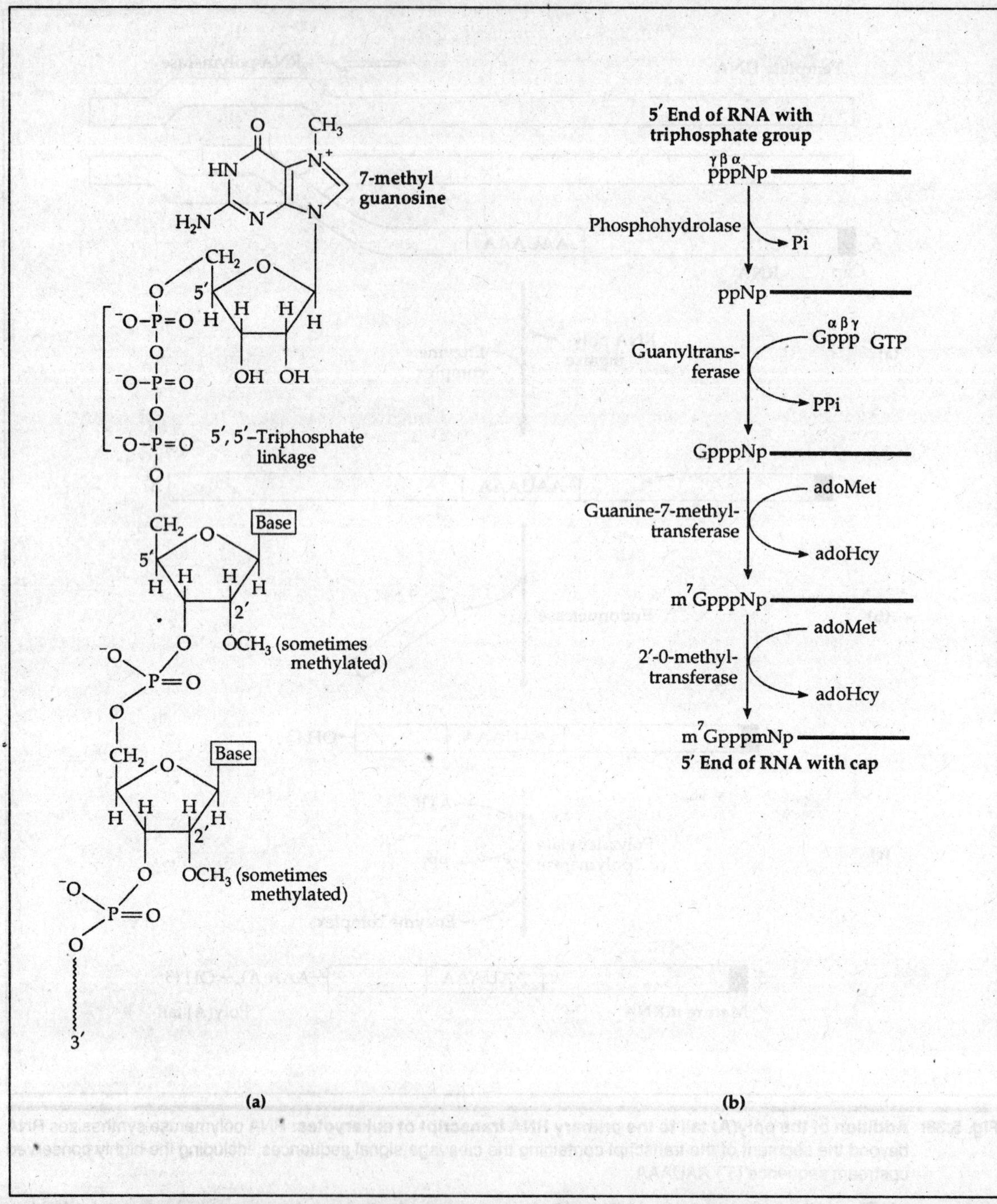

Fig. 5.39: The 5′ cap of mRNA:

(a) 7-Methylguanosine is joined to the 5′ end of almost all eukaryotic mRNAs in an unusual 5′, 5′-triphosphate linkage. Methyl groups are often found at the 2′ position of the first and second nucleotides. RNAs in yeast cells lack the 2′-methyl groups.

(b) Generation of the 5′ cap involves four to five separate steps (adoHcy → *S*-adenosylhomocysteine; adoMet → *S*-adenosylmethionine).

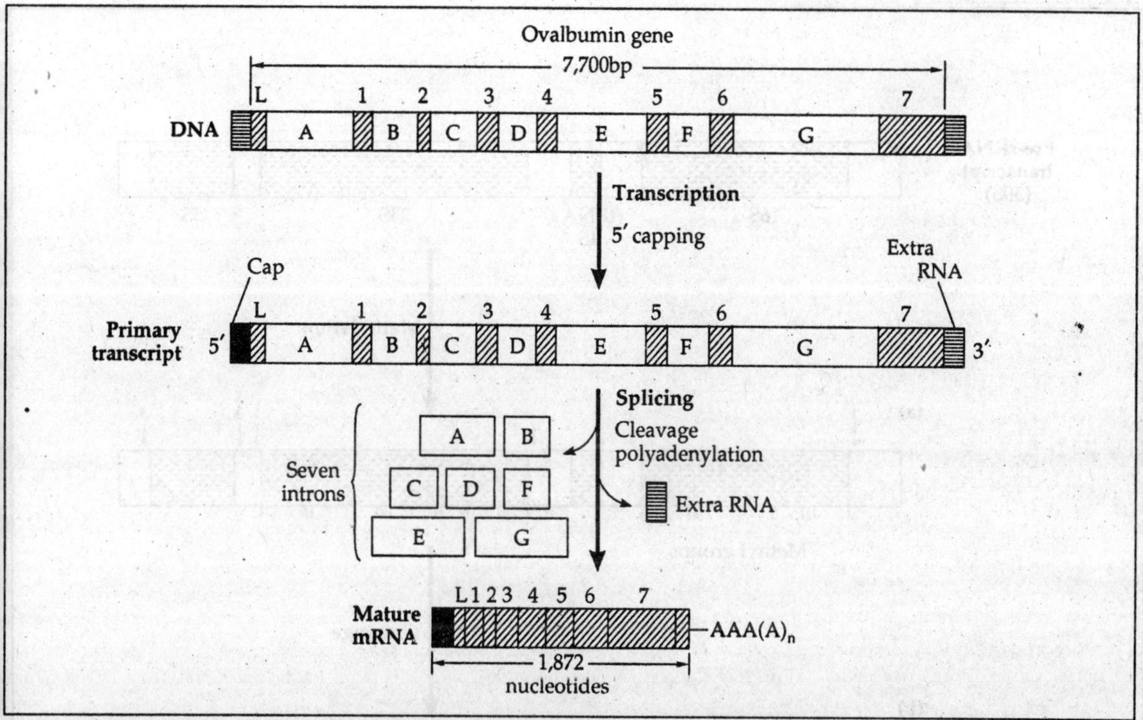

Fig. 5.40: Overview of the processing of eukaryotic mRNA: The ovalbumin gene contains seven introns (A-G) and exons (1-7, L). About three-quarters of the RNA is removed during processing. RNA polymerase II extends the primary transcript well beyond the cleavage and polyadenylation site (extra RNA) before terminating transcription. Termination signals for RNA polymerase II have not yet been defined.

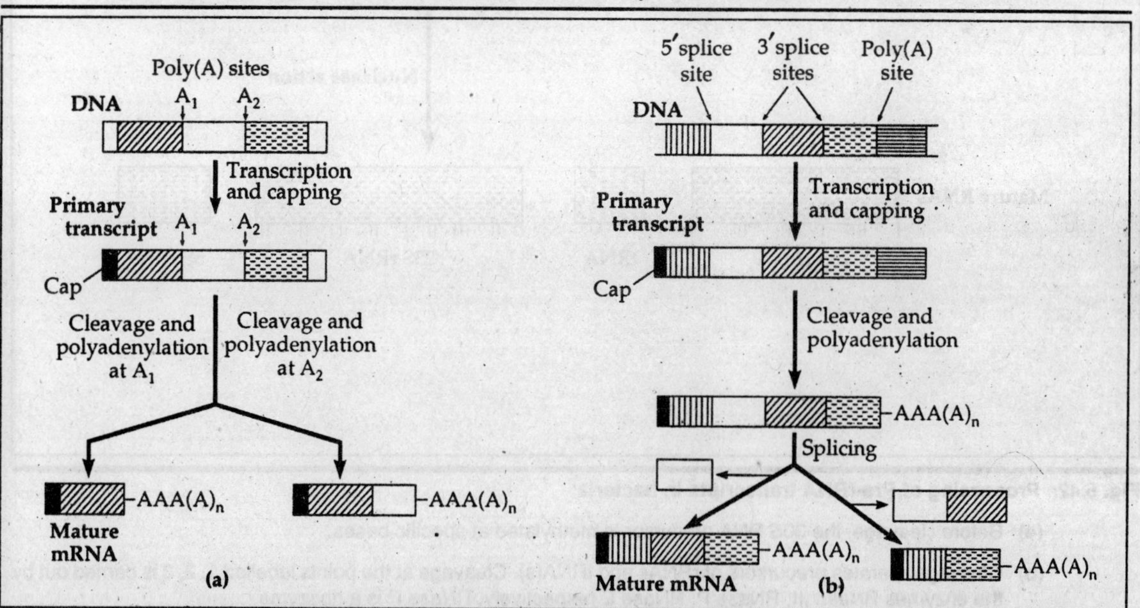

Fig. 5.41: Two mechanisms for the alternative processing of complex transcripts in eukaryotes: (a) Alternative cleavage and polyadenylation patterns showing two poly(A) sites—A_1 and A_2. **(b)** Alternative splicing patterns showing two different 3′ splice sites. In both mechanisms, different mature mRNAs are produced from the same primary transcript.

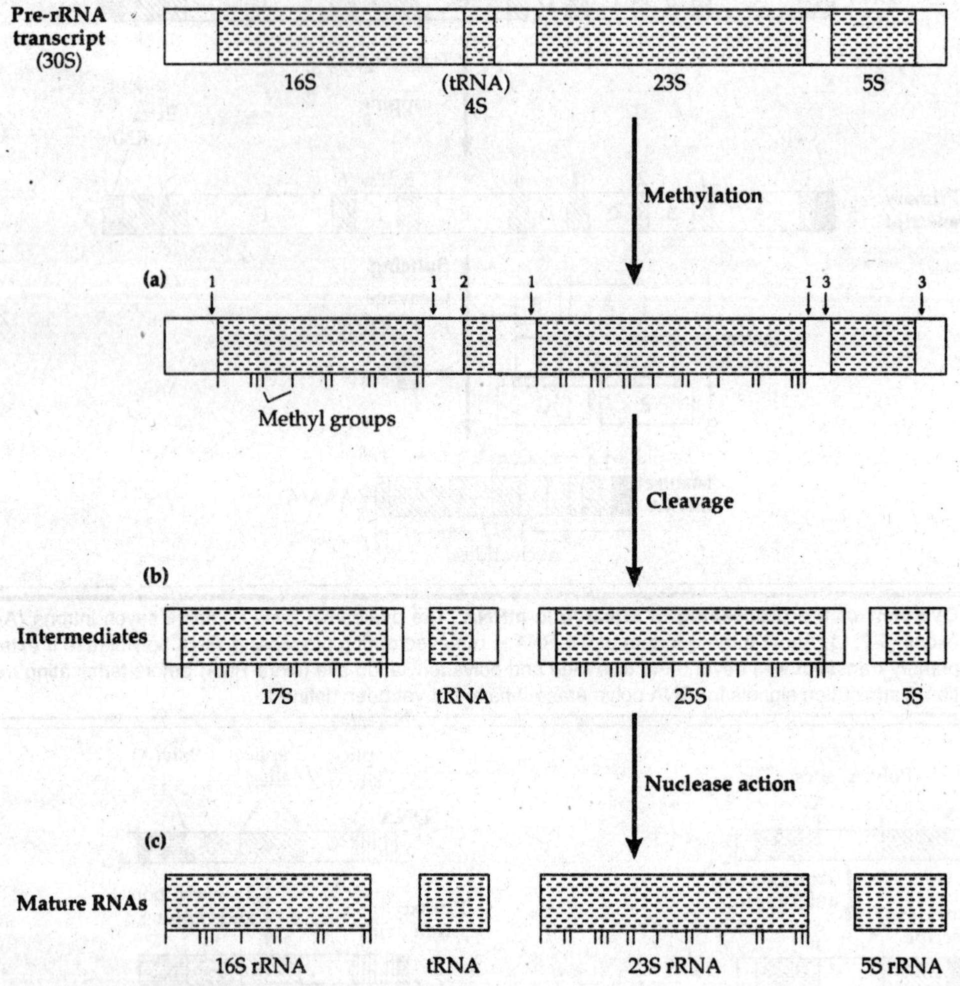

Fig. 5.42: Processing of Pre-rRNA transcripts in bacteria:

(a) Before cleavage, the 30S RNA precursor is methylated at specific bases.

(b) Cleavage liberates precursors of rRNAs and tRNA(s). Cleavage at the points labelled 1, 2, 3 is carried out by the enzymes RNase III, RNase P, RNase E respectively. RNase P is a ribozyme.

(c) The final 16S, 23S, 5S rRNA products result from the action of a variety of specific nucleases. The seven copies of the gene for pre-rRNA in the *E. coli* chromosome differ in the number, location and identity of tRNAs included in the primary transcript. Some copies of the gene have additional tRNA gene segments between the 16S and 23S rRNA segments and at the far 3′ end of the primary transcript.

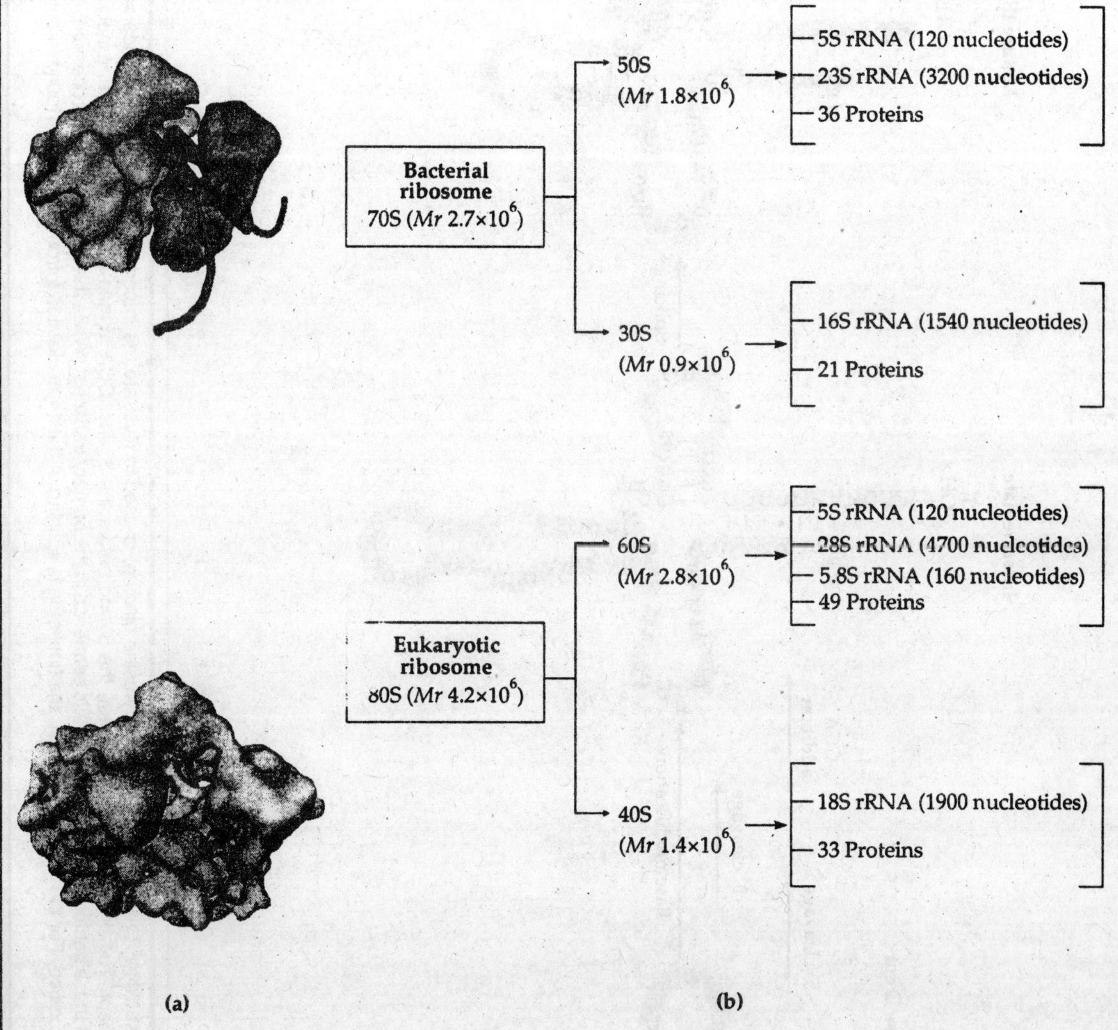

Fig. 5.43: Ribosomes:

(a) Structure of the bacterial ribosome at near-molecular resolution. The model shows the locations of mRNA and the bound tRNAs. The large subunit is on the left, the small subunit on the right. The image at the bottom is the complete ribosome at about 15 Å resolution. The subunits are not distinguished, but the orientation is similar to that of the image at the top.

(b) Ribosomal subunits are identified by their S(Svedverg unit) values, sedimentation coefficients that refer to their rate of sedimentation in a centrifuge. The S values are not necessarily additive when subunits are combined, because rates of sedimentation are affected by shape as well as mass.

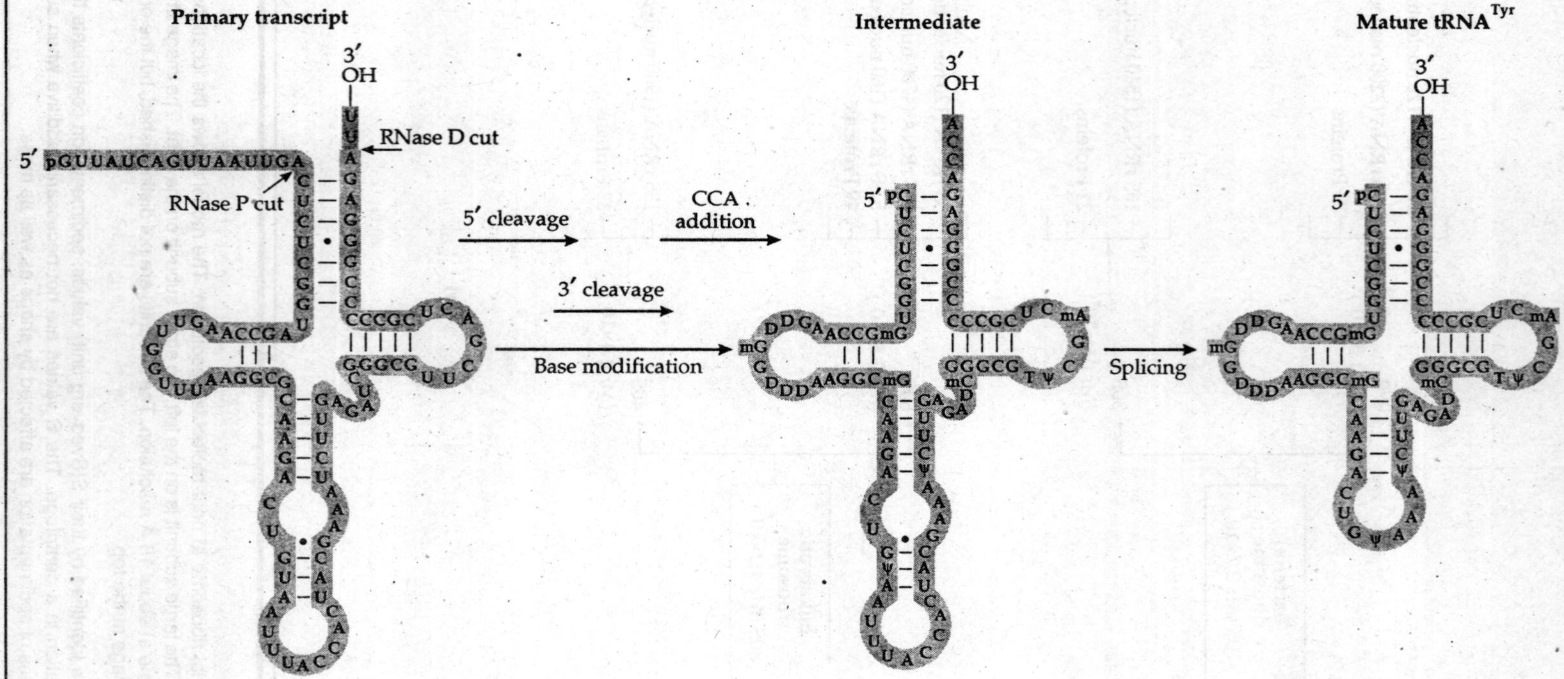

Fig. 5.44: Processing of tRNAs: The yeast tRNA^{Tyr} (the tRNA specific for tyrosine binding) is used to illustrate the important steps. The nucleotide sequences are removed from the primary transcript. The ends are processed first, the 5′ end before the 3′ end. CCA is then added to the 3′ end, a necessary step in processing eukaryotic tRNAs and those bacterial tRNAs that lack this sequence in the primary transcript. While the ends are being processed, specific bases in the rest of the transcript are modified. For the eukaryotic tRNA, the final step is splicing of the 14 nucleotide intron. Introns are found in some eukaryotic tRNAs but not in bacterial tRNAs.

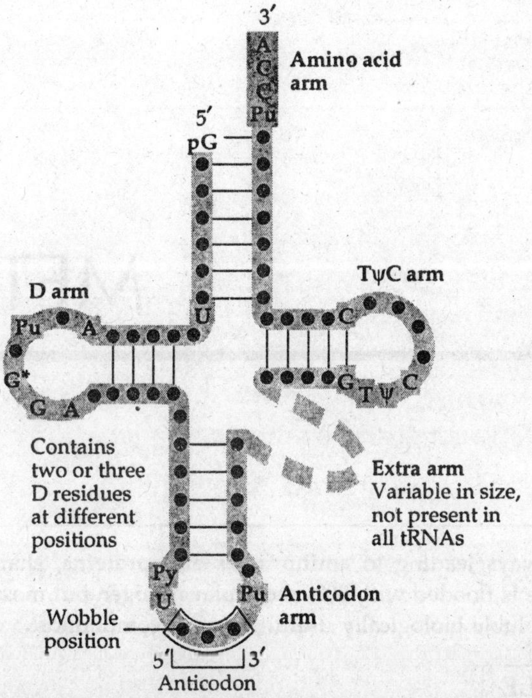

Fig. 5.45: General cloverleaf secondary structure of all tRNAs: The large dots on the backbone represent nucleotide residues and the lines represent base pairs. Transfer RNAs vary in length from 73 to 93 nucleotides. Extra nucleotides occur in the extra arm or in the D arm. At the end of the anticodon arm is the anticodon loop, which always contains seven unpaired nucleotides. The D arm contains two or three D residues, depending on the tRNA. In some tRNAs, the D arm has only three hydrogen-bonded base pairs. [Pu–Purine nucleotide; Py–Pyrimidine nucleotide; G*–Guanylate or 2′ -O-methylguanylate].

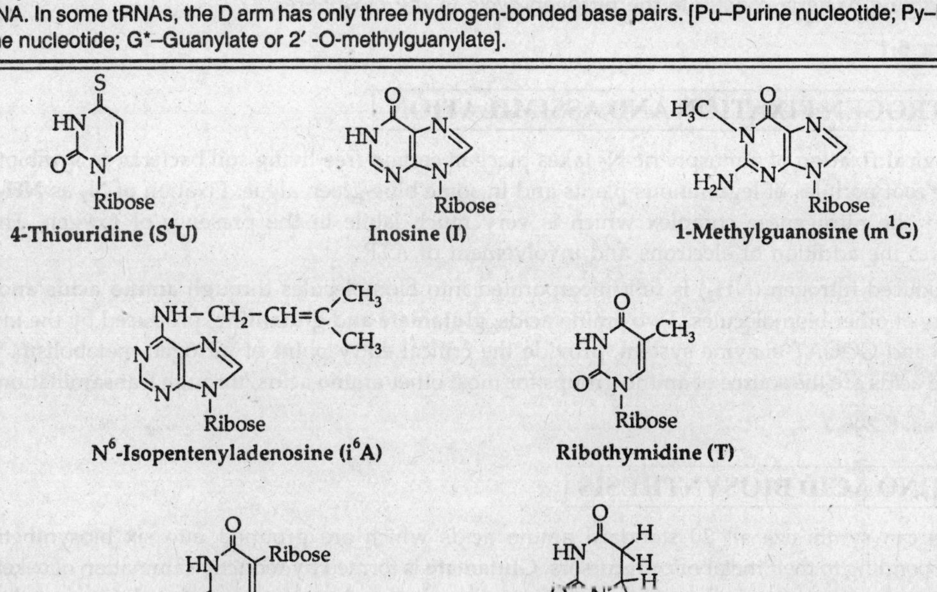

Fig. 5.46: Some modified bases of tRNAs, produced in post-transcriptional reactions.

CHAPTER

6

PROTEIN METABOLISM

The biosynthetic pathways leading to **amino acids** and **proteins**, share a requirement for nitrogen. Though the atmosphere is flooded with 80% molecular nitrogen but most living organisms cannot avail until it is reduced to soluble biologically useful nitrogen compounds.

NITROGEN CYCLE

Nitrogen cycle maintains a pool of biologically available nitrogen. Formation of ammonia by **biological fixation of N_2**, **nitrification** of ammonia to nitrate by soil microbes, **reduction** of nitrate to ammonia by higher plants, **synthesis of amino acids** from ammonia, **denitrification** of nitrate to molecular N_2 by denitrifying bacteria constitute the nitrogen cycle in the biosphere.

Figure: 6.1

NITROGEN FIXATION AND ASSIMILATION

Biological fixation of atmospheric N_2 takes place in certain free living soil bacteria, in symbiotic bacteria in the root nodules of leguminous plants and in some blue-green algae. Fixation of N_2 as NH_3 is carried out by the nitrogenase complex which is very much labile in the presence of oxygen. The reaction requires the addition of electrons and involvement of ATP.

Reduced nitrogen (NH_3) is first incorporated into biomolecules through **amino acids** and then into variety of other biomolecules. Two amino acids, **glutamate** and **glutamine**, produced by the involvement of GS and GOGAT enzyme system, provide the critical entry point of nitrogen metabolism. These two amino acids are the source of amino groups for most other amino acids, through transamination reactions.

Figures: 6.2–6.7

AMINO ACID BIOSYNTHESIS

Plants can synthesize all 20 standard amino acids which are grouped into six biosynthetic families corresponding to their metabolic precursors. **Glutamate** is formed by reductive amination of α-ketoglutarate and is the precursor of glutamine, proline and arginine. **Aspartate**, produced from oxaloacetate by transamination, is the precursor of lysine, methionine, asparagine and threonine. **Pyruvate** is the precursor of leucine, isoleucine, valine and alanine. **Serine** whose carbon chain is derived from 3-phosphoglycerate, is the precursor of glycine; the β-carbon atom of serine is transferred to tetrahydrofolate. **Cysteine** is produced from serine and sulphide, the later formed by the reduction of environmental sulfate. **Aromatic amino acids**, phenylalanine, tyrosine and tryptophan are synthesized from PEP and erythrose-4-P through an intermediate chorismate. **Histidine** is produced from ribose-5-P through phosphoribosyl pyrophosphate.

Many other important biomolecules are derived from amino acids—**porphyrins** from glycine; **auxin** from tryptophan; **cinnamate** from phenylalanine; **glutathione** from three amino acids, an important cellular reducing agent; **purines** from glutamine, glycine, aspartate and **pyrimidines** from aspartate.

Figures: **6.8–6.27**

AMINO ACID OXIDATION

Amino acid oxidation in plants occurs mainly to produce the metabolites for other biosynthetic pathways, rarely to provide energy as in animals. Catabolism of amino acids involves the separation of the amino group from the carbon skeleton and its transfer mostly to α-ketoglutarate to form glutamate. After removal of amino groups, the carbon skeleton of amino acids undergoes oxidation to compounds that can enter the **citric acid cycle** for oxidation to CO_2 and H_2O.

They enter the cycle through five intermediates—acetyl-CoA, α-ketoglutarate, succinyl-CoA, fumarate and oxaloacetate. Alanine, cysteine, glycine, tryptophan and serine yield **acetyl-CoA** via pyruvate. Leucine, lysine, phenylalanine, tyrosine and tryptophan yield acetoacetyl-CoA via **acetoacetyl-CoA**. Leucine, isoleucine and tryptophan also form acetyl-CoA directly. Proline, histidine, arginine, glutamine and glutamate produce **α-ketoglutarate**. Threonine, methionine, isoleucine and valine produce **succinyl-CoA**. Phenylalanine and tyrosine give rise to **fumarate**; aspartate and asparagine produce **oxaloacetate**.

Figures: **6.28–6.35**

PROTEIN SYNTHESIS

Proteins are synthesized with a particular amino acid sequence through the **translation** of information encoded in mRNA carried out by supramolecular machine ribosome. Amino acids are specified by **mRNA codons** consisting of nucleotide **triplet**. Translation requires adaptor molecule, the **tRNAs** which recognize codons by their anticodon, insert amino acids into their appropriate sequential positions in the polypeptide. The **genetic code dictionary**, prepared using synthetic mRNAs of known composition and sequence, reveals that codon is **non-ambiguous** but **degenerate**, because the third base in each codon is much less specific and is said to **wobble**. The standard genetic code words are **universal** in all species; and **non-overlapping**, **commaless** in nature. The **initiation codon** is AUG for methionine and UAA, UAG, UGA act as **termination codon**.

Protein Synthesis occurs in **five stages**:

Stage I : Activation of amino acids occurs by aminoacyl-tRNA synthetases in the cytosol forming aminoacyl-tRNAs with the involvement of ATP. The anticodon of a tRNA is responsible for the specificity of interaction between the aminoacyl-tRNA and the complementary mRNA codon.

Stage 2 : Initiation of protein synthesis involves the formation of a complex between the ribosome subunits, mRNA, GTP, specific aminoacyl-tRNA (met-tRNA), initiation factors. The process differs between prokaryotes and eukaryotes.

Stage 3 : Elongation of polypeptide involves the binding of incoming aminoacyl-tRNA to the A site of ribosome with the involvement of GTP and elongation factors followed by peptidyl transfer reaction of the amino acid residue of P site of ribosome with the amino group of the incoming aminoacyl-tRNA. Movement of the ribosome along the mRNA then translocates peptidyl-tRNA from A site to P site, a process requiring hydrolysis of GTP. Deacylated tRNA dissociates from the ribosomal P site.

Stage 4 : Termination of protein synthesis occurs after many cycles of elongation with the aid of release factors.

Stage 5 : Modification of released polypeptide occurs by folding into biologically active three-dimensional forms along with processing through several post-translational modification reactions at amino-terminal, carboxyl-terminal of individual amino acid; attachment of carbohydrate side chain; addition of isoprenyl and prosthetic groups; proteolytic processing; formation of disulphide cross links.

At least four high-energy phosphate bonds are required to generate one peptide bond. Large clusters of ribosomes called **polysomes** use the single molecule of mRNA to translate protein with maximum efficiency within short lifetime of mRNAs.

Figures: 6.36–6.48

PROTEIN TARGETTING

After synthesis, many proteins are directed to particular locations in the cell. One targetting mechanism involves a peptide **signal sequence**, generally found at the amino terminus of a newly synthesized protein. In eukaryotic cells, one class of signal sequences is recognized and bound by a large protein-RNA complex called the **signal recognition particle** (SRP). SRP binds the signal sequence as soon as it appears on the ribosome and transfers the entire ribosome and incomplete polypeptide to the endoplasmic reticulum. In ER lumen, the proteins are further modified—removal of signal sequences, folding of polypeptides, formation of disulfide bonds and glycosylation. Next moved to the Golgi complex, then sorted and sent to lysosomes, the plasma membrane or transport vesicles. Proteins targetted to mitochondria and chloroplasts in eukaryotic cells, and for export in bacteria, also make use of an amino-terminal signal sequence. Other known targetting signals include carbohydrates and **signal patches**. Some proteins are imported into the cell by receptor-mediated **endocytosis**.

Proteins are eventually degraded by specialized **proteolytic systems** present in all cells. Proteins are first tagged to a highly conserved protein, ubiquitin, which is critical to the regulation of many cellular processes.

Figures: 6.49–6.53

REGULATION OF PROTEIN SYNTHESIS

The expression of genes is regulated by a number of processes that affect the rates at which proteins are synthesized by the gene. The regulation mostly occurs at the level of initiation of transcription, mediated by regulatory proteins which recognize specific DNA sequences. The effect of repressors is called **negative regulation** and of activators is called **positive regulation.**

In **prokaryotes**, a single transcriptional unit with independent function is called an **operon**. Transcription of **Lactose** (*lac*) **operon** in *E. coli* is blocked by binding of a repressor protein at specific site of DNA called **operator** (negative regulation), dissociation of which is mediated by **inducer**, allolactose. Positive regulation of *lac* operon is carried out by cAMP receptor protein (CRP) which binds a specific site on the DNA, stimulating transcription. CRP decreases with increasing concentration of glucose and depresses expression of *lac* genes.

In **arabinose** (*ara*) **operon**, the AraC protein acts as both activator and repressor. Repressors regulate their own synthesis (autoregulation) and some regulatory proteins bind DNA sites many base pairs apart and interact by DNA looping.

Tryptopha (*trp*) **operon** that produces the enzymes of amino acid tryptophan synthesis, has a regulatory circuit called **attenuation** that uses a transcription termination site (attenuator). Formation of the attenuator structure in the mRNA is modulated by a mechanism that couples transcription and translation while responding to small changes in tryptophan concentration.

In the **SOS system**, multiple unlinked genes that are repressed by a single type of repressor protein, are induced simultaneously when DNA damage triggers RecA protein-facilitated autocatalytic proteolysis of the repressor.

Translational regulation can also occur in prokaryotes. In the synthesis of ribosomal proteins, one protein in each r-protein operon acts as a translational repressor. The mRNA is bound by the repressor and translation thus blocked only when r-protein is present in excess of available rRNA.

Eukaryotic cells use some regulatory schemes with more common positive regulation. Promoters for RNA polymerase II typically have **TATA box** and Inr sequence, as well as multiple binding sites for DNA-binding transactivators called **enhancers**. The effects of DNA-binding transactivators on RNA polymerase II are mediated by coactivator protein complexes called **mediator**. Other protein complexes including histone acetyl transferases and ATP-dependent complexes reversibly remodel chromatin structure for transcriptional regulation. **Translational regulation** in eukaryotes involves the binding of repressors to specific sites in mRNA which interact with initiation factors or with ribosome to prevent translation.

Figures: 6.54–6.68

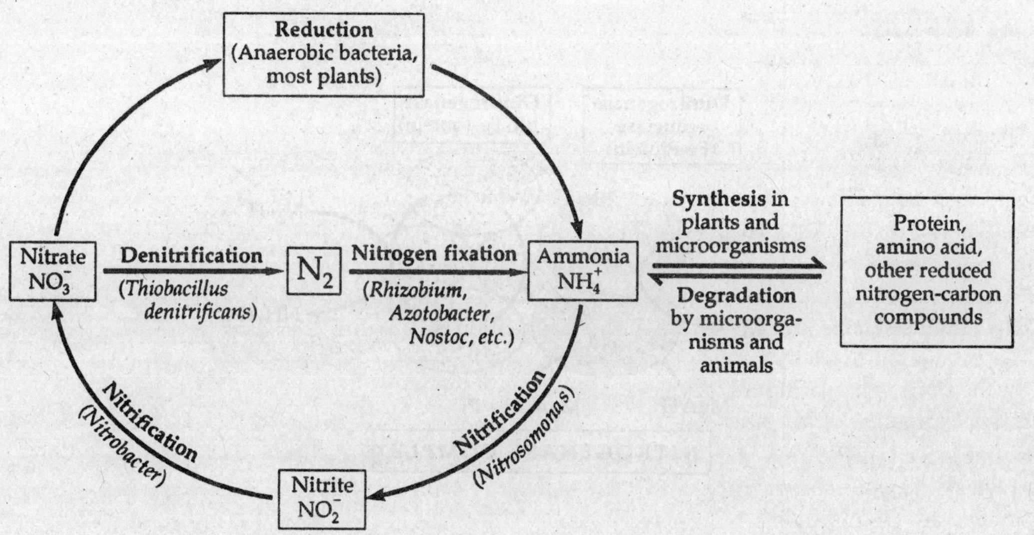

Fig. 6.1: Nitrogen cycle: The total amount of nitrogen fixed annually in the biosphere exceeds 10^{11} kg.

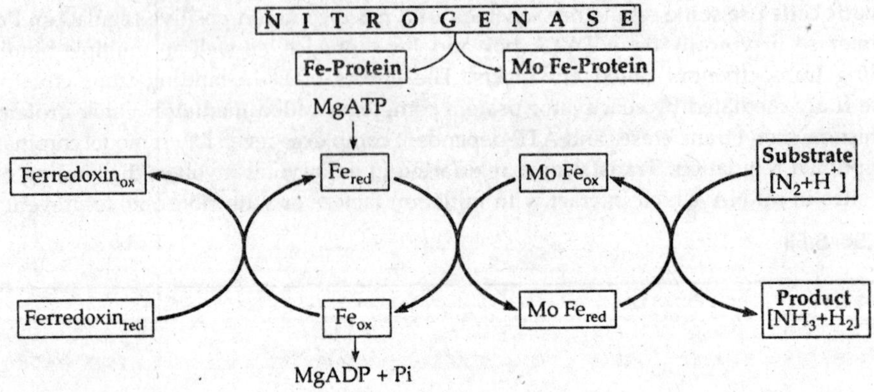

Fig. 6.2: The reaction catalyzed by nitrogenase: Ferredoxin reduces the Fe-protein of nitrogenase. Binding and hydrolysis of ATP to Fe-protein facilitates redox reactions. The Fe-protein reduces the Mo Fe-protein which in turn reduces N_2.

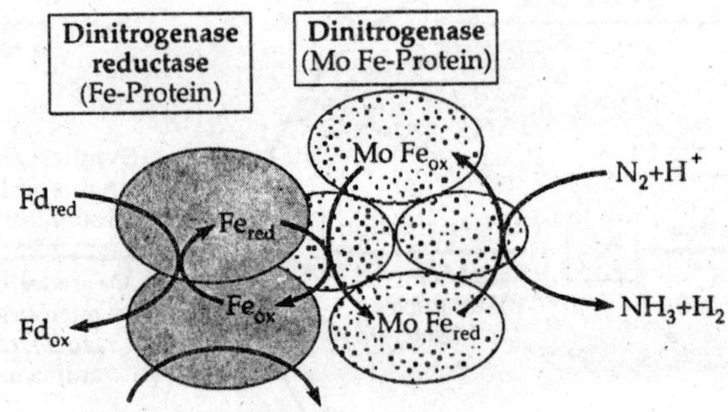

Fig. 6.3: Structure of nitrogenase enzyme complex showing the flow of electron to the substrate in biological N_2 fixation: The Fe-protein, a homodimer encoded by *nifH*, accepts electrons from ferredoxin and transfers it to Mo Fe-protein accompanied by hydrolysis of ATP. The Mo Fe-protein, a heterotetramer encoded by *nifD* and *nifK*, reduces N_2 and H^+ to ammonia and H_2.

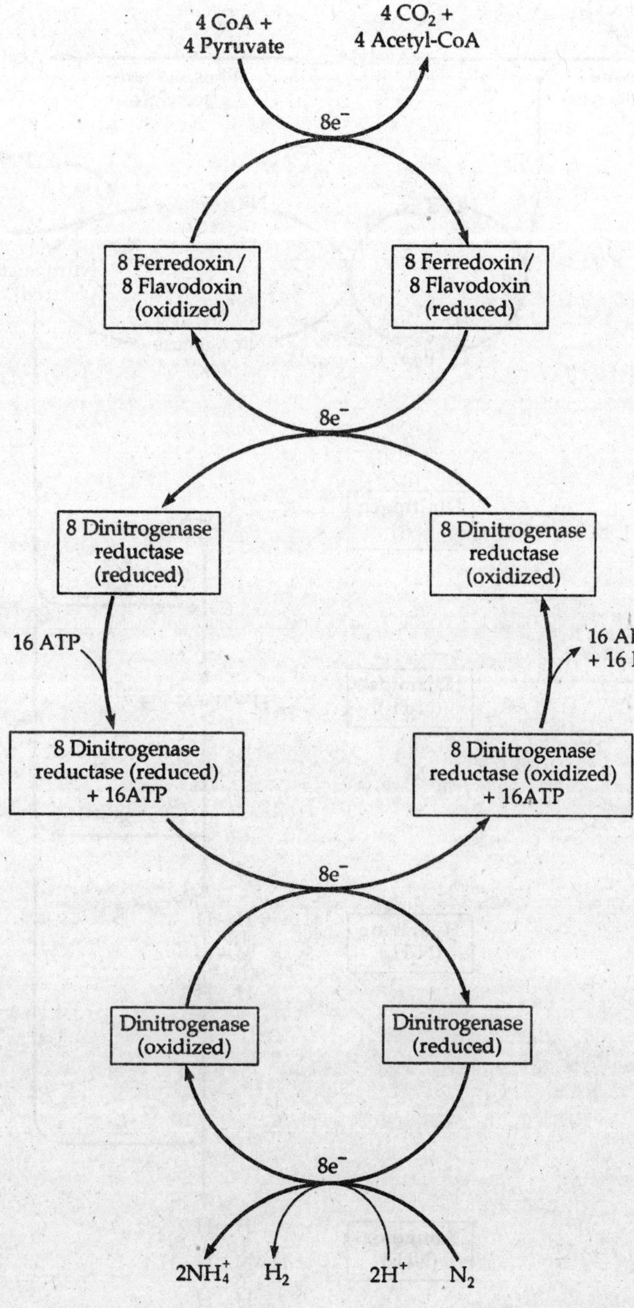

Fig. 6.4: Nitrogen fixation by the nitrogenase complex: Electrons are transferred from pyruvate to dinitrogenase via ferredoxin or flavodoxin and dinitrogenase reductase. Six electrons required to fix one molecule N_2 and an additional two electrons are used to reduce $2H^+$ to H_2 in a process that obligatorily accompanies nitrogen fixation, making a total of eight electrons required per N_2 molecule fixed.

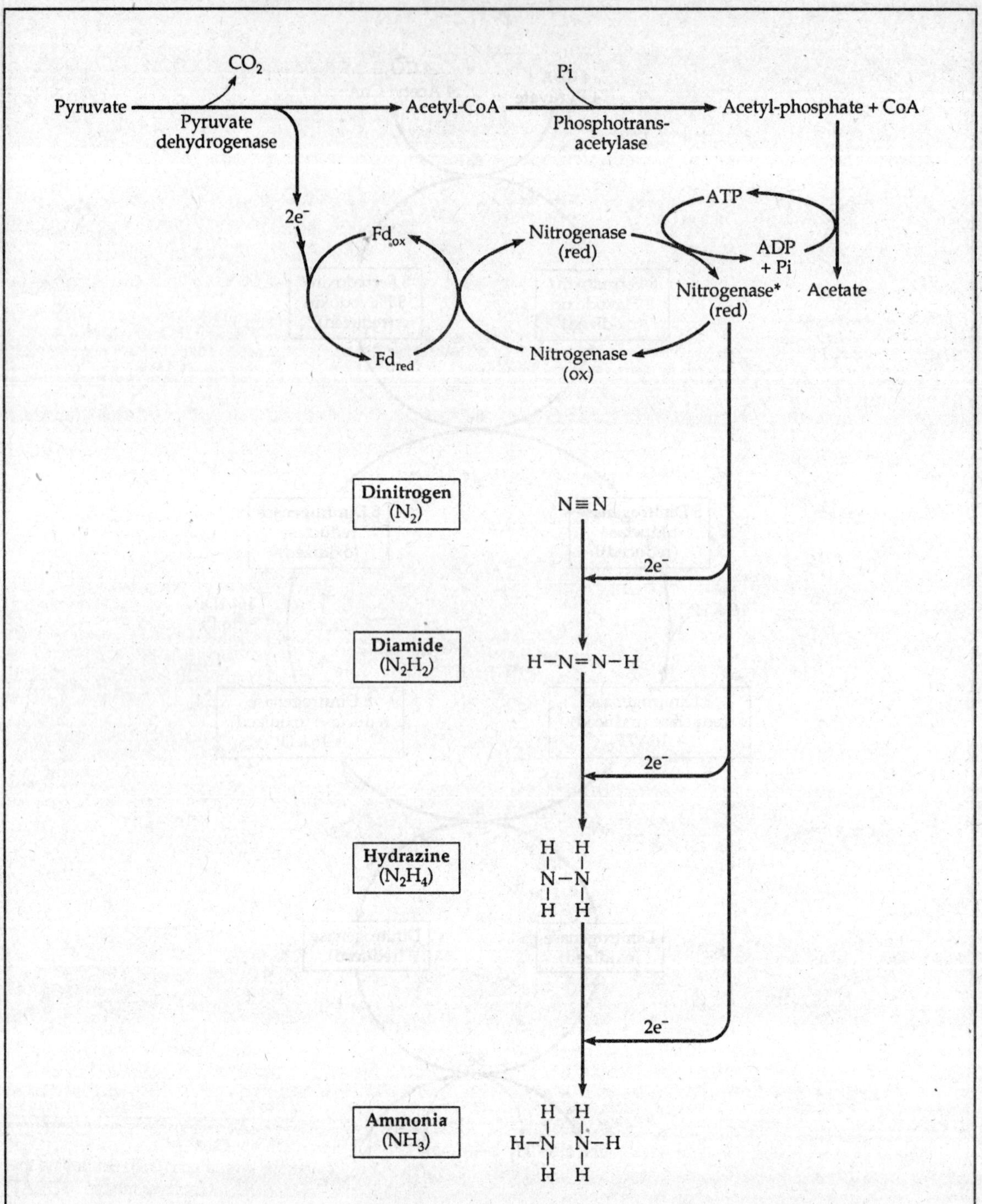

Fig. 6.5: Conversion of dinitrogen to ammonia by the nitrogenase enzyme system in *Clostridium pasteurianum*.

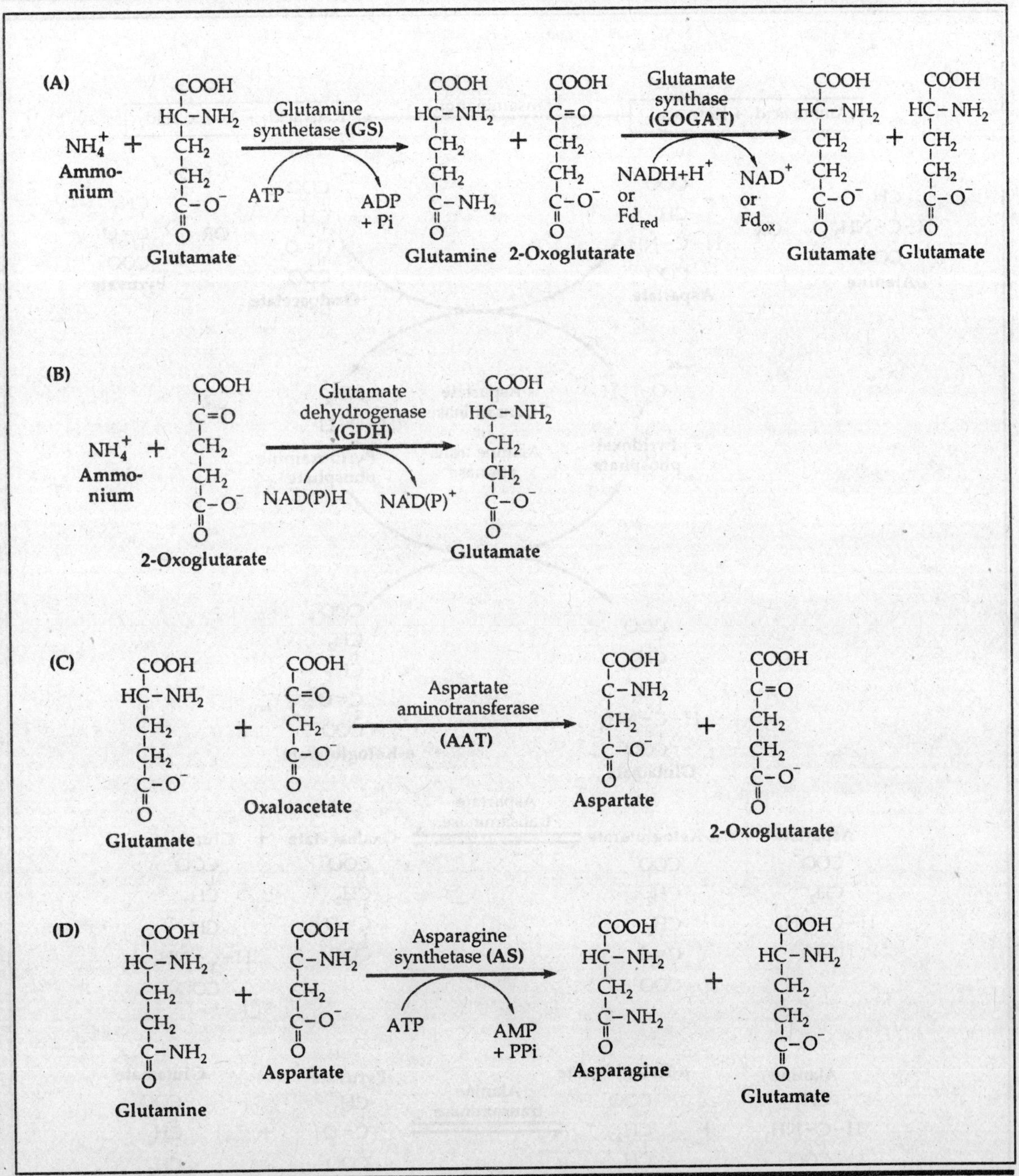

Fig. 6.6: **Ammonium metabolism**: Ammonium may be assimilated by one of several processes—

(A) Combine with glutamate to form amide (glutamine) catalyzed by glutamine synthetase (GS). Glutamine then combines with 2-oxoglutarate to form two molecules of glutamate catalyzed by glutamate synthase (GOGAT). [A reduced co-factor is required for the reaction: ferredoxin in green leaves and NADH in non-photosynthetic tissue.]

(B) Combine with 2-oxoglutarate to form glutamate catalyzed by glutamate dehydrogenase (GDH).

(C) Transfer of amino group from glutamate to aspartate catalyzed by aspartate aminotransferase (AAT).

(D) Transfer of amino group from glutamine to aspartate to synthesize asparagine catalysed by asparagine synthetase (AS).

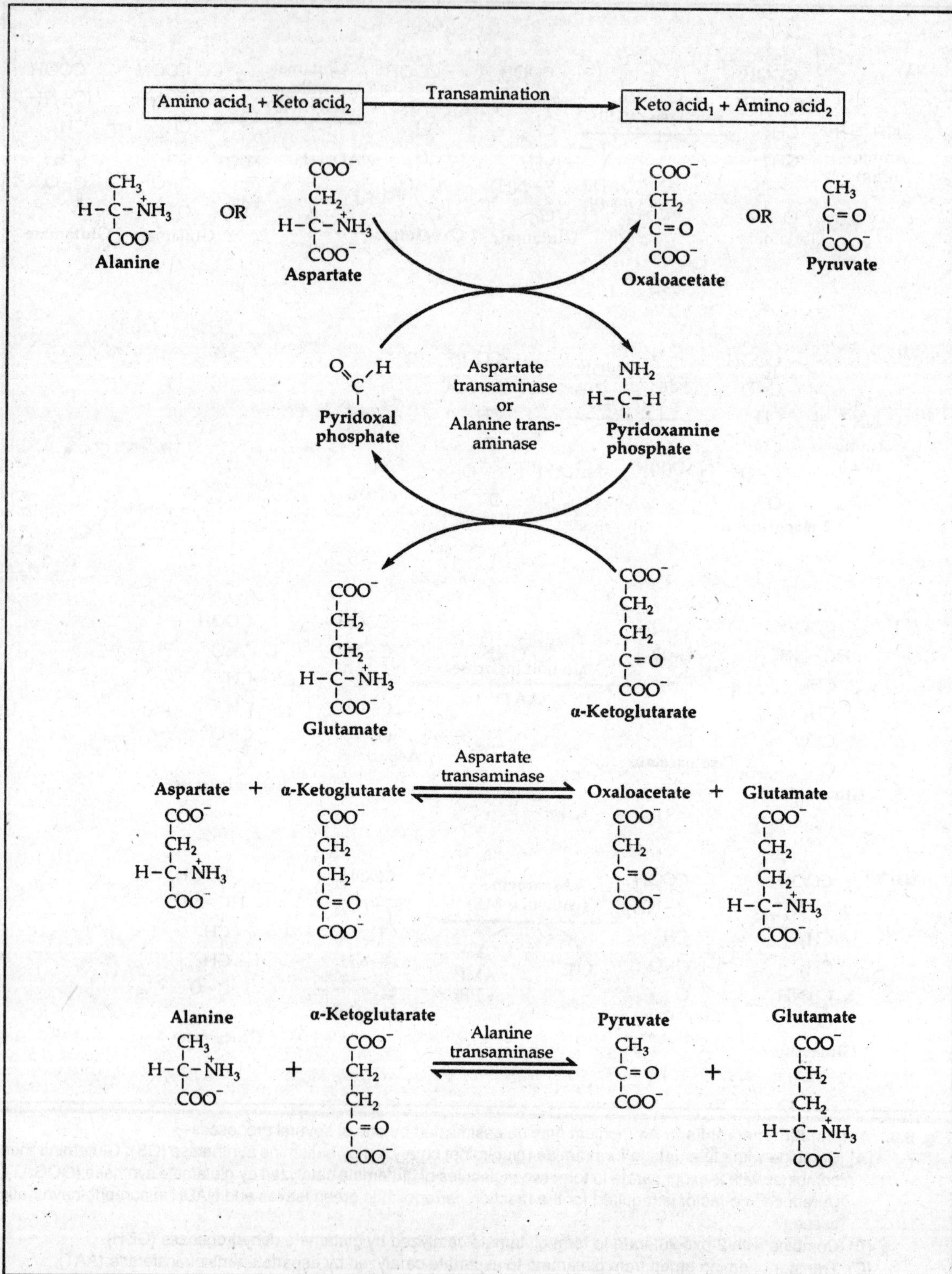

Fig. 6.7a: Transamination reaction catalyzed by aspartate transaminase or alanine transaminase.

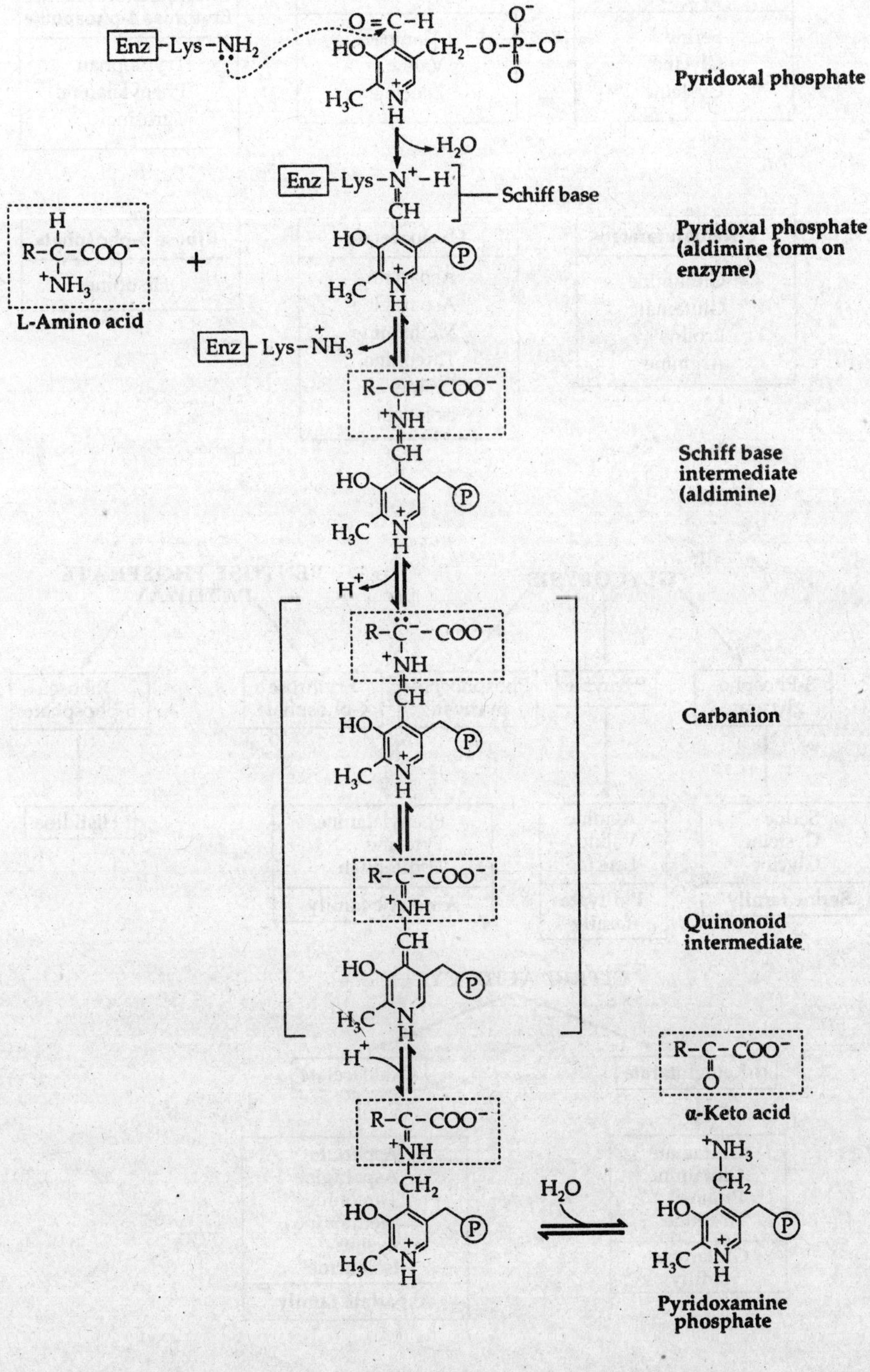

Fig. 6.7b: Mechanism of transamination facilitated by pyridoxal phosphate.

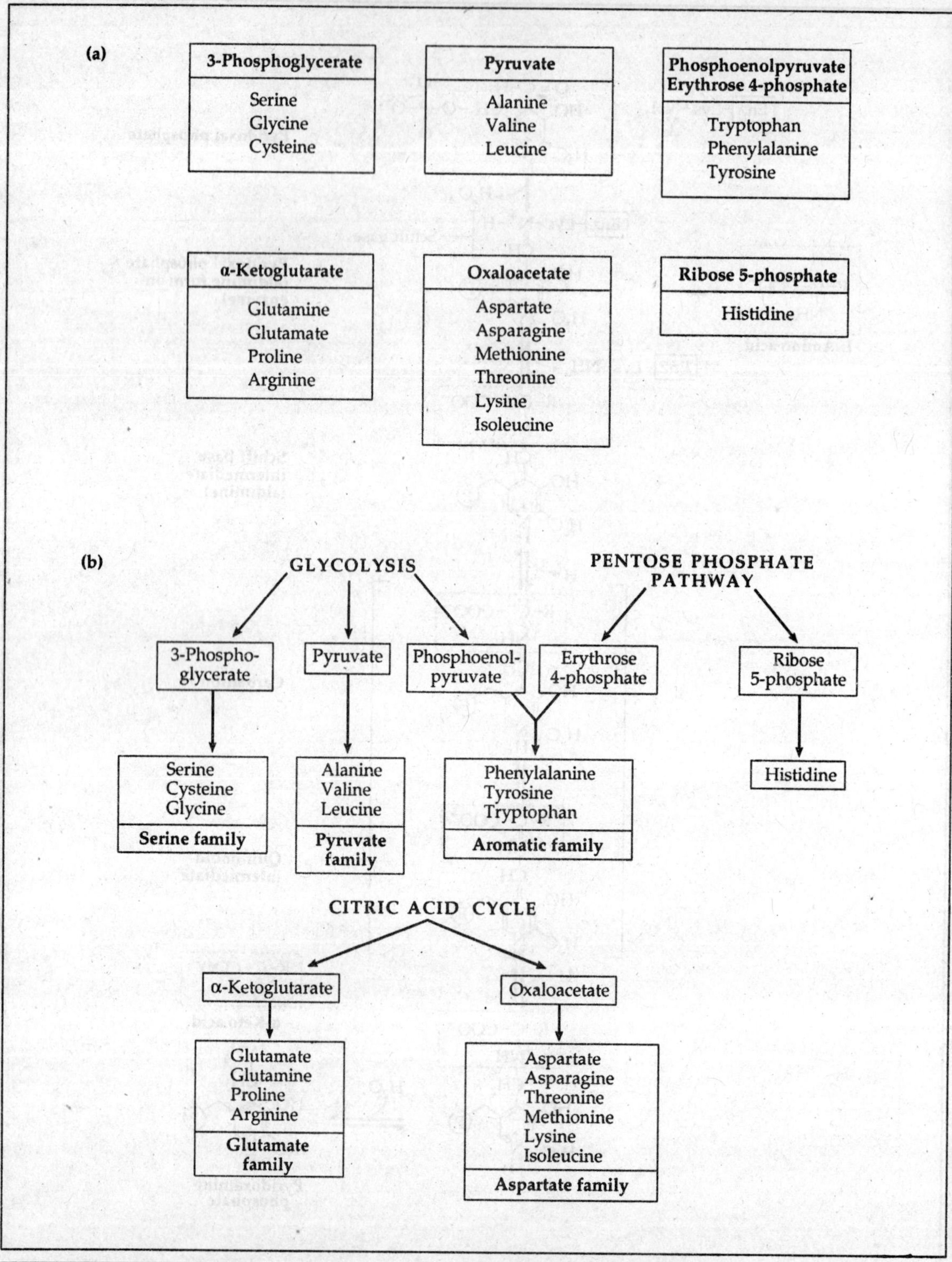

Fig. 6.8: **Amino acid biosynthetic families: (a)** Grouped by metabolic precursor, **(b)** Precursors are derived from glycolytic pathway, pentose phosphate pathway and citric acid cycle.

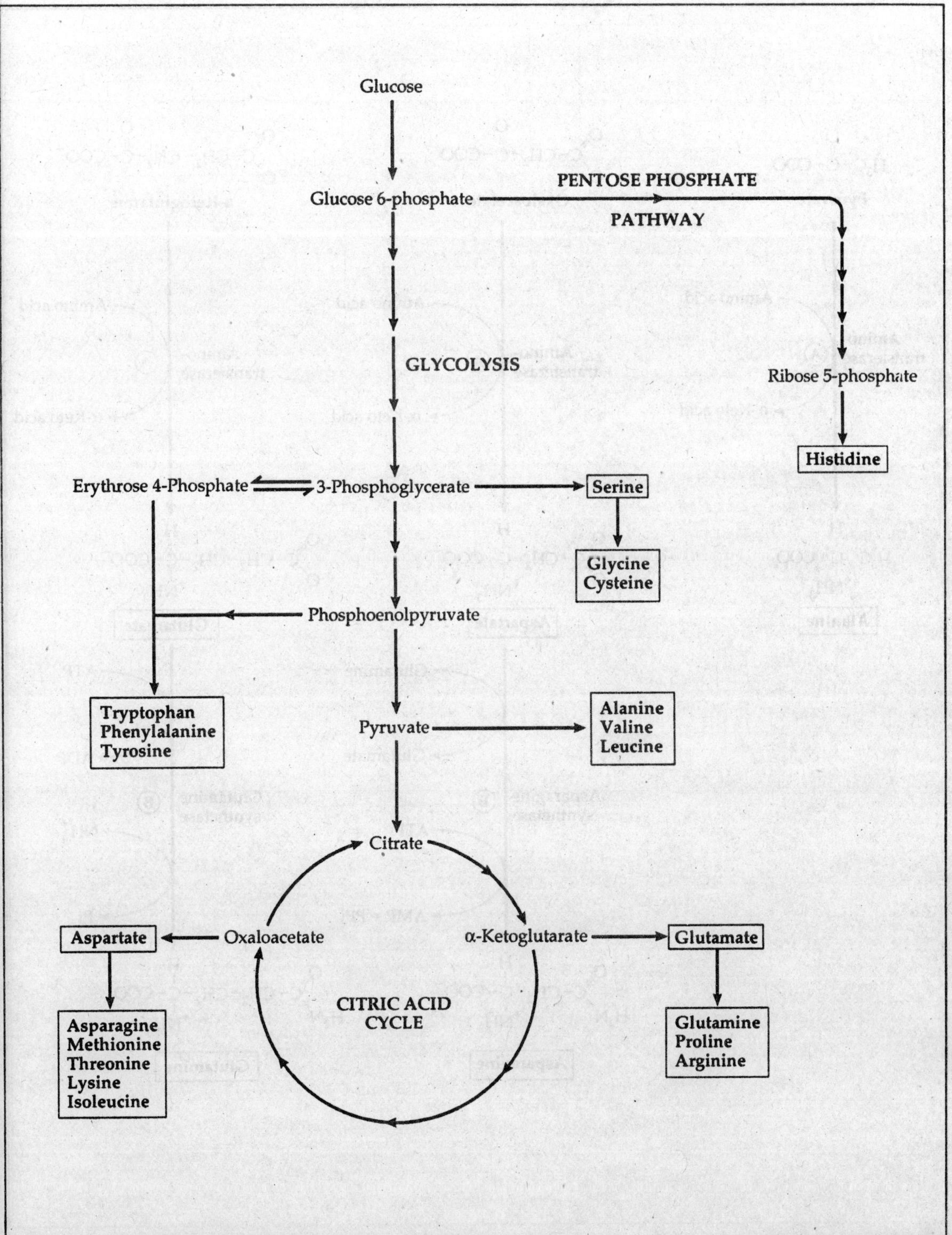

Fig. 6.9: Overview of amino acid biosynthesis: Precursors from glycolysis, citric acid cycle and pentose phosphate pathway are used to derive the amino acids.

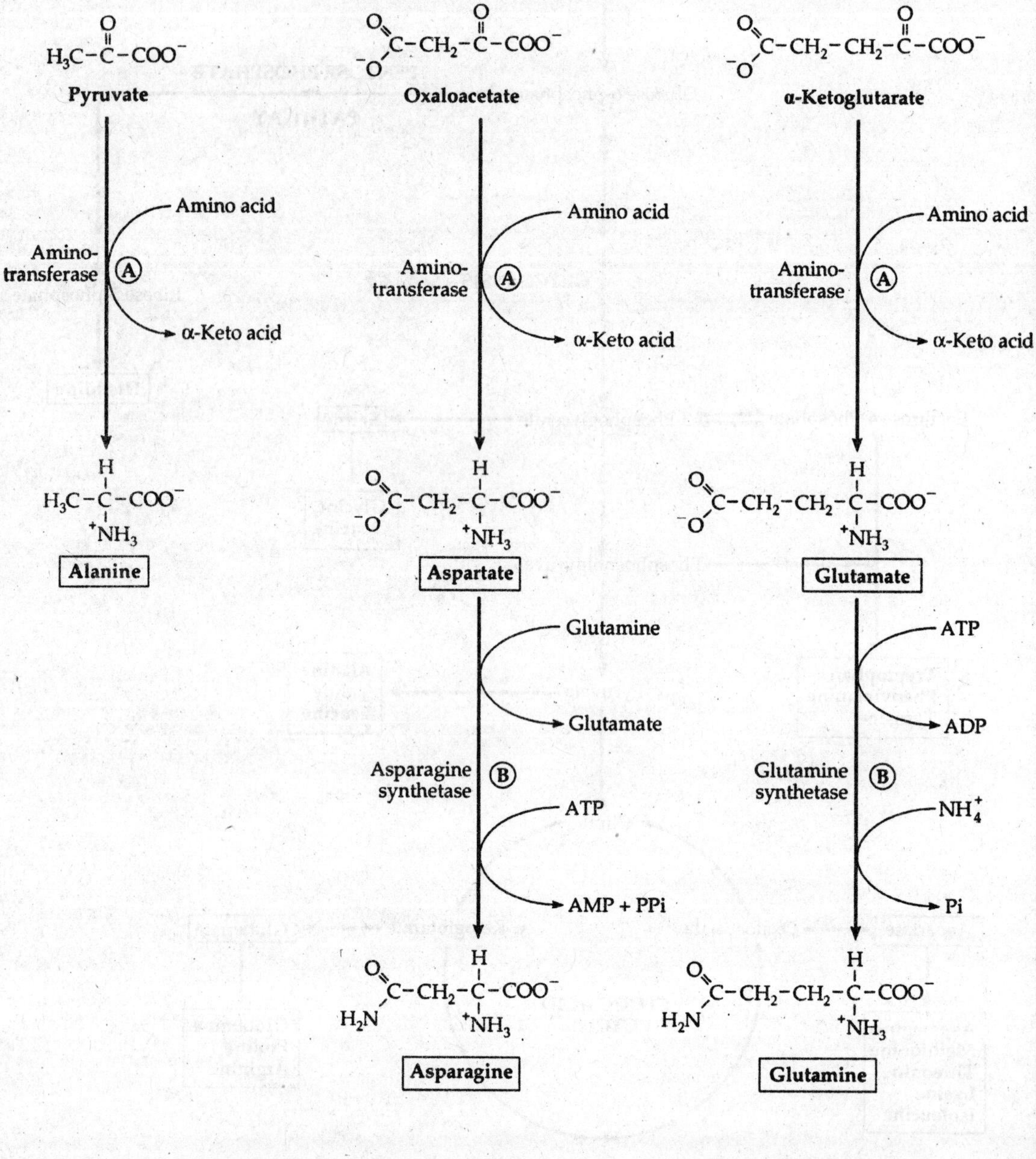

Fig. 6.10: The syntheses of alanine, aspartate, asparagine, glutamate, glutamine involve **(A)** transamination of pyruvate, oxaloacetate and α-ketoglutarate and **(B)** amidation of aspartate and glutamate.

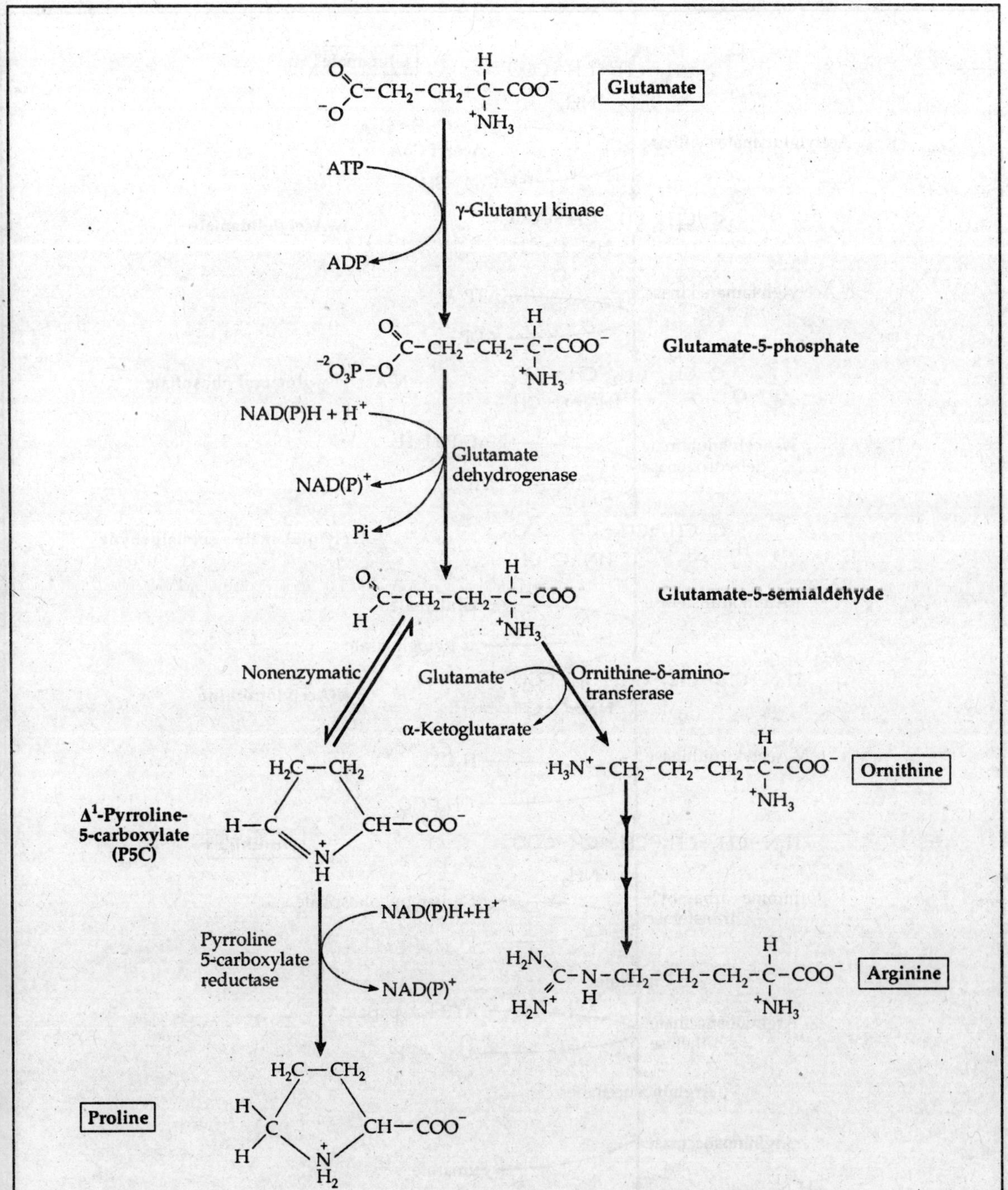

Fig. 6.11: Biosynthesis of glutamate family of amino acids—Arginine, Ornithine and Proline: All five carbon atoms of proline arise from glutamate. The γ-semialdehyde in the proline pathway undergoes a rapid, reversible cyclization to Δ^1-pyrroline-5-carboxylate (P5C), with the equilibrium favoring P5C formation. Cyclization is averted in the ornithine/arginine pathway.

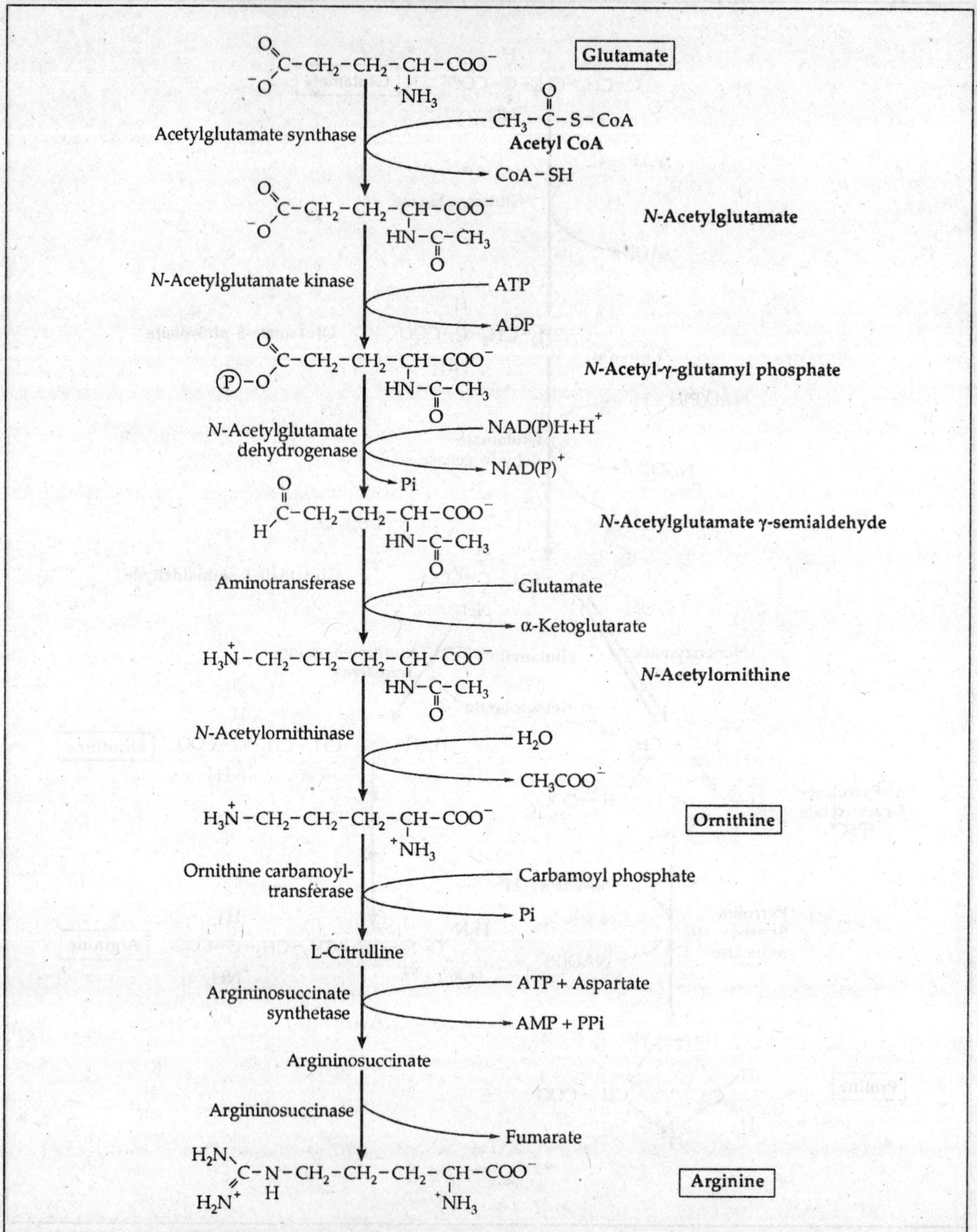

Fig. 6.12: Biosynthesis of arginine from glutamate in bacteria: Acetylation of amino group of glutamate and removal of acetyl group after transamination occur in ornithine/arginine biosynthetic pathway. The synthesis of arginine from ornithine occurs with citrulline and arginosuccinate as intermediates.

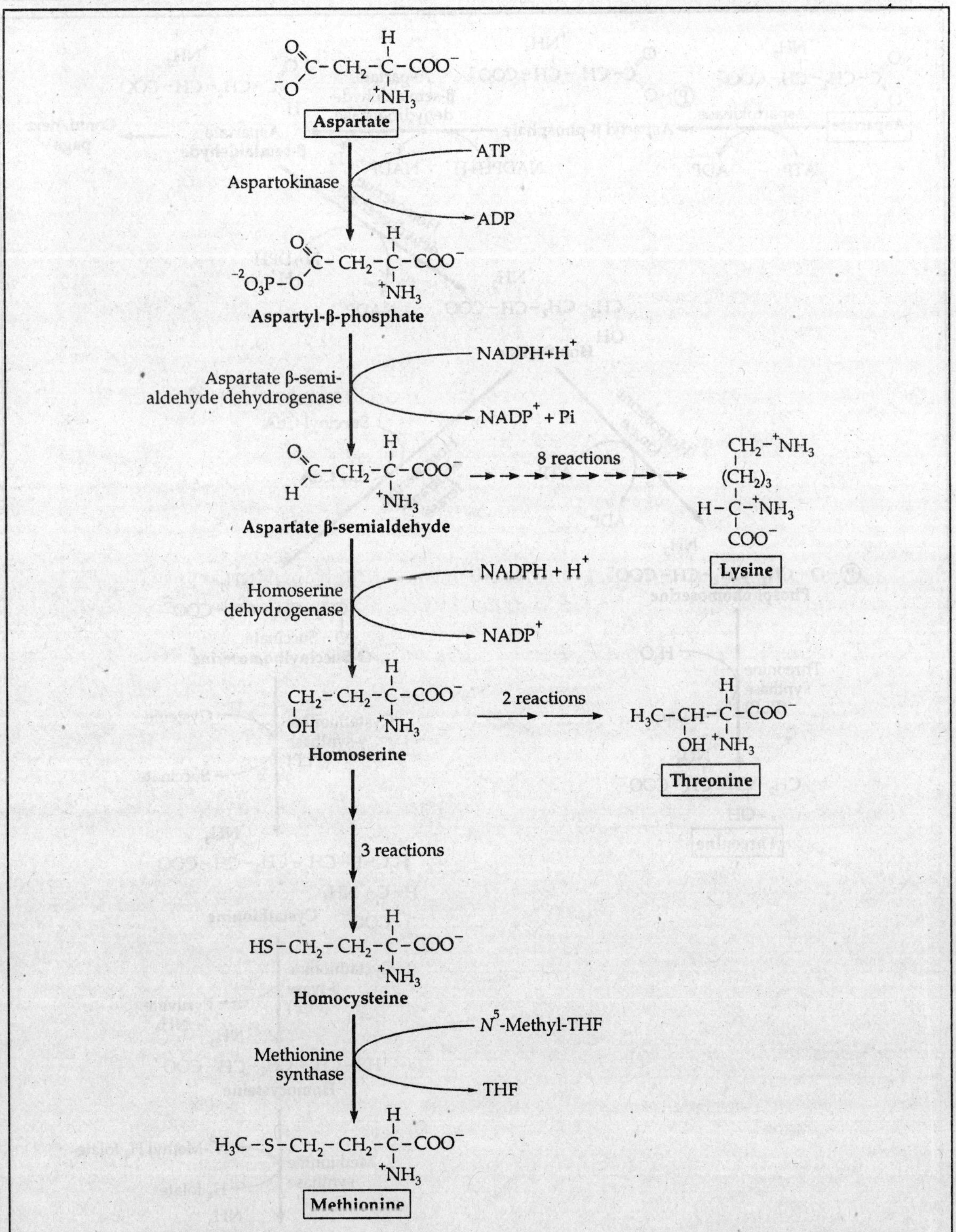

Fig. 6.13: Biosynthesis of the aspartate family amino acids–Lysine, Methionine and Threonine (in brief).

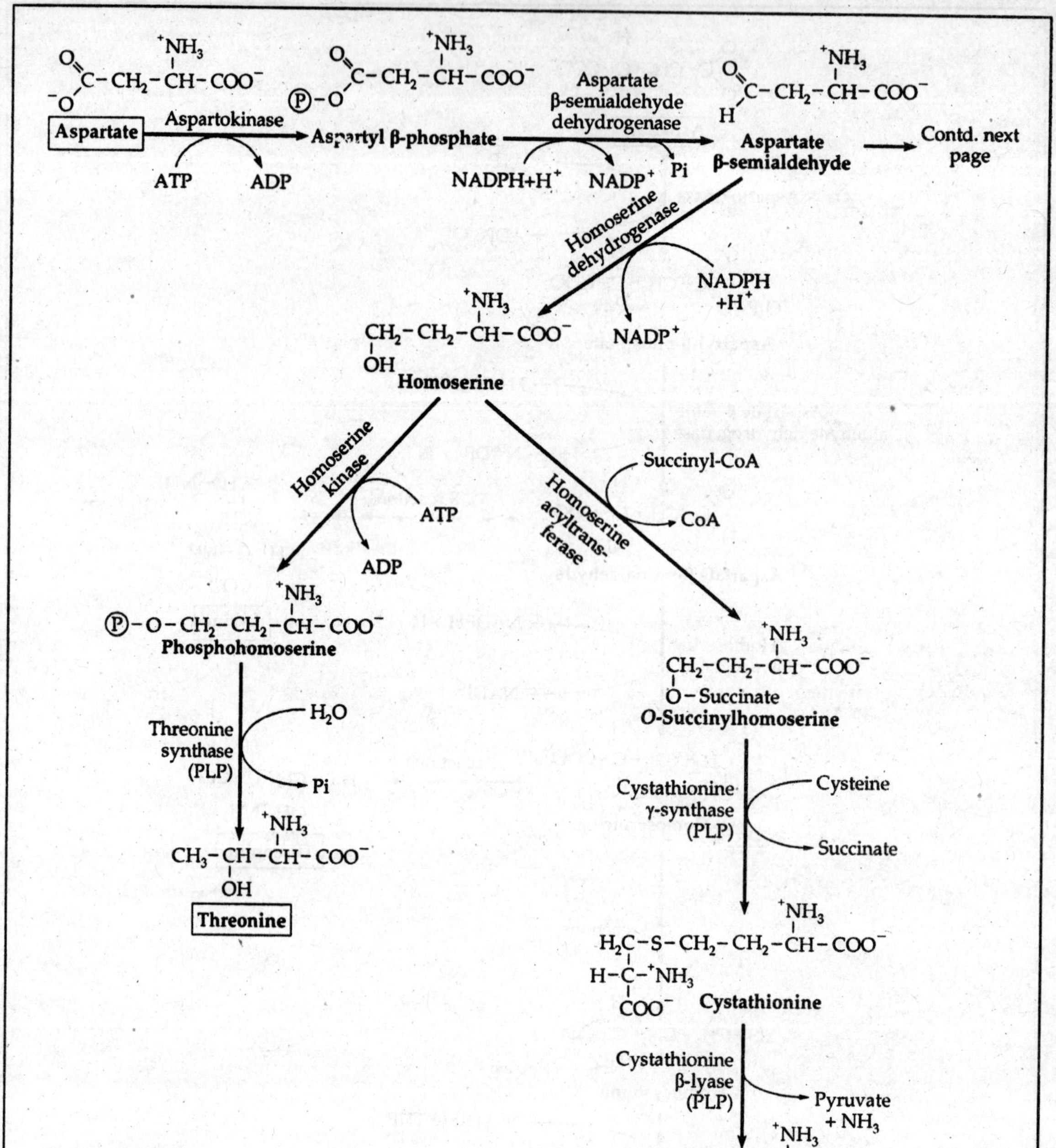

Aspartate
Aspartokinase
ATP
ADP
Aspartyl β-phosphate
Aspartate β-semialdehyde dehydrogenase
NADPH+H⁺
NADP⁺
Pi
Aspartate β-semialdehyde
Contd. next page
Homoserine dehydrogenase
NADPH +H⁺
NADP⁺
Homoserine
Homoserine kinase
ATP
ADP
Phosphohomoserine
Threonine synthase (PLP)
H₂O
Pi
Threonine
Homoserine acyltransferase
Succinyl-CoA
CoA
O-Succinylhomoserine
Cystathionine γ-synthase (PLP)
Cysteine
Succinate
Cystathionine
Cystathionine β-lyase (PLP)
Pyruvate + NH₃
Homocysteine
Methionine synthase
N⁵-Methyl H₄ folate
H₄ folate
Methionine

Cont. previous page

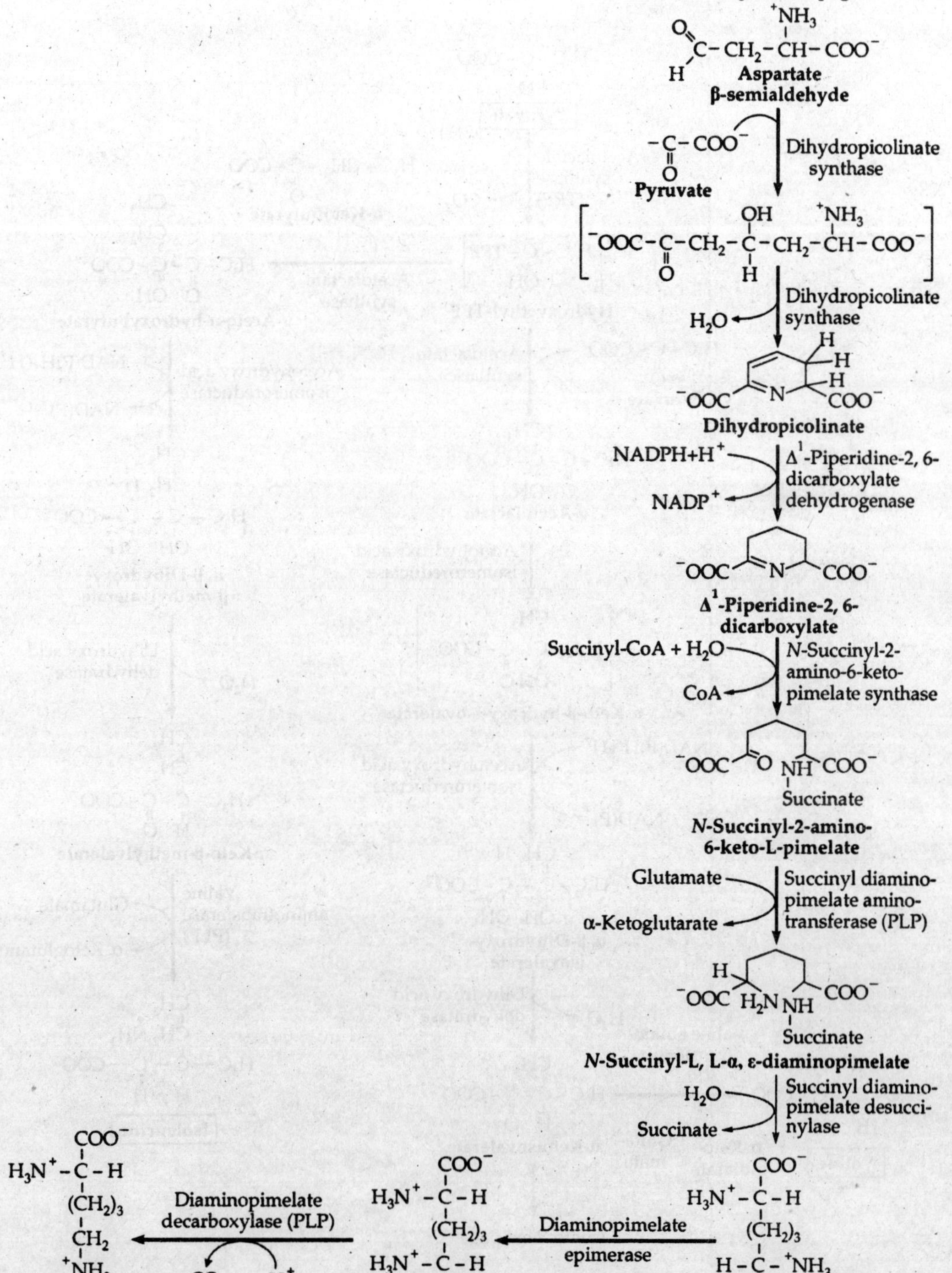

Fig. 6.14: Biosynthesis of aspartate family amino acids: Threonine, Methionine, Lysine (in details).

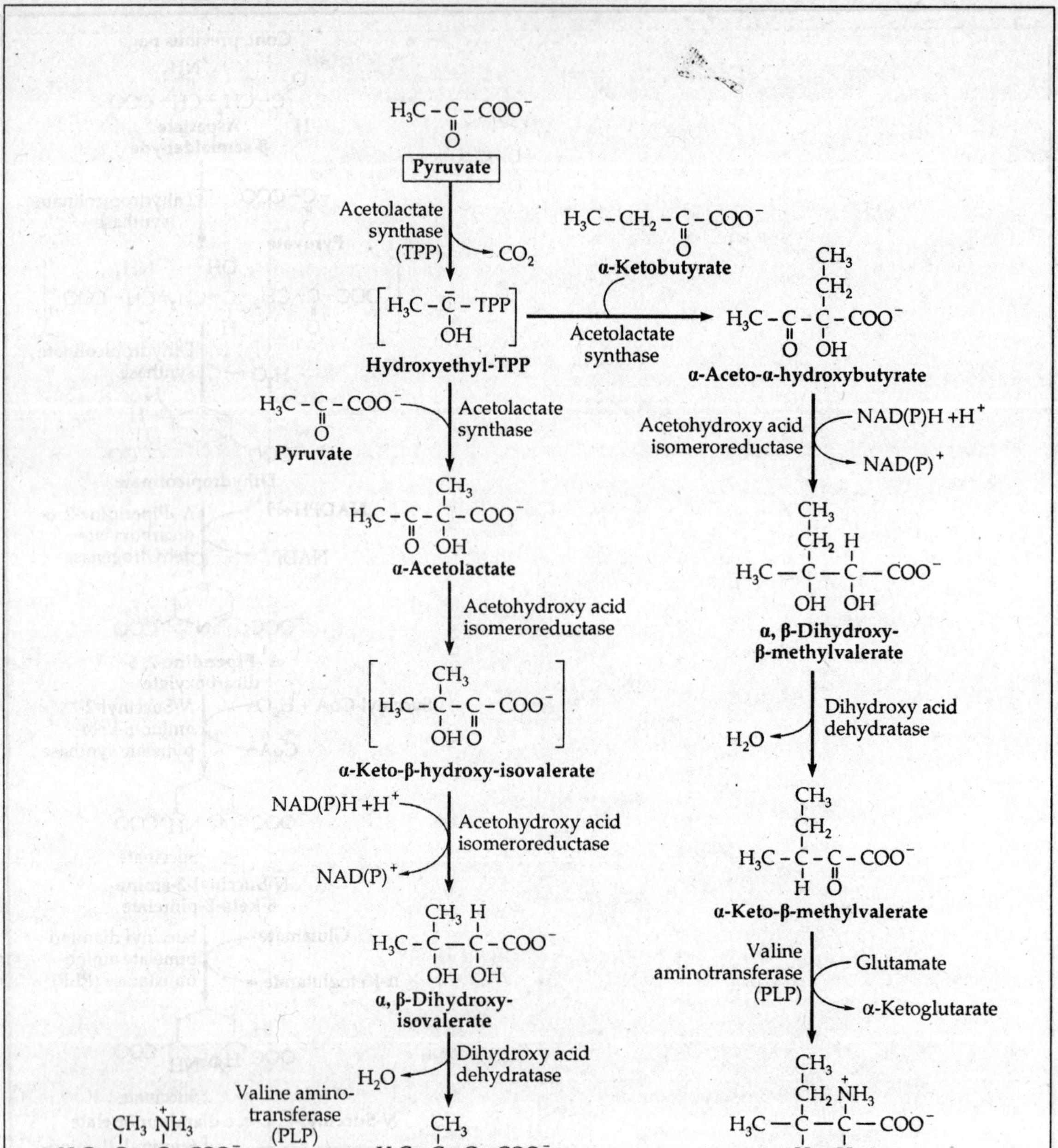
Pyruvate
Acetolactate synthase (TPP)
CO_2
Hydroxyethyl-TPP
α-Ketobutyrate
Acetolactate synthase
α-Aceto-α-hydroxybutyrate
Pyruvate
Acetolactate synthase
α-Acetolactate
Acetohydroxy acid isomeroreductase
NAD(P)H +H⁺
NAD(P)⁺
α, β-Dihydroxy-β-methylvalerate
α-Keto-β-hydroxy-isovalerate
Dihydroxy acid dehydratase
H_2O
α-Keto-β-methylvalerate
Valine aminotransferase (PLP)
Glutamate
α-Ketoglutarate
α, β-Dihydroxy-isovalerate
Valine amino-transferase (PLP)
α-Keto-glutarate
Gluta-mate
α-Ketoisovalerate
Valine
Isoleucine
Contd. next page

Contd. previous page

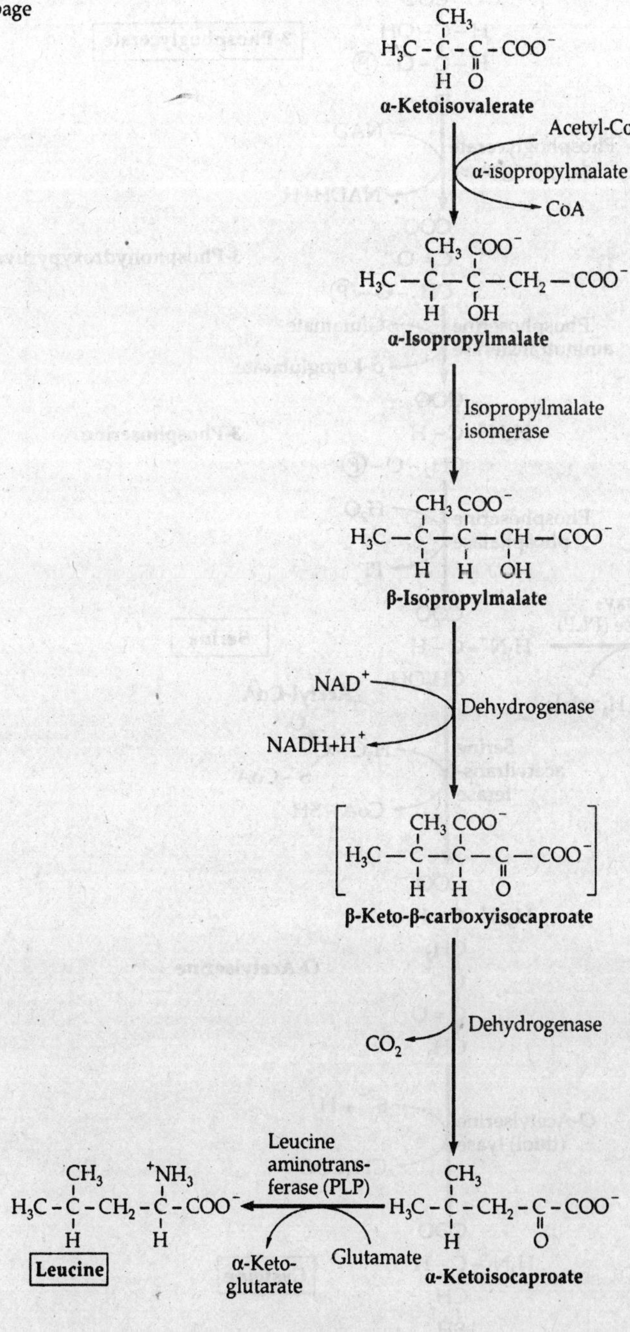

Fig. 6.15: Biosynthesis of pyruvate family amino acids—Isoleucine, Valine, Leucine: The first enzyme, acetolactate synthase, catalyzes two reactions—one leading to valine and leucine, and other to isoleucine.

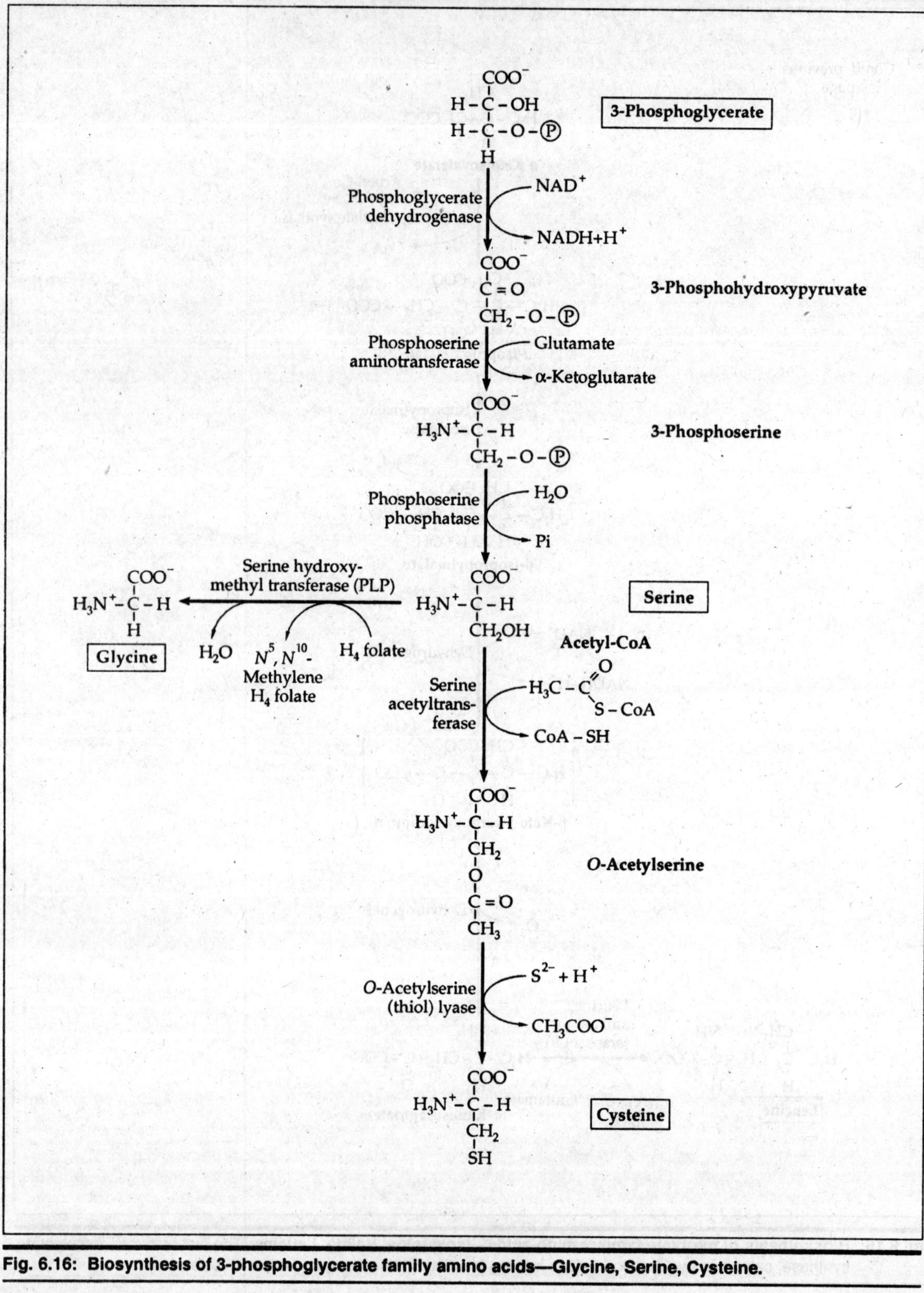

Fig. 6.16: Biosynthesis of 3-phosphoglycerate family amino acids—Glycine, Serine, Cysteine.

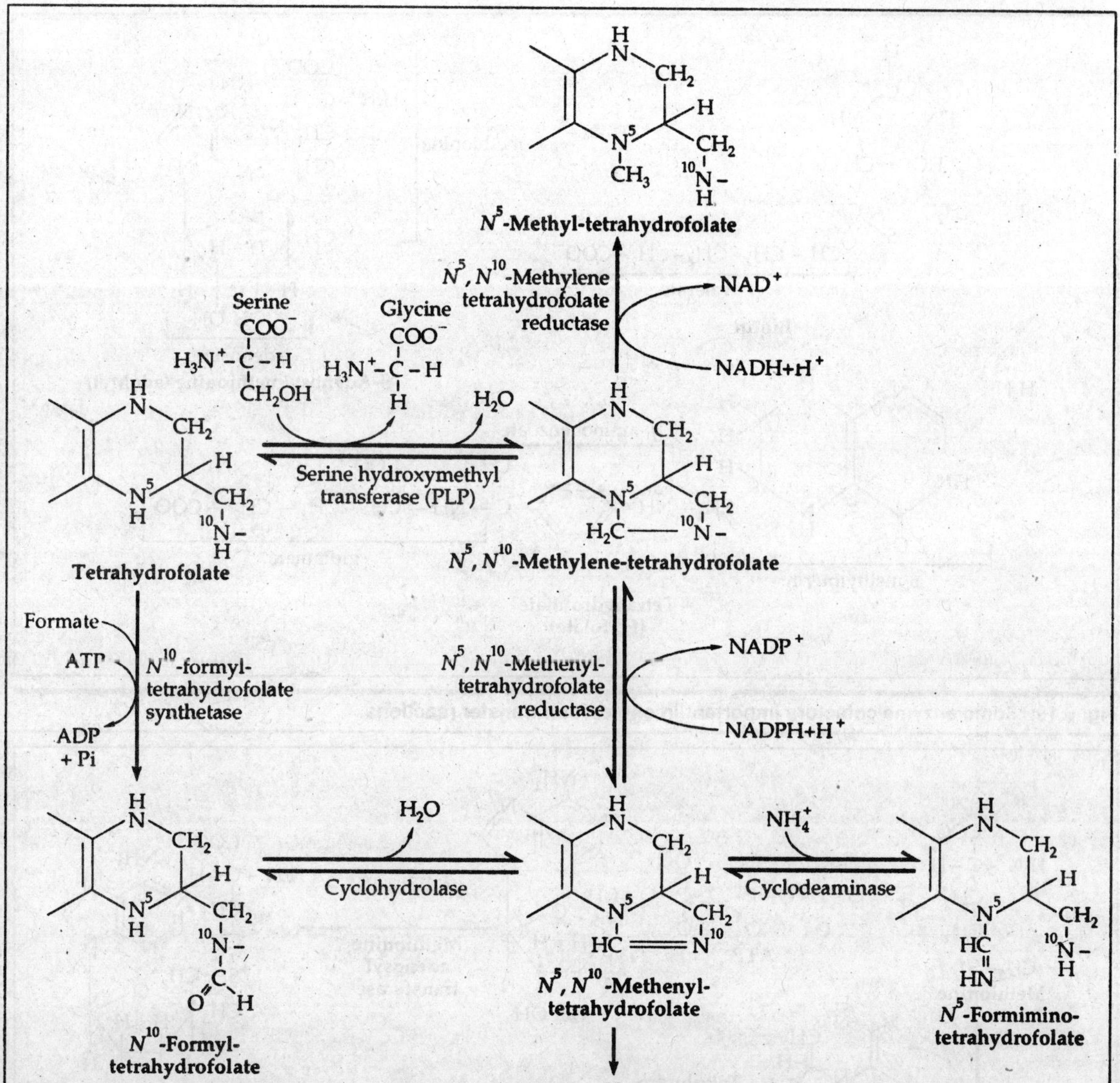

Fig. 6.17: Conversions of one-carbon units on tetrahydrofolate: The different molecular species are shown. The enzymatic transfer of formyl groups (as in purine biosynthesis or in the formation of formylmethionine) generally uses N^{10}-formyltetrahydrofolate rather than N^5-formyltetrahydrofolate because the latter species is significantly more stable and therefore a weaker donor of formyl groups. The equilibria of the reactions that interconnect these molecular species favour formation of N^5-formyltetrahydrofolate. Conversion of N^5-to N^{10}-formyltetrahydrofolate requires ATP.

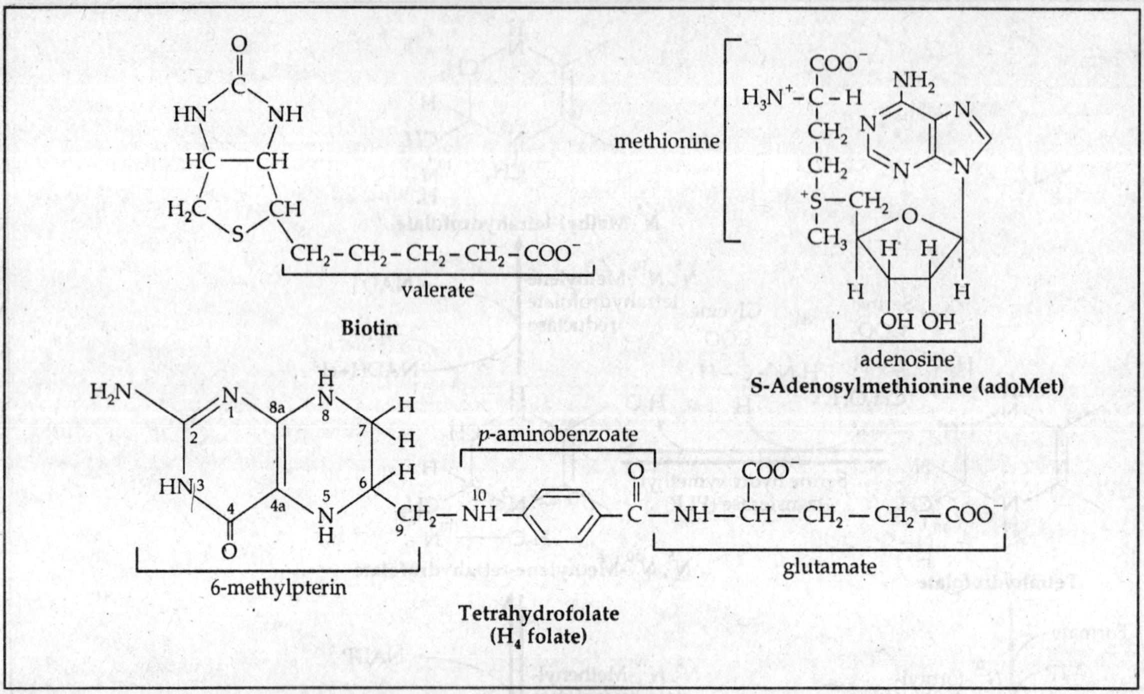

Fig. 6.18: Some enzyme cofactors important in one-carbon transfer reactions.

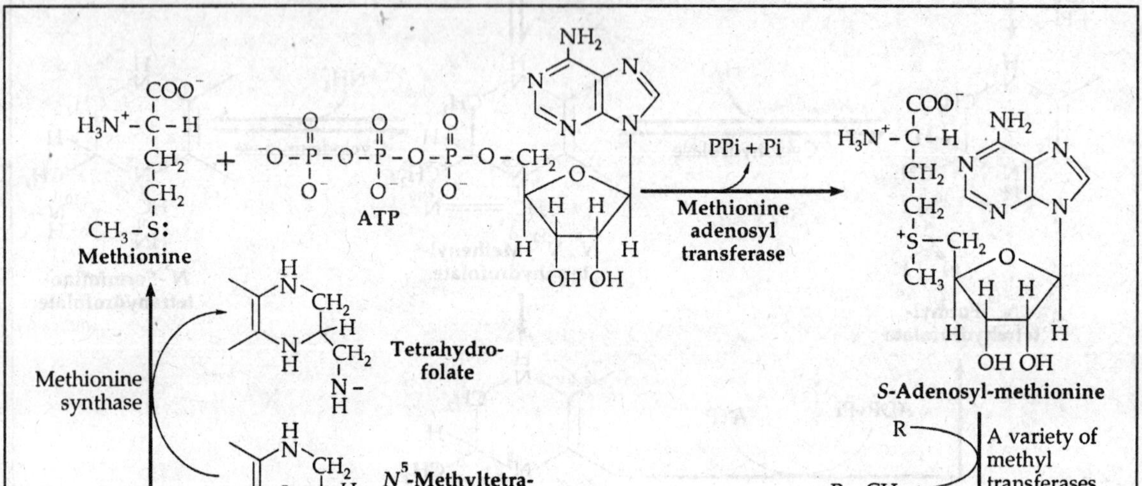

Fig. 6.19: Synthesis of methionine and S-adenosylmethionine in an activated-methyl cycle: S-Adenosylmethionine, which has positively charged sulfur, is a powerful methylating agent in a number of biosynthetic reactions.

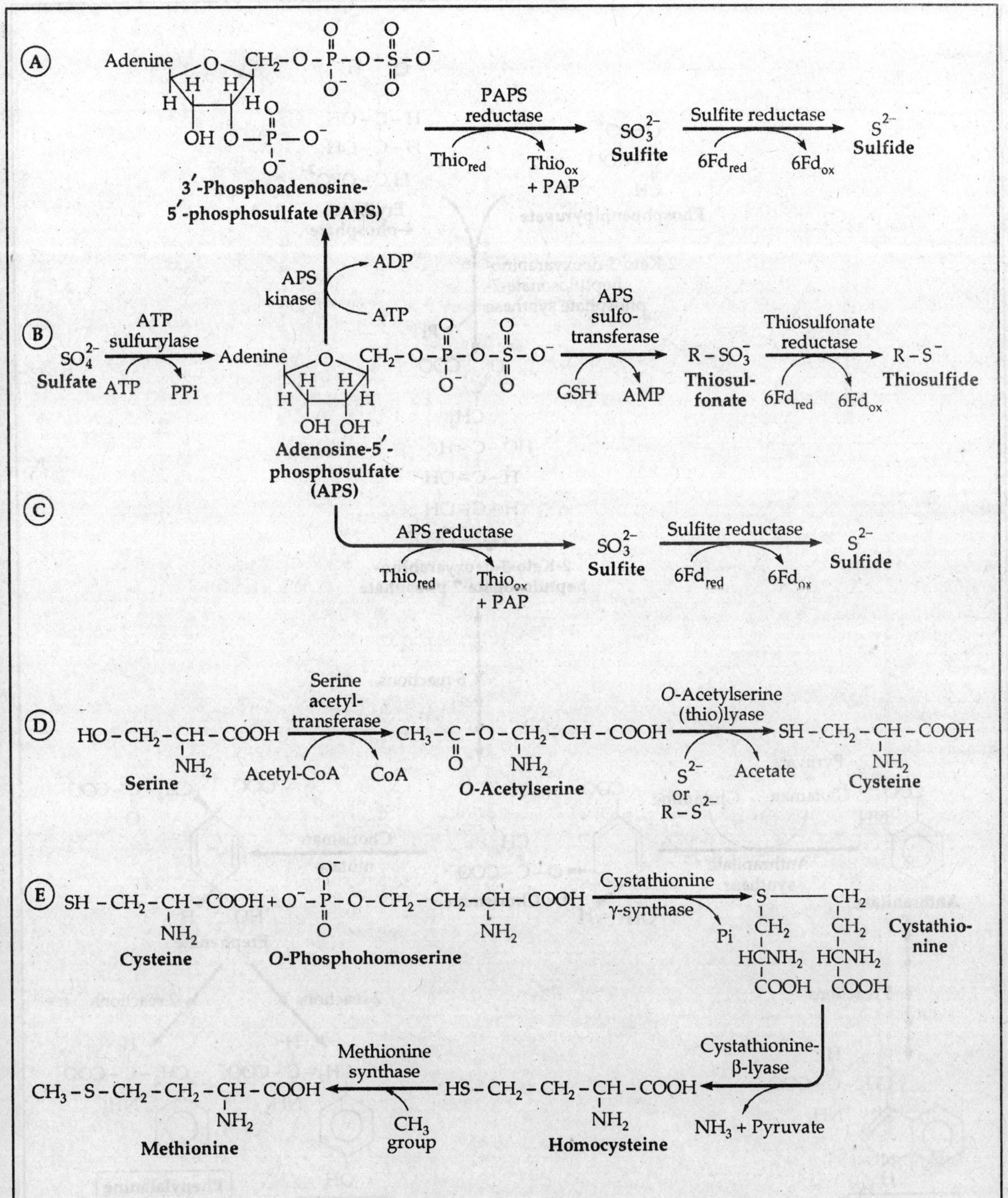

Fig. 6.20: Sulfur assimilation pathways in amino acid synthesis: **(A–C)** Sulfide or thiosulfide is produced from adenosine-5′-phosphosulfate (APS) through one of three pathways. **(B)** The enzyme ATP sulfurylase cleaves pyrophosphate from ATP and replaces it with sulfate in the pathway. **(D)** Serine reacts with acetyl-CoA to produce *O*-acetylserine and the sulfide or thiosulfide produced reacts with *O*-acetylserine to form cysteine. **(E)** Cysteine may be further modified to produce methionine. [Fd–ferredoxin, GSH–glutathione, Thio–thioredoxin]

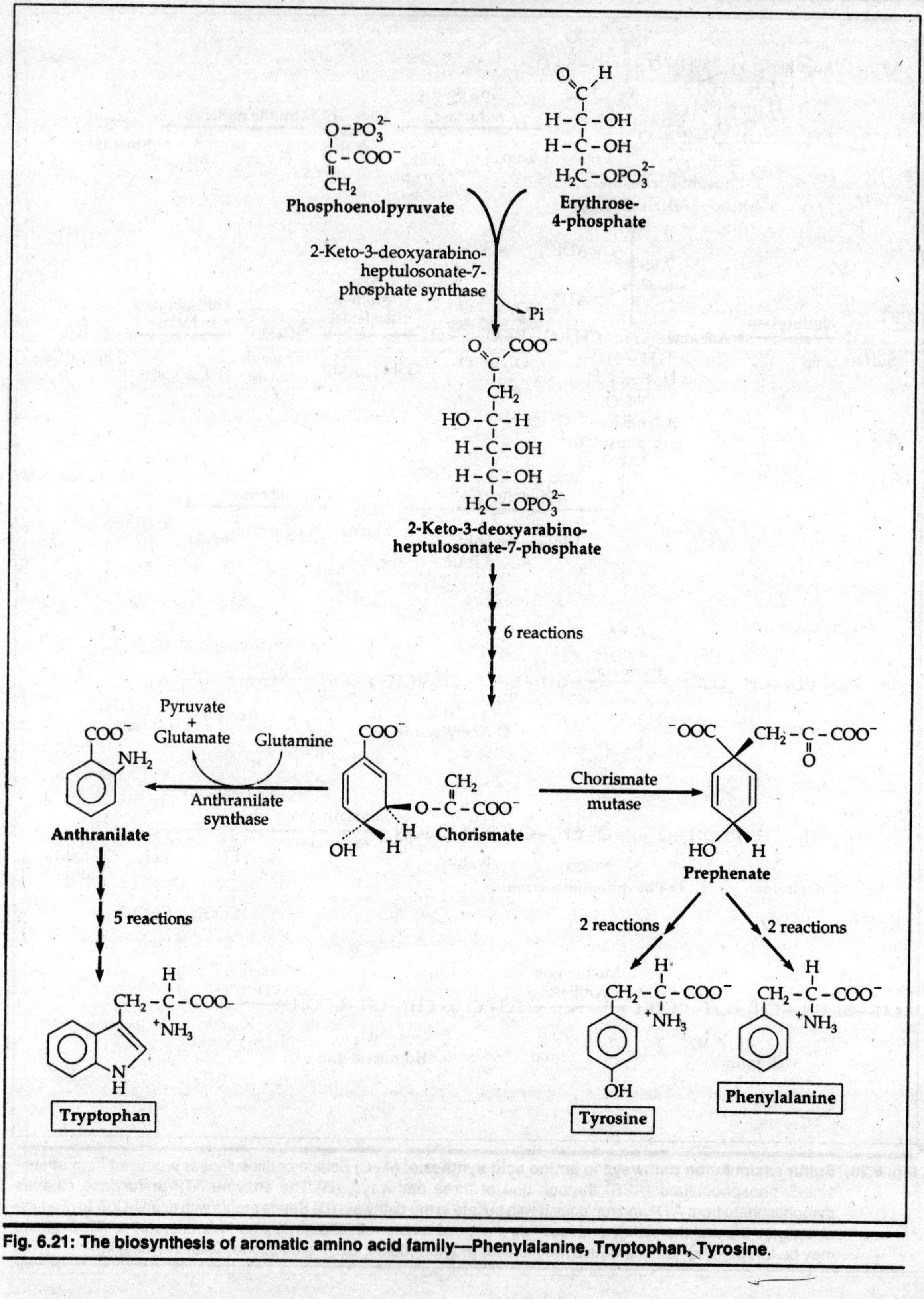

Fig. 6.21: The biosynthesis of aromatic amino acid family—Phenylalanine, Tryptophan, Tyrosine.

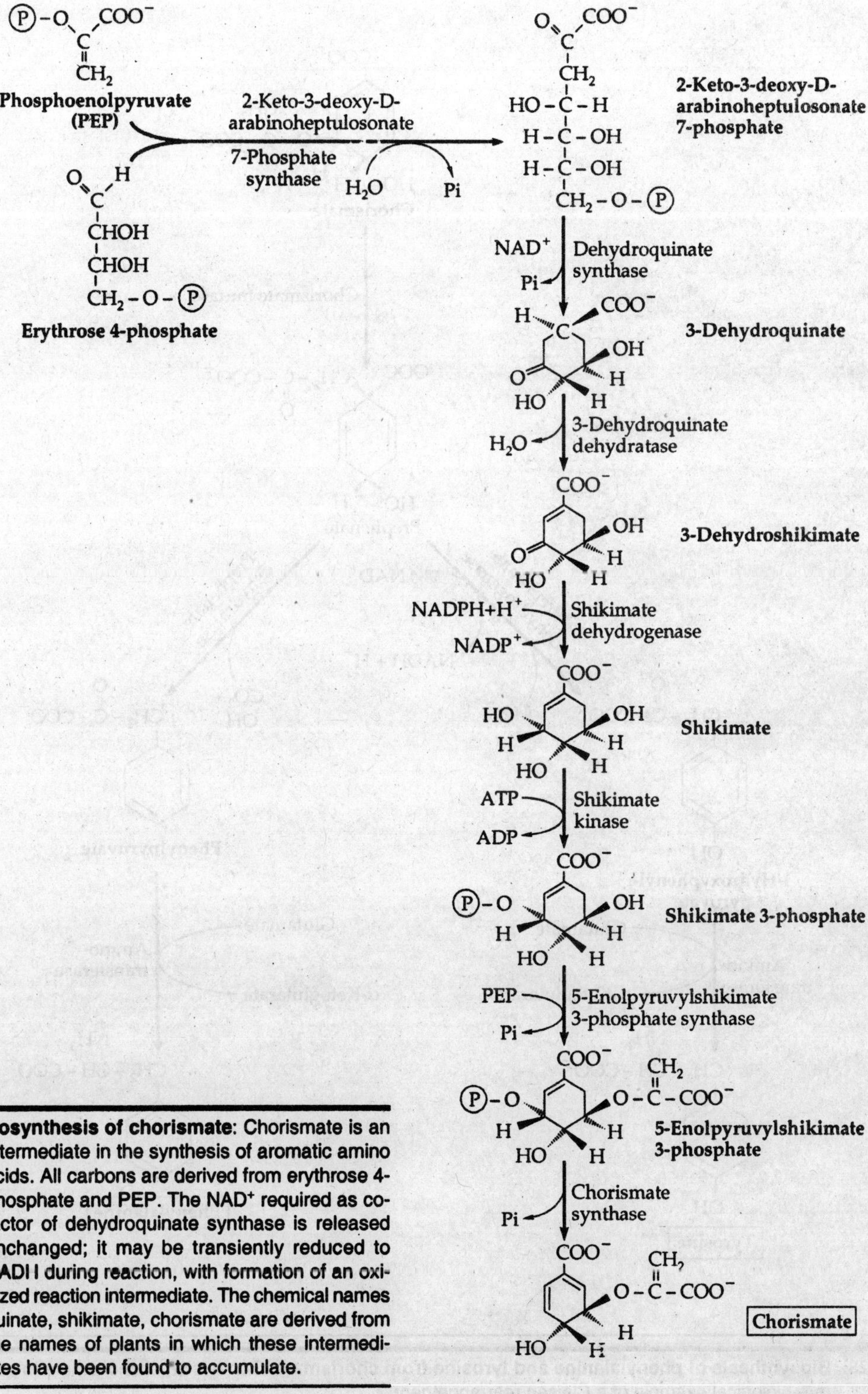

Fig. 6.22: Biosynthesis of chorismate: Chorismate is an intermediate in the synthesis of aromatic amino acids. All carbons are derived from erythrose 4-phosphate and PEP. The NAD^+ required as cofactor of dehydroquinate synthase is released unchanged; it may be transiently reduced to NADI I during reaction, with formation of an oxidized reaction intermediate. The chemical names quinate, shikimate, chorismate are derived from the names of plants in which these intermediates have been found to accumulate.

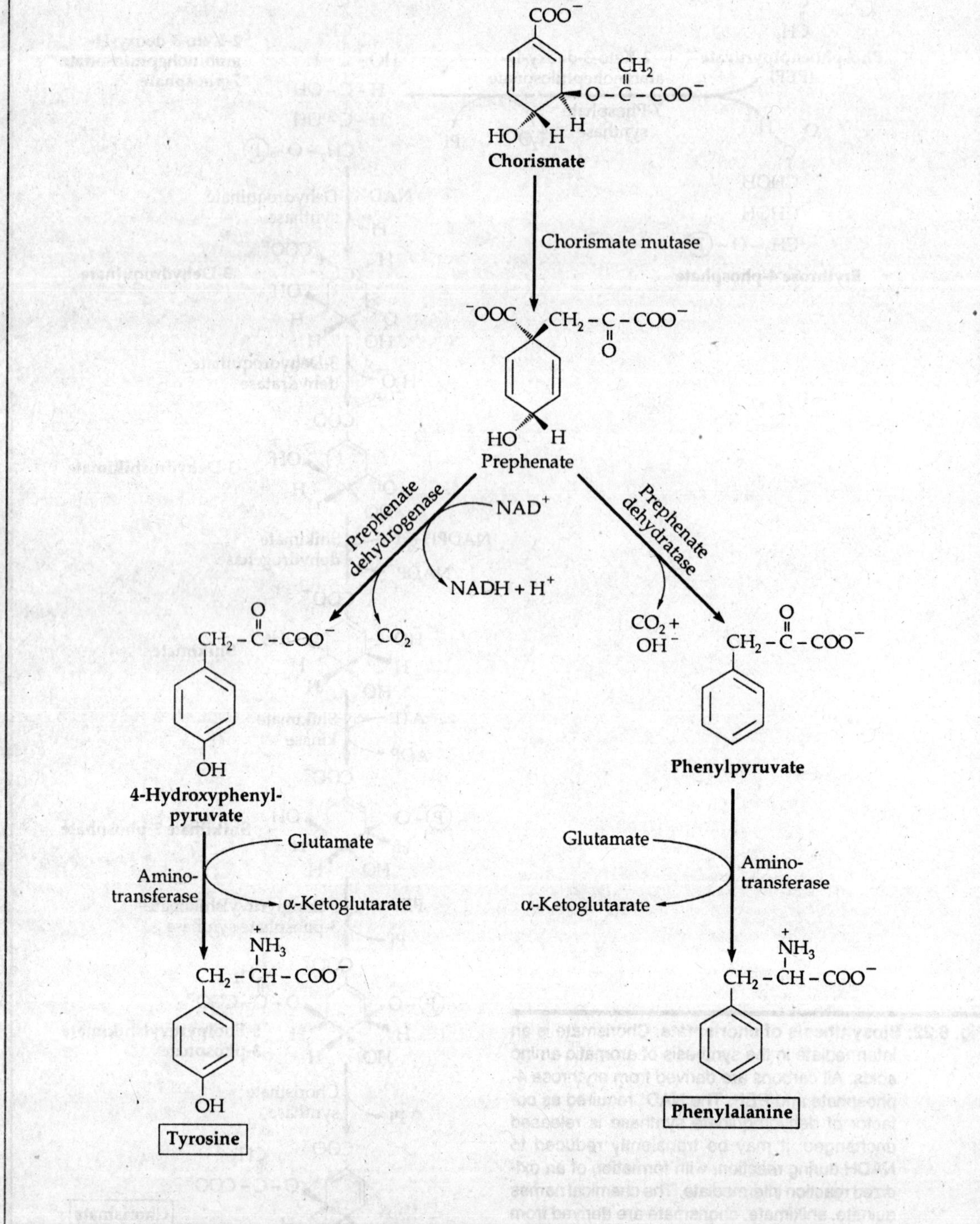

Fig. 6.23: Biosynthesis of phenylalanine and tyrosine from chorismate: Conversion of chorismate to prephenate is a rare biological example of a Claisen rearrangement.

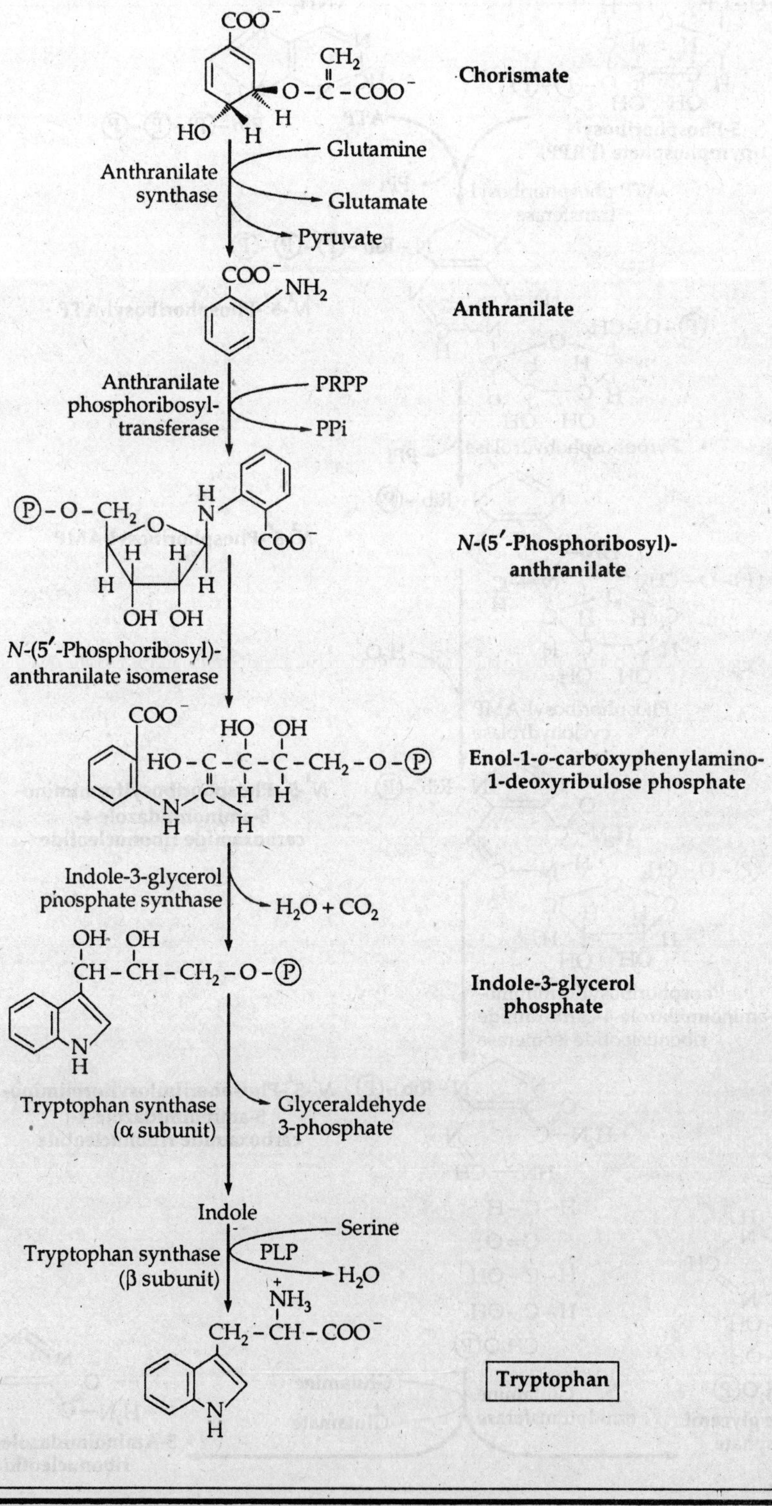

Fig. 6.24: Biosynthesis of tryptophan from chorismate.

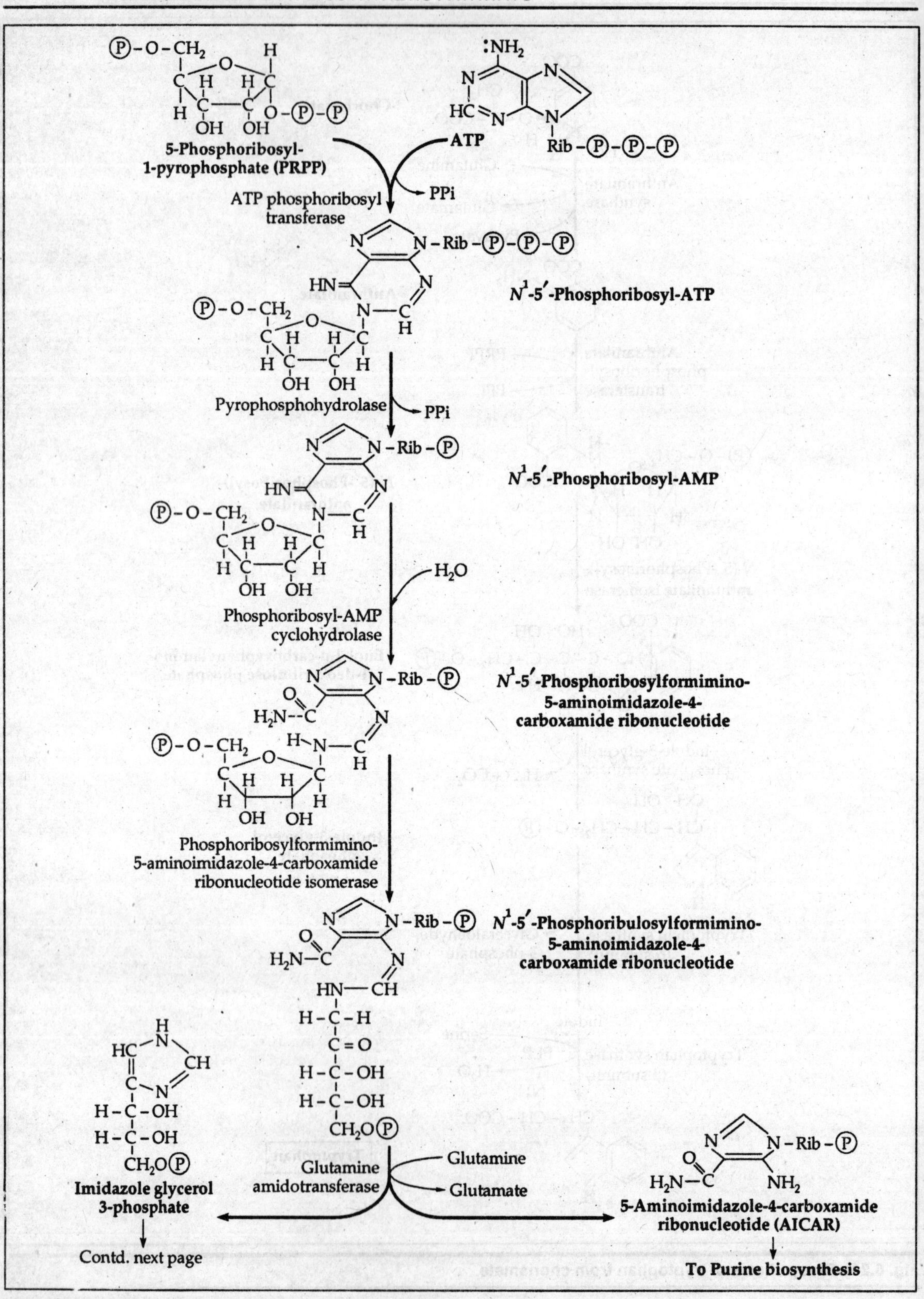
5-Phosphoribosyl-
1-pyrophosphate (PRPP)
ATP
PPi
ATP phosphoribosyl
transferase
N^1-5′-Phosphoribosyl-ATP
Pyrophosphohydrolase
PPi
N^1-5′-Phosphoribosyl-AMP
H_2O
Phosphoribosyl-AMP
cyclohydrolase
N^1-5′-Phosphoribosylformimino-
5-aminoimidazole-4-
carboxamide ribonucleotide
Phosphoribosylformimino-
5-aminoimidazole-4-carboxamide
ribonucleotide isomerase
N^1-5′-Phosphoribulosylformimino-
5-aminoimidazole-4-
carboxamide ribonucleotide
Glutamine
Glutamine
amidotransferase
Glutamate
Imidazole glycerol
3-phosphate
Contd. next page
5-Aminoimidazole-4-carboxamide
ribonucleotide (AICAR)
To Purine biosynthesis

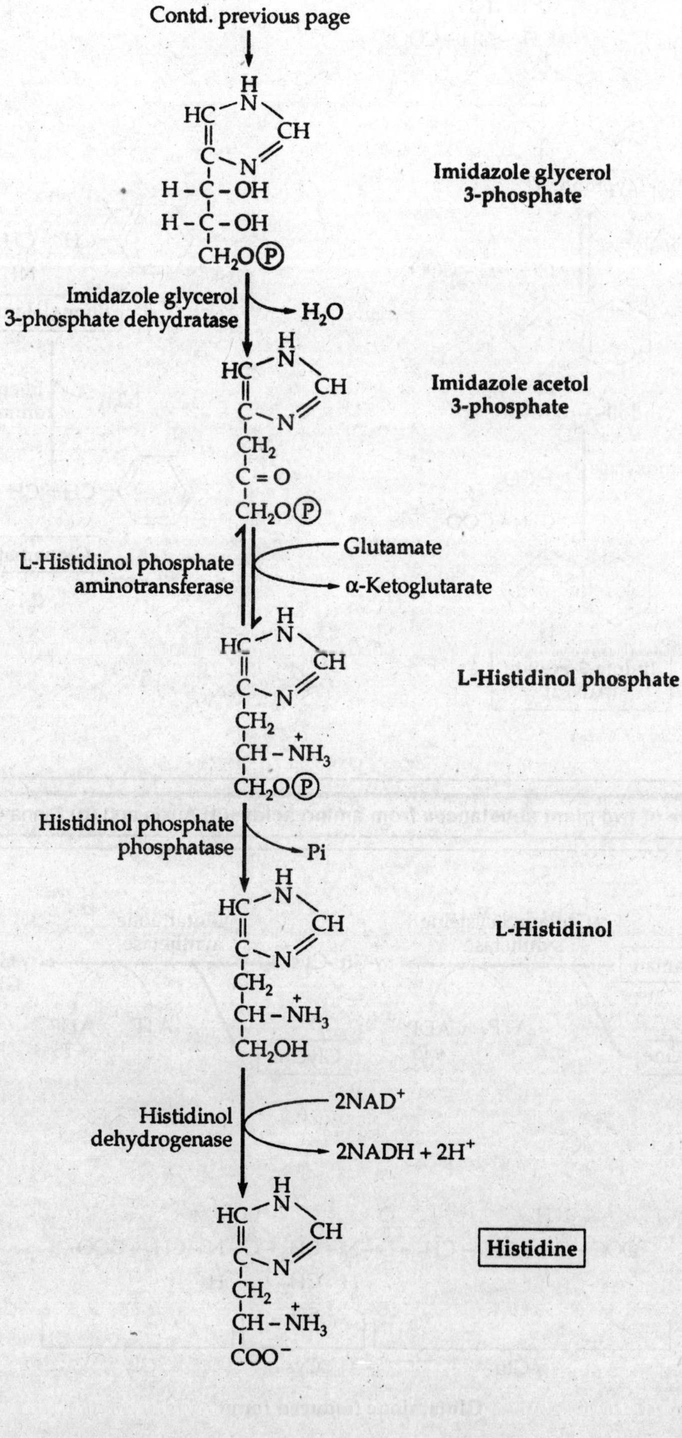

Fig. 6.25: Biosynthesis of Histidine: Nitrogens of histidine are derived one each from glutamine, glutamate, ATP. Derivative of ATP, i.e., AICAR is an intermediate in purine biosynthesis, so ATP is rapidly regenerated.

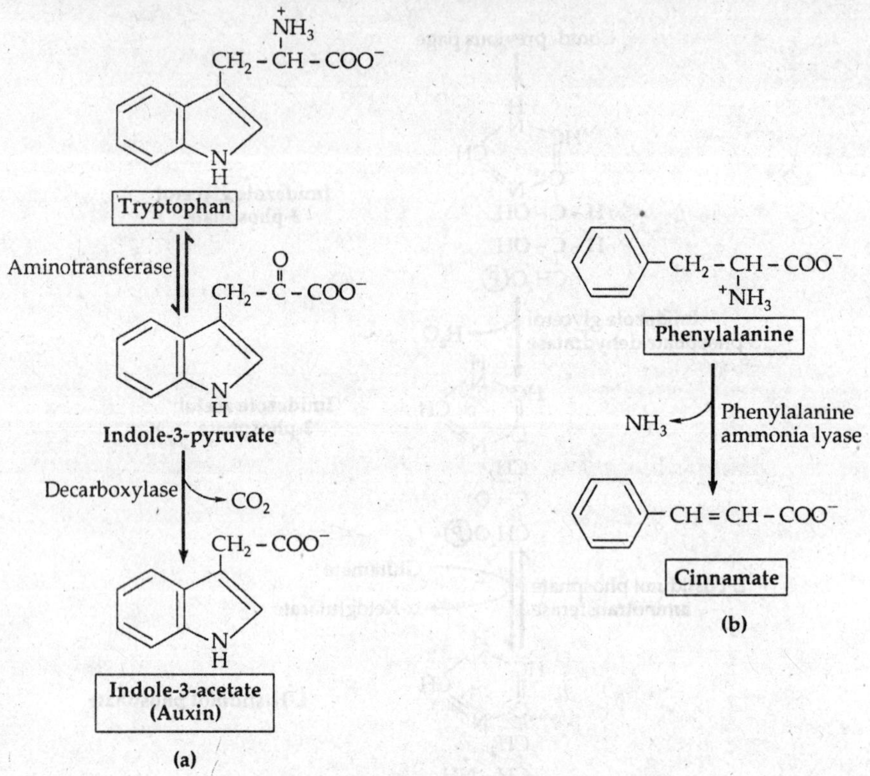

Fig. 6.26: Biosynthesis of two plant substances from amino acids: (a) Auxin and **(b)** Cinnamate.

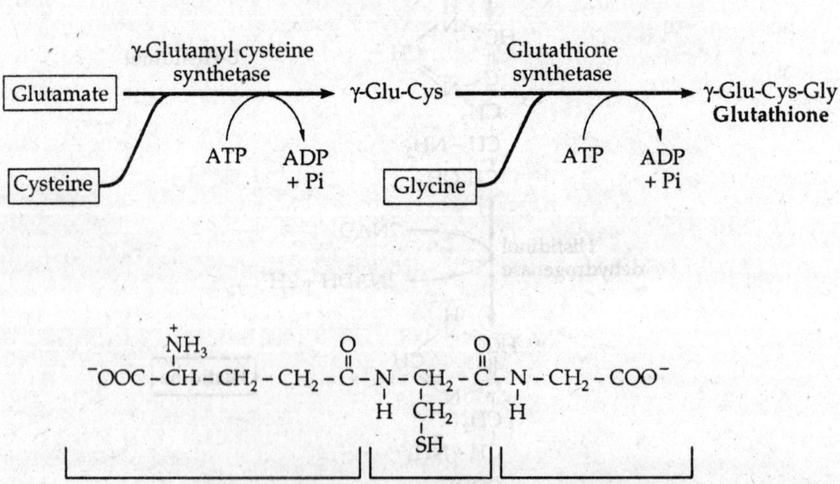

Fig. 6.27: Biosynthesis of glutathione.

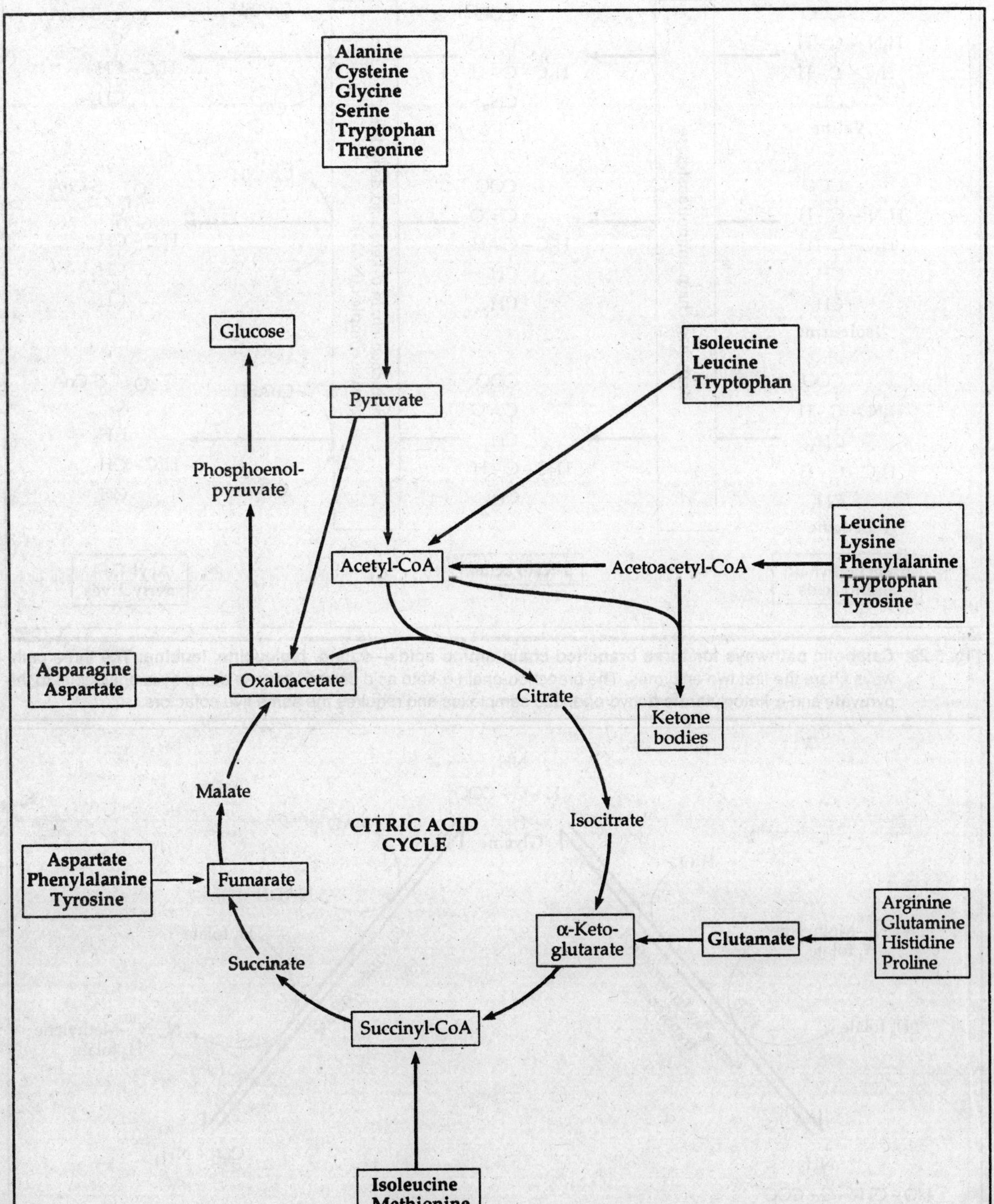

Fig. 6.28: Fate of carbon skeleton of amino acids.

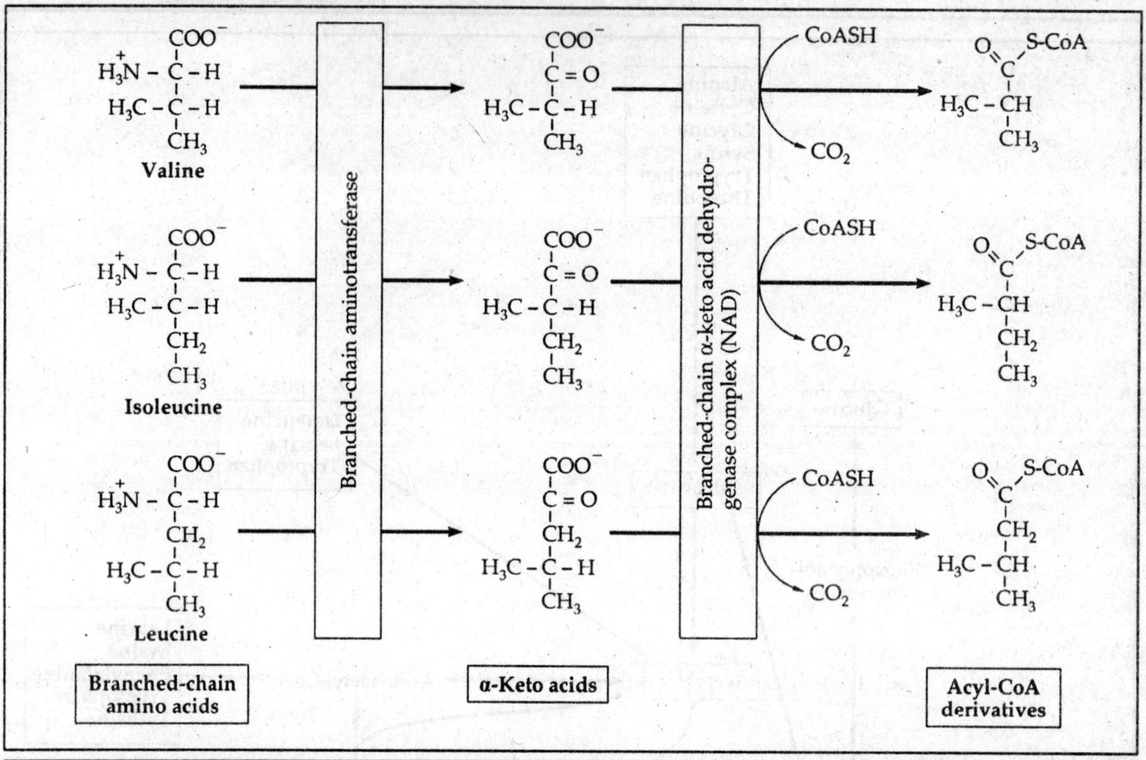

Fig. 6.29: Catabolic pathways for three branched-chain amino acids—valine, isoleucine, leucine: The three pathways share the first two enzymes. The branched-chain α-keto acid dehydrogenase complex is analogous to the pyruvate and α-ketoglutarate dehydrogenase complexes and requires the same five cofactors.

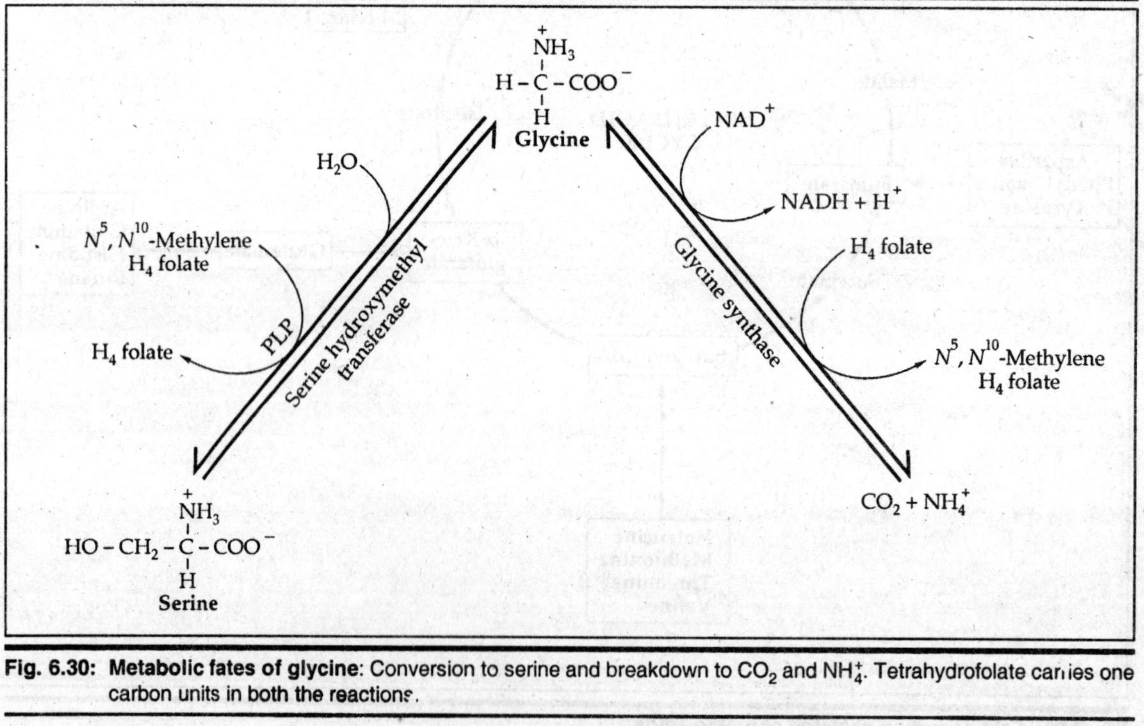

Fig. 6.30: Metabolic fates of glycine: Conversion to serine and breakdown to CO_2 and NH_4^+. Tetrahydrofolate carries one carbon units in both the reactions.

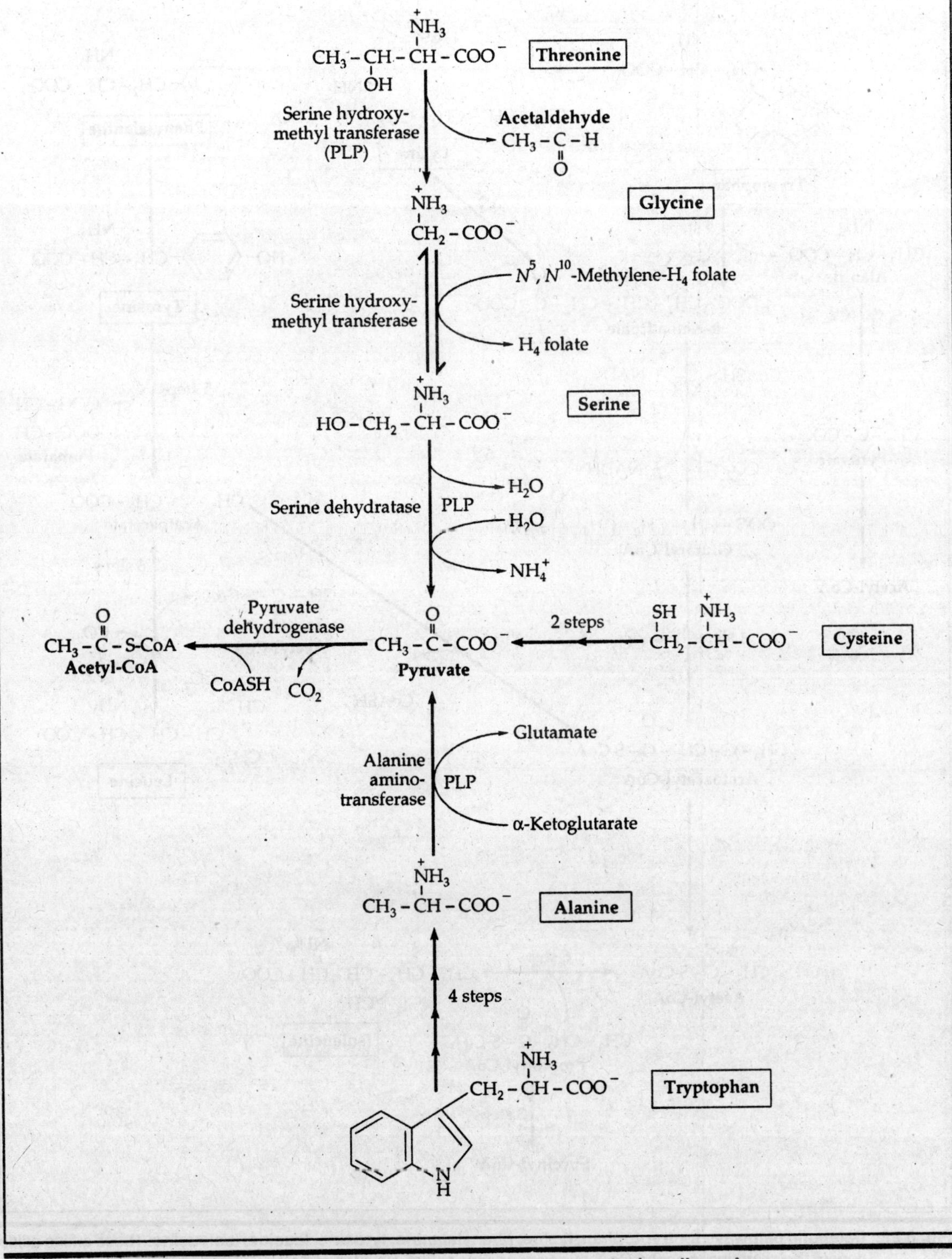

Fig. 6.31: Catabolic pathways for alanine, glycine, serine, cysteine, tryptophan, threonine.

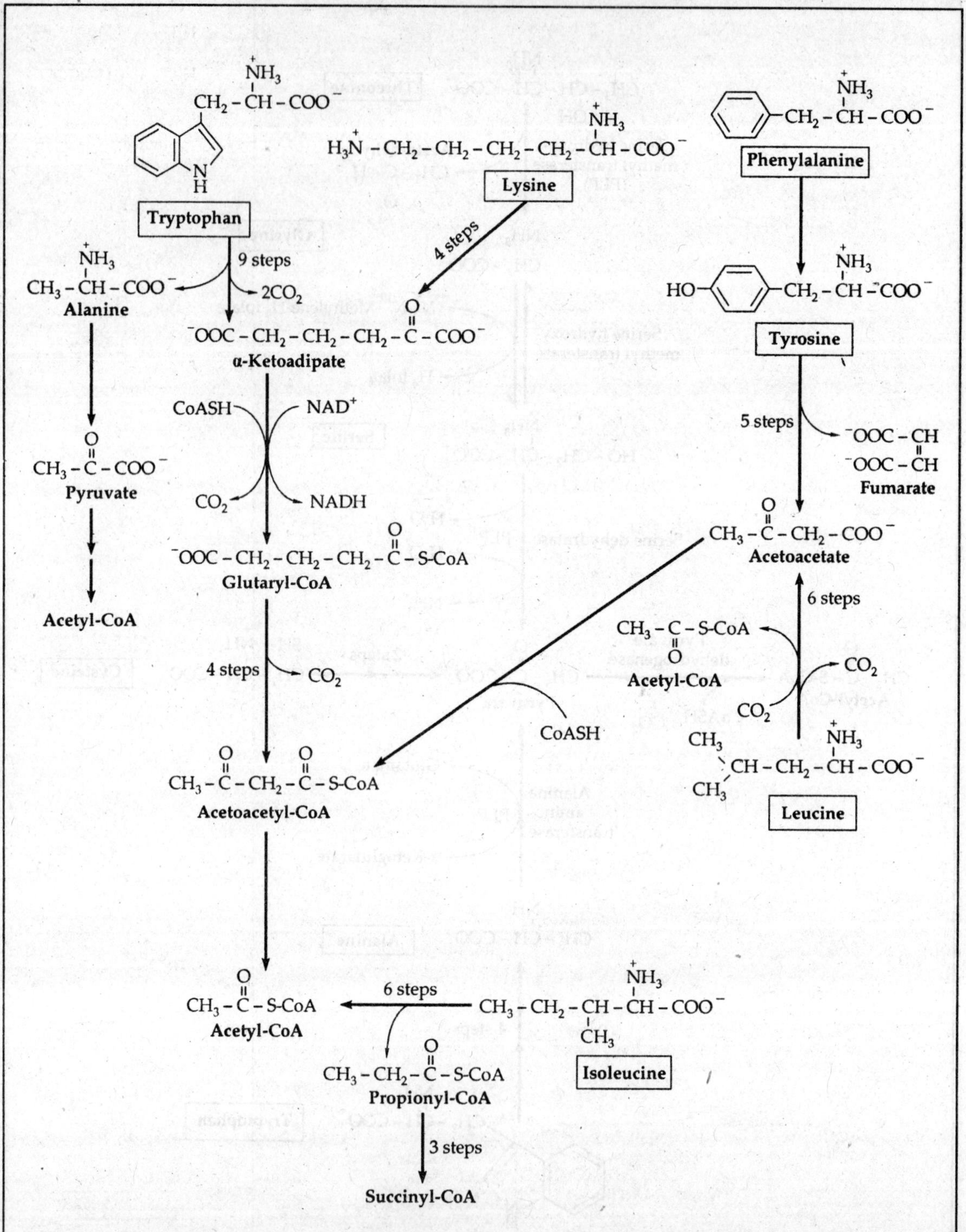

Fig. 6.32: Catabolic pathways for tryptophan, lysine, phenylalanine, tyrosine, leucine, isoleucine: These amino acids donate their carbons to acetyl-CoA, pyruvate or citric acid cycle intermediates. Nitrogen atoms are transferred in most cases to α-ketoglutarate to form glutamate.

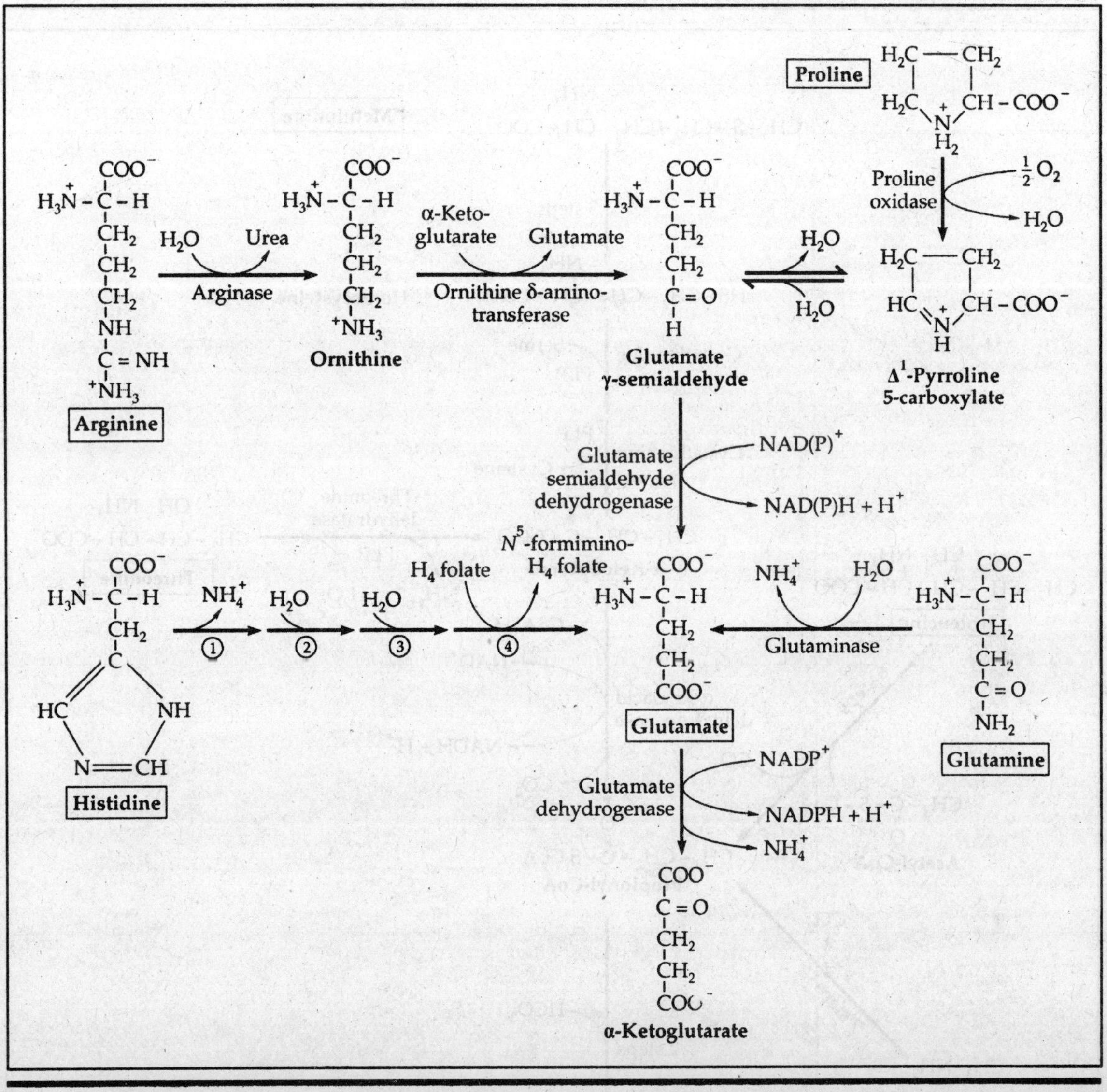

Fig. 6.33: Catabolic pathways for arginine, histidine, glutamate, glutamine, proline: These amino acids are converted to α-ketoglutarate. [(1) Histidine ammonia lyase, (2) Urocanate hydratase, (3) Imidazolonepropionase, (4) Glutamate formimino transferase.]

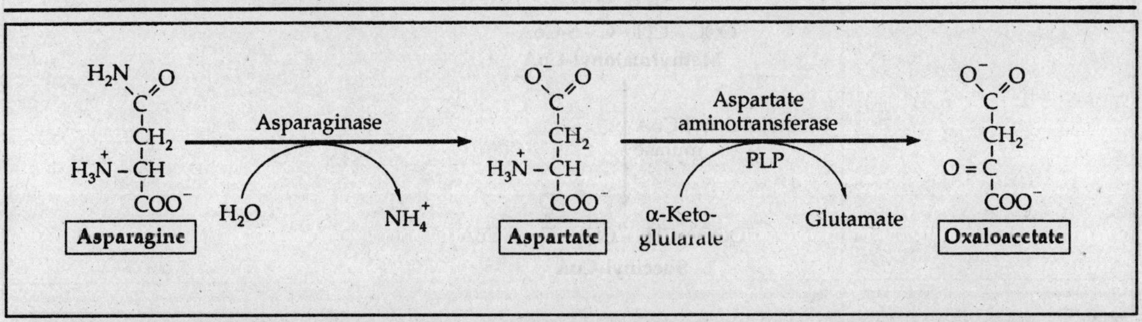

Fig. 6.34: Catabolic pathway for asparagine and aspartate: Both amino acids are converted to oxaloacetate.

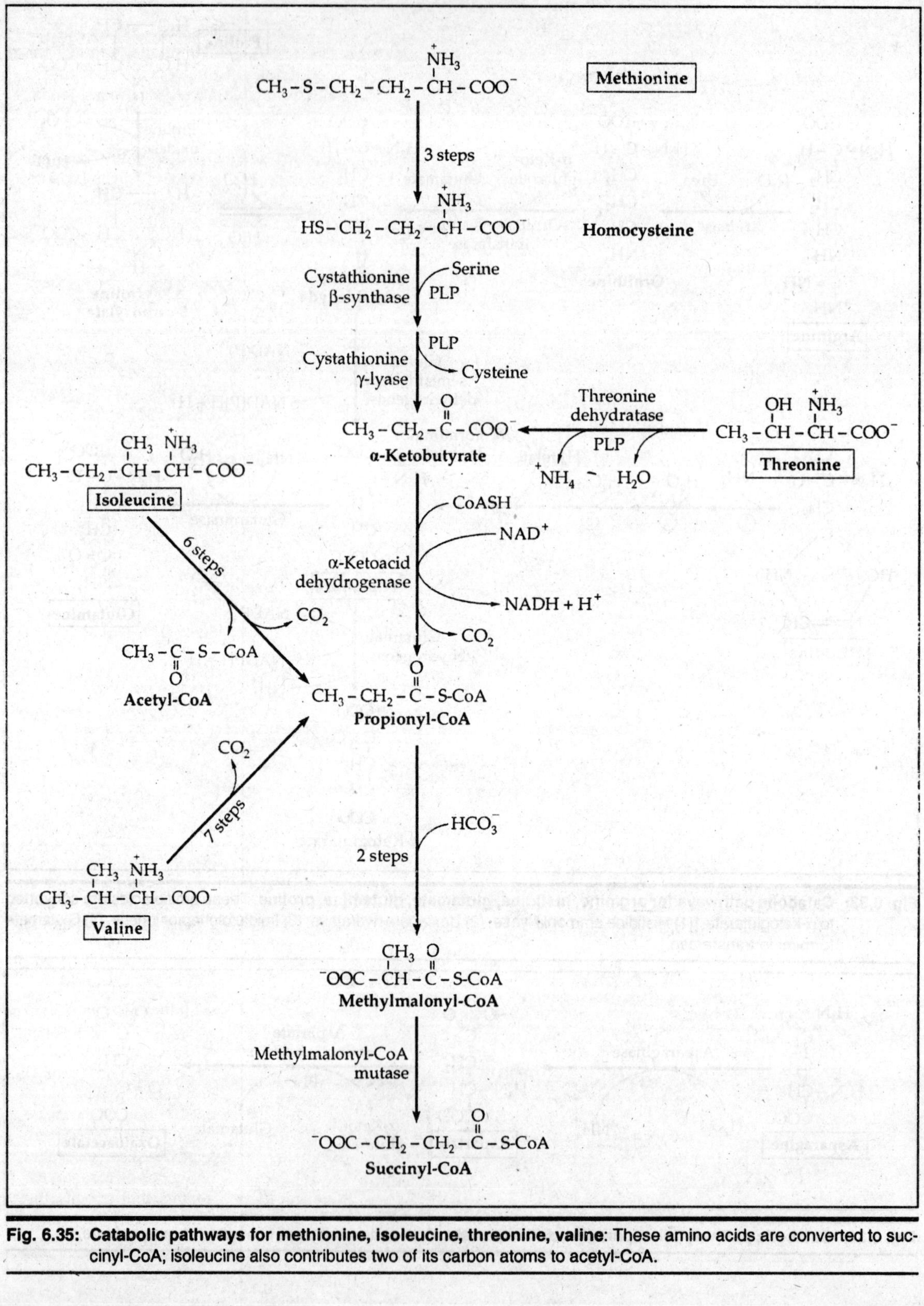

Fig. 6.35: Catabolic pathways for methionine, isoleucine, threonine, valine: These amino acids are converted to succinyl-CoA; isoleucine also contributes two of its carbon atoms to acetyl-CoA.

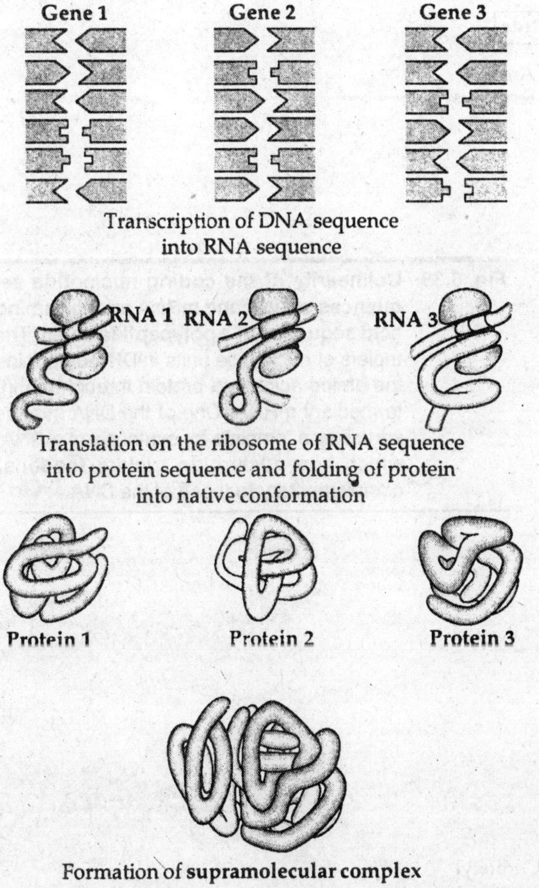

Fig. 6.36: DNA, RNA, Protein relationship: Linear sequences of deoxyribonucleotides in **DNA**, arranged into units known as **genes**, are transcribed into **RNA** molecules with complementary ribonucleotide sequences. The RNA sequences are then translated into linear **protein** chains, which fold into their native three-dimensional shapes. Individual proteins commonly associate with other proteins to form supramolecular complexes.

First letter of codon (5′ end)	Second letter of codon: U		C		A		G		Third letter of codon (3′ end)
U	UUU	Phe	UCU	Ser	UAU	Tyr	UGU	Cys	U
U	UUC	Phe	UCC	Ser	UAC	Tyr	UGC	Cys	C
U	UUA	Leu	UCA	Ser	**UAA**	Stop	**UGA**	Stop	A
U	UUG	Leu	UCG	Ser	**UAG**	Stop	UGG	Trp	G
C	CUU	Leu	CCU	Pro	CAU	His	CGU	Arg	U
C	CUC	Leu	CCC	Pro	CAC	His	CGC	Arg	C
C	CUA	Leu	CCA	Pro	CAA	Gln	CGA	Arg	A
C	CUG	Leu	CCG	Pro	CAG	Gln	CGG	Arg	G
A	AUU	Ile	ACU	Thr	AAU	Asn	AGU	Ser	U
A	AUC	Ile	ACC	Thr	AAC	Asn	AGC	Ser	C
A	AUA	Ile	ACA	Thr	AAA	Lys	AGA	Arg	A
A	**AUG**	Met	ACG	Thr	AAG	Lys	AGG	Arg	G
G	GUU	Val	GCU	Ala	GAU	Asp	GGU	Gly	U
G	GUC	Val	GCC	Ala	GAC	Asp	GGC	Gly	C
G	GUA	Val	GCA	Ala	GAA	Glu	GGA	Gly	A
G	GUG	Val	GCG	Ala	GAG	Glu	GGG	Gly	G

Fig. 6.37: Dictionary of amino acid code words as they occur in mRNAs: The codons are written in the 5′ → 3′ direction. The third base of each codon plays a lesser role in specifying an amino acid than the first two. The three termination codons (UAA, UAG, UGA) and the initiation codon (AUG) are in bold. All the amino acids except methionine and tryptophan have more than one codon. In most cases, codons that specify the same amino acid differ only at the third base.

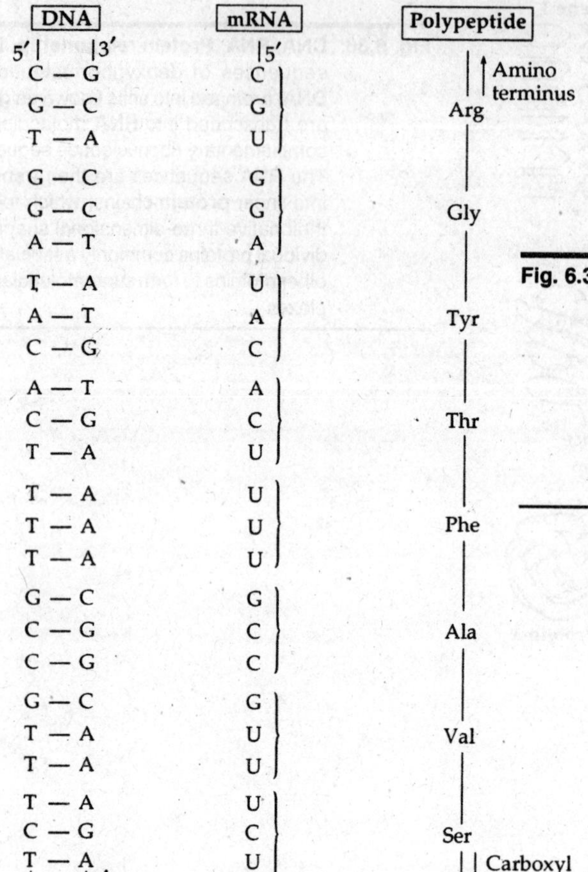

Fig. 6.38: Colinearity of the coding nucleotide sequences of DNA and mRNA and the amino acid sequence of a polypeptide chain: The triplets of nucleotide units in DNA determine the amino acids in a protein through the intermediary mRNA. One of the DNA strands serves as a template for synthesis of mRNA, which has nucleotide triplets (codons) complementary to those of the DNA.

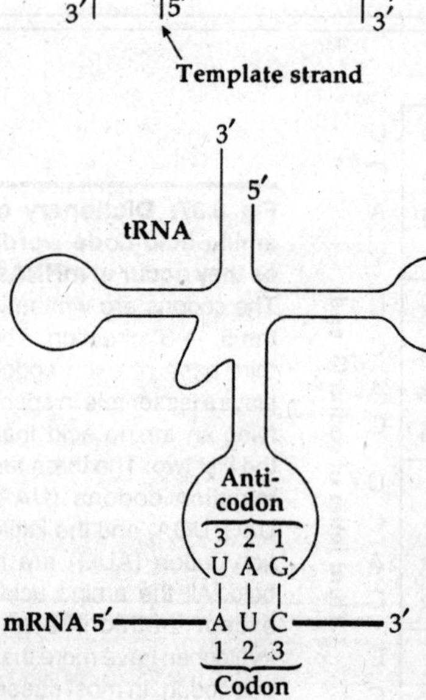

Fig. 6.39: Pairing relationship of codon and anticodon: Alignment of the two RNAs is antiparallel.

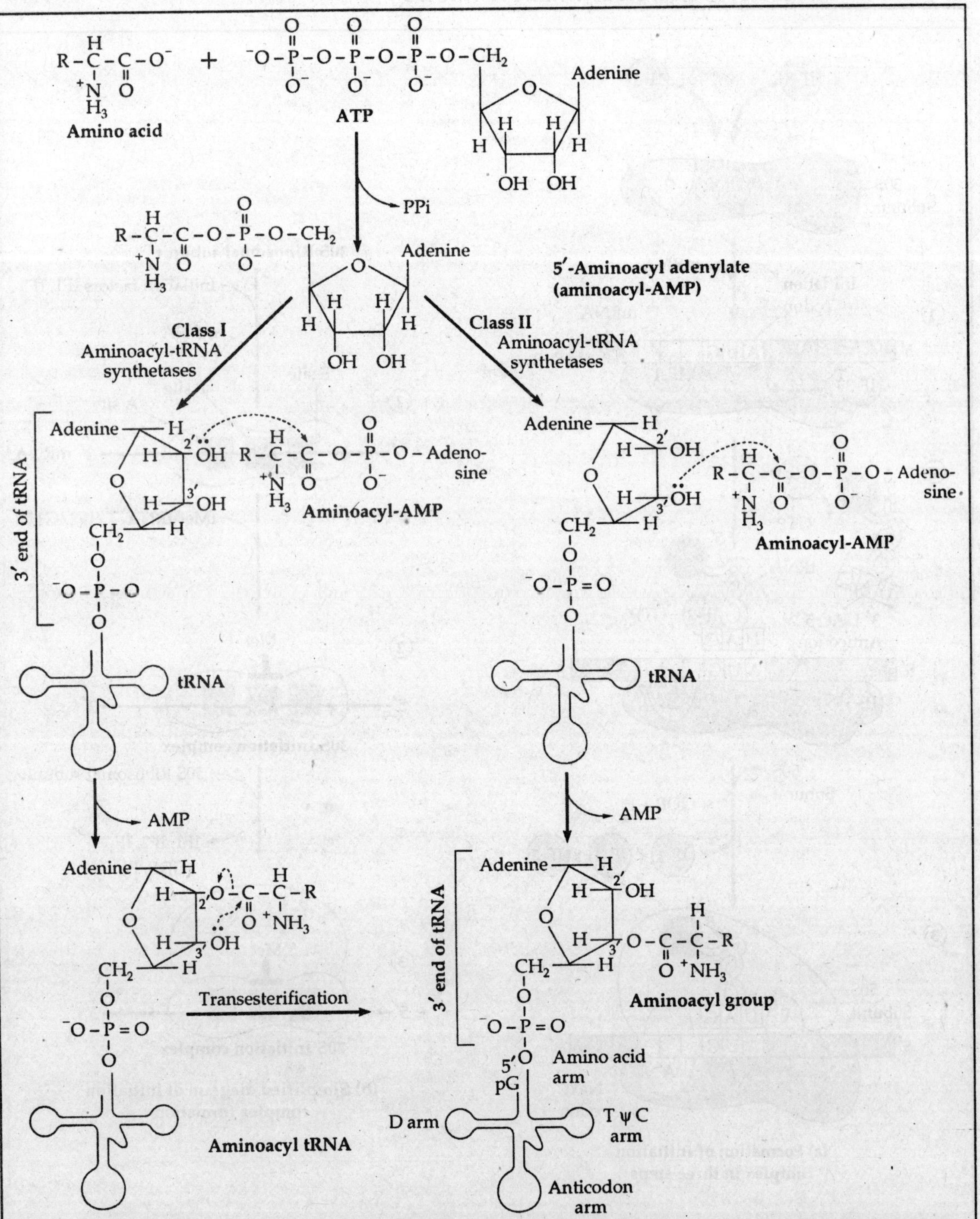

Fig. 6.40: Aminoacylation of tRNA by aminoacyl-tRNA synthetase: The **first step** is the formation of aminoacyl adenylate, which remains bound to active site. In **second step**, aminoacyl group is transferred to the tRNA, which vary in the mechanism for two classes of enzymes. For **class I enzymes**, the aminoacyl group is transferred initially to the 2′ hydroxyl of the 3′-terminal adenylate residue, then moves to the 3′ hydroxyl by a transesterification reaction. For **class II enzymes**, the aminoacyl group is transferred directly to the 3′ hydroxyl of the terminal adenylate.

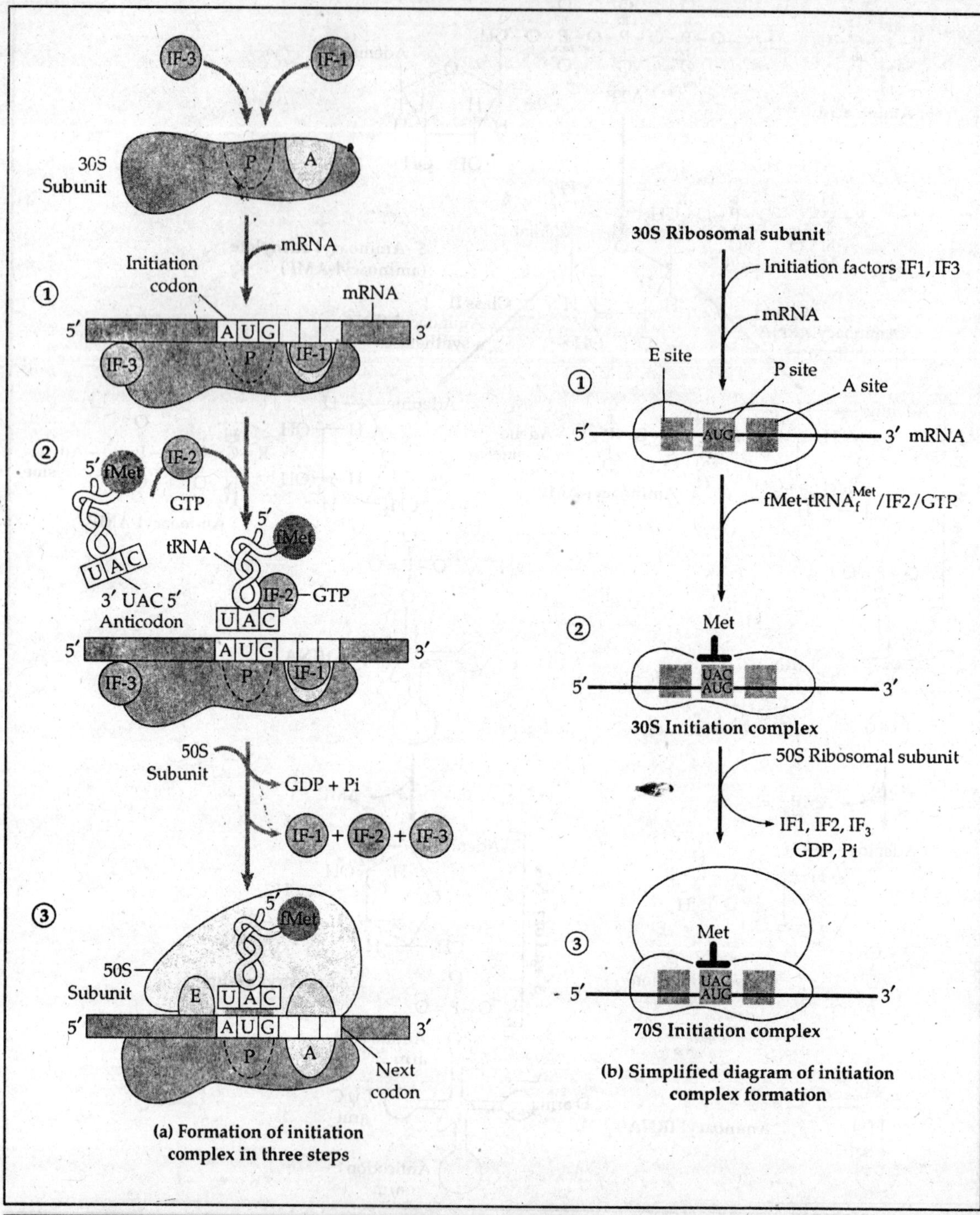

Fig. 6.41: Initiation of protein synthesis in bacteria: **(1)** Initiation begins with the binding of IF1 and IF3 to the small 30S ribosomal subunit. The small subunit then binds to the mRNA via the Shine-Dalgarno sequence and moves 3′ along the mRNA until it locates the AUG initiation codon. **(2)** The initiator tRNA charged with N-formylmethionine and in a complex with IF2 and GTP (fMet-tRNA$_i^{Met}$/IF2/GTP) then binds to form a complex of mRNA, fMet-tRNA$_i^{Met}$, IF1, IF2, IF3 and 30S ribosomal subunit. **(3)** The large 50S ribosomal subunit then binds, with the release of IF1, IF2, IF3 and hydrolysis of GTP, to form 70S initiation complex.

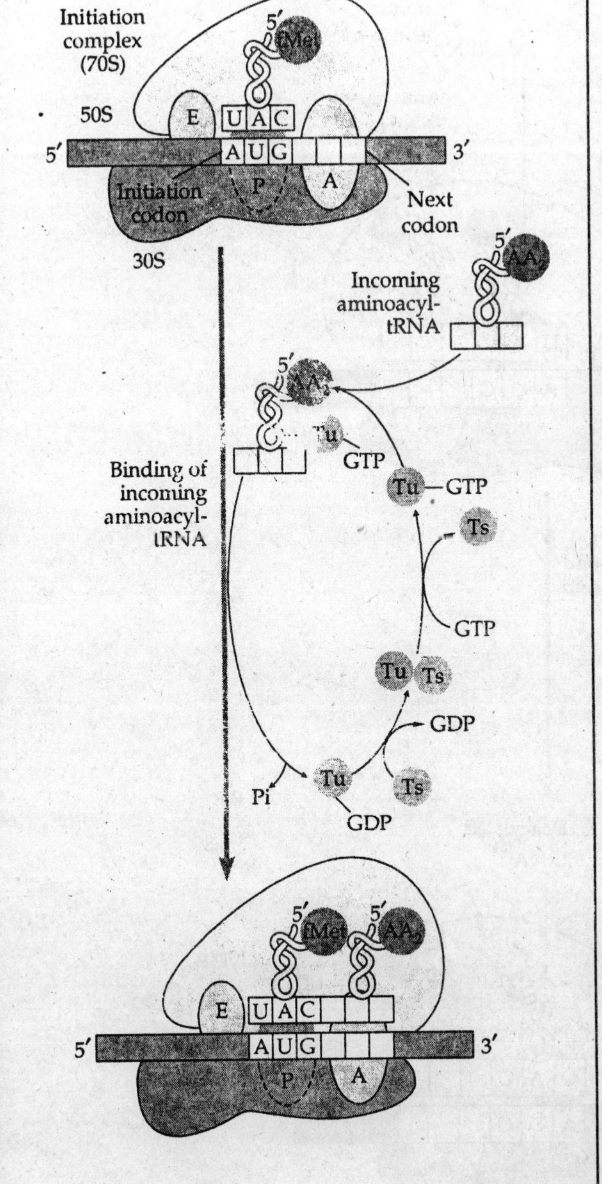

Fig. 6.42: First step in elongation—Binding of the next aminoacyl-tRNA: The next aminoacyl-tRNA enters A site of the ribosome bound to EF-Tu containing GTP. The process is accompanied by hydrolysis of GTP to GDP and Pi, and release of Tu.

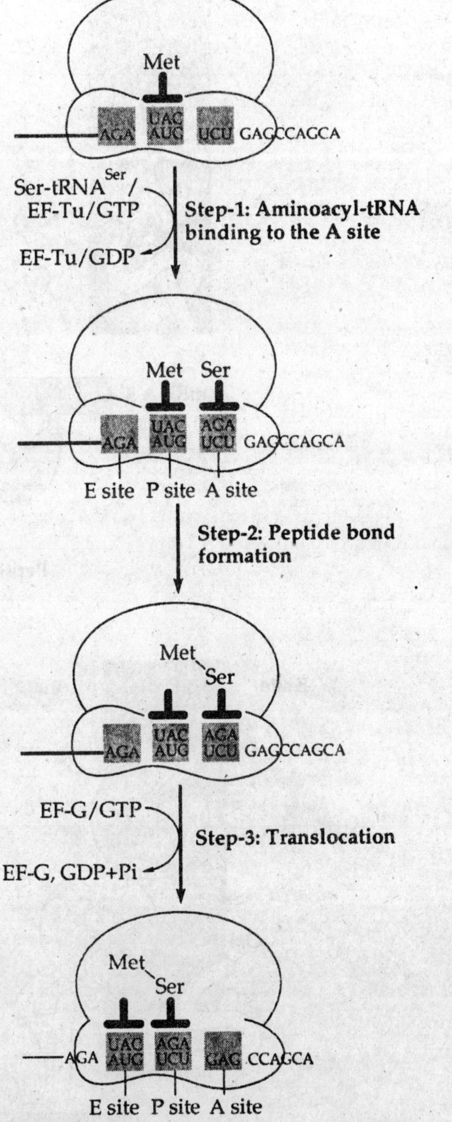

Fig. 6.43: The elongation phase of protein synthesis in bacteria: It occurs in three steps—

Step-1: Binding of aminoacyl-tRNA to A site of ribosome.

Step-2: Peptide bond formation.

Step-3: Translocation.

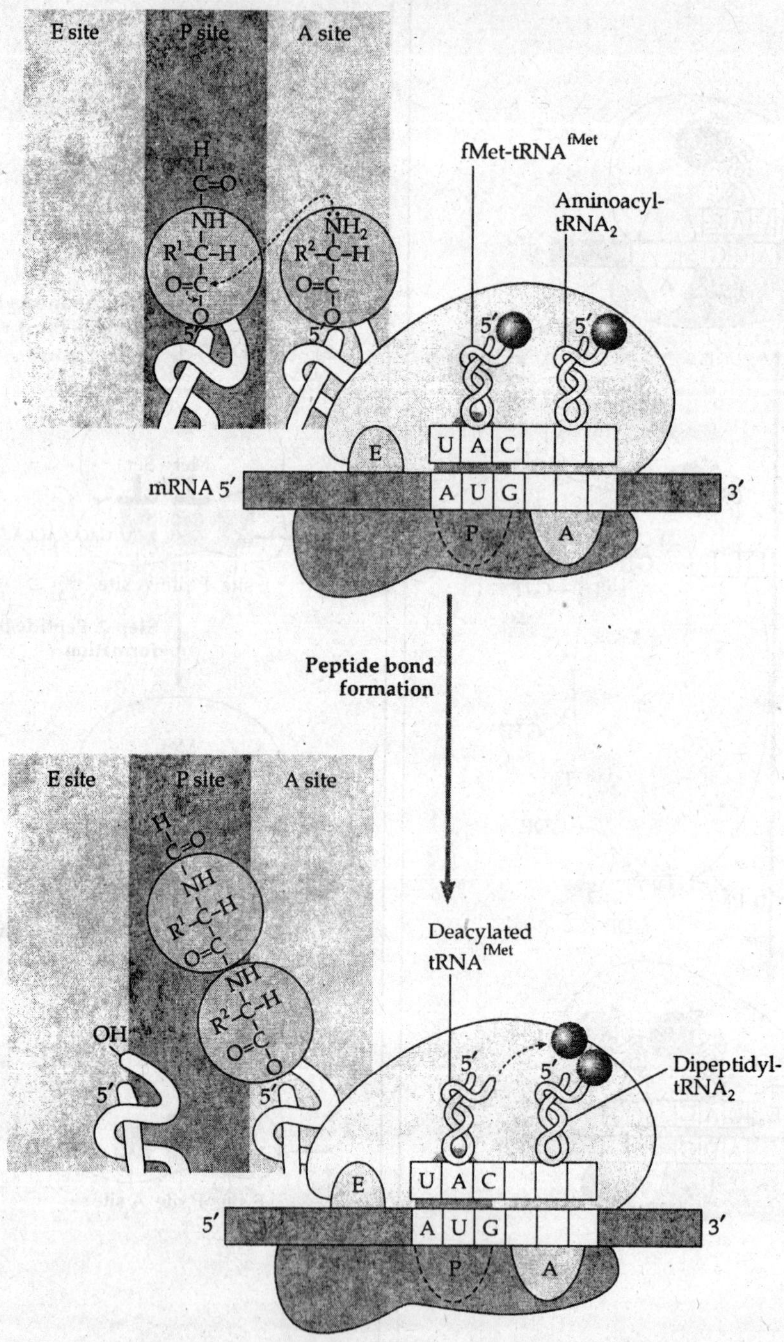

Fig. 6.44: Second step in elongation—Formation of peptide bond: The N-formylmethionyl group is transferred to the amino group of the second aminoacyl-tRNA in the A site, forming a depeptidyl-tRNA with the help of peptidyl transferase. The uncharged tRNA shifts so that its 3′ and 5′ ends are in the E site. Similarly, the 3′ and 5′ ends of the peptidyl tRNA shift to the P site. The anticodons remain in the A and P sites.

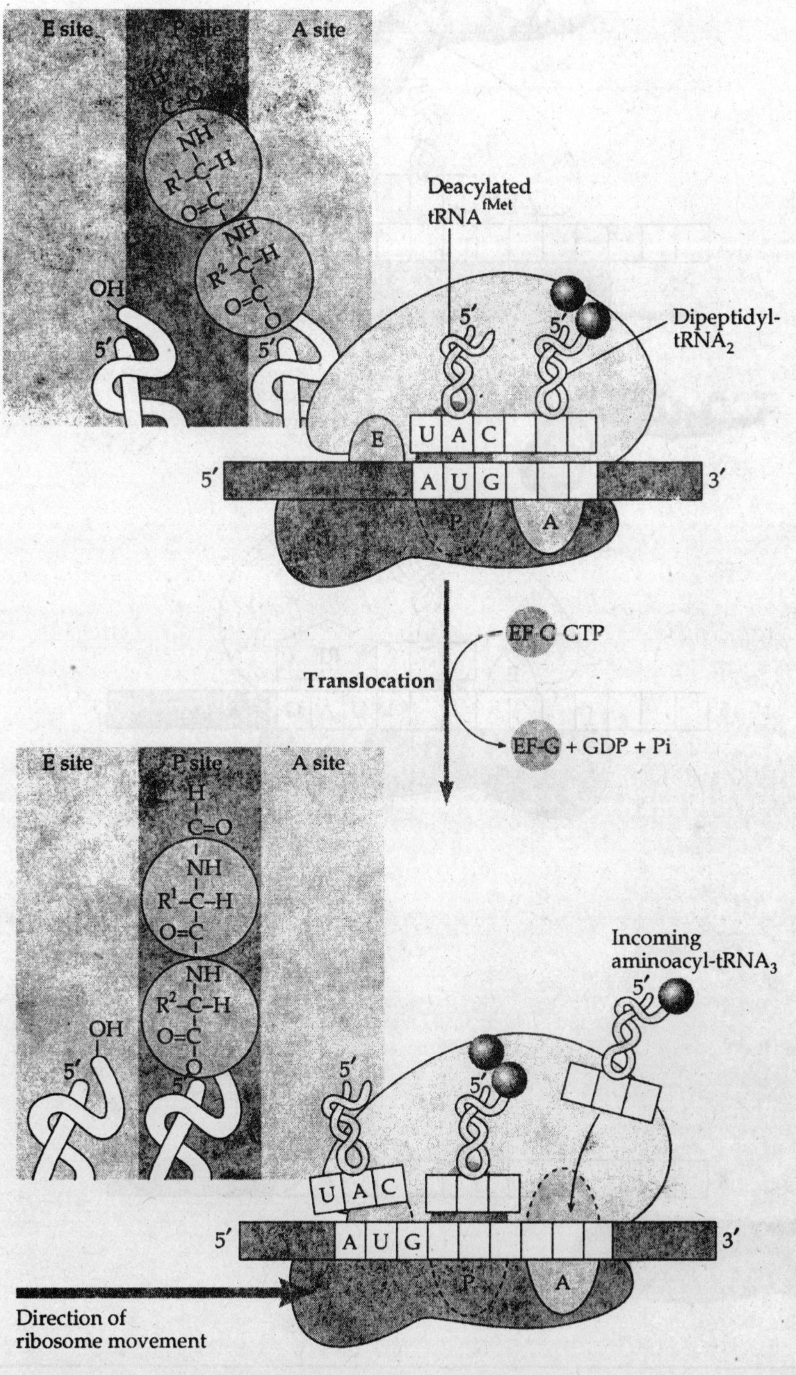

Fig. 6.45: Third step in elongation—Translocation: The ribosome moves one codon towards the 3′ end of mRNA, using energy provided by hydrolysis of GTP bound to EF-G (translocase). The dipeptidyl-tRNA is now entirely in the P site, leaving the A site open for the next incoming aminoacyl-tRNA. The uncharged tRNA dissociates from the E site, and the elongation cycle begins again.

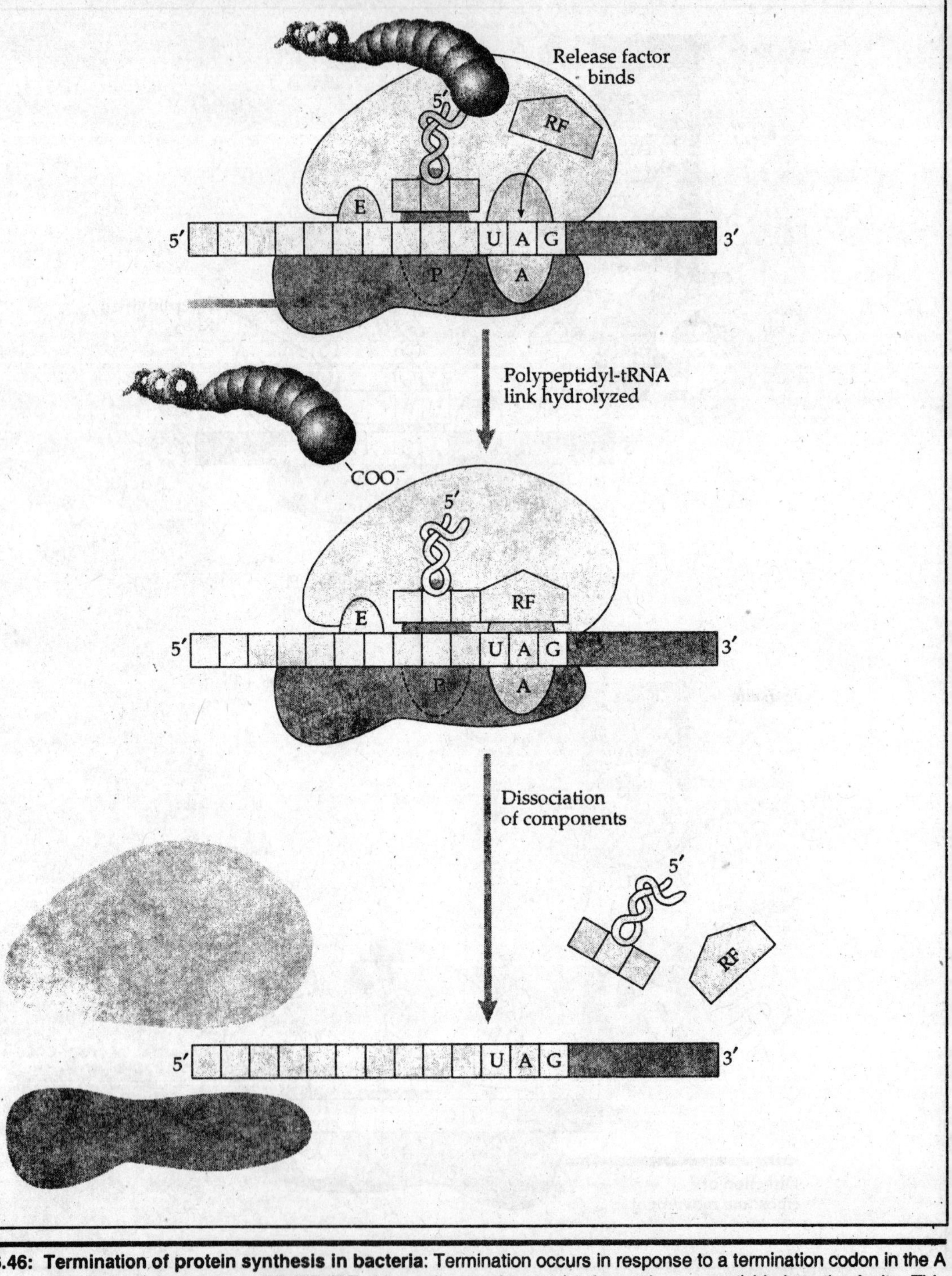

Fig. 6.46: Termination of protein synthesis in bacteria: Termination occurs in response to a termination codon in the A site. First, a release factor $RF_1/RF_2/RF_3$ (depending on the termination codon present) binds to the A site. This leads to hydrolysis of the ester linkage between the nascent polypeptide and the tRNA in the P site and release of the completed polypeptide. Finally, the mRNA, deacylated tRNA and release factor leave the ribosome and ribosome dissociated into its 30S and 50S subunits.

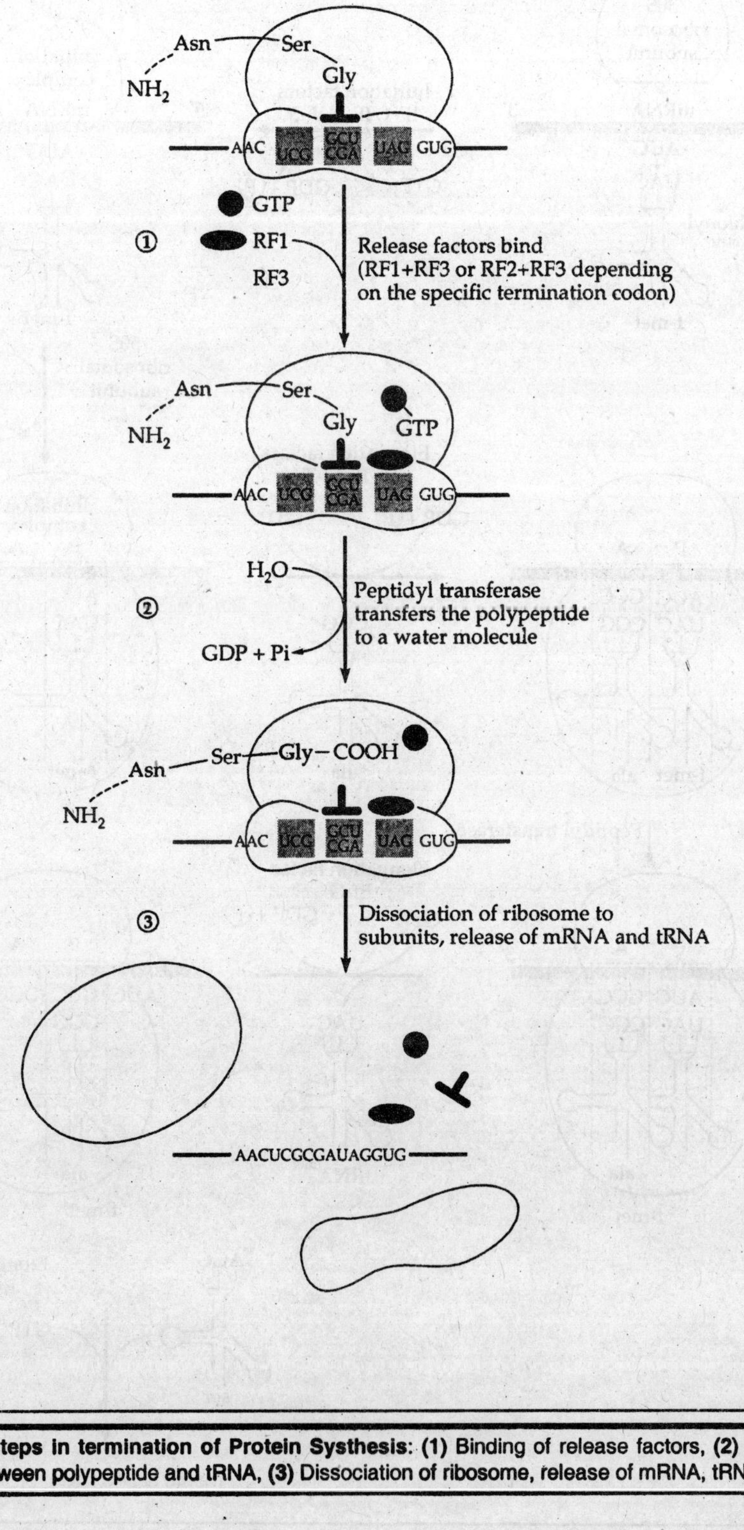

Fig. 6.47: Different steps in termination of Protein Systhesis: (1) Binding of release factors, **(2)** Hydrolysis of ester linkage between polypeptide and tRNA, **(3)** Dissociation of ribosome, release of mRNA, tRNA.

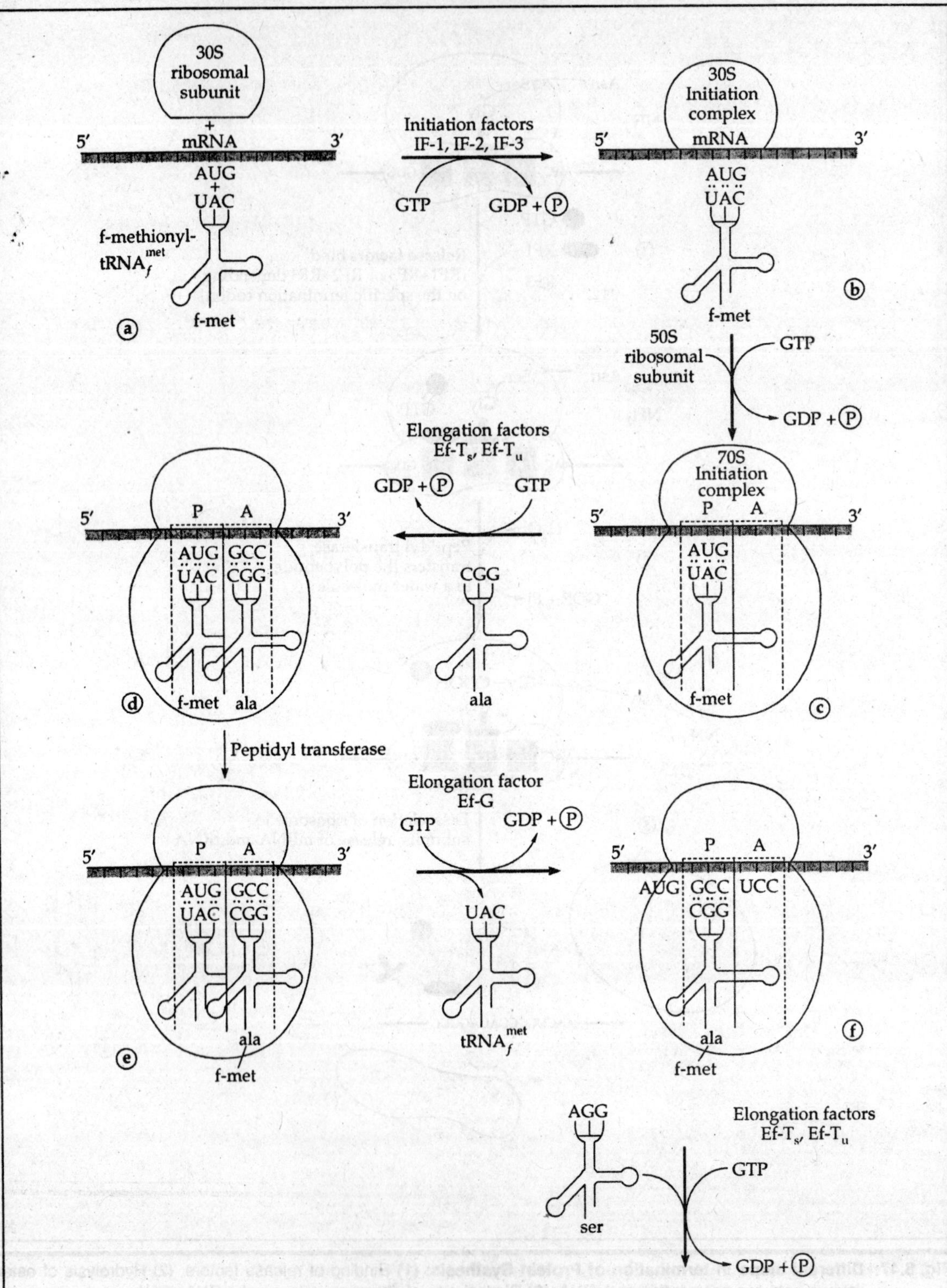

(contd. to next page)

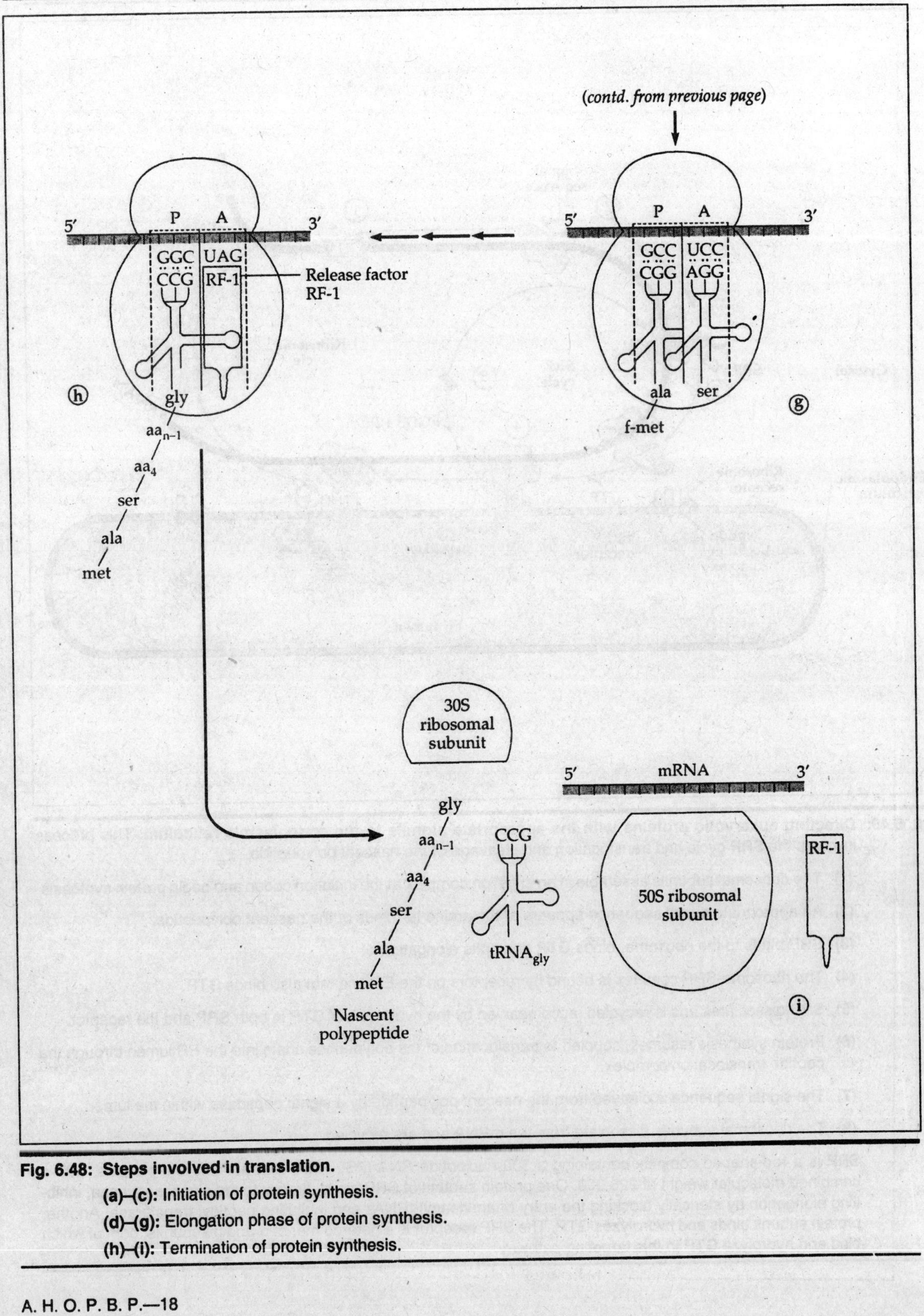

Fig. 6.48: Steps involved in translation.

(a)–(c): Initiation of protein synthesis.

(d)–(g): Elongation phase of protein synthesis.

(h)–(i): Termination of protein synthesis.

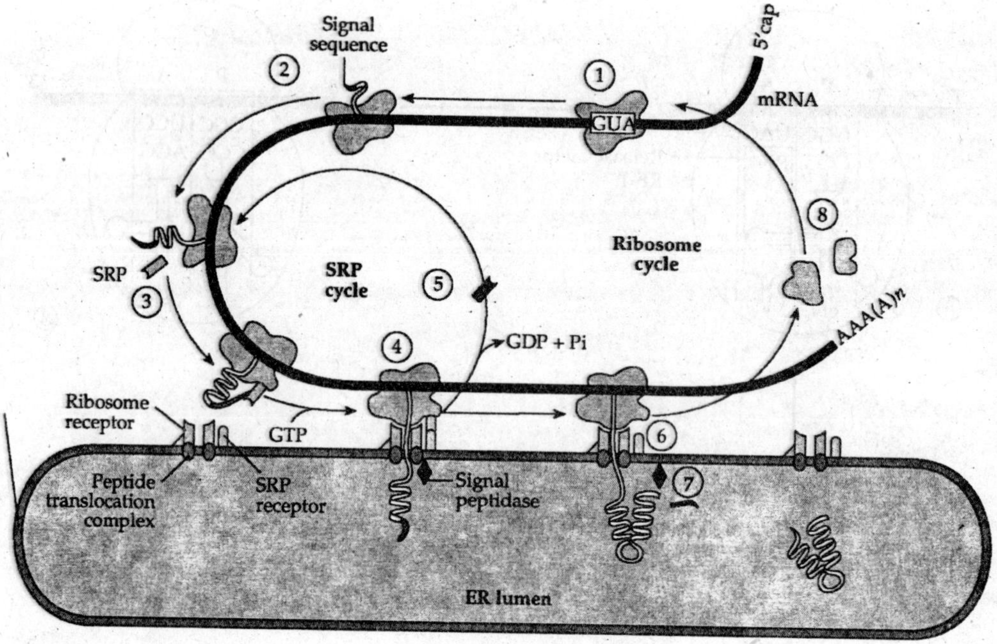

Fig. 6.49: Directing eukaryotic proteins with the appropriate signals to the endoplasmic reticulum: This process involves the SRP cycle and translocation and cleavage of the nascent polypeptide.

(1) The ribosomal subunits assemble in an initiation complex at the initiation codon and begin protein synthesis.

(2) An appropriate signal sequence appears at the amino terminus of the nascent polypeptide.

(3) SRP binds to the ribosome, binds GTP and halts elongation.

(4) The ribosome-SRP complex is bound by receptors on the ER and that also binds GTP.

(5) SRP dissociates and is recycled, accompanied by the hydrolysis of GTP in both SRP and the receptor.

(6) Protein synthesis resumes, coupled to translocation of the polypeptide chain into the ER lumen through the peptide translocation complex.

(7) The signal sequence is cleaved from the nascent polypeptide by a signal peptidase within the lumen.

(8) The ribosomal subunits dissociate from the mRNA and are recycled.

SRP is a rod-shaped complex containing a 300 nucleotide RNA (7SL-RNA) and six different proteins, with a combined molecular weight of 325,000. One protein subunit of SRP binds directly to the signal sequence, inhibiting elongation by sterically blocking the entry of aminoacyl-tRNAs and inhibiting peptidyl transferase. Another protein subunit binds and hydrolyzes GTP. The SRP receptor is a heterodimer of α and β subunits, both of which bind and hydrolyze GTP in this targeting pathway.

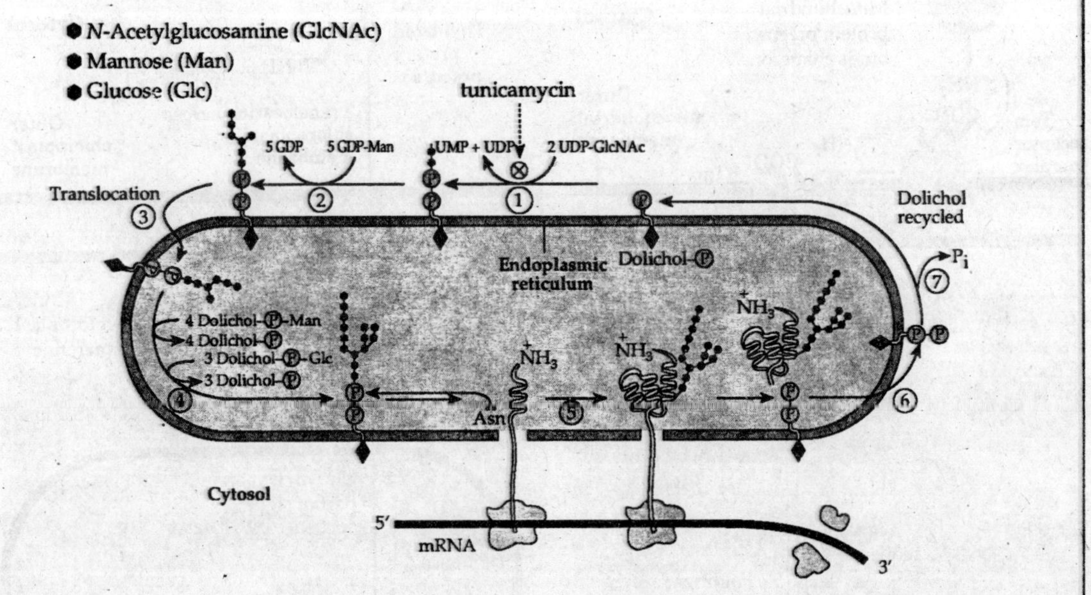

Fig. 6.50: Synthesis of the core oligosaccharide of glycoproteins: The core oligosaccharide is built up by the successive addition of monosaccharide units.

(1) and **(2)** The first two steps occur on the cytosolic face of the ER.

(3) Translocation moves the incomplete oligosaccharide across the membrane.

(4) Completion of the core oligosaccharide occurs within the lumen of the ER. The synthetic precursors that contribute additional mannose and glucose residues to the growing oligosaccharide in the lumen are dolichol phosphate derivatives.

(5) In the construction of the *N*-linked oligosaccharide moiety of a glycoprotein, the core oligosaccharide is transferred from dolichol phosphate to an Asn residue of the protein within the ER lumen. The core oligosaccharide is then further modified in the ER and the Golgi complex in pathways that differ for different proteins. The five sugar residues are retained in the final structure of all *N*-linked oligosaccharides.

(6) The released dolichol pyrophosphate is again translocated so that the pyrophosphate is on the cytosolic face of the ER.

(7) Then a phosphate is hydrolytically removed to regenerate dolichol phosphate.

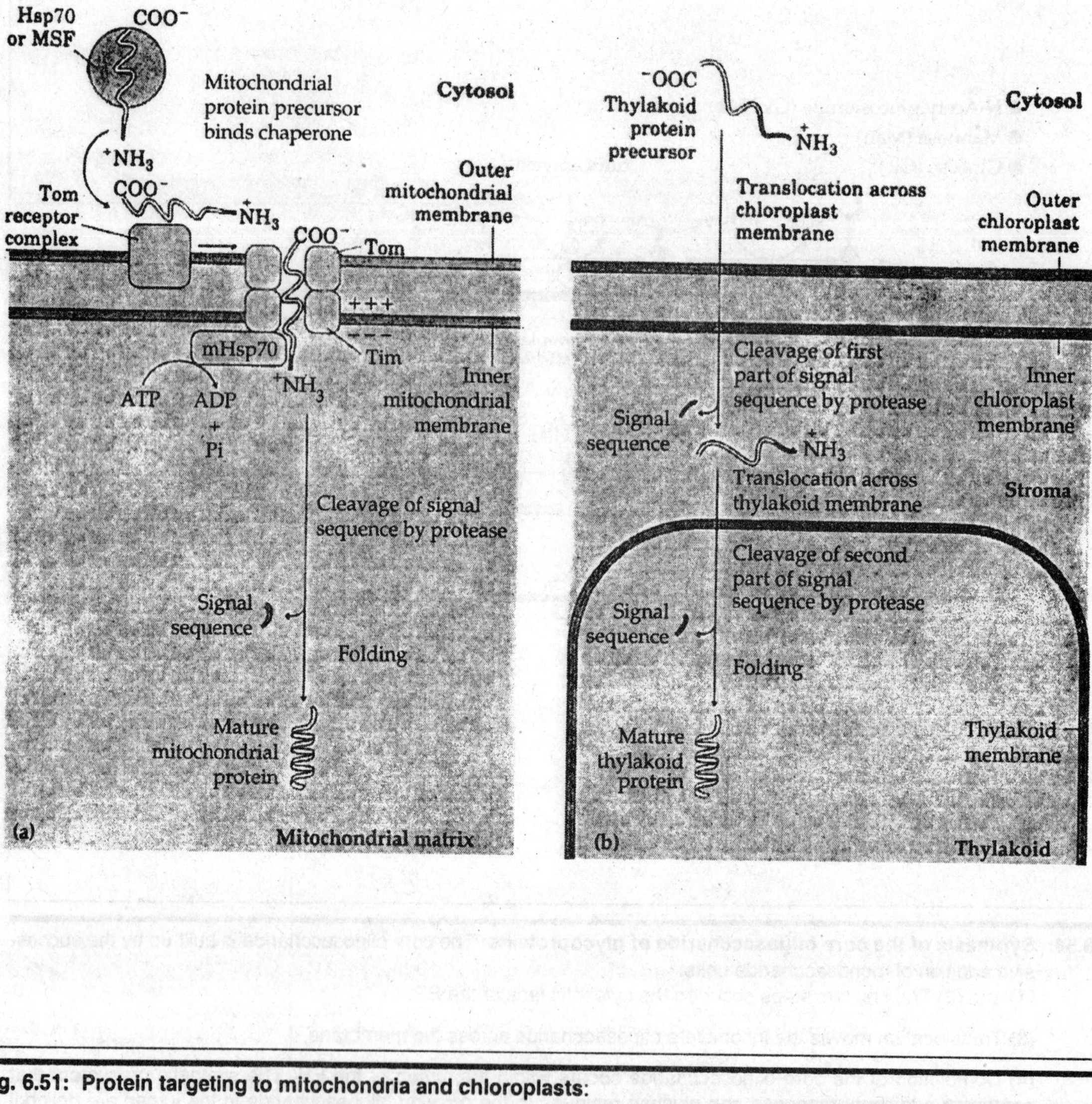

Fig. 6.51: Protein targeting to mitochondria and chloroplasts:

(a) Targeting to the mitochondrial matrix: A mitochondrial protein precursor is bound by a cytosolic chaperon (Hsp 70 or MSF) and delivered to a Tom receptor complex, then to a membrane channel made up of specialised Tom and Tim channel complexes. Translocation of the precursor protein is driven both by the ATPase of mitochondrial Hsp 70 protein (mHsp 70) on the matrix side of the inner mitochondrial membrane and by the electrochemical potential across the inner membrane. The signal sequence is cleaved in the mitochondrial matrix, and the protein folds into its functional conformation.

(b) Proteins targeted to the chloroplast: Thylakoids have bipartite signal sequences. The first part of the sequence mediates targeting to the chloroplast stroma, where it is removed, the second part mediates translocation into the thylakoid, where it, in turn, is removed.

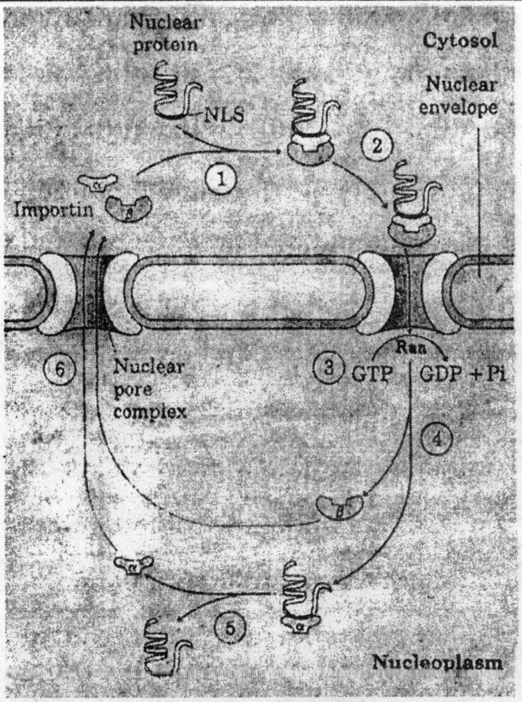

Fig. 6.52: Targeting of nuclear proteins:

(1) A protein with an appropriate nuclear localisation signal is bound by a complex of importing α and β.

(2) The resulting complex binds to a nuclear pore.

(3) Translocation is mediated by the Ran GTPase.

(4) Inside the nucleus importing β dissociates from importing α.

(5) Importing α then dissociates from the nuclear protein.

(6) Importing α and β are transported out of the nucleus and recycled.

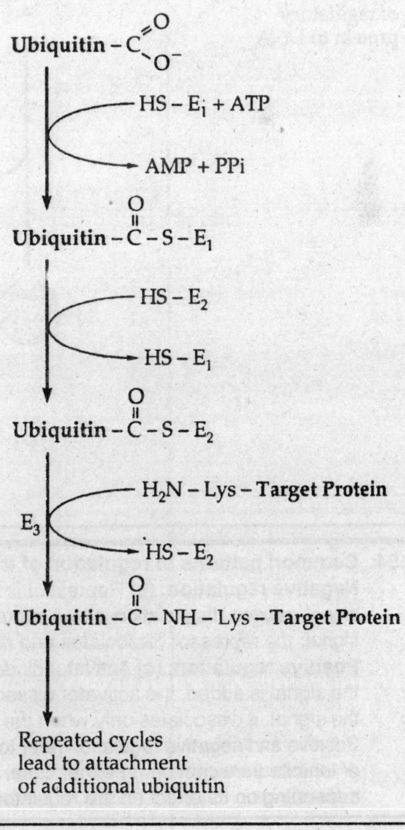

Fig. 6.53: Three-step cascade pathway by which ubiquitin is attached to a protein: Two different enzyme-ubiquitin intermediates are involved. The free carboxyl group of ubiquitin's carboxyl-terminal Gly residue is ultimately linked through an amide bond to an ε-amino group of a Lys residue of the target protein. Additional cycles produce polyubiquitin, a covalent polymer of ubiquitin subunits that targets the attached protein for destruction in eukaryotes.

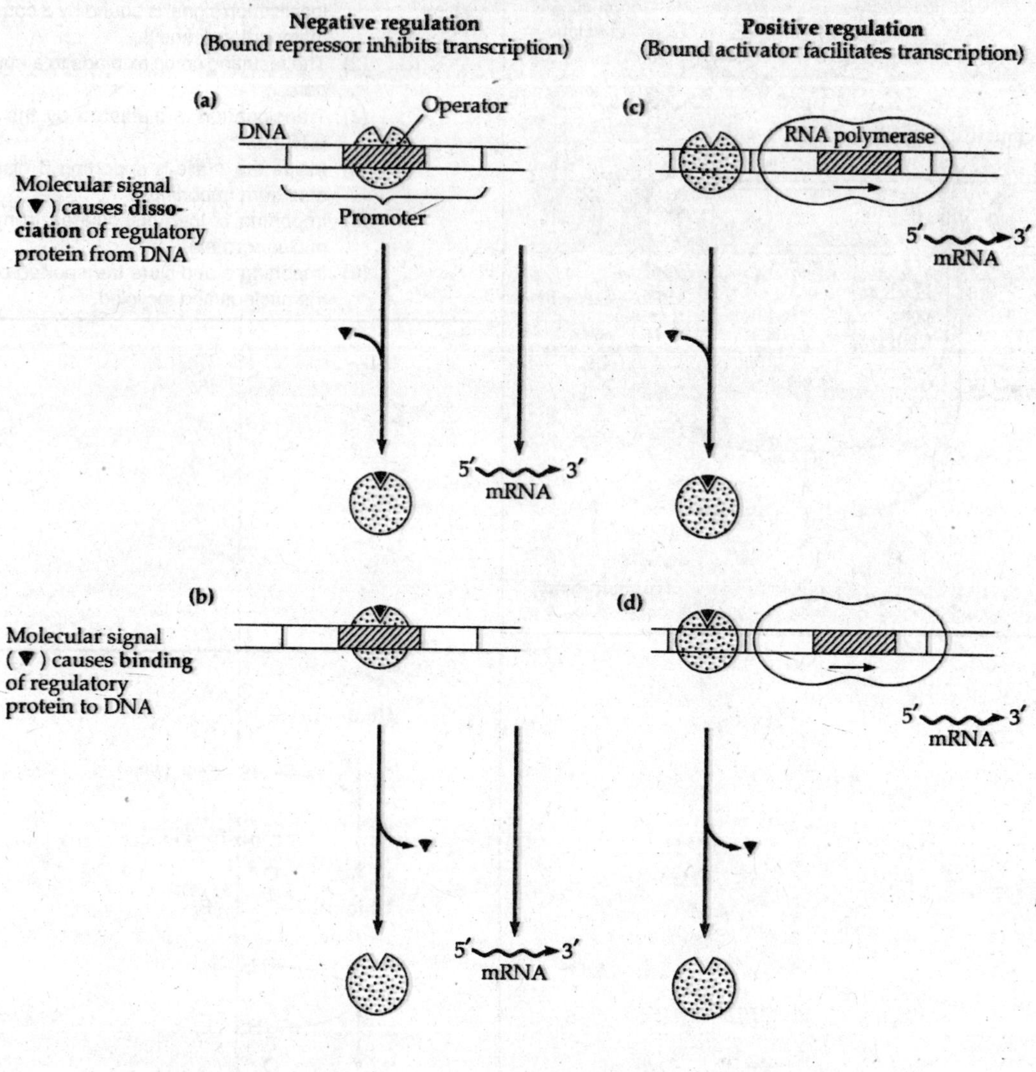

Fig. 6.54: Common patterns of regulation of transcription initiation:
Negative regulation: (a) Repressor is bound to the operator in the absence of the molecular signal; the external signal causes dissociation of the repressor to permit transcription. **(b)** Repressor is bound in the presence of signal; the repressor dissociates and transcription ensues when the signal is removed.
Positive regulation: (c) Activator binds in the absence of the molecular signal and transcription proceeds, when the signal is added, the activator dissociates and transcription is inhibited. **(d)** Activator binds in the presence of the signal; it dissociates only when the signal is removed.
Positive and negative regulation refer to the type of regulatory protein involved: the bound protein either facilitates or inhibits transcription. In either case, addition of the molecular signal may increase or decrease transcription, depending on its effect on the regulatory protein.

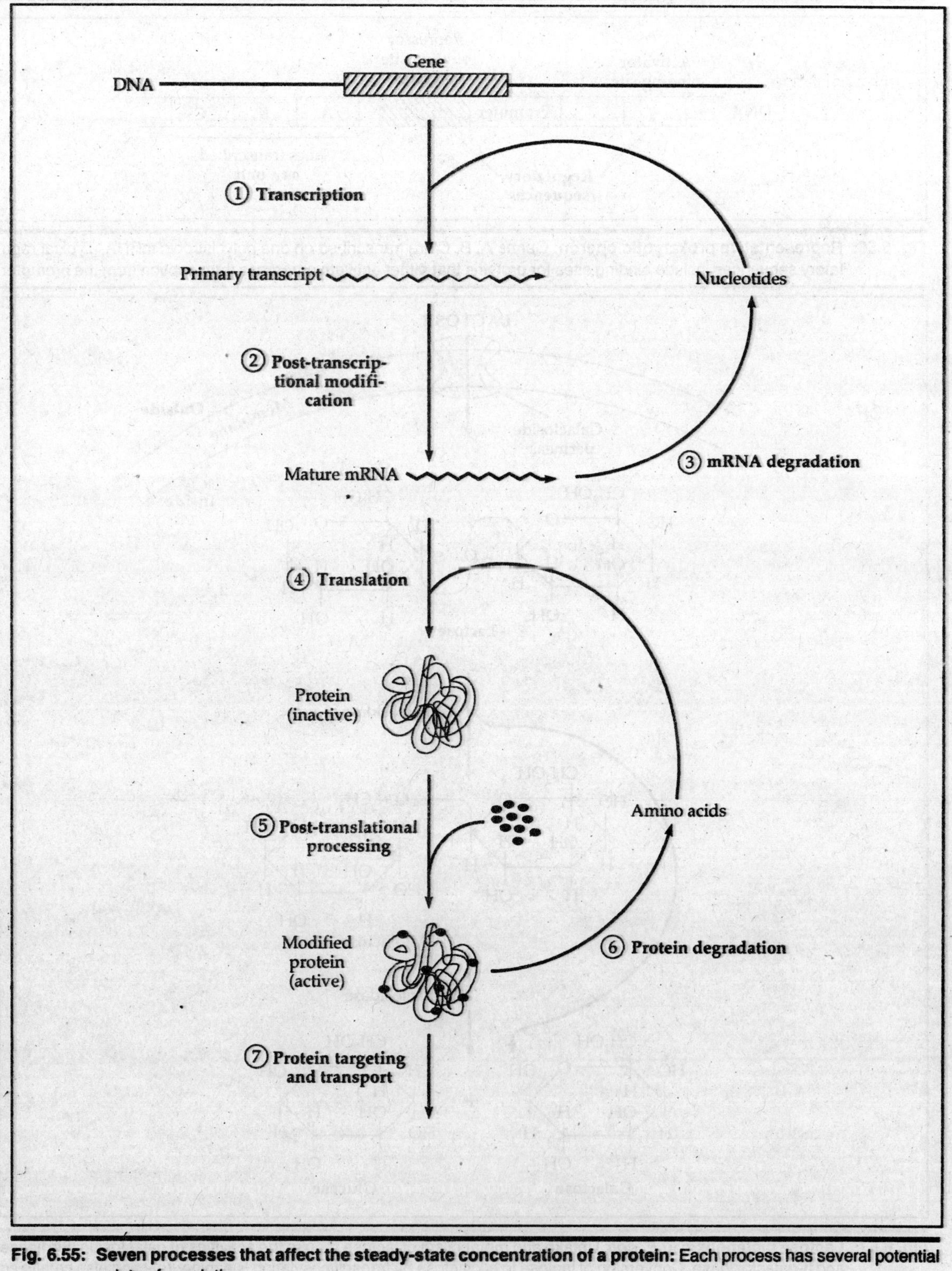

Fig. 6.55: Seven processes that affect the steady-state concentration of a protein: Each process has several potential points of regulation.

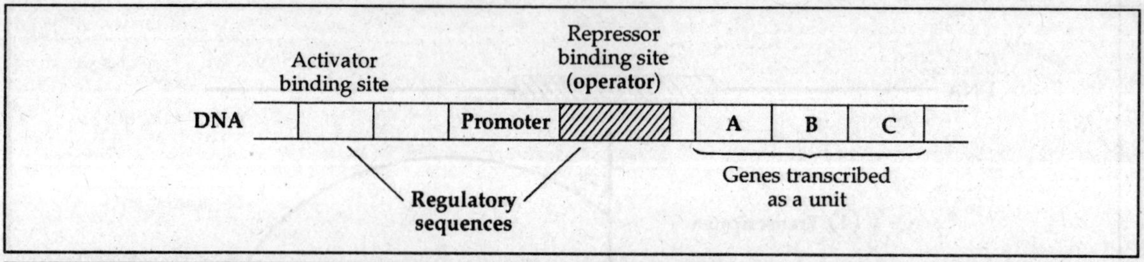

Fig. 6.56: Representative prokaryotic operon: Genes A, B, C are transcribed on one polycistronic mRNA. Typical regulatory sequences include binding sites for proteins that either activate or repress transcription from the promoter.

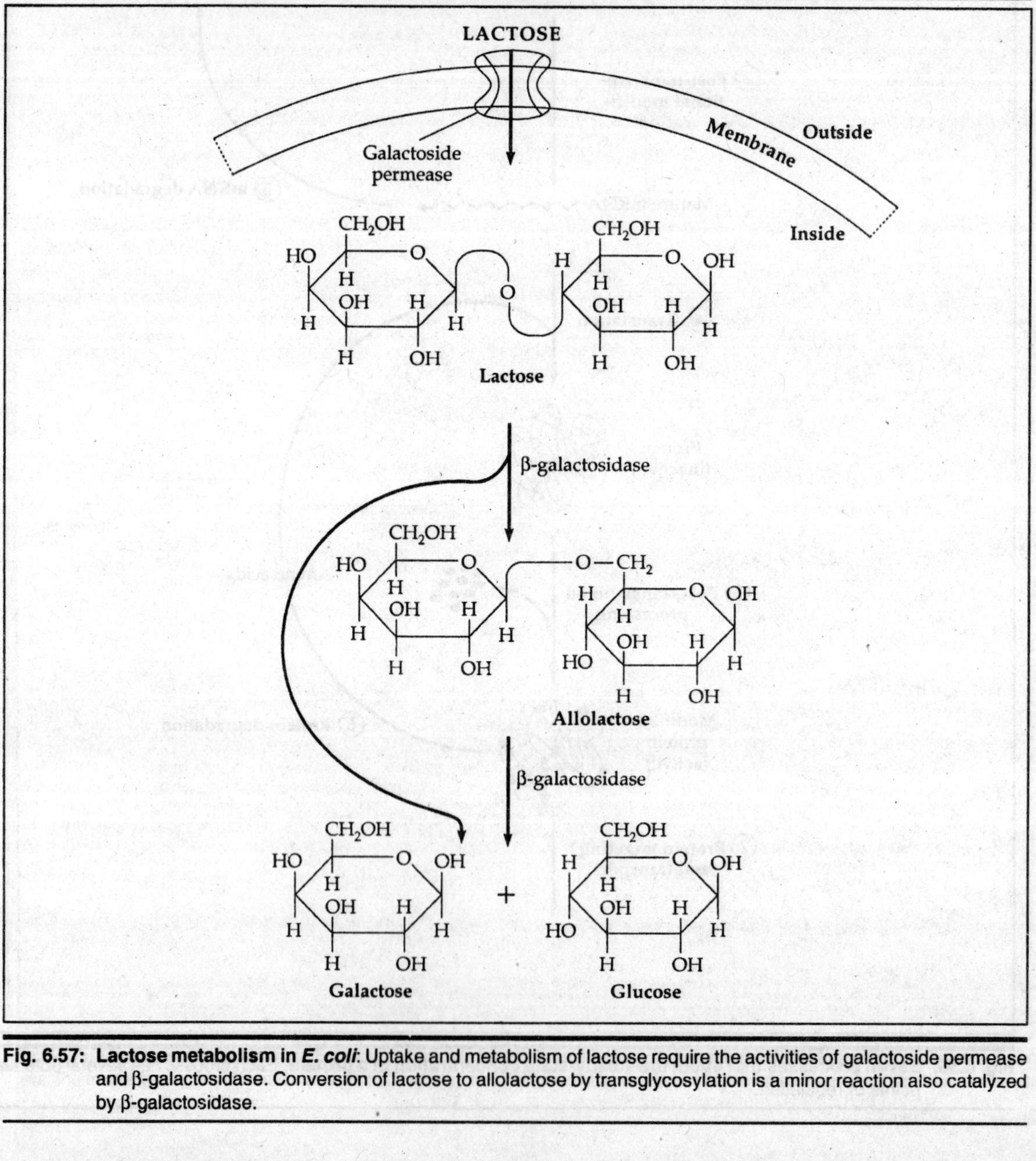

Fig. 6.57: Lactose metabolism in *E. coli*: Uptake and metabolism of lactose require the activities of galactoside permease and β-galactosidase. Conversion of lactose to allolactose by transglycosylation is a minor reaction also catalyzed by β-galactosidase.

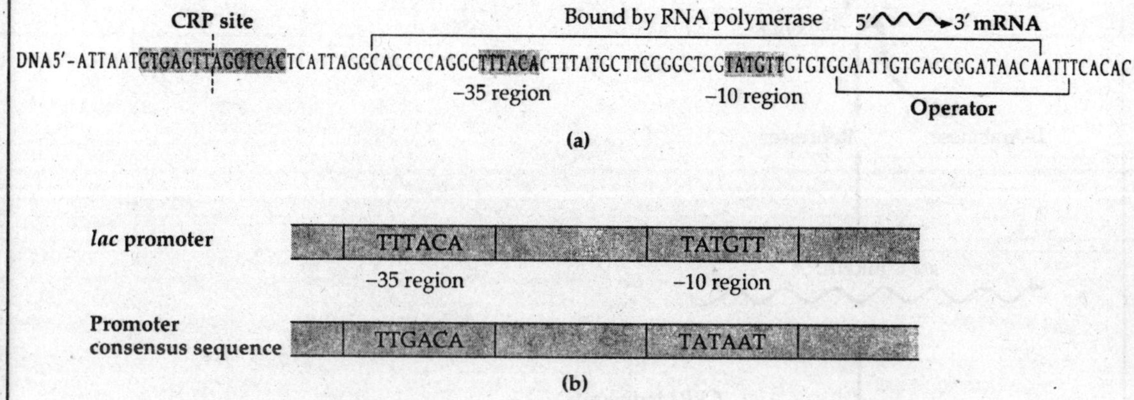

Fig. 6.58: Activation of transcription of the *lac* operon by CRP: (a) The binding site for CRP is near the promoter. As in the case of *lac* operator, the CRP site has two-fold symmetry about the axis indicated by the dashed line. **(b)** Sequence of the *lac* promoter compared with the promoter consensus sequence. The differences mean that RNA polymerase binds relatively weakly to the *lac* promoter until the polymerase is activated by CRP.

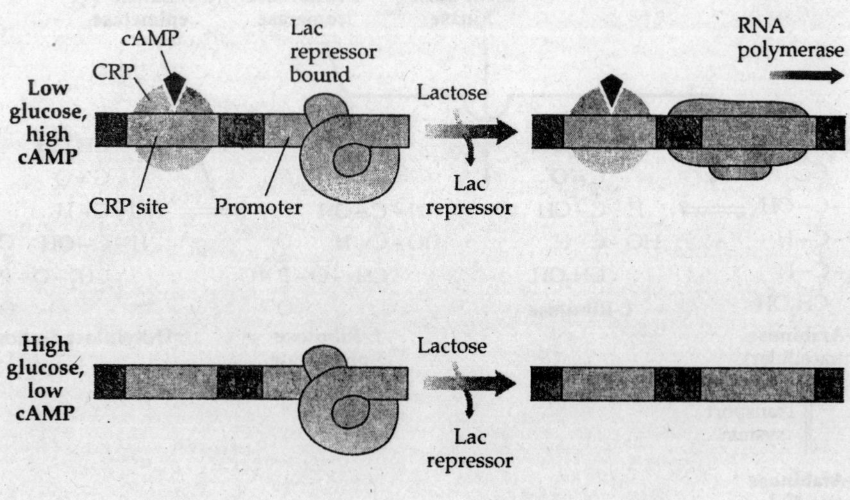

Fig. 6.59: Combined effects of glucose and lactose on expression of the *lac* operon: Transcription occurs only when glucose concentrations are low (so that cAMP levels are high and CRP is bound) and lactose concentrations are high (so that the Lac repressor is not bound).

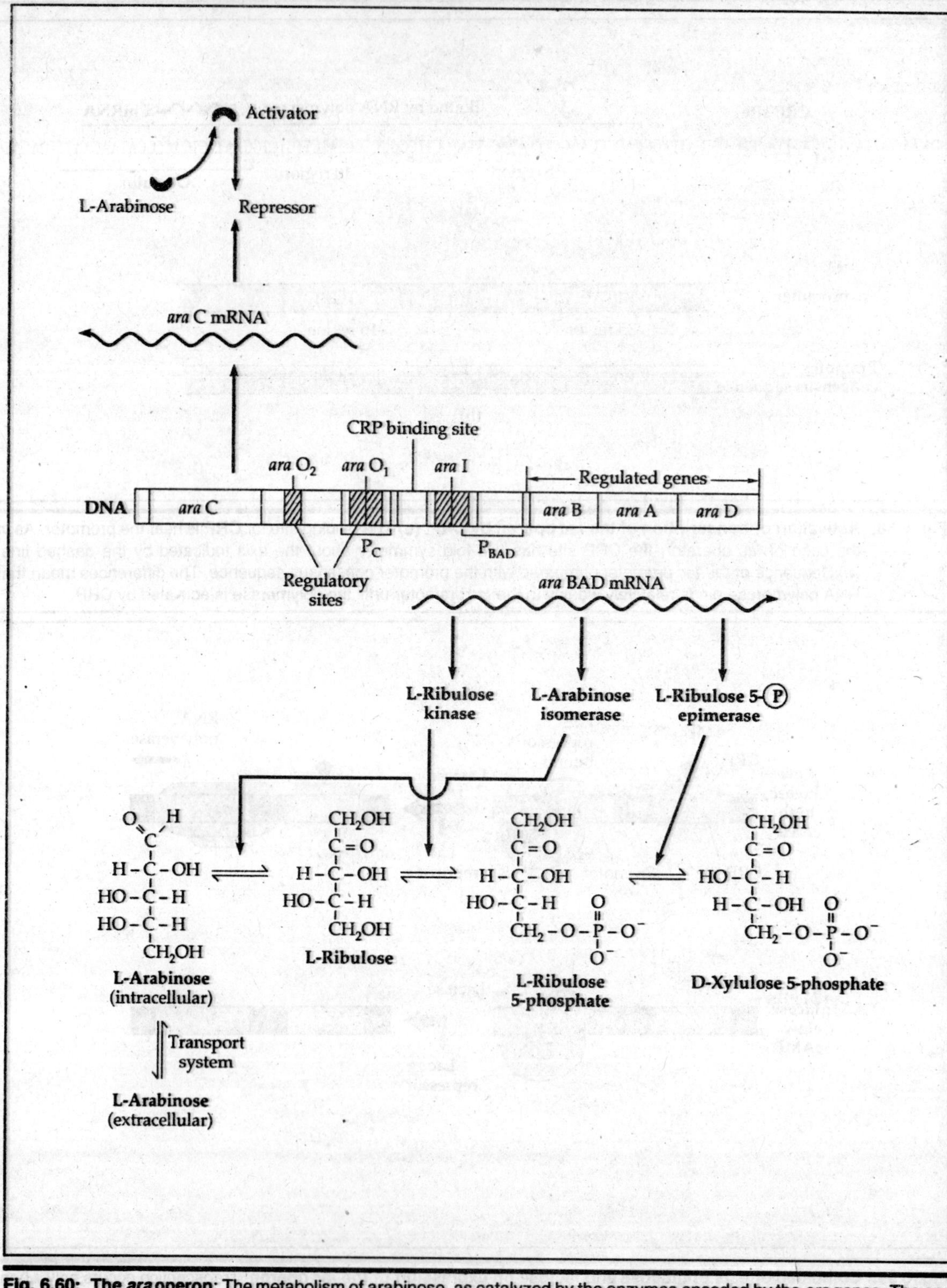

Fig. 6.60: The *ara* operon: The metabolism of arabinose, as catalyzed by the enzymes encoded by the *ara* genes. The end product of this pathway, D-xylulose 5-phosphate, is an intermediate in the pentose phosphate pathway.

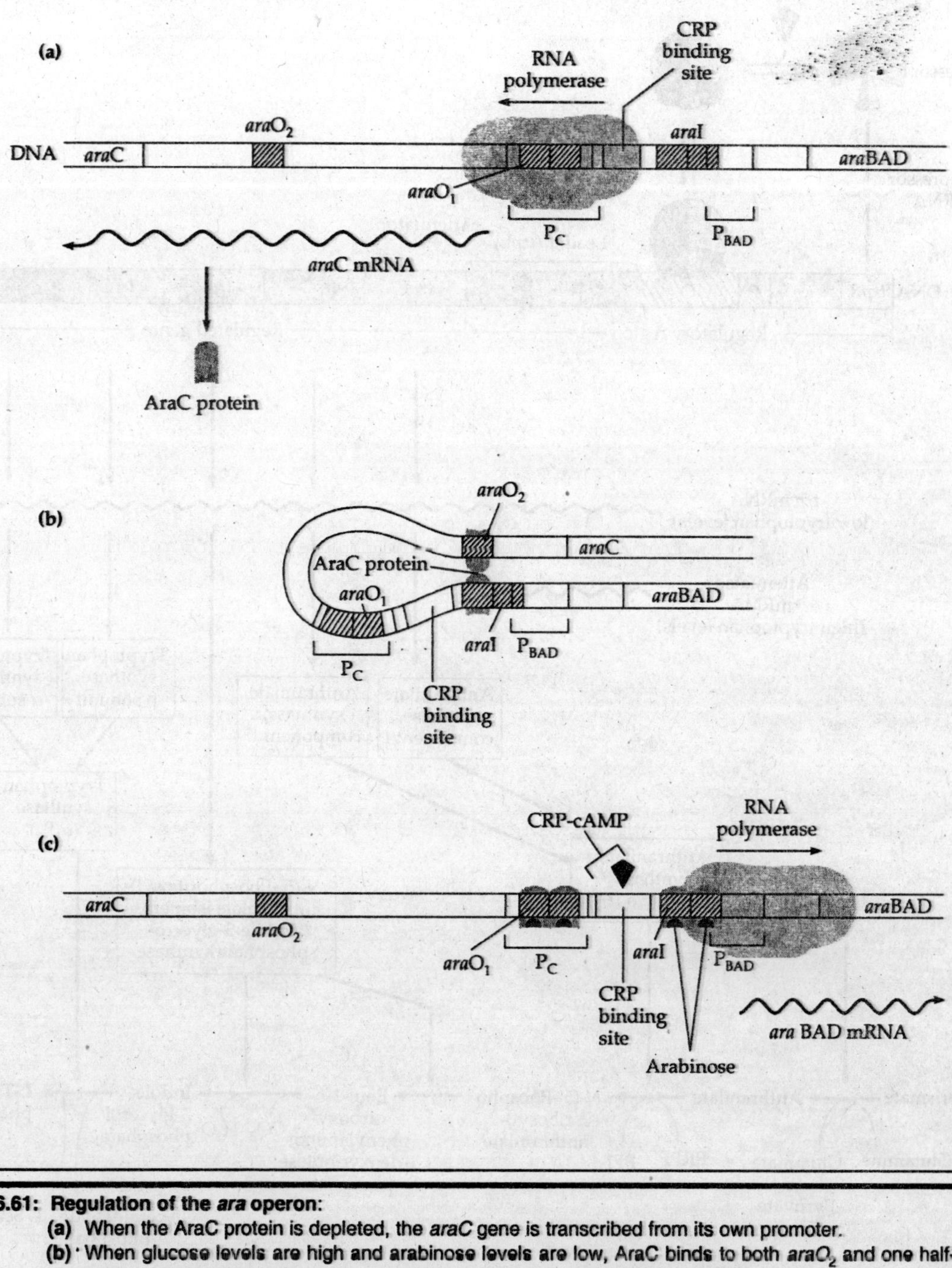

Fig. 6.61: Regulation of the *ara* operon:

(a) When the AraC protein is depleted, the *araC* gene is transcribed from its own promoter.

(b) When glucose levels are high and arabinose levels are low, AraC binds to both *araO*$_2$ and one half-site of *araI* and brings these sites together to form a DNA loop, repressing *araBAD*.

(c) When arabinose is present and the glucose concentration is low, AraC binds arabinose and changes conformation to become an activator. The DNA loop is opened, and AraC binds to each half-site of *araI* and *araO*$_1$. The tandem proteins interact with each other and act in concert with CRP-cAMP to facilitate transcription of the *araBAD* genes.

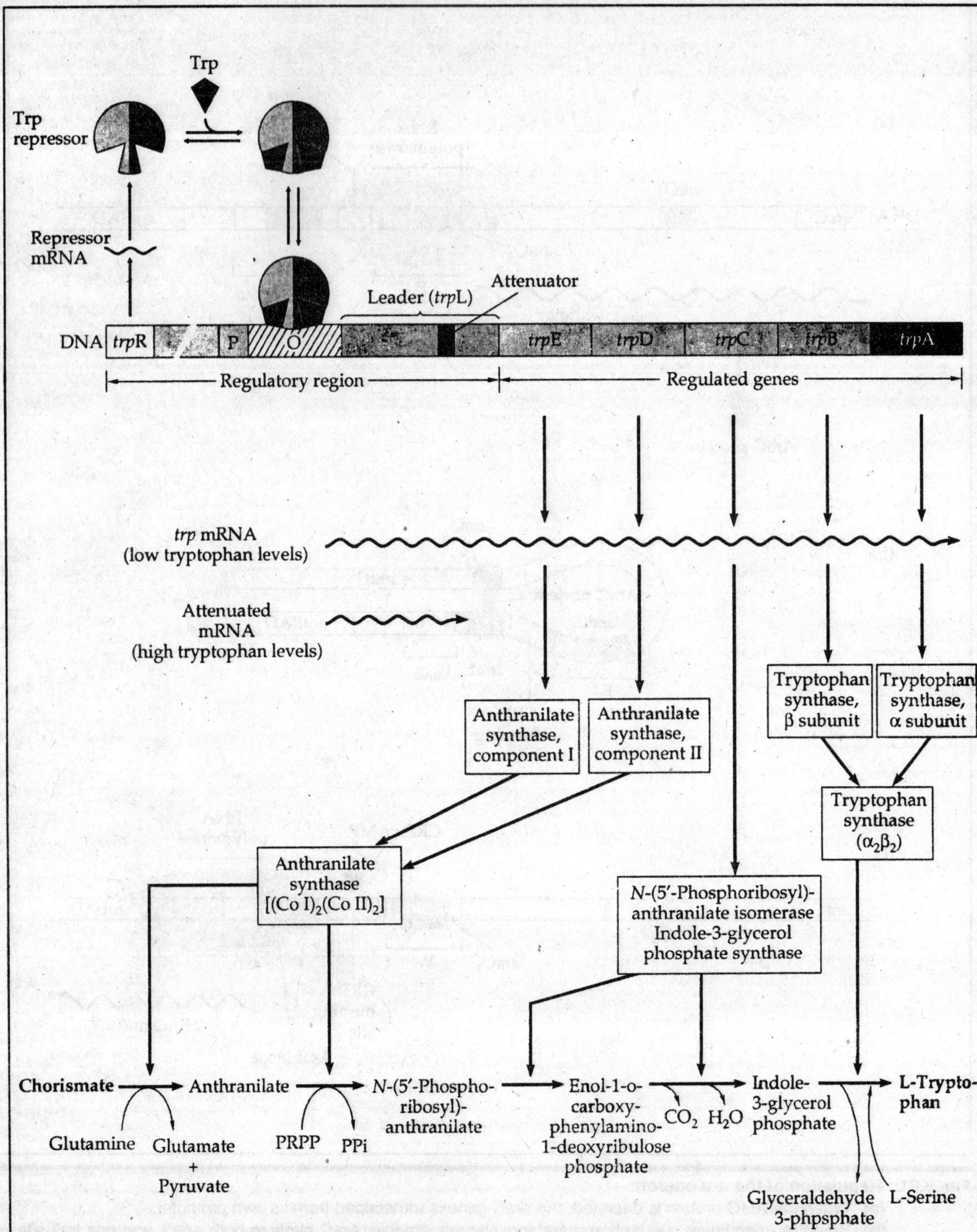

Fig. 6.62: The *trp* operon: This operon is regulated by two mechanisms. When tryptophan levels are high, (1) the repressor (upper left) binds its operator and (2) transcription of *trp* mRNA is attenuated. The biosynthesis of tryptophan by the enzymes encoded in the *trp* operon has shown at the bottom

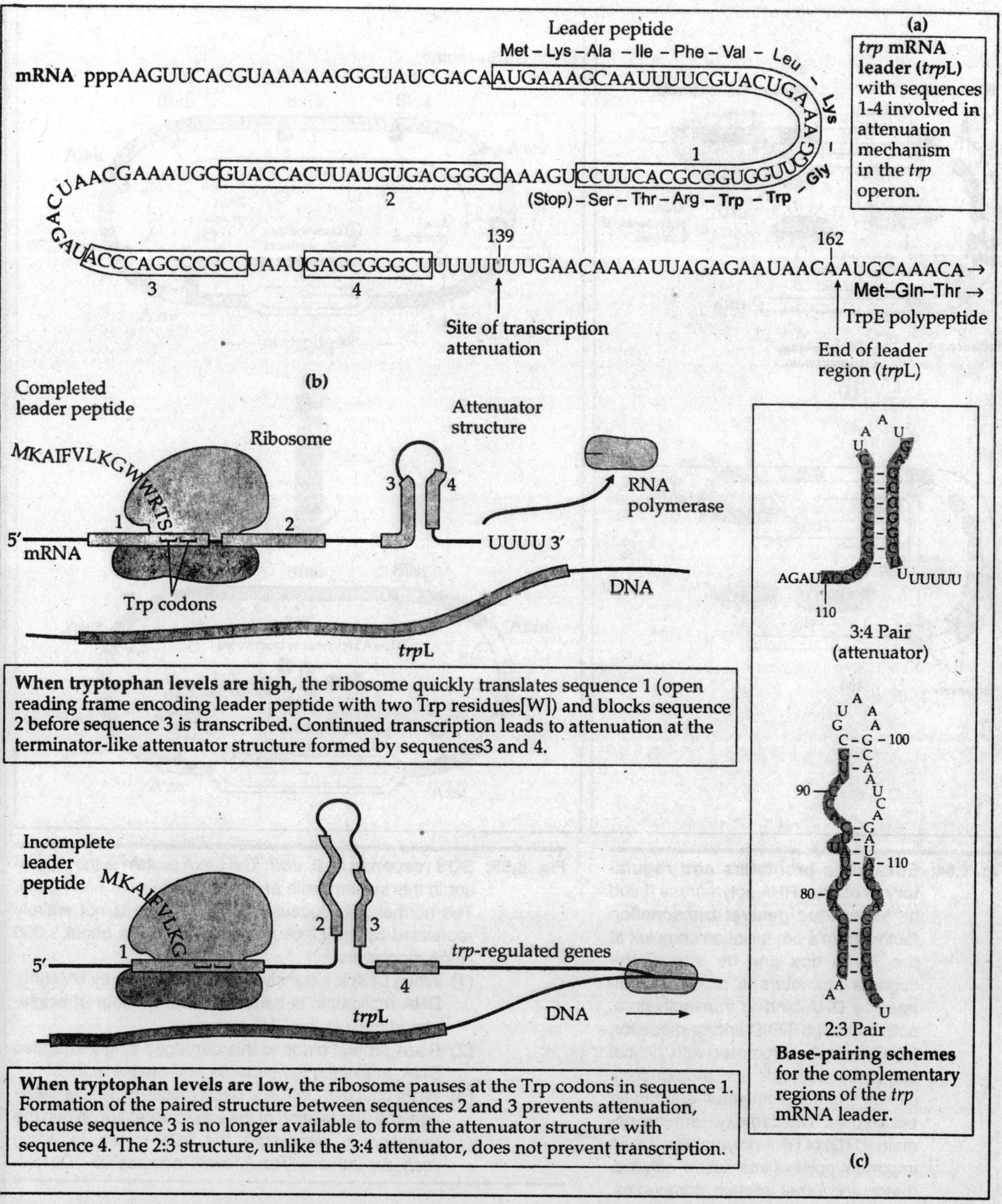

Fig. 6.63 : Transcriptional attenuation in the *trp* operon: Transcription is initiated at the beginning of the 162 nucleotide mRNA leader encoded by a DNA region called *trpL*. A regulatory mechanism determines whether transcription is attenuated at the end of the leader or continues into the structural genes. As sequences 3 and 4 are complementary, attenuator structure is formed by pairing whose structure and function is like transcription terminator. As sequences 2 and 3 are also complementary, pairing between them prevents the formation of attenuator structure. Translation of open reading frame of the leader peptide has a purely regulatory function that determines which complementary sequences are paired.

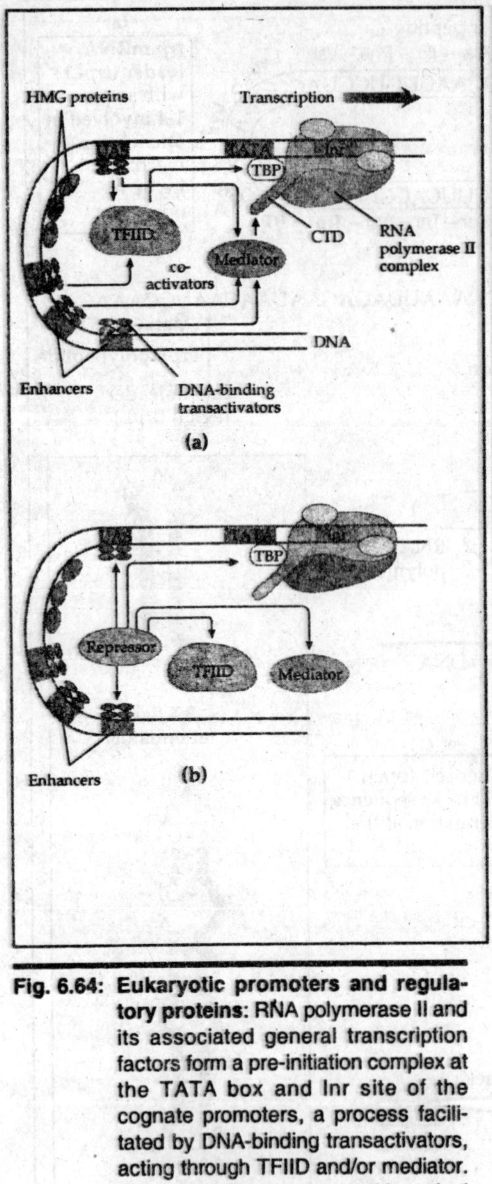

Fig. 6.64: Eukaryotic promoters and regulatory proteins: RNA polymerase II and its associated general transcription factors form a pre-initiation complex at the TATA box and Inr site of the cognate promoters, a process facilitated by DNA-binding transactivators, acting through TFIID and/or mediator. **(a)** Composite promoter with typical sequence elements and protein complexes found in both yeast and higher eukaryotes. The carboxyl-terminal domain (CTD) of RNA polymerase II is an important point of interaction with mediator and other protein complexes. **(b)** A wide variety of eukaryotic transcriptional repressors function by a range of mechanisms. Some bind directly to DNA, displacing a protein complex required for activation; others interact with various parts of the transcription or activation complexes to prevent activation.

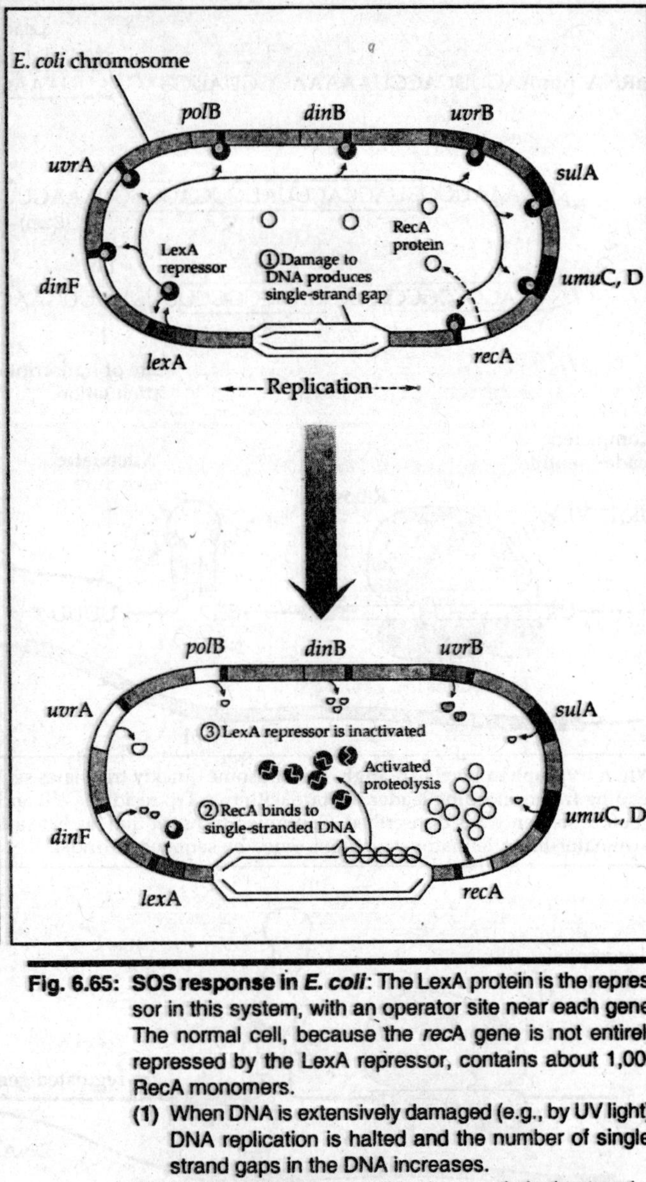

Fig. 6.65: SOS response in *E. coli*: The LexA protein is the repressor in this system, with an operator site near each gene. The normal cell, because the *rec*A gene is not entirely repressed by the LexA repressor, contains about 1,000 RecA monomers.

(1) When DNA is extensively damaged (e.g., by UV light), DNA replication is halted and the number of single-strand gaps in the DNA increases.

(2) RecA protein binds to this damaged single-stranded DNA, activating the protein's coprotease activity.

(3) While bound to DNA, the RecA protein facilitates cleavage and inactivation of the LexA repressor. When the repressor is inactivated, the SOS genes, including *rec*A, are induced; RecA levels increase 50–100 fold.

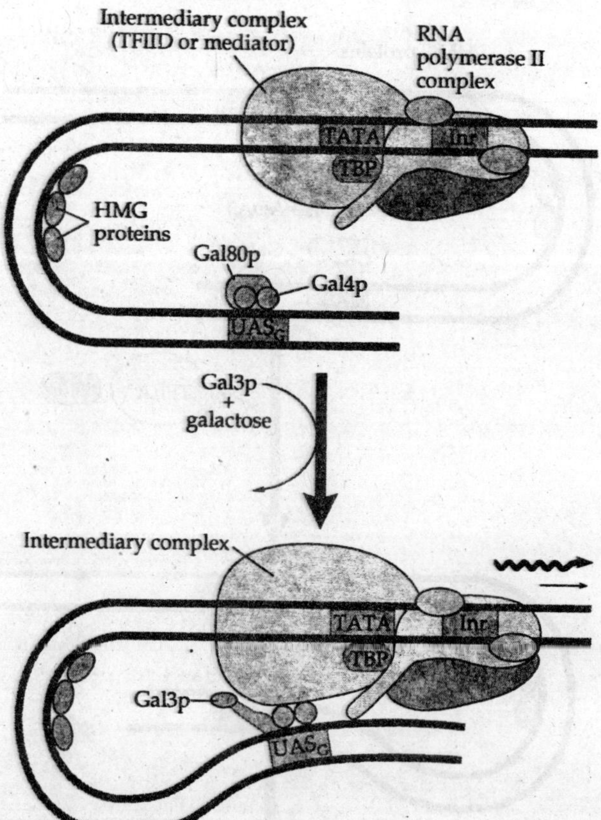

Fig. 6.66: Regulation of transcription at genes of galactose metabolism in yeast: Galactose is imported into the cell and converted to galactose 6-phosphate by a pathway involving six enzymes whose genes are scattered over three chromosomes. Transcription of these genes is regulated by the combined actions of the proteins Gal4p, Gal80p and Gal3p with Gal4p playing the central role of DNA-binding transactivator. The Gal4p/Gal80p complex is inactive in gene inactivation. Binding of galactose to Gal3p and its interaction with Gal80p produce a conformational change in Gal80p that allows Gal4p to function in transcription activation.

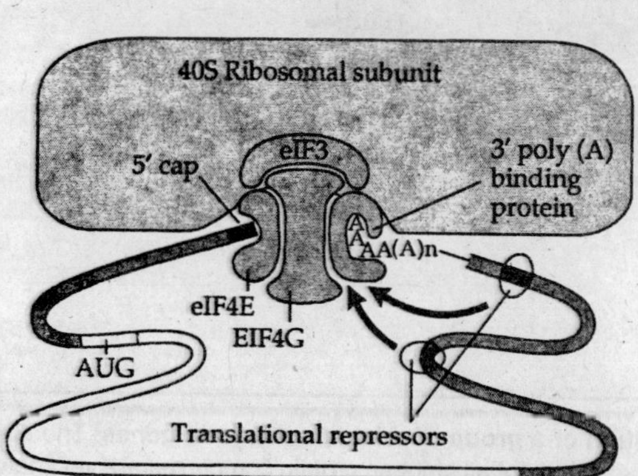

Fig. 6.67: Translational regulation of eukaryotic mRNA: A major mechanism for translational regulation in eukaryotes involves the binding of translational repressors (RNA-binding proteins) to specific sites in the 3′ untranslated region (3′ UTR) of the mRNA. These proteins interact with eukaryotic initiation factors or with the ribosome to prevent or slow translation.

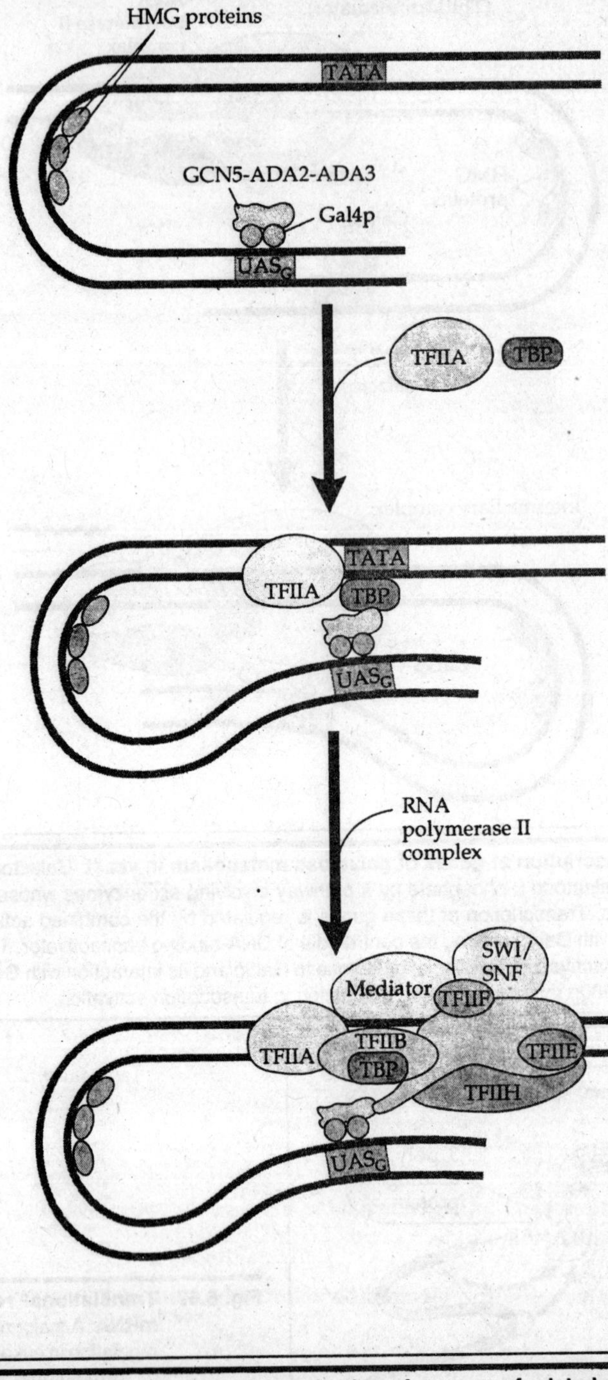

Fig. 6.68: Protein complexes involved in transcription activation of a group of related eukaryotic genes: The *GAL* system is shown to illustrate the complexity of this process, but not all of these protein complexes are yet known to affect *GAL* gene transcription. Many of the complexes (such as SWI/SNF, GCN5-ADA2-ADA3, mediator) affect the transcription of many genes. The complexes assemble stepwise. DNA-binding transactivators bind first, then the additional protein comlexes needed to remodel the chromatin and allow transcription to begin.

CHAPTER 7

SECONDARY METABOLISM

The **basic metabolic pathways** constitute the origins of secondary metabolism in plants and give rise to a vast array of compounds (**secondary metabolites**). Some of these are responsible for characteristic odours, pungencies and colours of plants, others give a particular plant its culinary, medicinal or poisonous virtues, and by far the **greatest numbers** are of obscure value to the plant, though modern authors suggest their role in helping producer's survival. A number of **biogenetic groups** are characterized by particular skeletal structures—aromatic compounds, phenols and phenolic glycosides, terpenes, steroids, alkaloids, volatile oils and resins, etc. **Shikimic acid pathway, malonic acid pathway, mevalonic acid pathway, pyruvate pathway** play central role in the synthesis of secondary metabolites.

Figures: 7.1–7.2

AROMATIC BIOSYNTHESIS

A large number of aromatic compounds including amino acids are synthesized from carbohydrate precursors derived from the pentose phosphate pathway and glycolysis through **shikimic acid pathway.** An important branching point arises at **chorismic acid**; which gives rise to anthranilic acid, a precursor of tryptophan; and also to prephenate, a precursor of phenylalanine and tyrosine. The shikimic acid pathway is important in the genesis of aromatic building blocks of lignin, tannins, valine, phenylpropane units of the flavones, coumarins, opium alkaloids, etc.

Figures: 7.3–7.4

PHENOLS AND PHENOLIC GLYCOSIDES

Phenols, having aromatic moieties, constitute the largest group of plant secondary metabolism. It includes simple phenolics, flavones, lignin, tannins, coumarins, anthraquinones and naphthaquinones, etc. The biosynthetic origin of most of these compounds involves shikimic acid pathway. Phenylalanine, a product of shikimic acid pathway, is the precursor of most phenolics. Deamination of phenylalanine to cinnamic acid, followed by hydroxylation to *p*-coumaric acid, is the first step in phenolic biosynthesis.

Simple phenolic compounds often possess alcoholic (e.g., catechol), aldehydic (e.g., vanillin) and carboxylic acid groups (salicylic, ferulic, caffeic acids). **Tannins** are polyphenols and glycosides illustrating particular aspects of gallic acid and flavan-3-ol phytochemistry. Two main groups of tannins are recognised—hydrolysable tannins (e.g., gallitannins, ellagitannins) and condensed tannins (e.g., proanthocyanidins). **Coumarins** (e.g., aesculetin, umbelliferone, scopoletin, etc.) are derivatives of benzo-α-pyrone, occur in plants both in free state and as glycosides, produced through shikimic acid pathway.

Anthraquinones like emodin, alizarin, etc., are synthesized either via the acetate-malonate pathway or from shikimate and mevalonate. **Naphthoquinones** are biosynthesized via a variety of pathways including acetate and malonate (e.g., plumbagin), shikimate/succinyl CoA combined pathway (e.g., lawsone) and shikimate/mevalonate combined pathway (e.g., alkannin). **Chromones** and **xanthones**, derivatives of benzo-γ-pyrone, are of mixed shikimate-acetate origin. **Flavonoids** are the largest naturally occurring phenols with γ-pyrone moiety, occur both in free state and as glycosides. The biosynthesis of flavonoid ring structure occurs from three acetate units and a phenylpropane unit. **Anthocyanidins** are structurally related to the flavones whose glycosides are anthocyanins. **Lignin** is a polymeric compound $(C_6\text{-}C_3)_n$ which vary in composition, depending on the particular monomeric units of which they are composed. **Lignans** are dimeric compounds formed by the union of two molecules of a phenylpropane derivative and are offshoots of the principal lignin biosynthetic pathway.

Figures: 7.5–7.9

ISOPRENOIDS–TERPENES

In nature there are tremendous range of isoprenoids. Basic five-carbon units of **terpenes** are synthesized from acetyl-CoA through mevalonic acid pathway. The major subclasses of terpenes are biosynthesized from the basic five-carbon units—IPP and DMAPP. Stereospecific reactions occur in terpenoid biogenesis. **Monoterpenes** (C_{10}), **sesquiterpenes** (C_{15}) and **diterpenes** (C_{20}) are produced by sequential addition of C_5 units. Monoterpenes (e.g., iridoids, valerenol) derived biogenetically from the C_{10} geranyl pyrophosphate. Sesquiterpenes (e.g., artemisinin, gossypol) are synthesized from farnesyl pyrophosphate. Diterpenoids comprise of structurally diverse range of compounds which may be acyclic or possess 1-5 ring systems. They include kaurane, forskolin, tigliane, ginkgolide, etc. **Triterpenes** are derived from two C_{15} units through squalene and **tetraterpenes** from two C_{20} units through phytoene. Triterpenoids are abundant in nature and occur as either esters or glycosides; may be aliphatic, tetracyclic or pentacyclic (e.g., quassin, neoquassin). Tetraterpenoids (carotenoids) are formed by the tail to tail union of two molecules of the C_{20} geranyl-geranyl diphosphate leading to lycopene synthesis, cyclization of one or both ends of which results various carotenes and derivatives. **Polyterpenoids** are composed of many isoprene units, e.g., rubber, latex.

Figures: 7.10–7.17

STEROIDS, SAPONINS, CARDIAC GLYCOSIDES

Cholesterol is biosynthesized from squalene through lanosterol and it is the precursor of a large number of steroids, sapogenins and cardiac glycosides. **Saponins** are glycosides of **sapogenins** (aglycone) and various sugars. Structurally, sapogenin may be steroidal tetracyclic or pentacyclic triterpenoids having a common biogenetic origin via mevalonic acid and isoprenoid units. Cholesterol, produced by cyclization of squalene, can be incorporated into a number of C_{27} sapogenins without side chain cleavage. Some steroidal sapogenins are diosgenin, sarsasapogenin, yamogenin, smilagenin, hecogenin. Pentacyclic saponins are aescin, aralin, glycyrrhizin, etc. **Cardiac glycosides** may be **cardenolides** (e.g., digitoxigenin) or **bufadienolides** (e.g., scillarenin). Aglycons are derived from mevalonic acid but the final molecules arise from a condensation of C_{21} steroid with a C_2 or C_3 unit. Other **phytosterols** (e.g., stigmasterol, sitosterol, sinosterol) are also being derived from cholesterol.

Figures: 7.18–7.21

VOLATILE OILS, GUMS, RESINS

Volatile oils are generally mixtures of hydrocarbons and their oxygenated compounds, frequently associated with gums and resins. Many oils are terpenoid in origin (e.g., pinenes, citral, carvone, geraniol, menthol, camphor, camphene), some are aromatic derivatives (e.g., eugenol) and few are structurally aromatic but terpenoid in origin (e.g., thymol, gingerol, carvacrol). Most of the oils are monoterpenoids, arise at the GPP level of terpenoid synthesis. They may be acyclic, monocyclic or bicyclic which are sequentially derived from limonene. **Resins** are complex mixtures of **resin** acids, resin alcohols (resinols), resin phenols (resinotannols), esters and resenes. Resins are often associated with volatile oils (**oleoresins**), with gums (**gum-resins**) or with oil and gum (**oleo-gum-resins**).

Figures: 7.22–7.23

ALKALOIDS

Alkaloids are nitrogenous basic compounds typically with heterocyclic ring (e.g., nicotine, quinine), though non-heterocyclic **protoalkaloids** (e.g., ephedrine) also exist. Amino acids, both aliphatic and aromatic, involve in biogenesis of most alkaloids. **Ornithine-derived alkaloids** are of tropane, ecgonine, nicotine, necine and stachydrine groups. **Lysine-derived alkaloids** are anabasine, lupinine, anaferine, piperine, etc. A number of alkaloids are **phenylalanine and tyrosine-derived**. Protoalkaloids like ephedrine is synthesized from phenylalanine through cathinone. Emetine alkaloids are being formed from two phenylalanine units and a C_9 terpenoid precursor, secologanin. A large number of benzylisoquinolines (e.g., thebaine, codeine, morphine) are biosynthesized from tyrosine. Reticuline derived from tyrosine, gives rise to a number of alkaloids—canadine, stylopine, naroein, etc. Phenethylisoquinoline like colchicine is derived from phenylalanine and tyrosine, later contributes tropolone moiety. **Tryptophan-derived** alkaloids constitute a large class of indole alkaloids. A key intermediate in the biogenesis of the monoterpene indole alkaloids is strictosidine. The non-tryptophan-derived portion of complex indole alkaloids (e.g., catharanthine, vindoline, reserpine, strychnine) are formed by mevalonic acid incorporation. Ergot alkaloids are derived from tryptophan and mevalonate via lysergic acid synthesis. Quinine having both indole and terpenoid moieties, is derived through strictosidine precursor. The ring formation of **purine alkaloids** (e.g., caffeine, theobromine) appears to follow the classical scheme for biosynthesis of purine nucleotides. **Pyridine alkaloids** (e.g., coniline, arecoline, ricinine) appear to be biosynthesized from acetate or succinate with the participation of nitrogen source. **Steroidal alkaloids** arise by the inclusion of a basic nitrogen at some point in the steroidal molecule, are either of C_{27} group (e.g., solasodine, solanidine) or C_{21} group (e.g., conessine, kurchine). Common **terpenoid alkaloid** is aconitine.

Figures: 7.24–7.44

GLYCOSIDES AND GLUCOSINOLATES

In addition to the glycosides discussed, there are a number of **cyanogenetic glycosides** (e.g., manihotoxin, amygdalin, linamarin, prunasin) and **glucosinolates** (e.g., sinigrin, sinalbin). Most of them are synthesized from aldoximes or nitriles of mandelic acid derived from amino acids. Sinigrin, a thioglucoside of mustard, derived from homomethionine.

Figures: 7.45–7.46

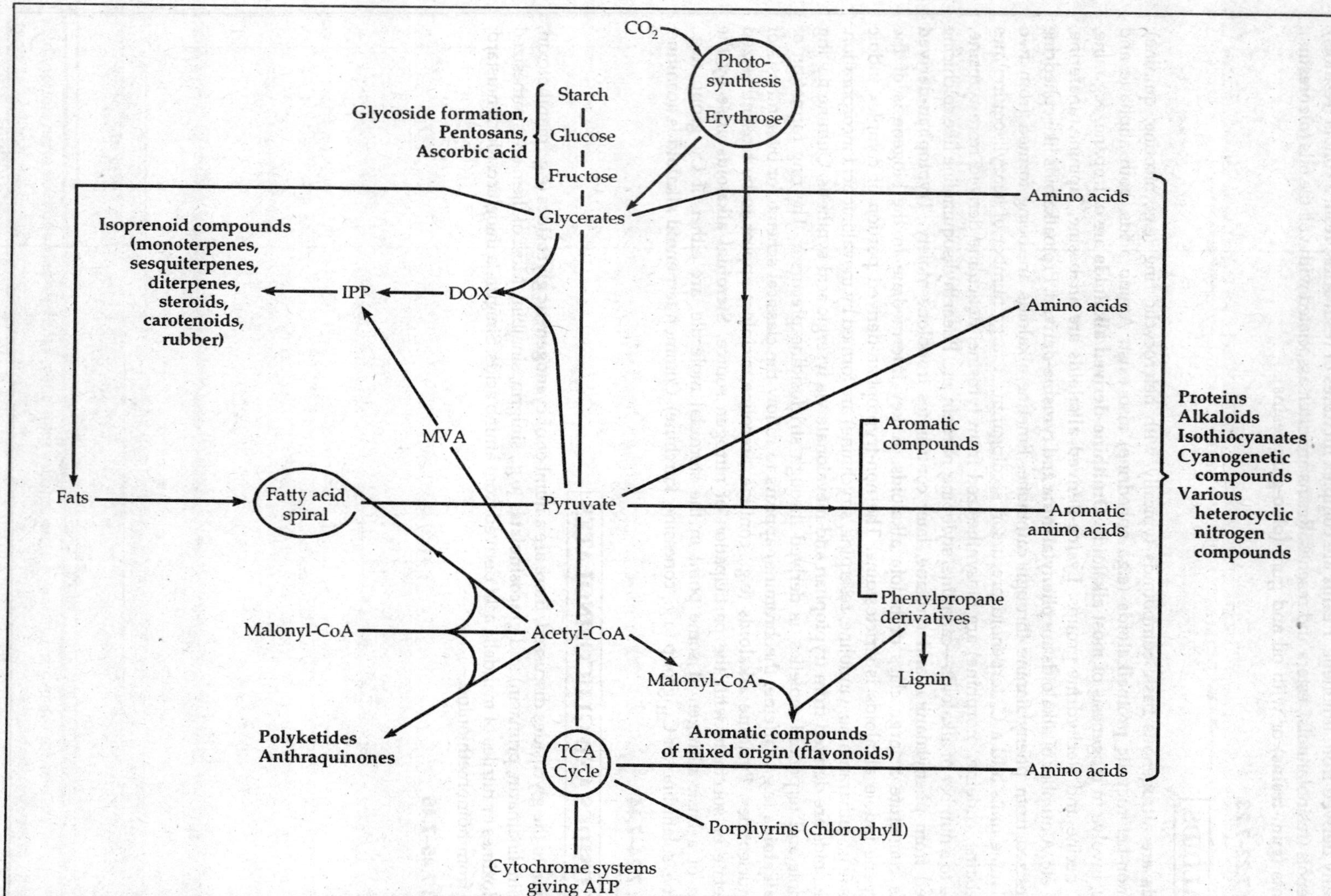

Fig. 7.1: Origins of some secondary metabolites in relation to the basic metabolic pathways of plants. [DOX = deoxyxylulose, IPP = isopentenyl diphosphate, MVA = mevalonic acid.]

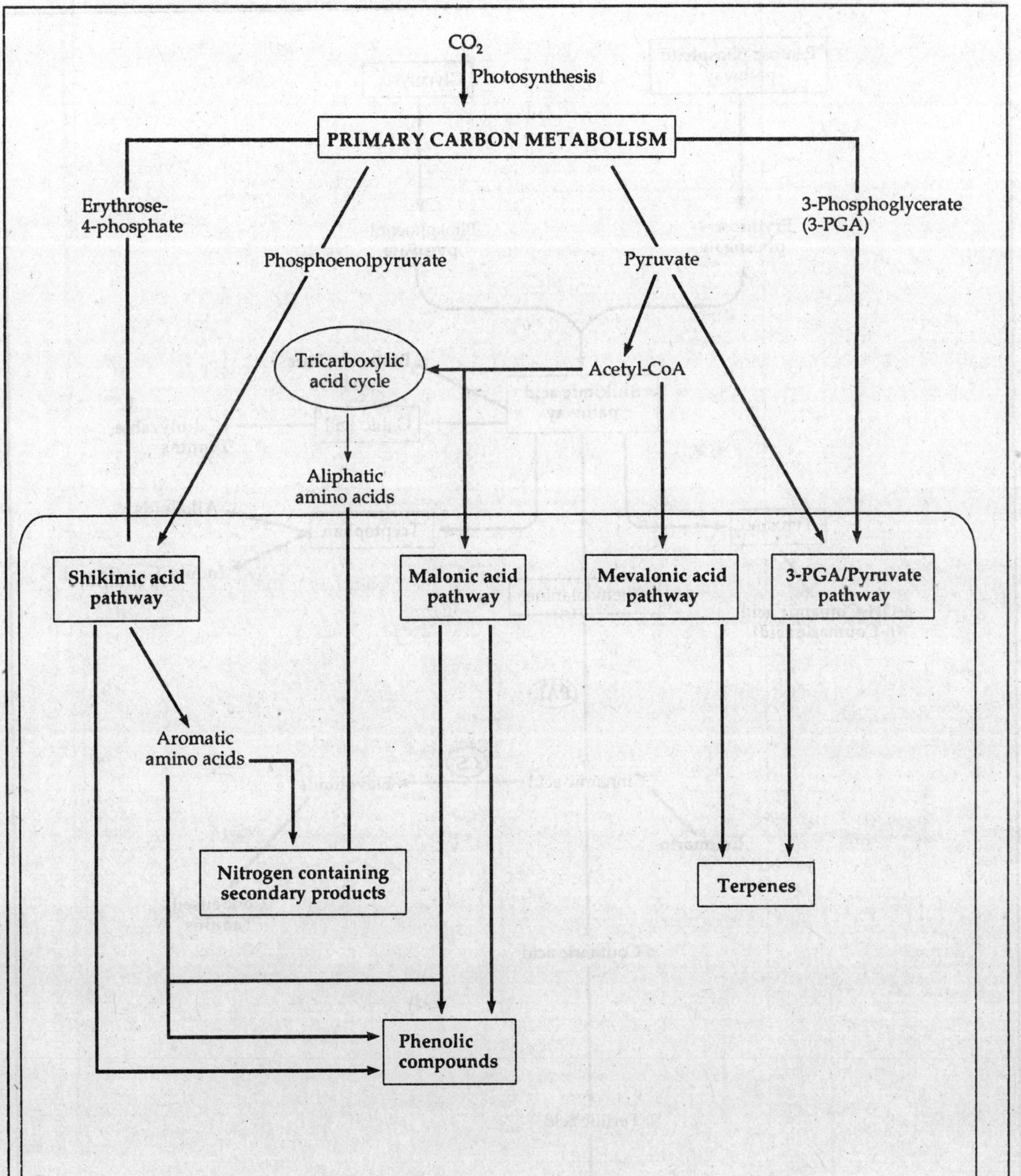

Fig. 7.2: A simplified view of the major pathways of secondary metabolite biosynthesis and their interrelationships with primary metabolism.

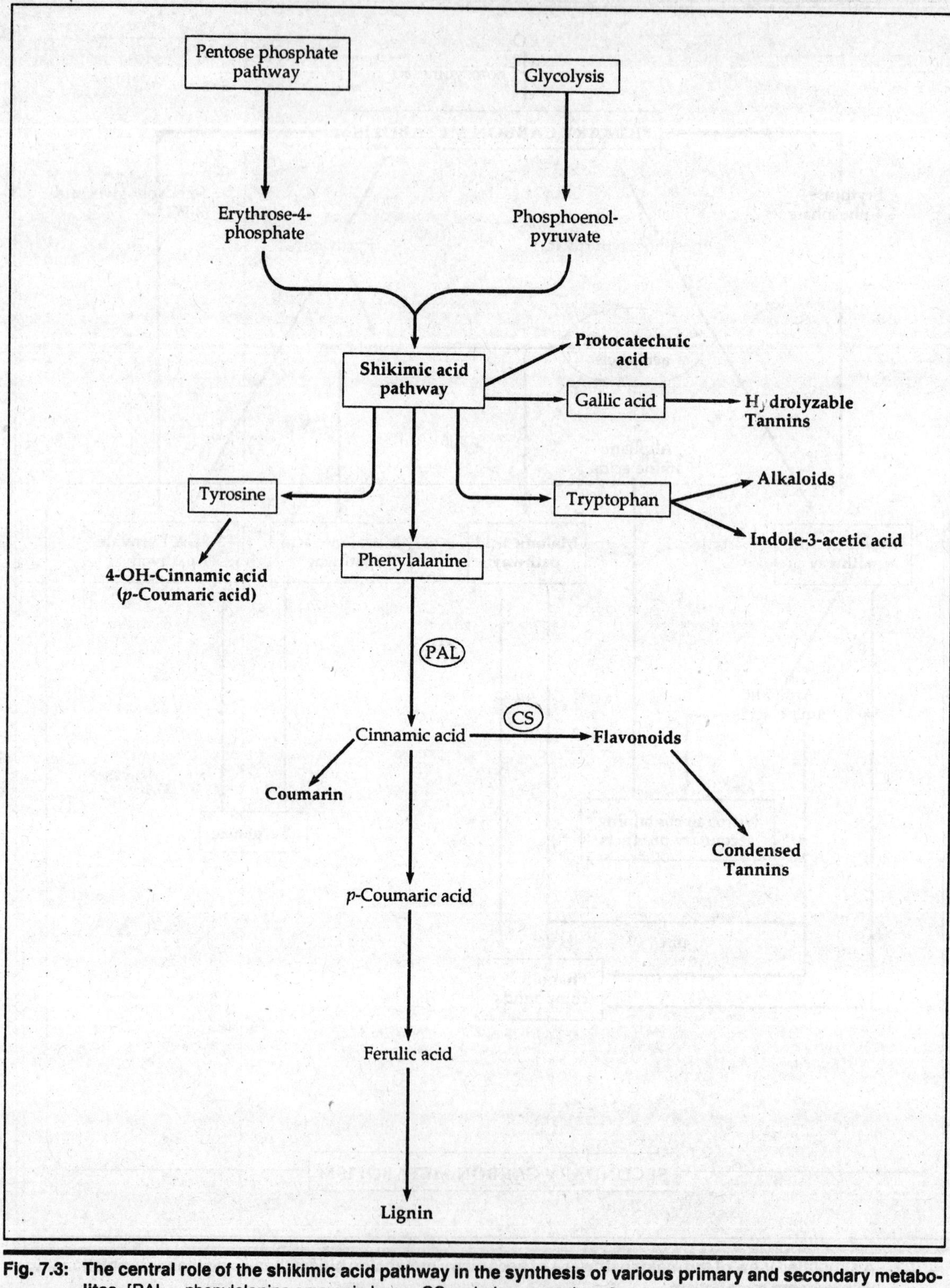

Fig. 7.3: The central role of the shikimic acid pathway in the synthesis of various primary and secondary metabolites. [PAL = phenylalanine ammonia lyase, CS = chalcone synthase]

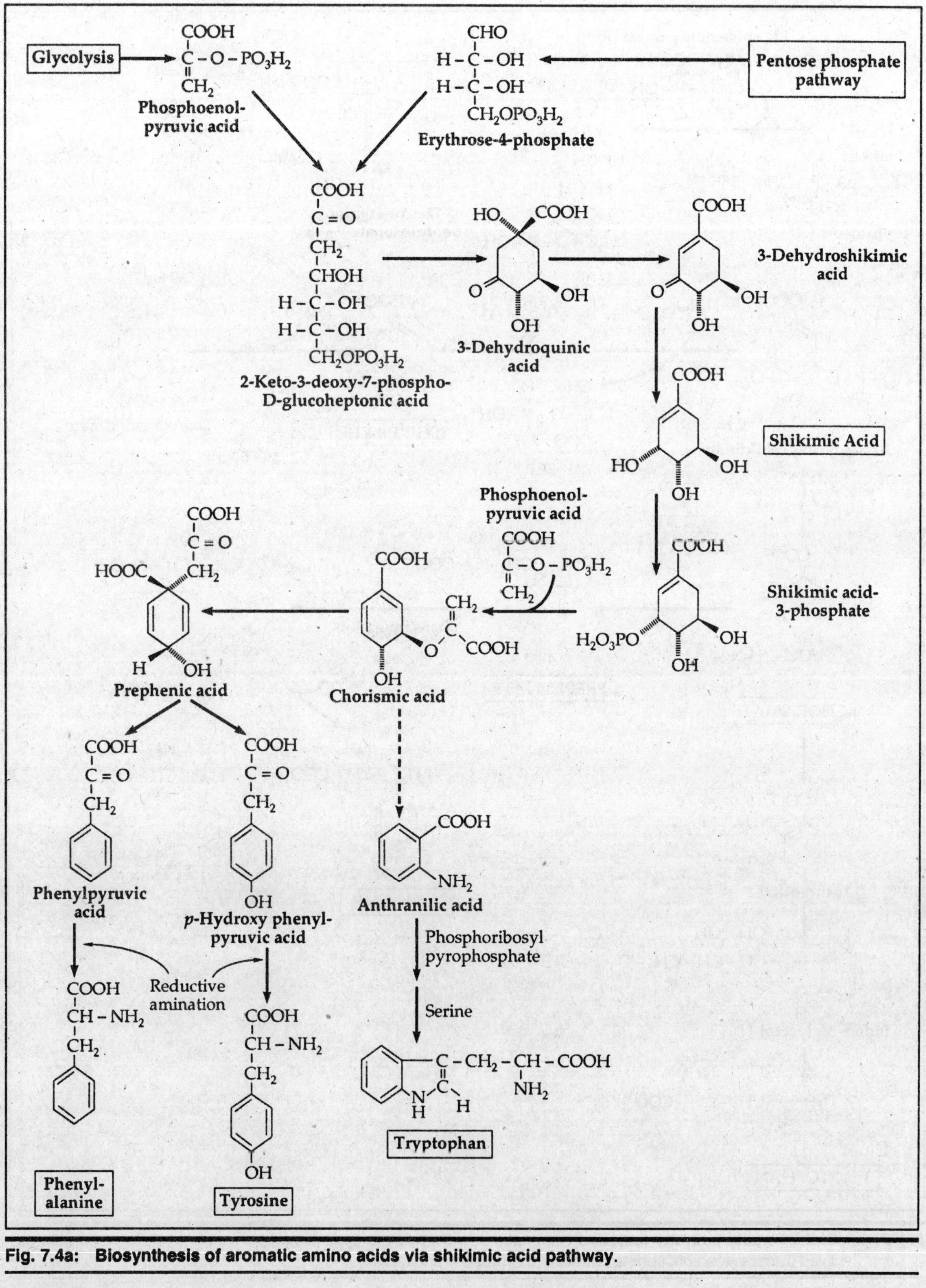

Fig. 7.4a: Biosynthesis of aromatic amino acids via shikimic acid pathway.

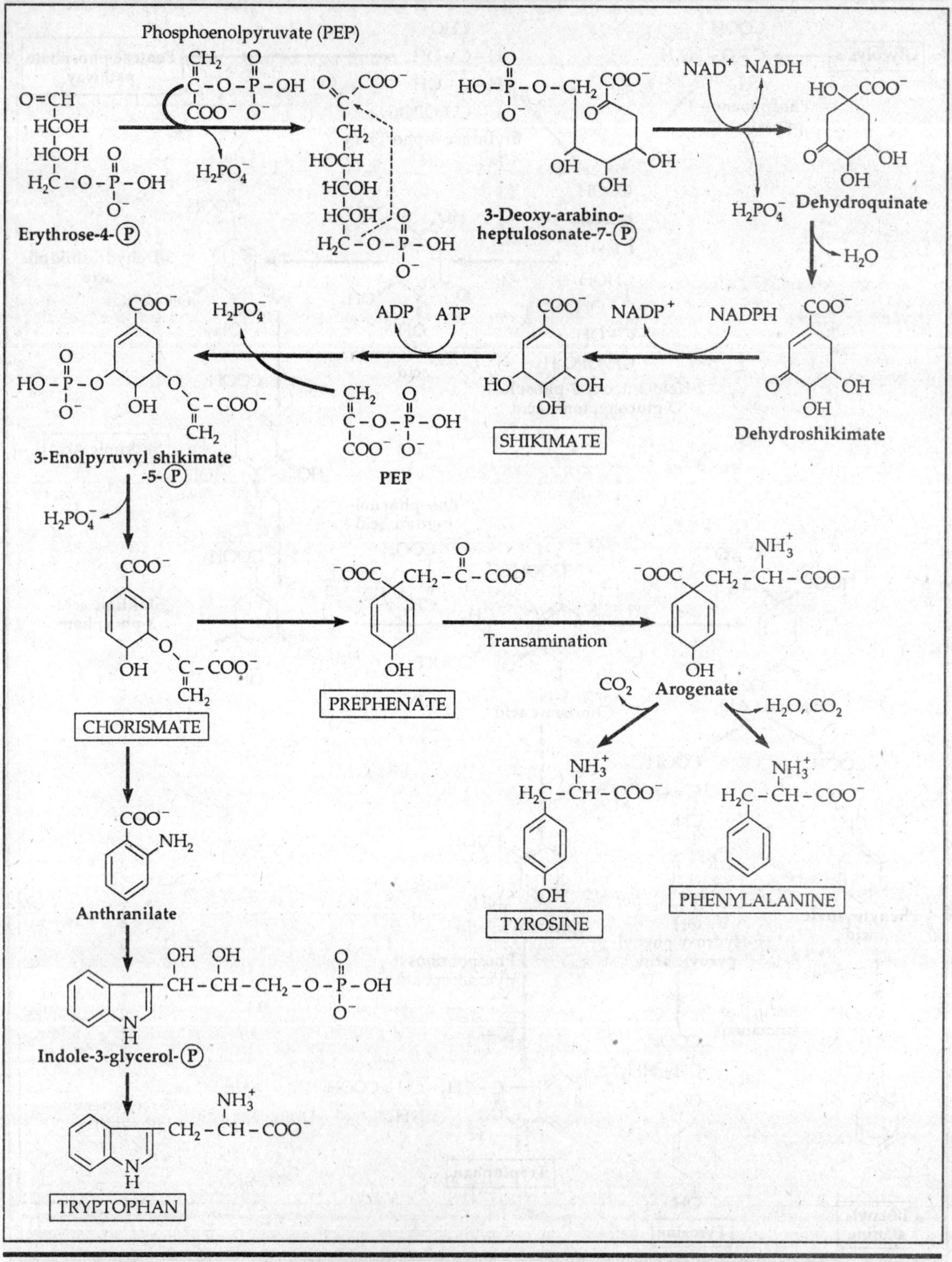

Fig. 7.4b: Shikimic acid pathway (detailed): Aromatic amino acids are synthesized from carbohydrate precursors derived from pentose phosphate pathway (Erythrose-4-phosphate) and glycolysis (Phosphoenolpyruvic acid).

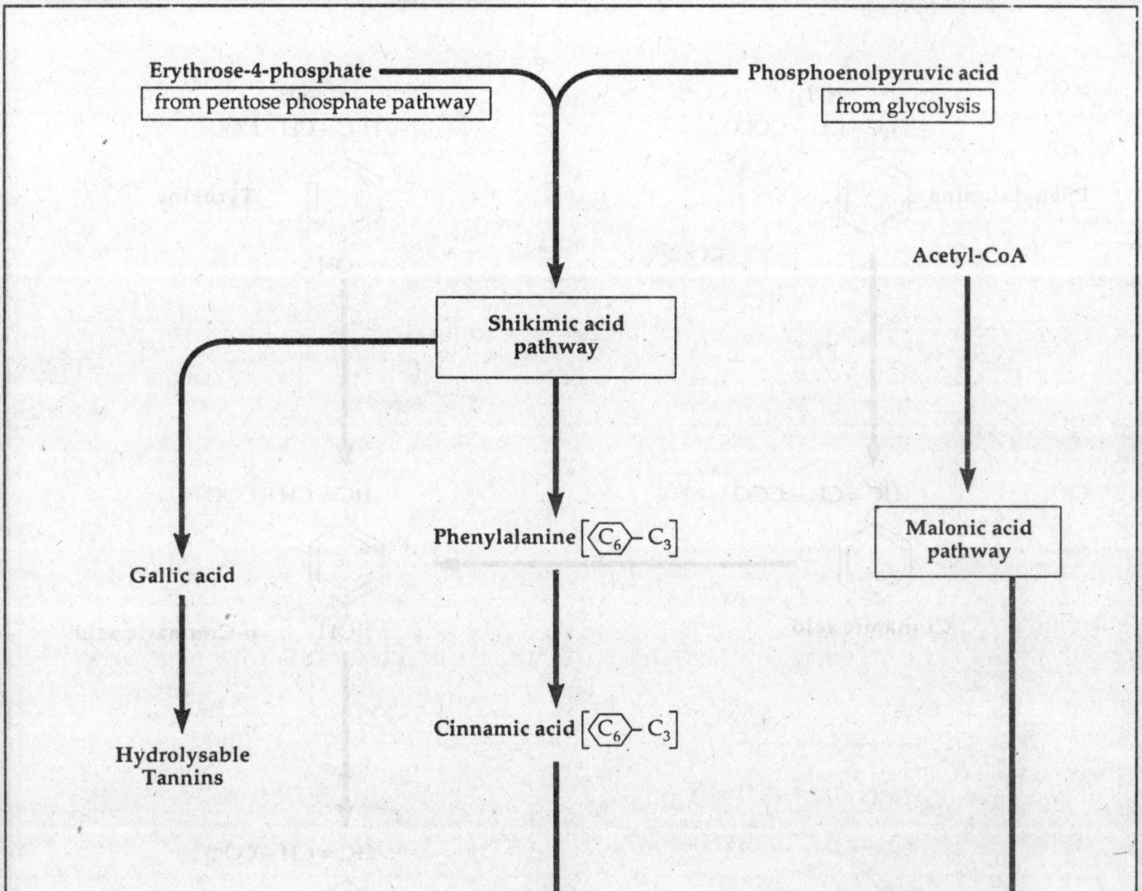

Fig. 7.5: Plant phenolics are biosynthesized in several different ways: In higher plants, most phenolics are derived at least in part from phenylalanine, a product of the shikimic acid pathway. Formulas in brackets indicate the basic arrangement of carbon skeletons: C_6 indicates a benzene ring, and C_3 is a three-carbon chain.

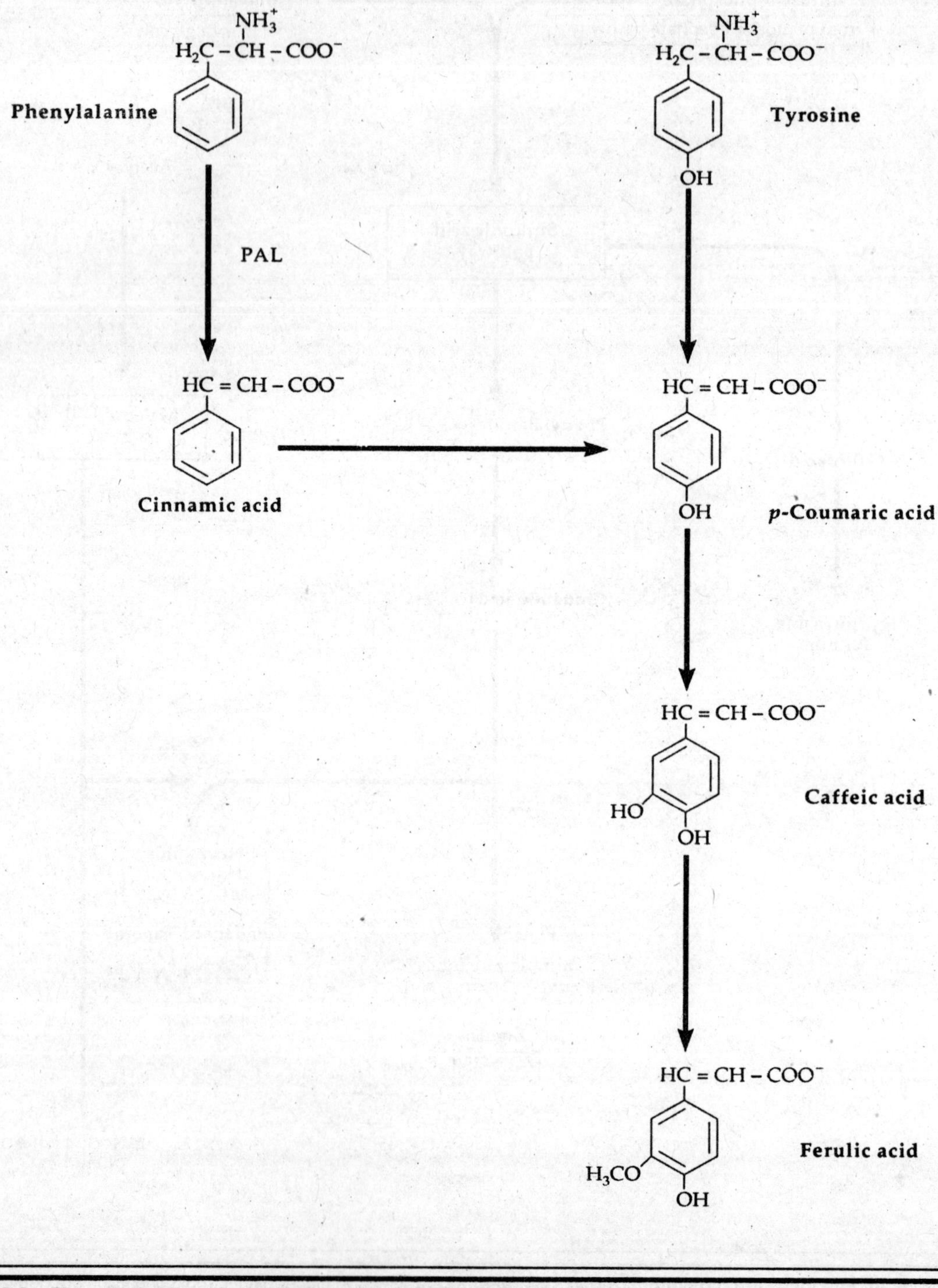

Fig. 7.6: Phenolic building blocks: Deamination of phenylalanine to cinnamic acid, followed by hydroxylation to *p*-coumaric acid, is the first step in phenolic biosynthesis. In members of the grass family, tyrosine can be deaminated directly to *p*-coumaric acid. [PAL = phenylalanine ammonia lyase]

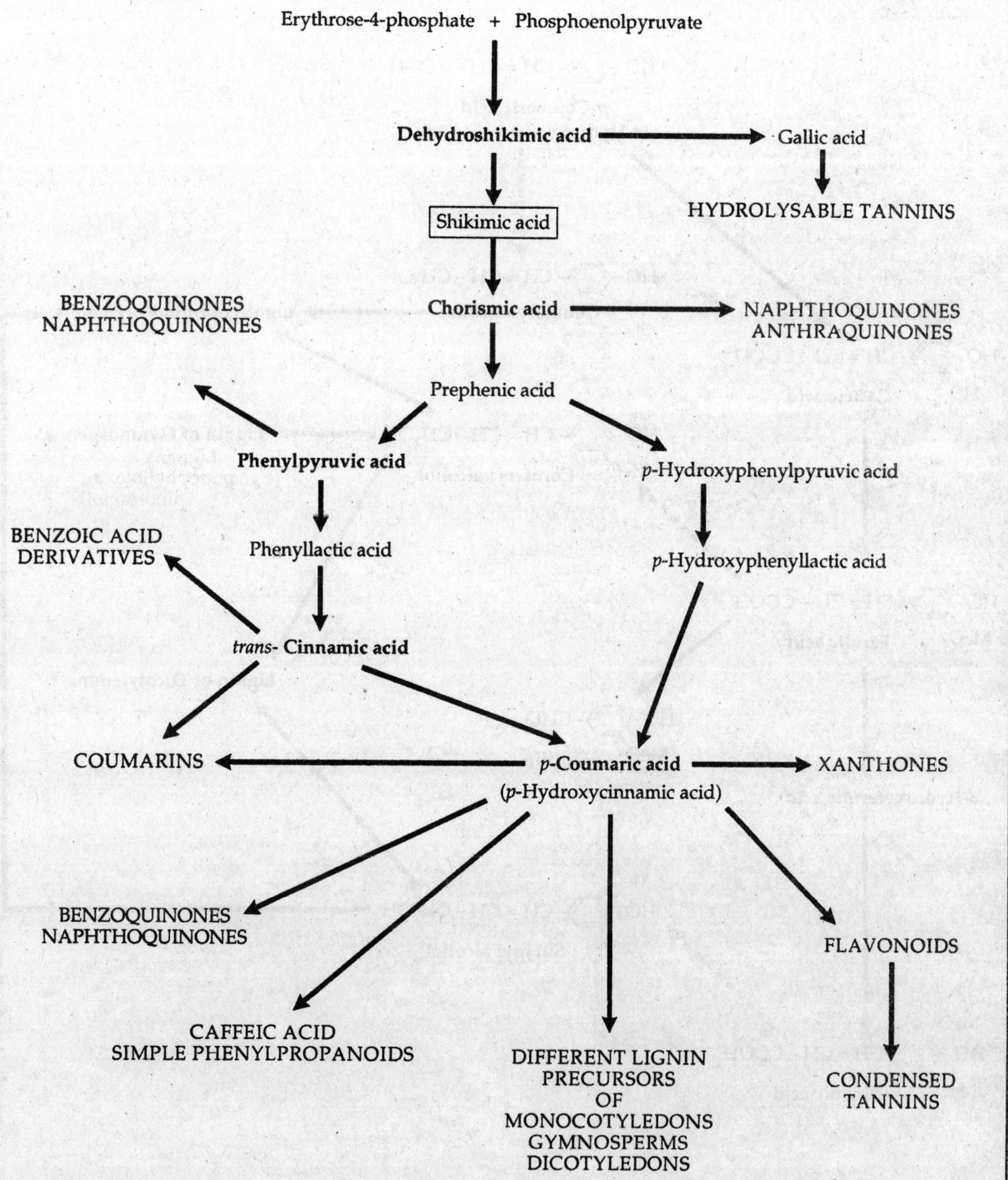

Fig. 7.7: Biosynthesis of different phenolic compounds originating from shikimic acid.

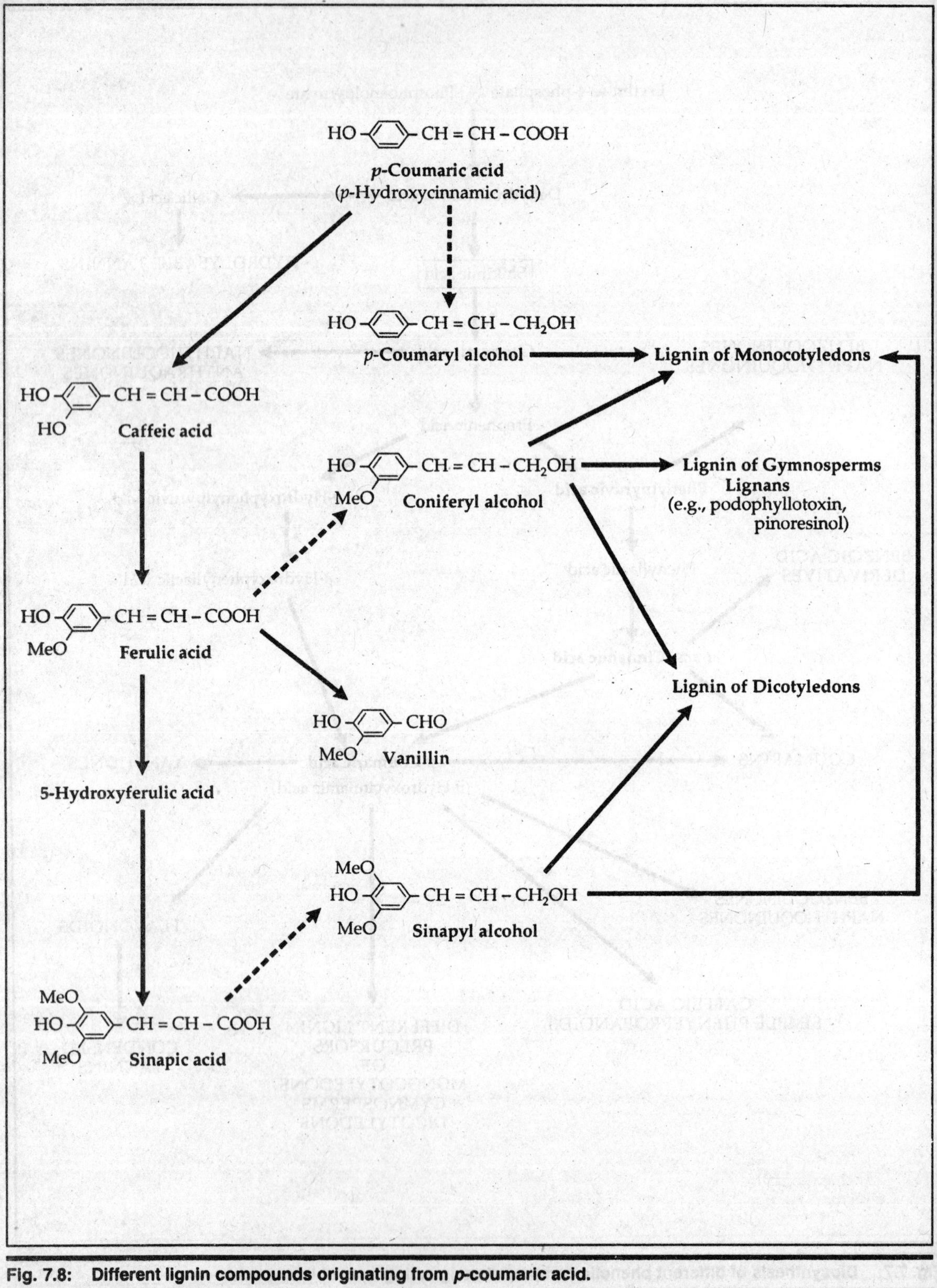

Fig. 7.8: Different lignin compounds originating from *p*-coumaric acid.

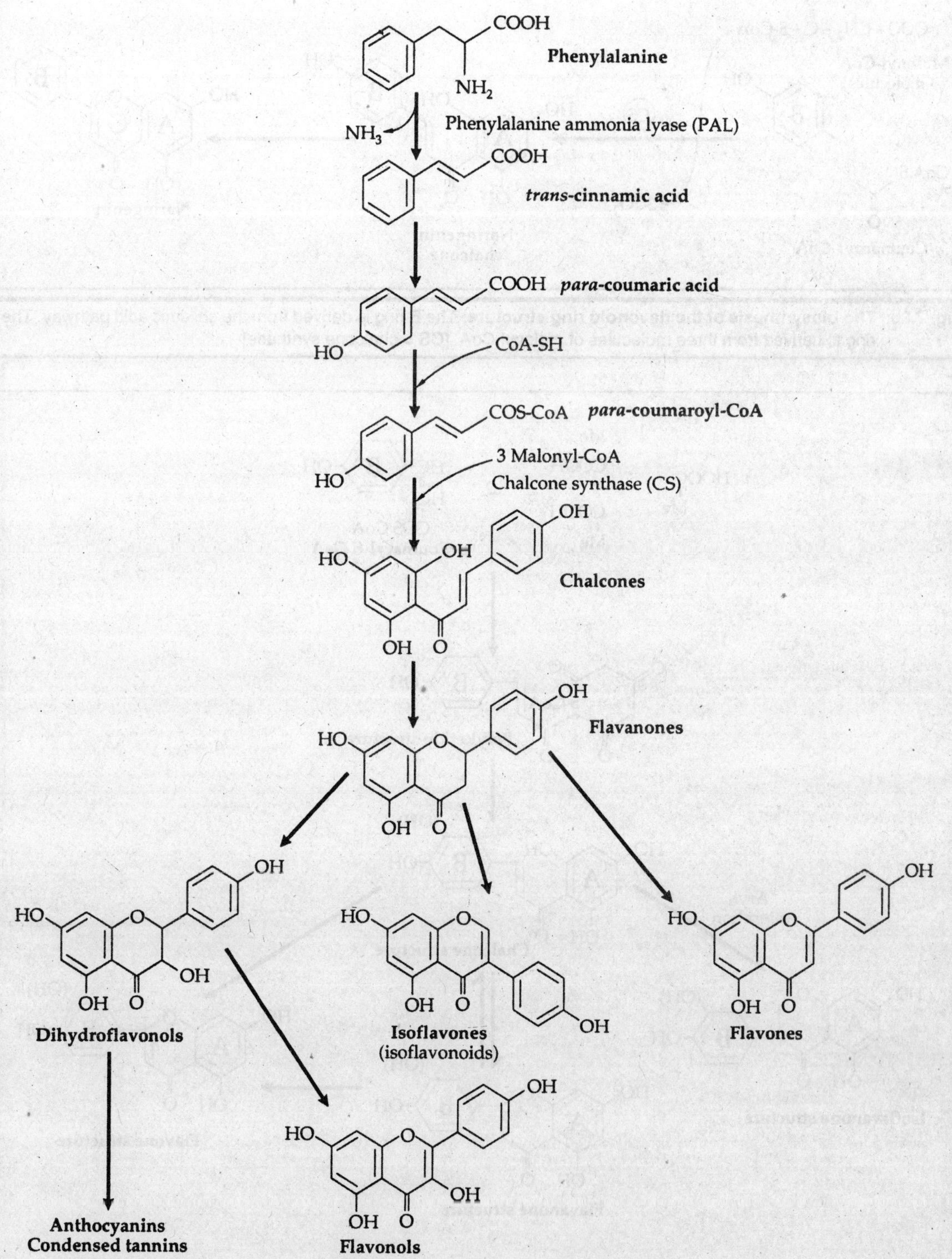

Fig. 7.9a: Biosynthesis of different flavone compounds.

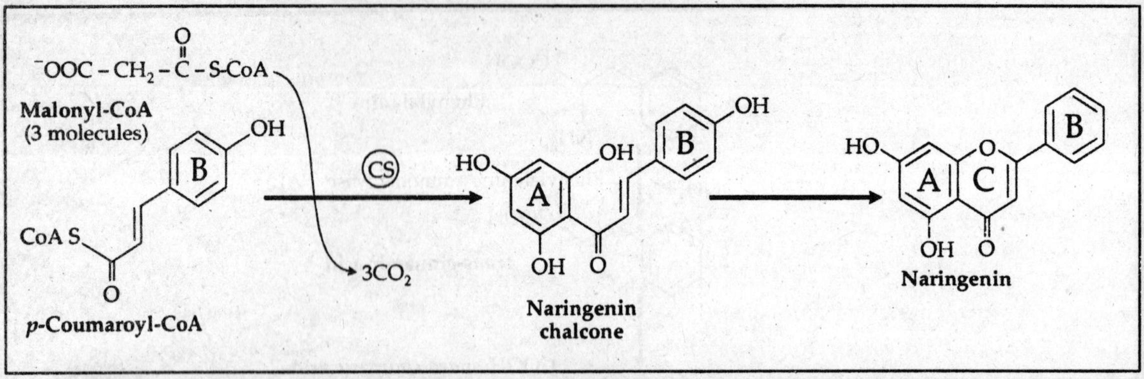

Fig. 7.9b: The biosynthesis of the flavonoid ring structure: The B ring is derived from the shikimic acid pathway. The A ring is derived from three molecules of malonyl-CoA. [CS = chalcone synthase]

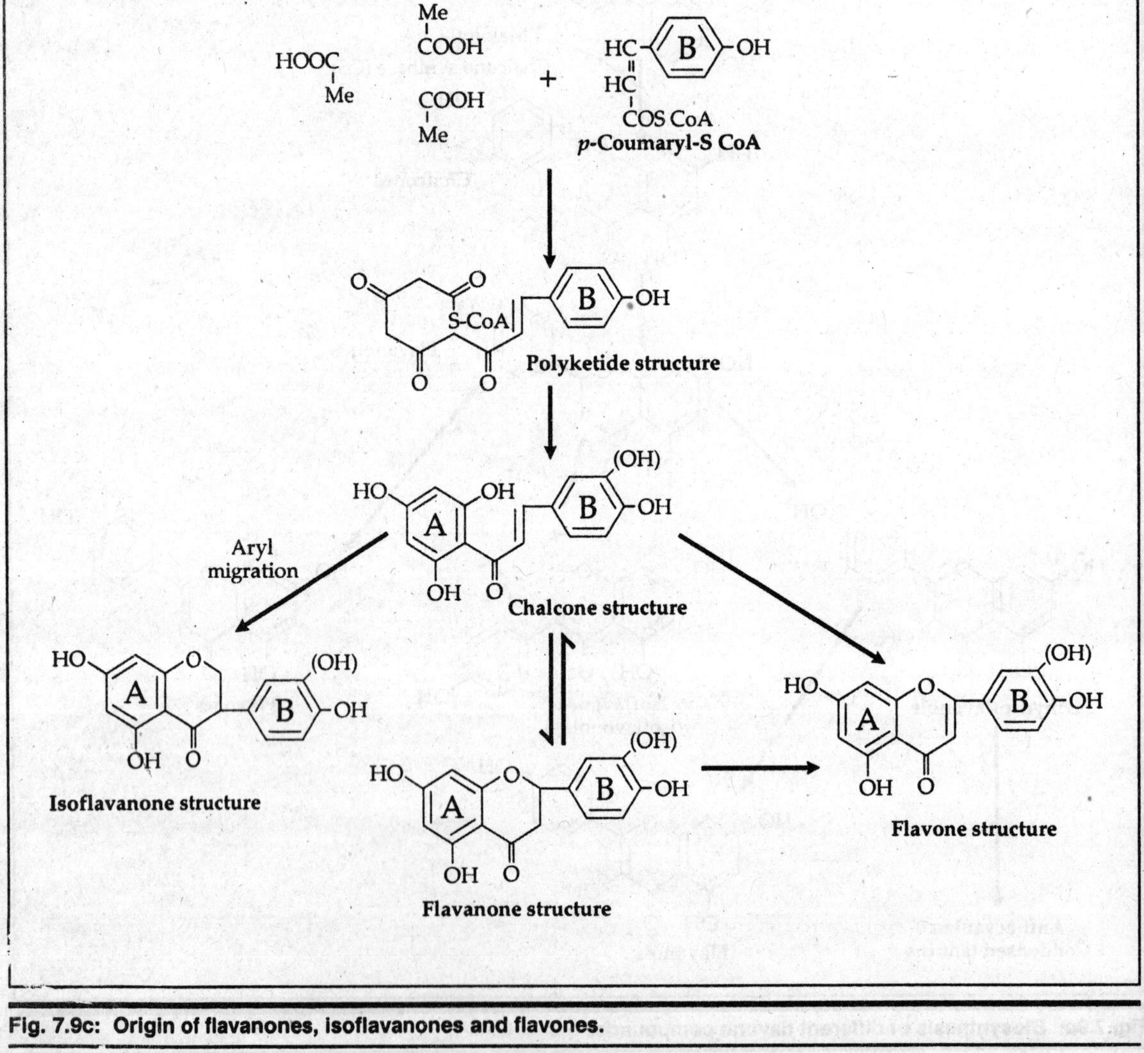

Fig. 7.9c: Origin of flavanones, isoflavanones and flavones.

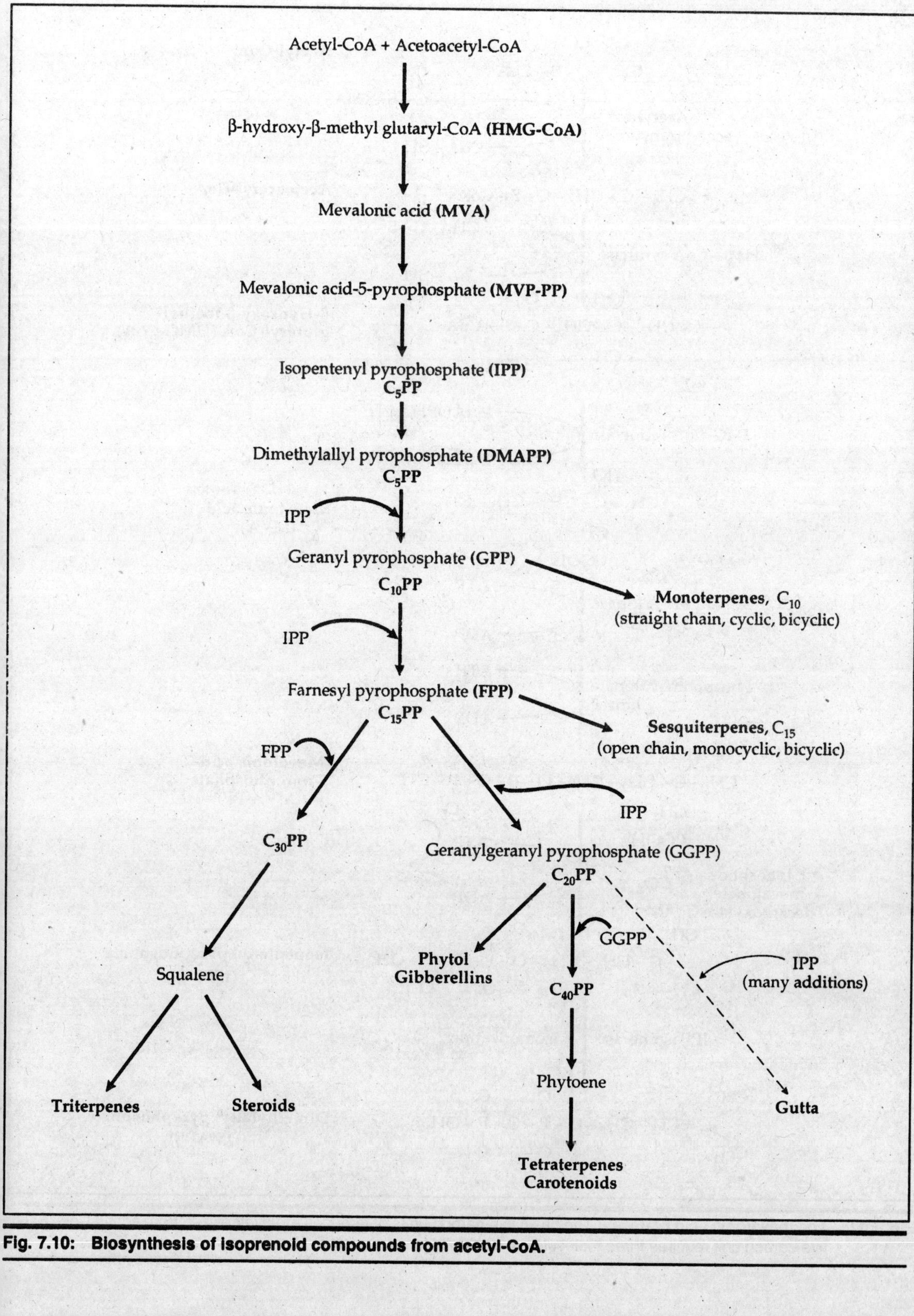

Fig. 7.10: Biosynthesis of isoprenoid compounds from acetyl-CoA.

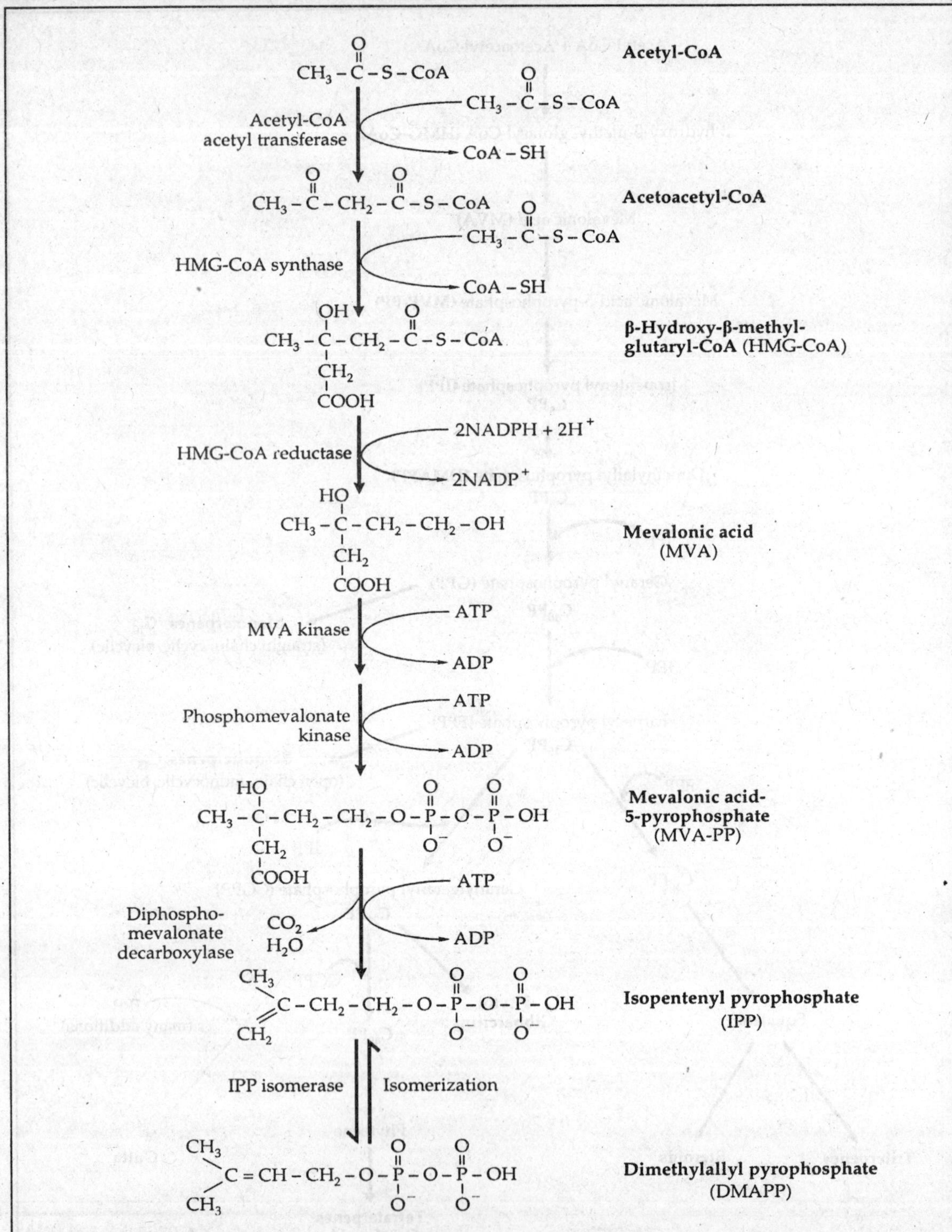

Fig. 7.11: The mevalonic acid pathway: The basic five-carbon units of terpenes are synthesized from acetyl-CoA. Each five-carbon unit requires three molecules of acetyl-CoA.

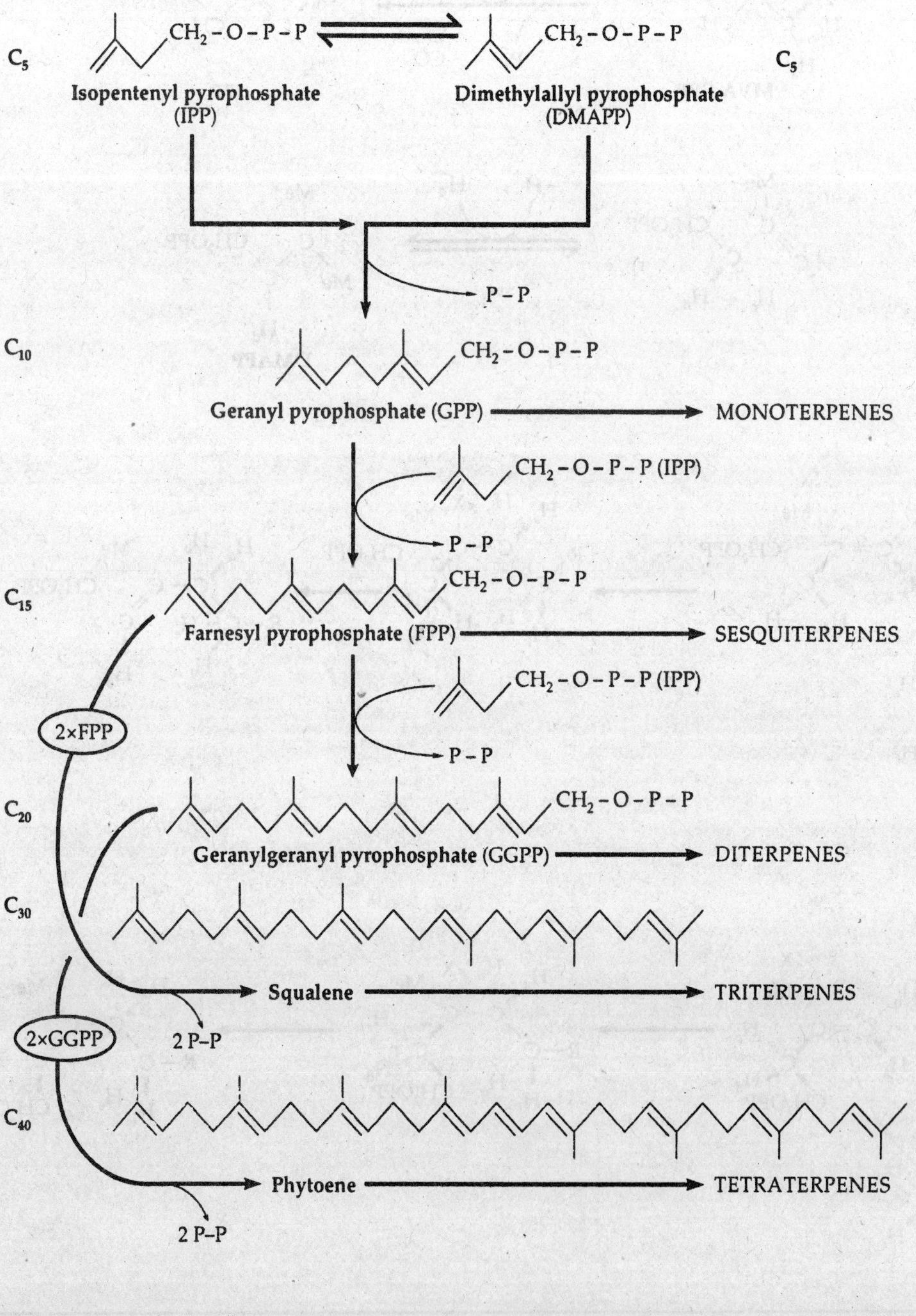

Fig. 7.12: The major subclasses of terpenes are biosynthesized from the basic five-carbon units IPP and DMAPP: Monoterpenes (C_{10}), sesquiterpenes (C_{15}) and diterpenes (C_{20}) are produced by the sequential addition of C_5 units. Triterpenes (C_{30}) are formed from two C_{15} units, tetraterpenes (C_{40}) from two C_{20} units. [PP = pyrophosphate]

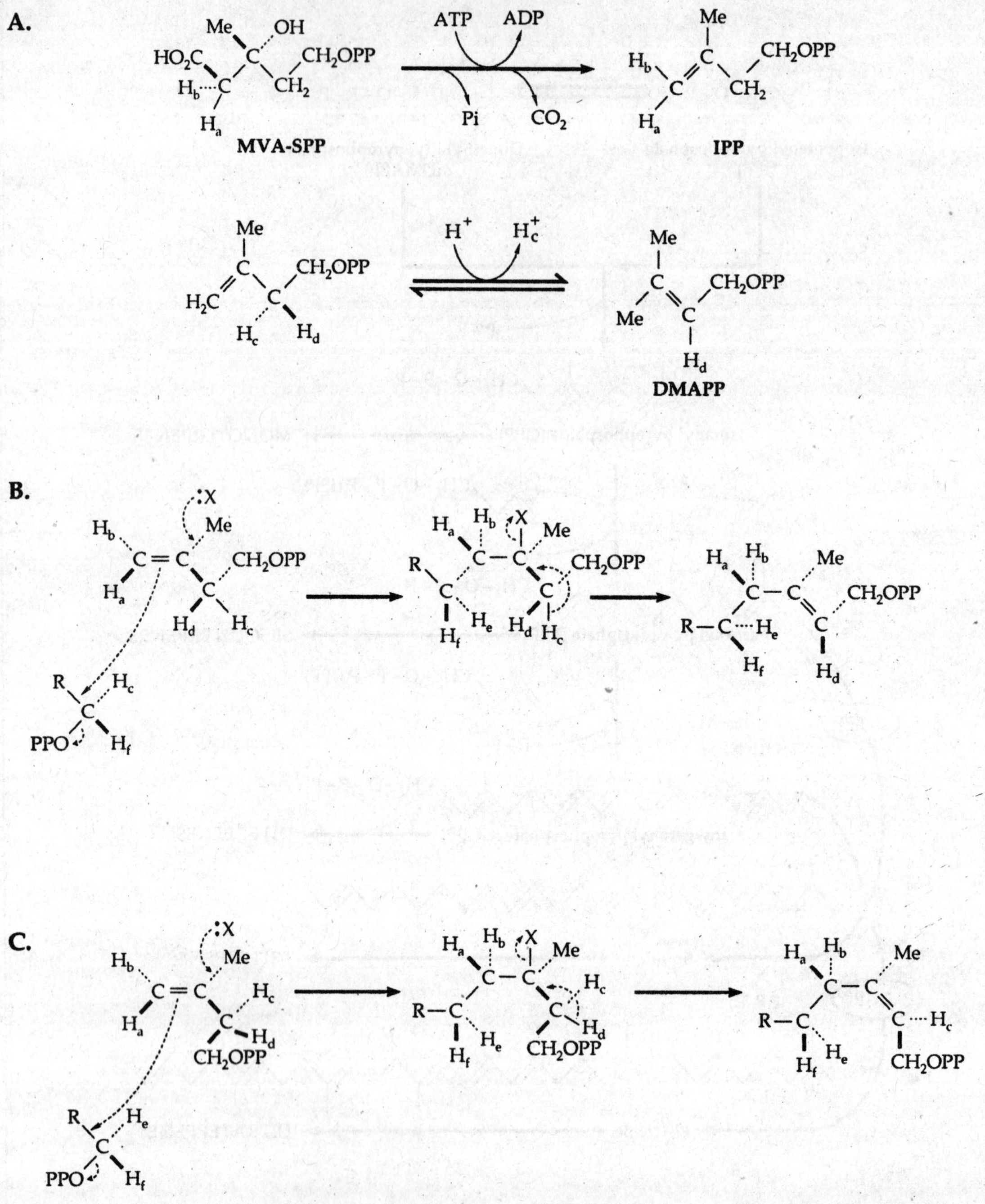

Fig. 7.13: Stereospecific reaction in terpenoid biogenesis:

A. Formation from mevalonic acid (MVA) of isopentenyl pyrophosphate (IPP) by *trans* elimination, isomerisation to dimethylallyl pyrophosphate (DMAPP).

B. Association of the two 5-C units with *trans* elimination of H_2.

C. As B but involving formation of *cis* double bonds as found in rubber.

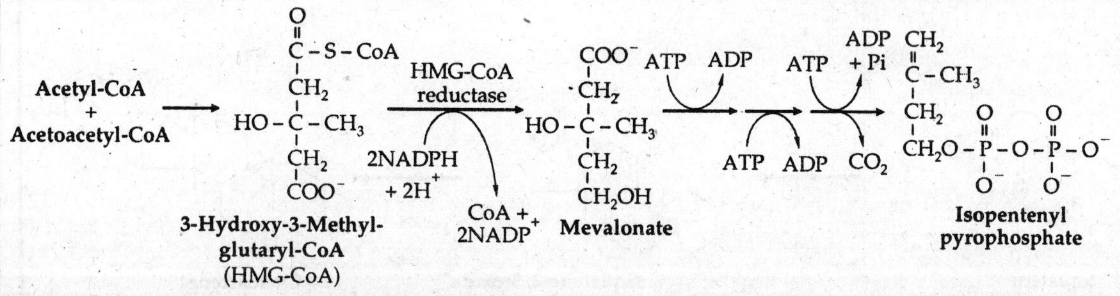

Fig. 7.14: Synthesis of isopentenyl pyrophosphate.

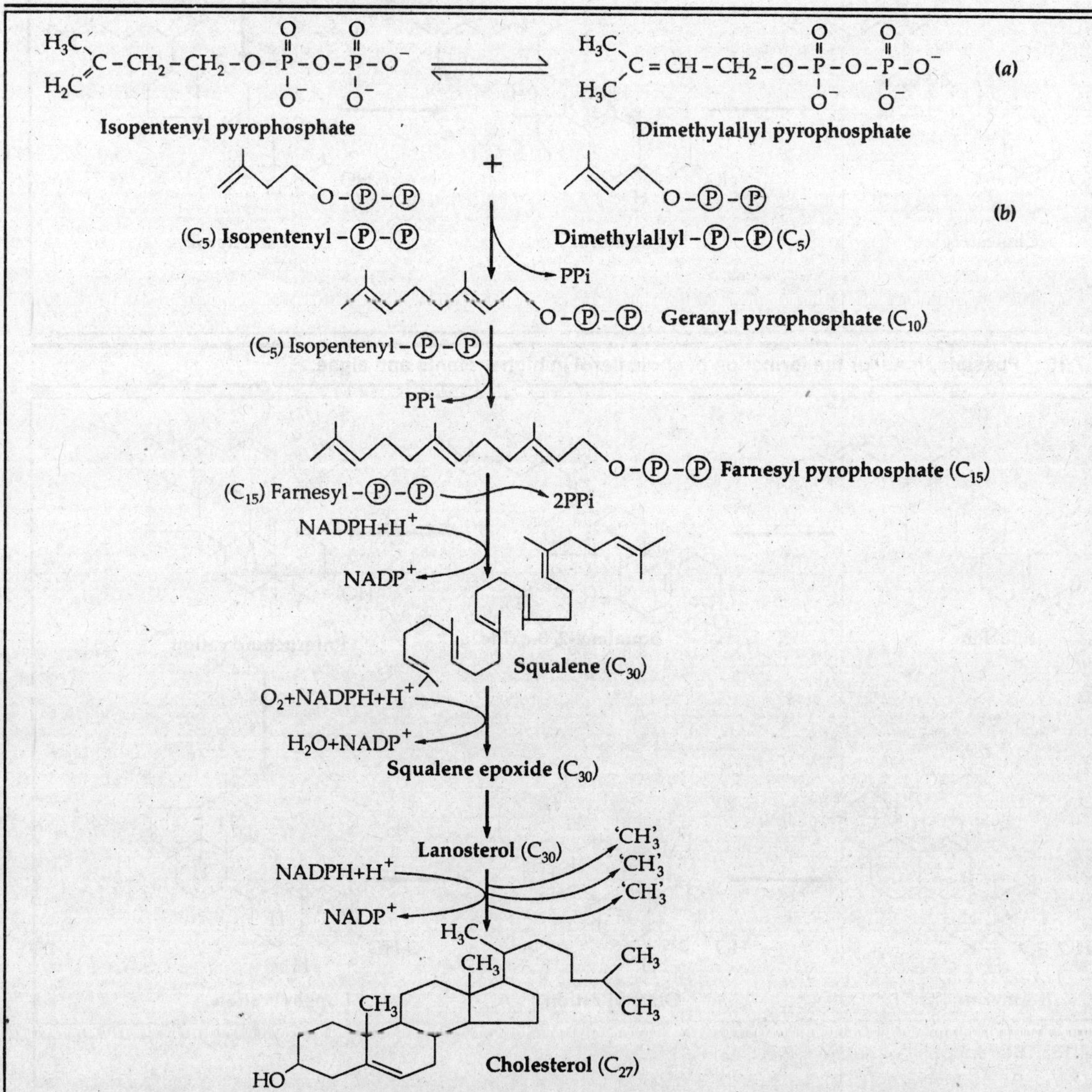

Fig. 7.15: Synthesis of squalene and cholesterol from isopentenyl pyrophosphate: (a) Isomerization of isopentenyl pyrophosphate to dimethylallyl pyrophosphate. **(b)** Synthesis of cholesterol.

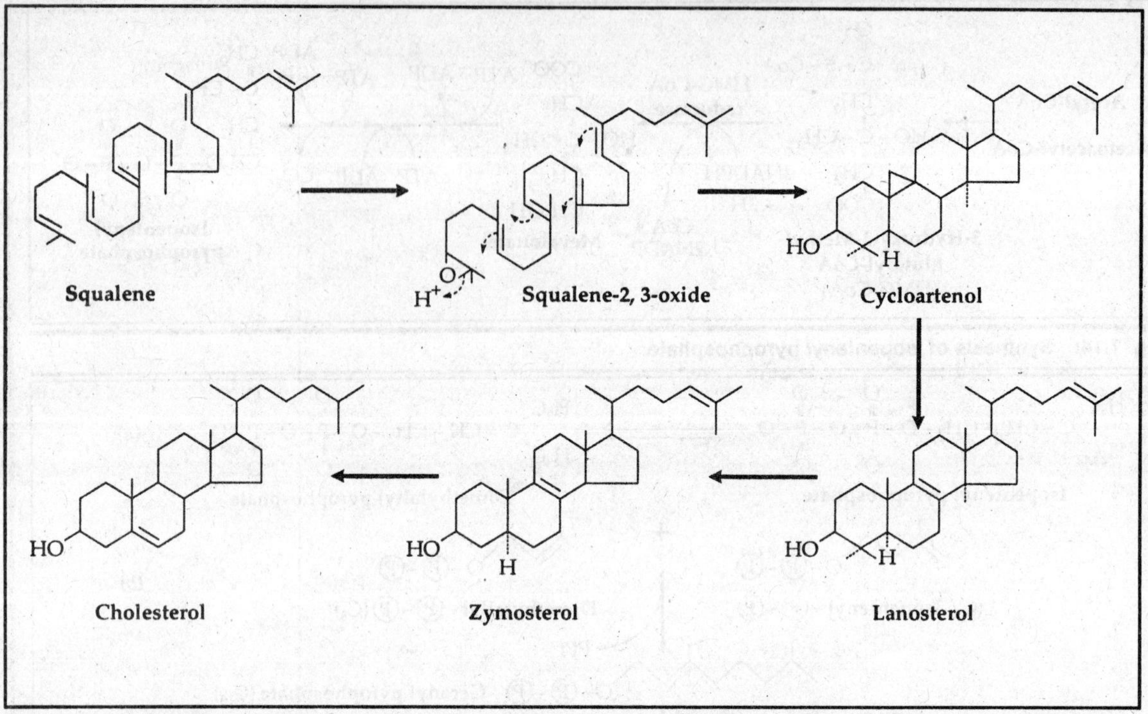

Fig. 7.16: Possible route for the formation of cholesterol in higher plants and algae.

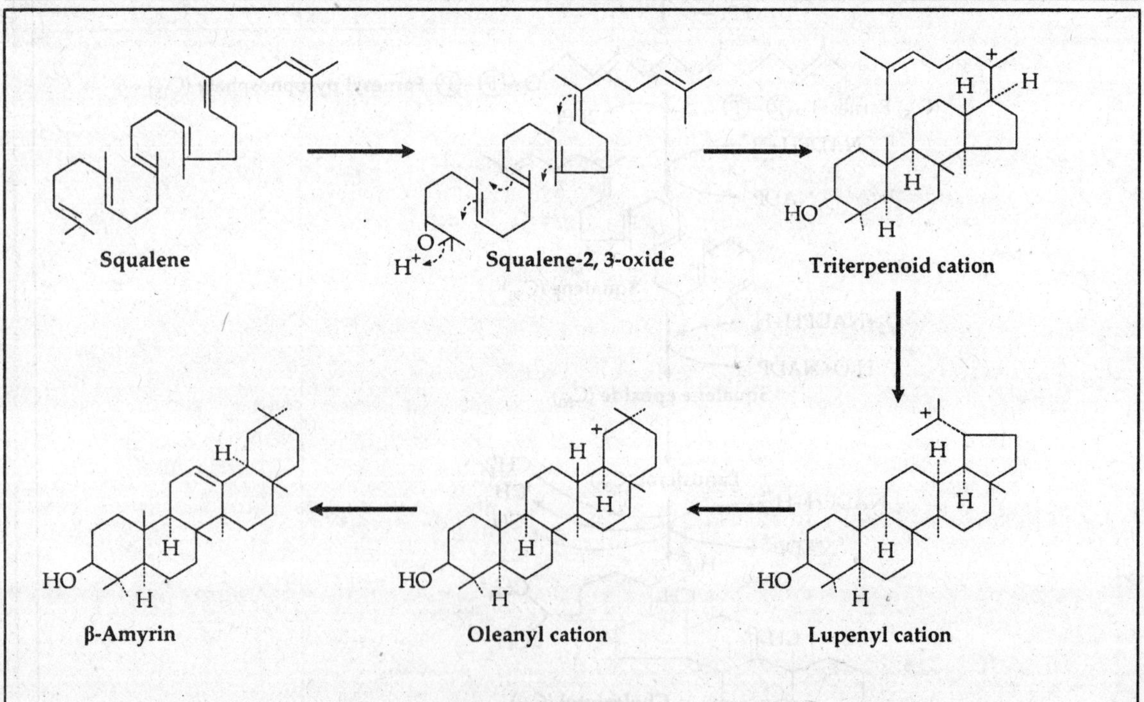

Fig. 7.17: Biosynthetic pathway of triterpenoids.

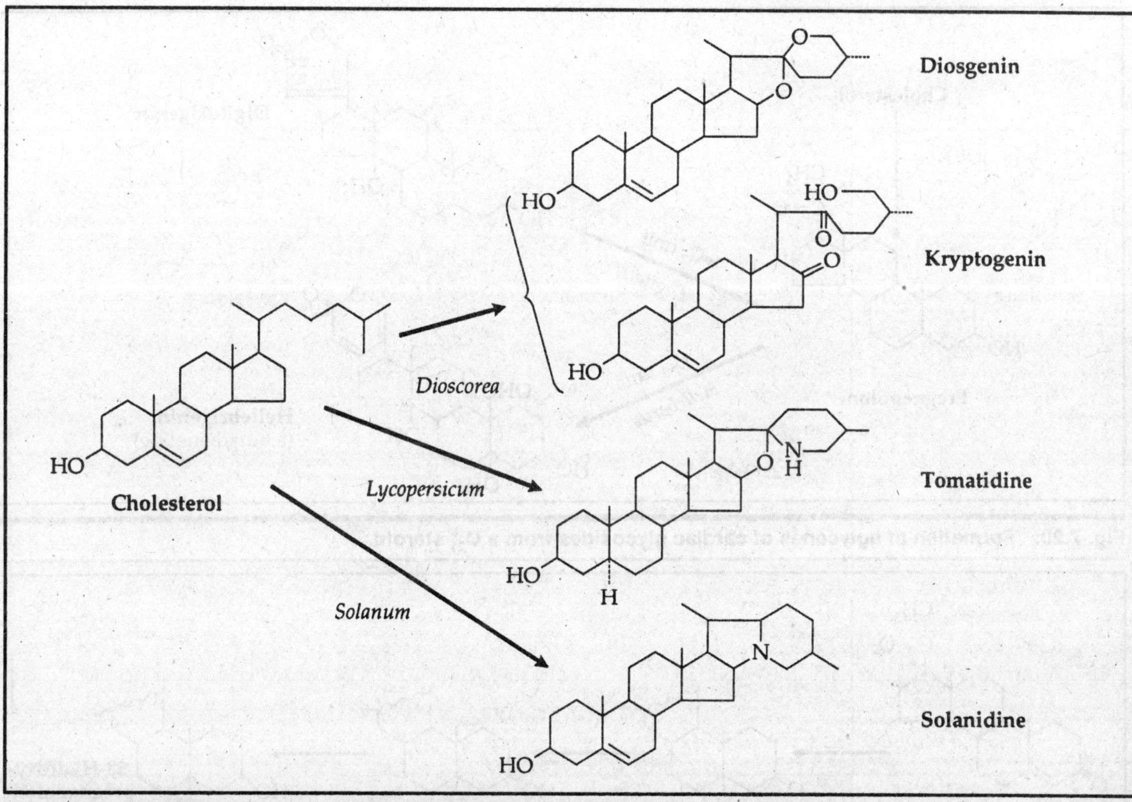

Fig. 7.18: Some plant metabolites from cholesterol.

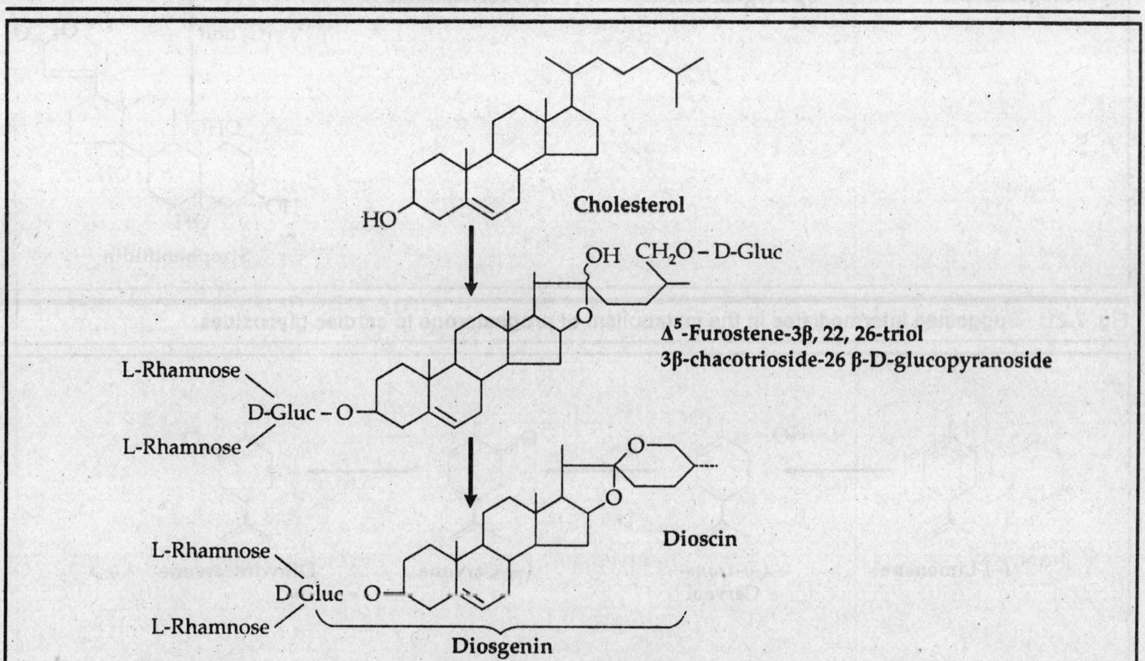

Fig. 7.19: Formation and metabolism of an open-chain saponin in *Dioscorea*.

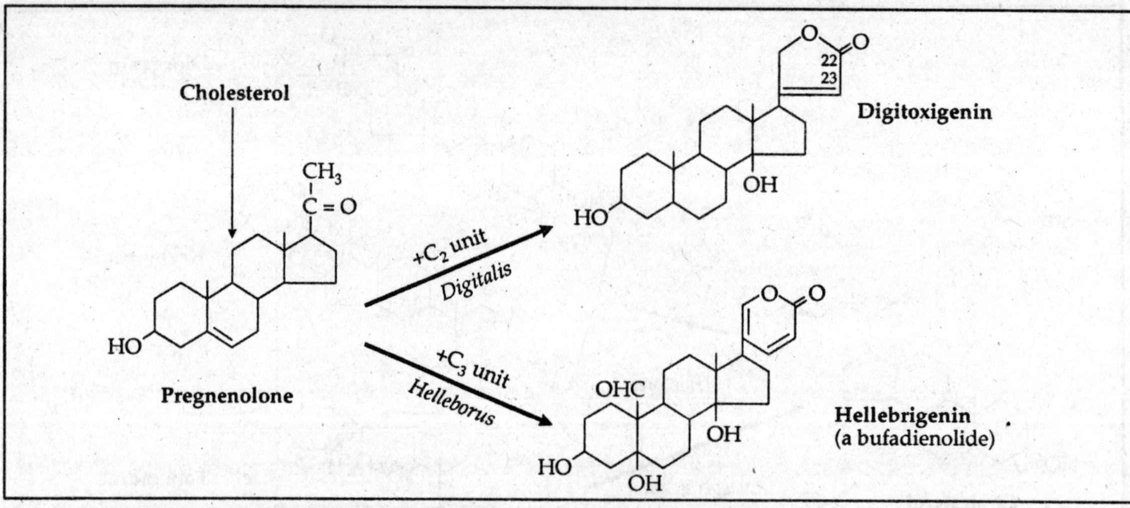

Fig. 7.20: Formation of aglycones of cardiac glycosides from a C_{21} steroid.

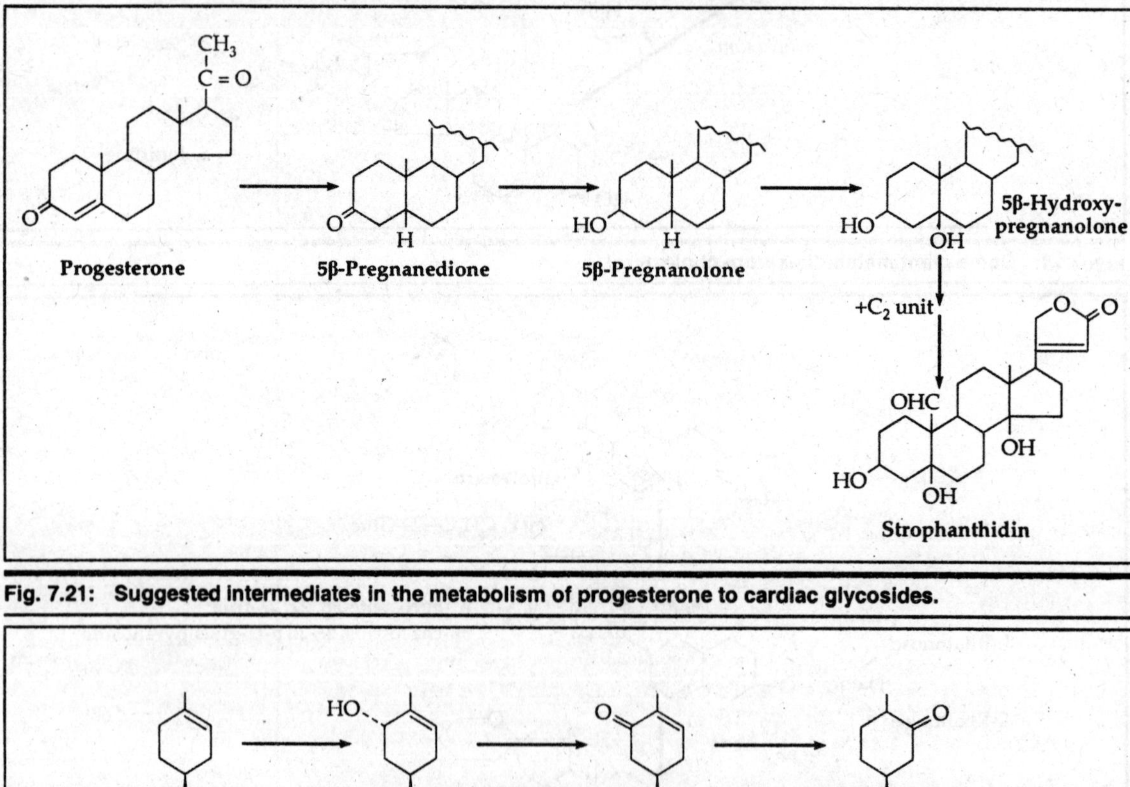

Fig. 7.21: Suggested intermediates in the metabolism of progesterone to cardiac glycosides.

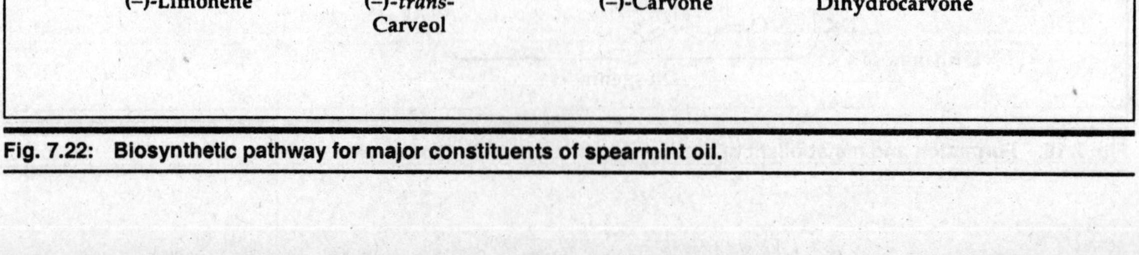

Fig. 7.22: Biosynthetic pathway for major constituents of spearmint oil.

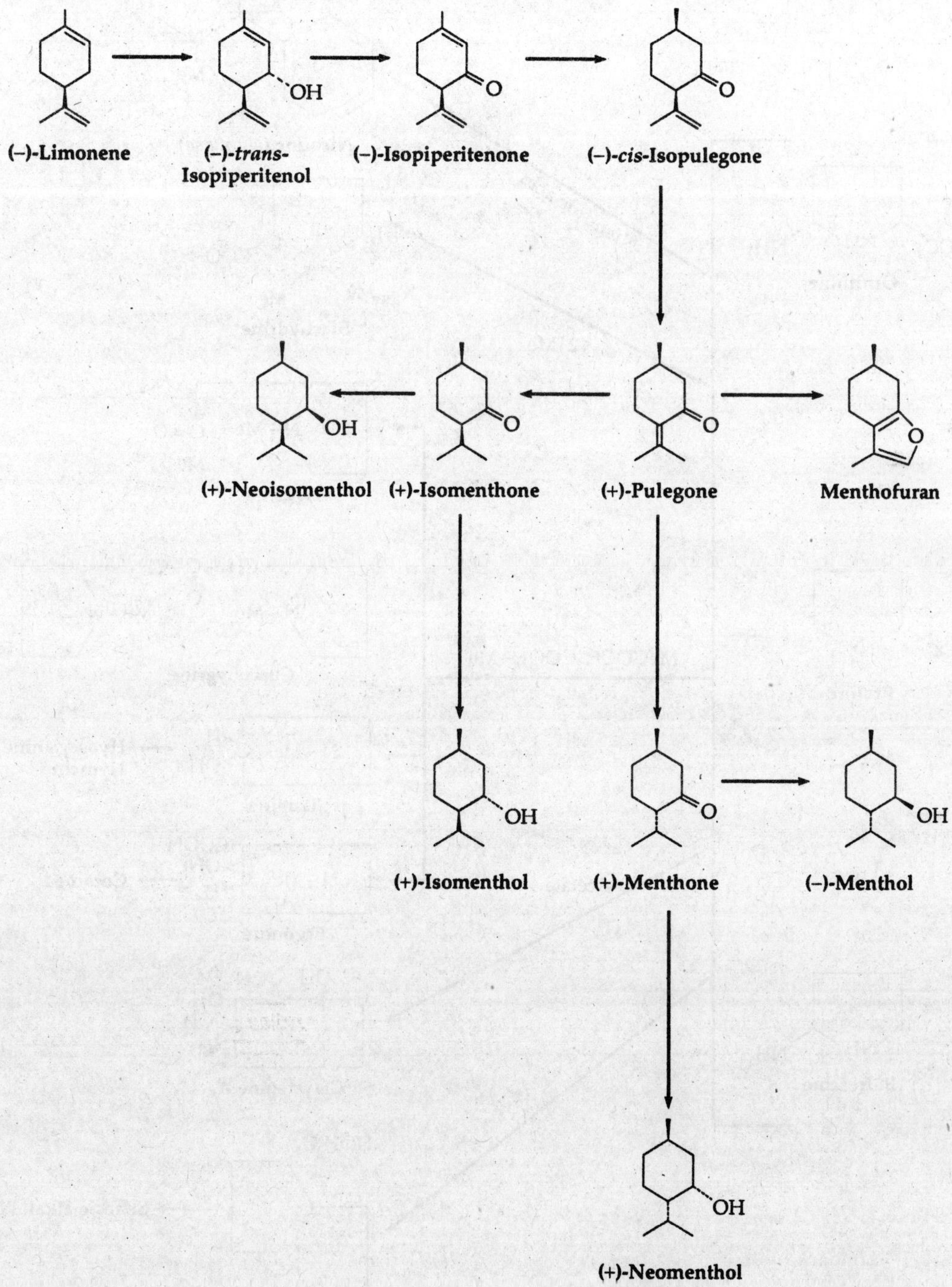

Fig. 7.23: A possible biogenetic scheme for some monoterpenoids of *Mentha piperita*.

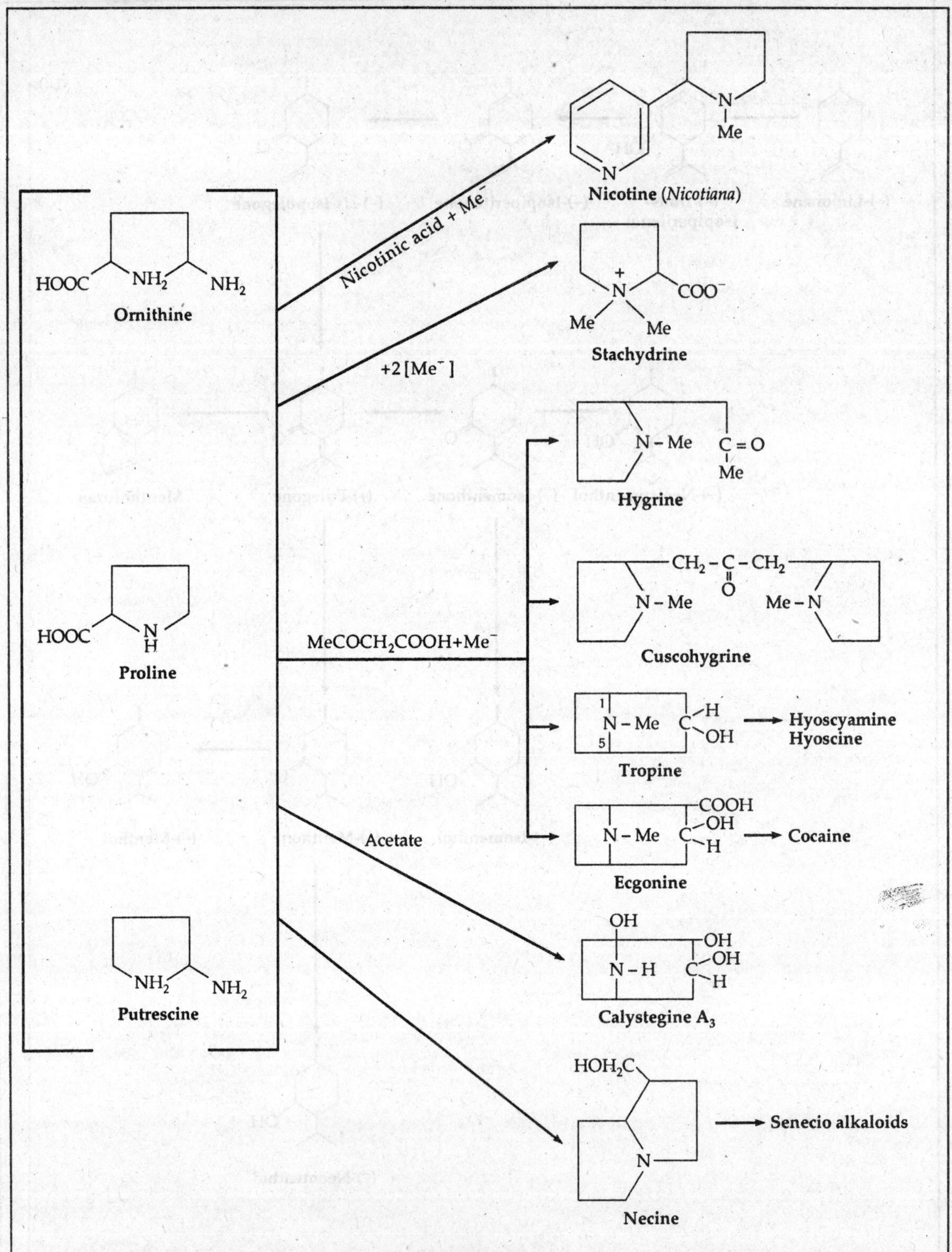

Fig. 7.24: Some ornithine-derived alkaloids.

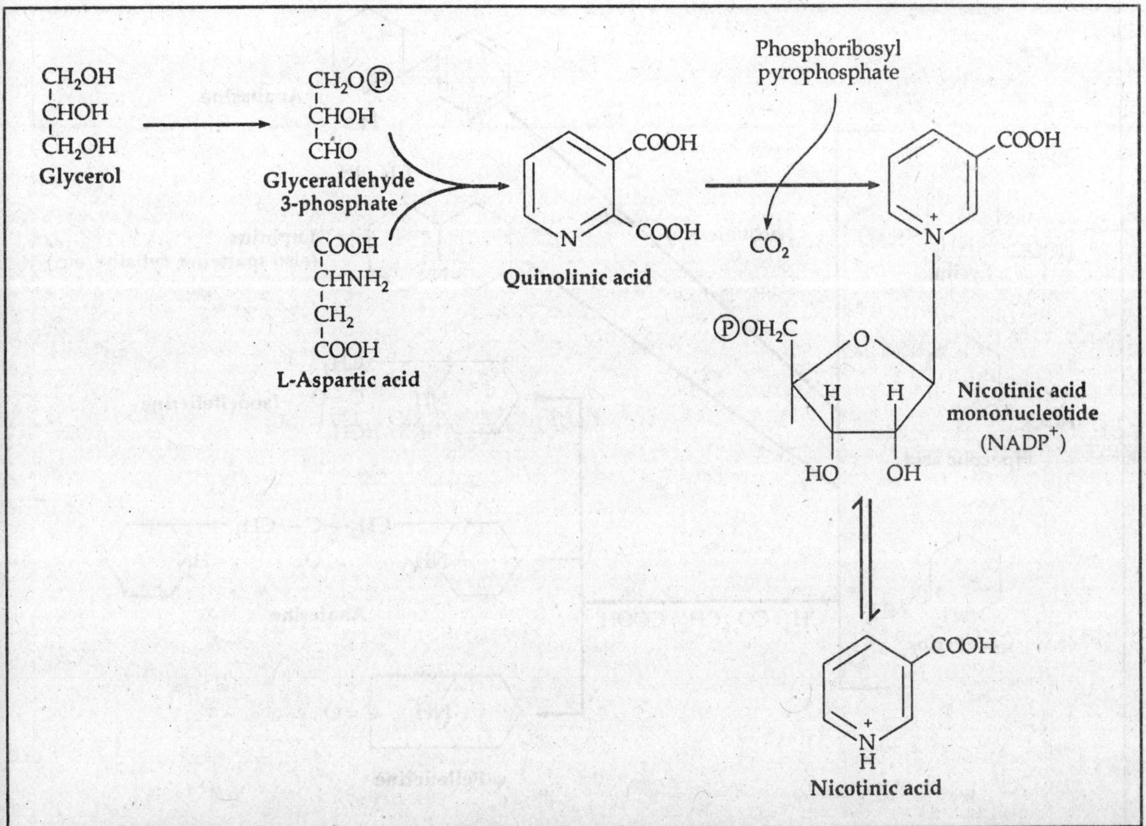

Fig. 7.25a: The biosynthesis of nicotinic acid, a precursor of the alkaloid nicotine. Nicotinic acid is also a component of NAD^+ and $NADP^+$, important participants in biological oxidation-reduction reactions. The pathway shown is thought to operate in both plants and bacteria.

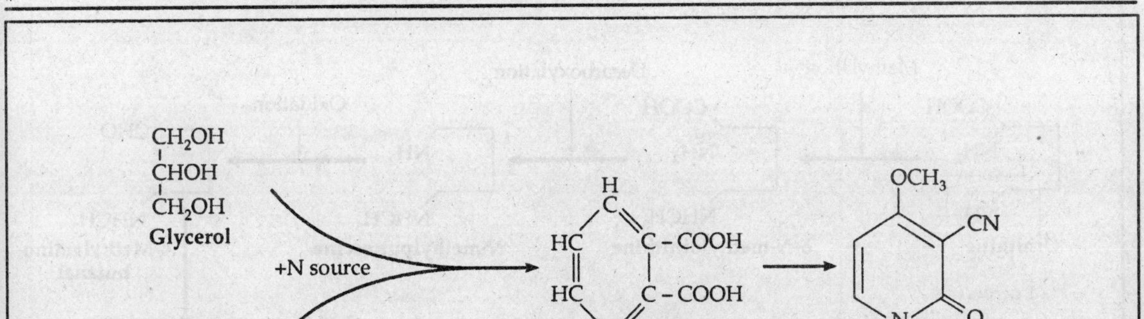

Fig. 7.25b: The biosynthesis of ricinine.

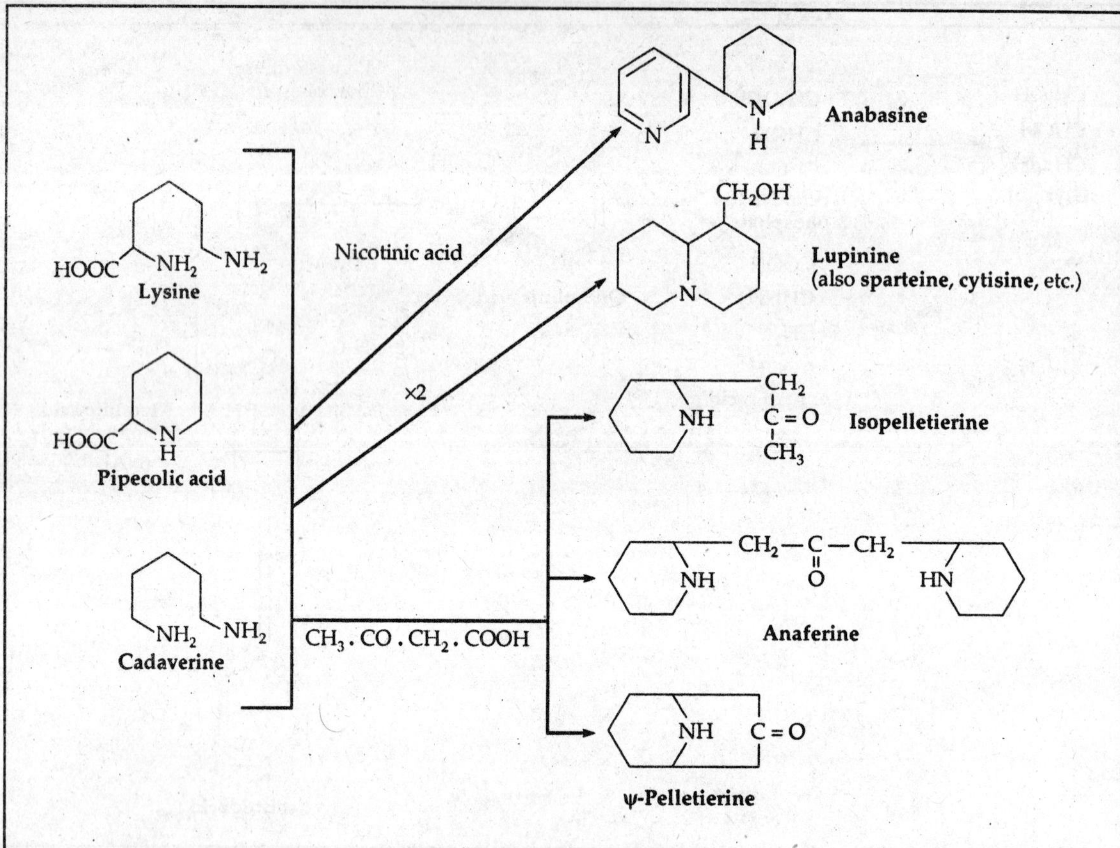

Fig. 7.26: Lysine as a precursor of different alkaloids.

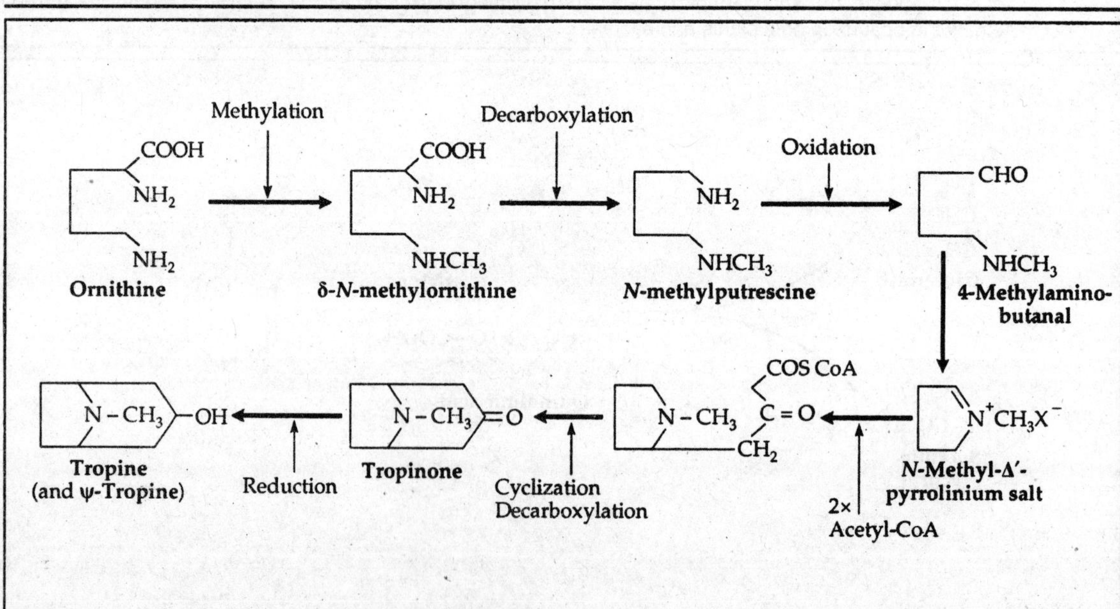

Fig. 7.27: Possible biogenetic routes for tropine and pseudotropine.

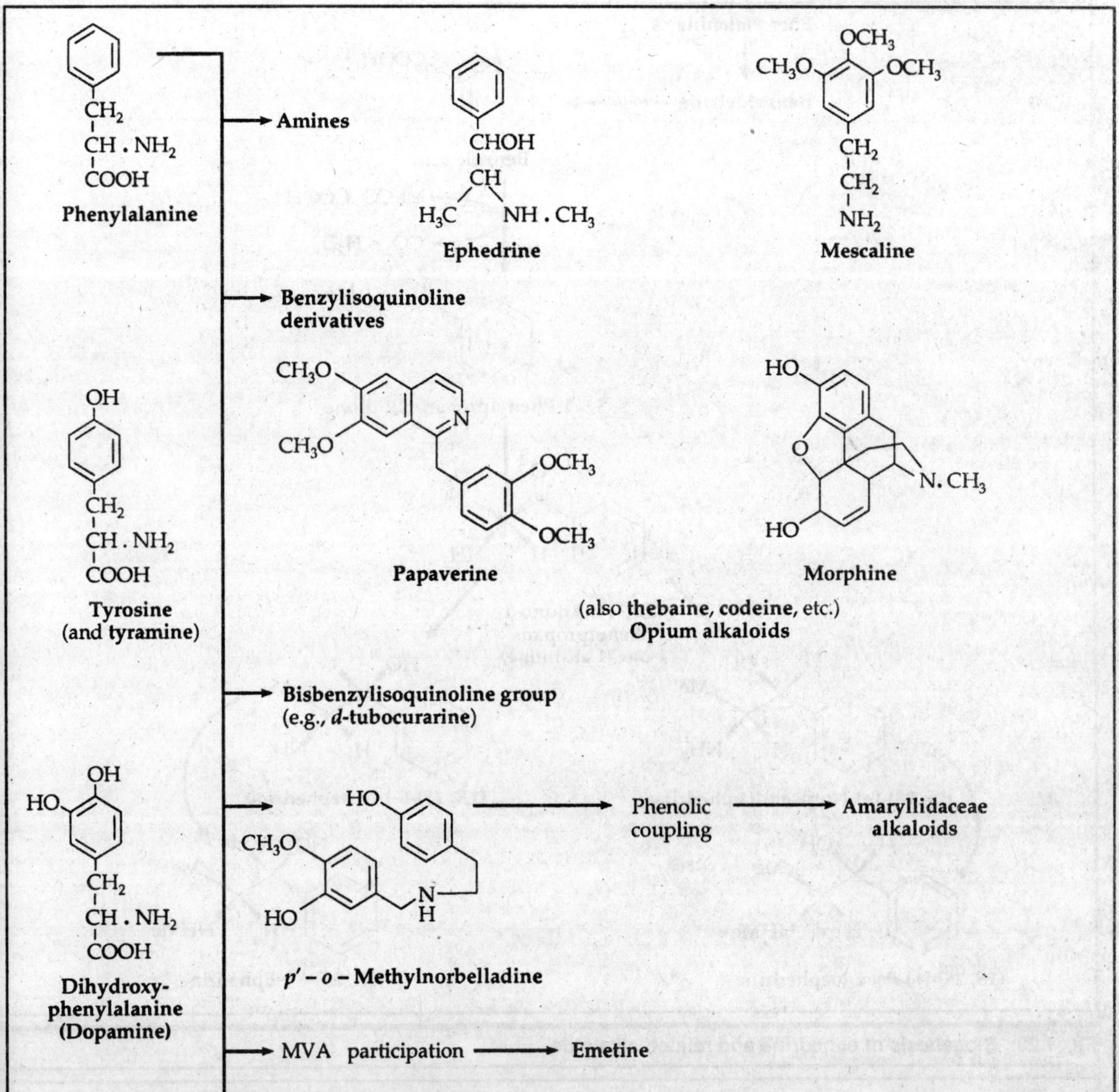

Fig. 7.28: Some bases derived from phenylalanine, tyrosine and dihydroxyphenylalanine.

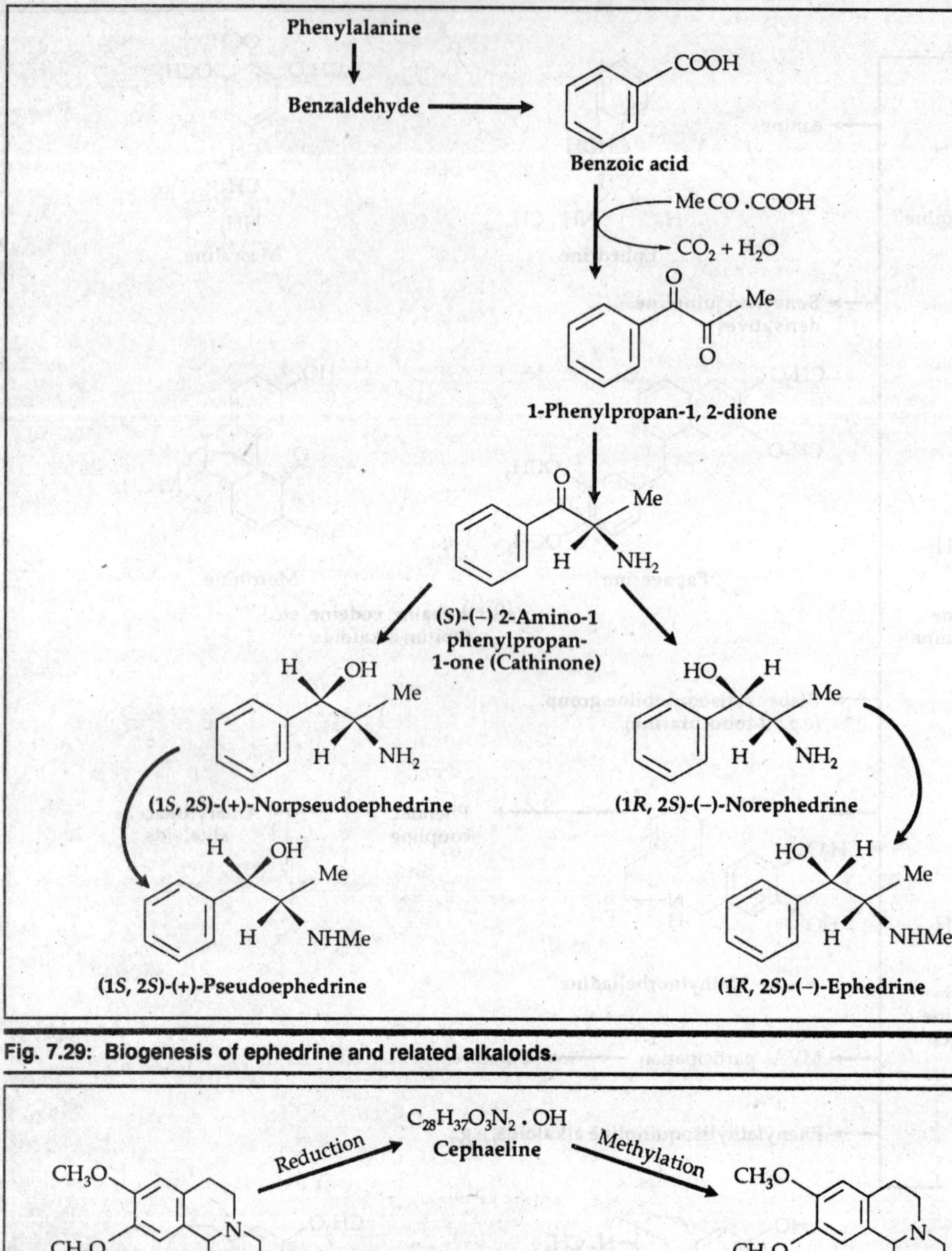

Fig. 7.29: Biogenesis of ephedrine and related alkaloids.

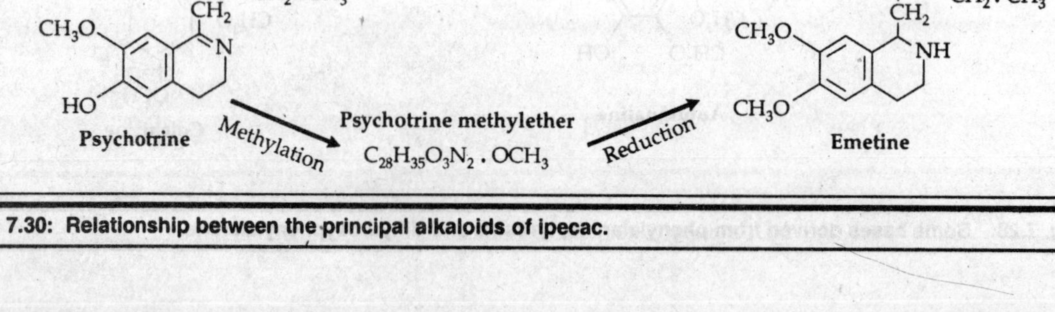

Fig. 7.30: Relationship between the principal alkaloids of ipecac.

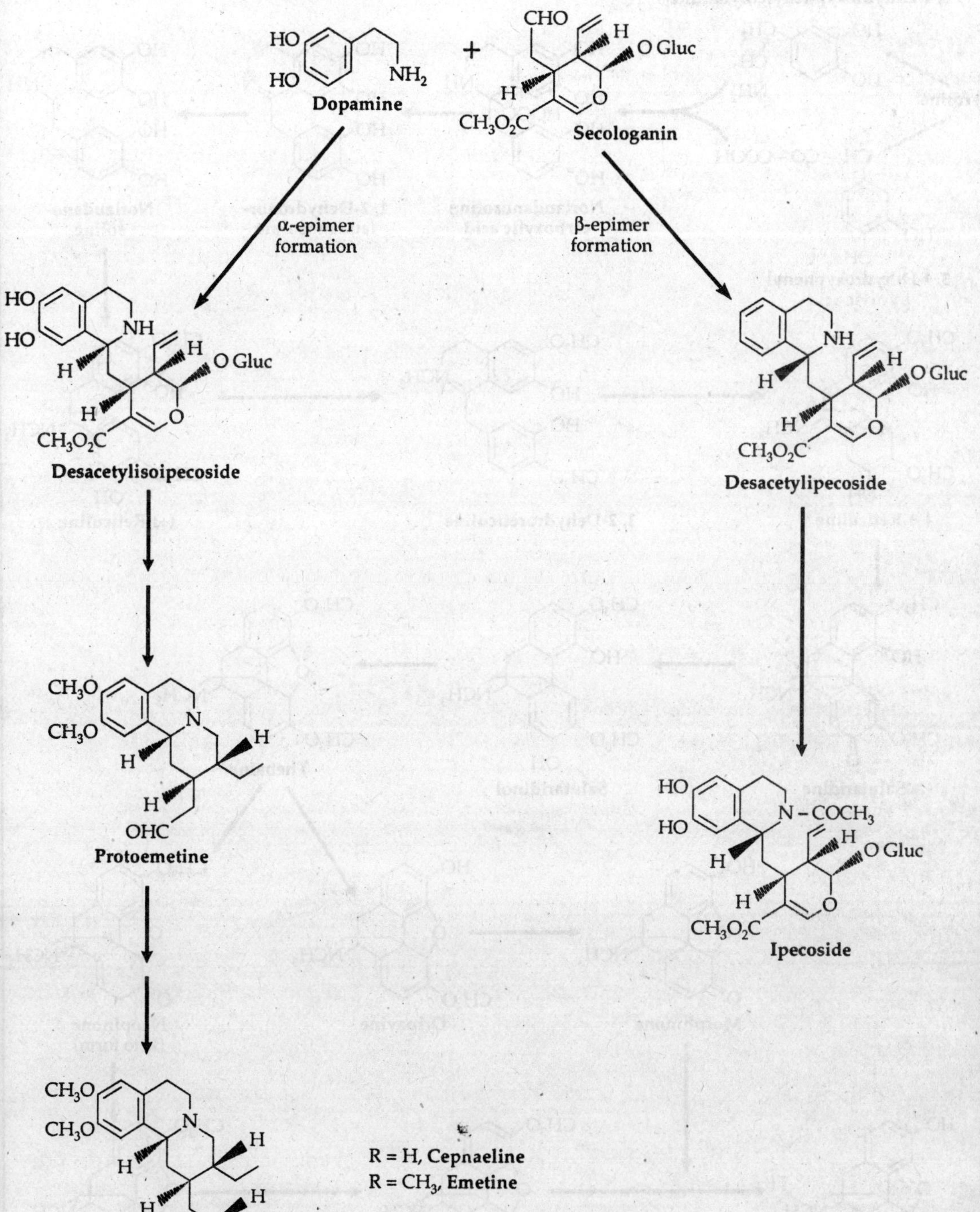

Fig. 7.31: Proposed biosynthetic sequence for the biosynthesis of cephaeline, emetine and the alkaloidal glucoside ipecoside.

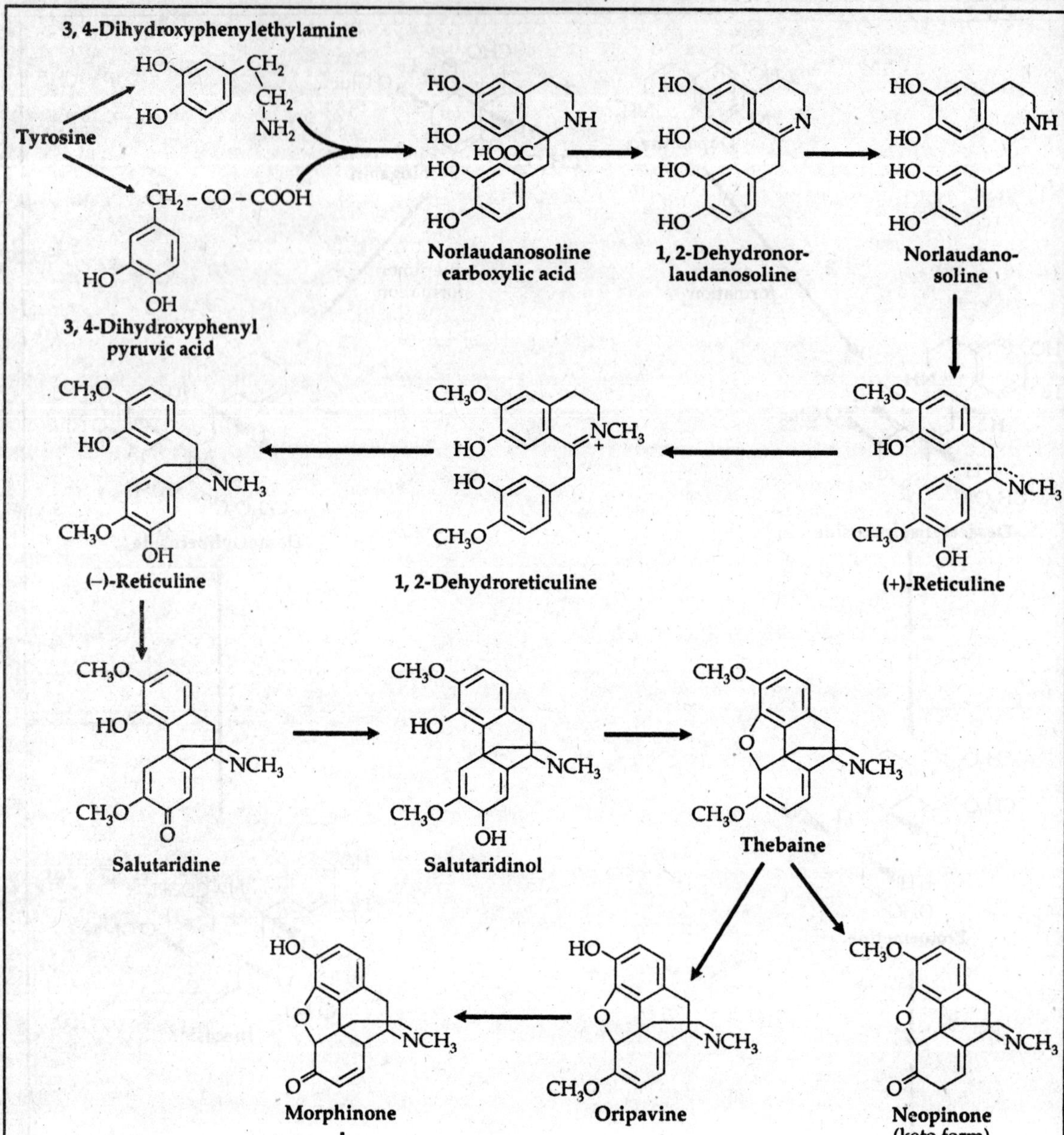

Fig. 7.32: Biogenesis of thebaine, codeine and morphine.

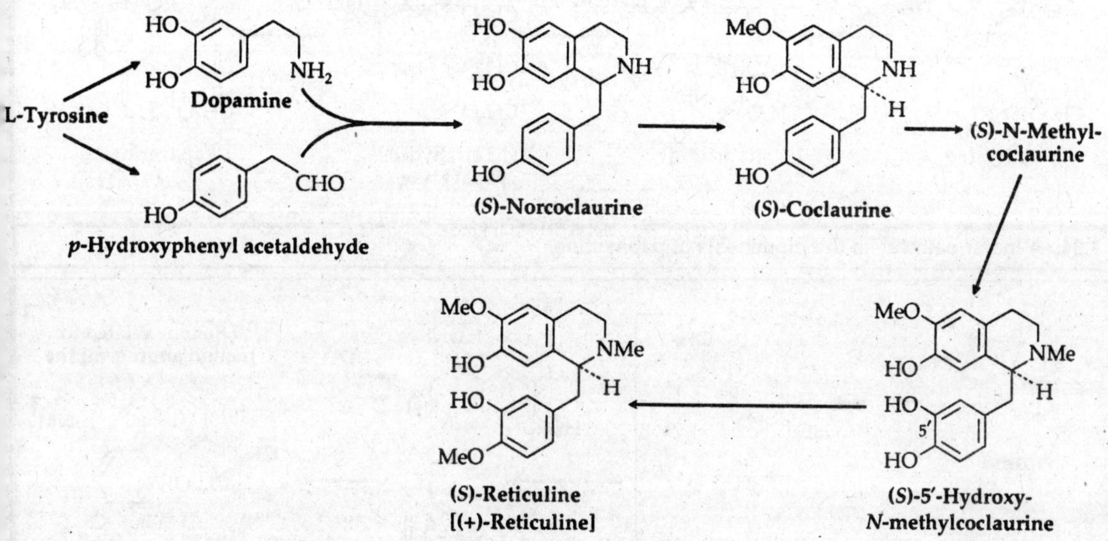

g. 7.33a: A revised biogenetic route from L-tyrosine to (S)-reticuline.

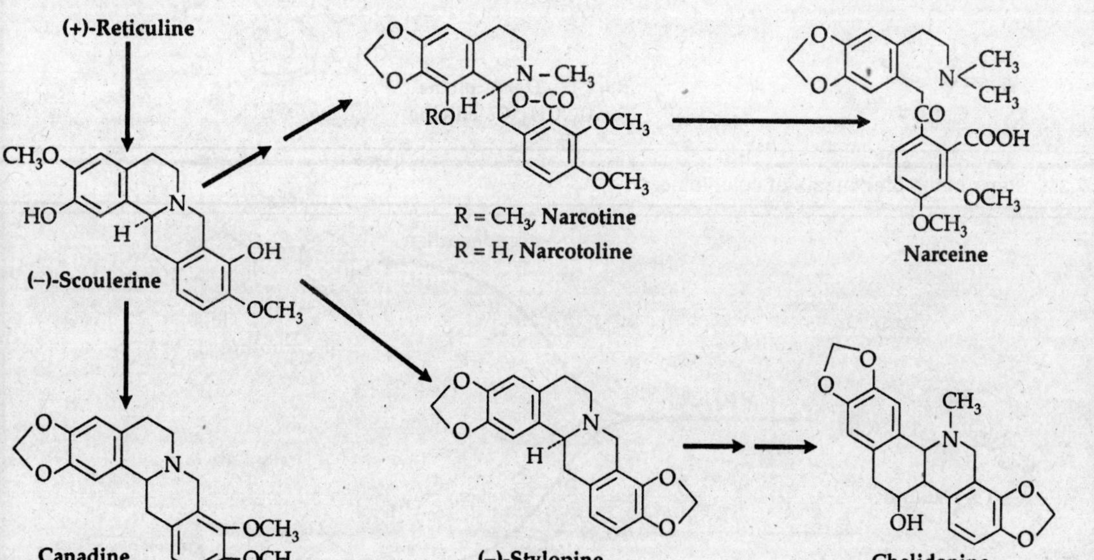

g. 7.33b: Some alkaloids derived from (+)-reticuline.

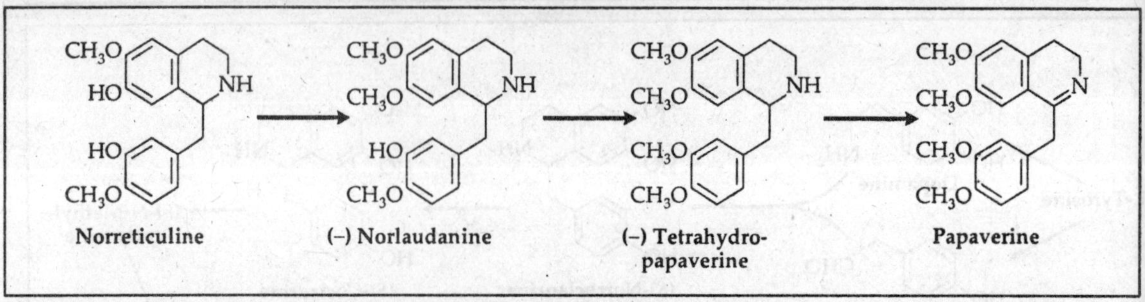

Fig. 7.34: A major pathway in the biogenesis of papaverine.

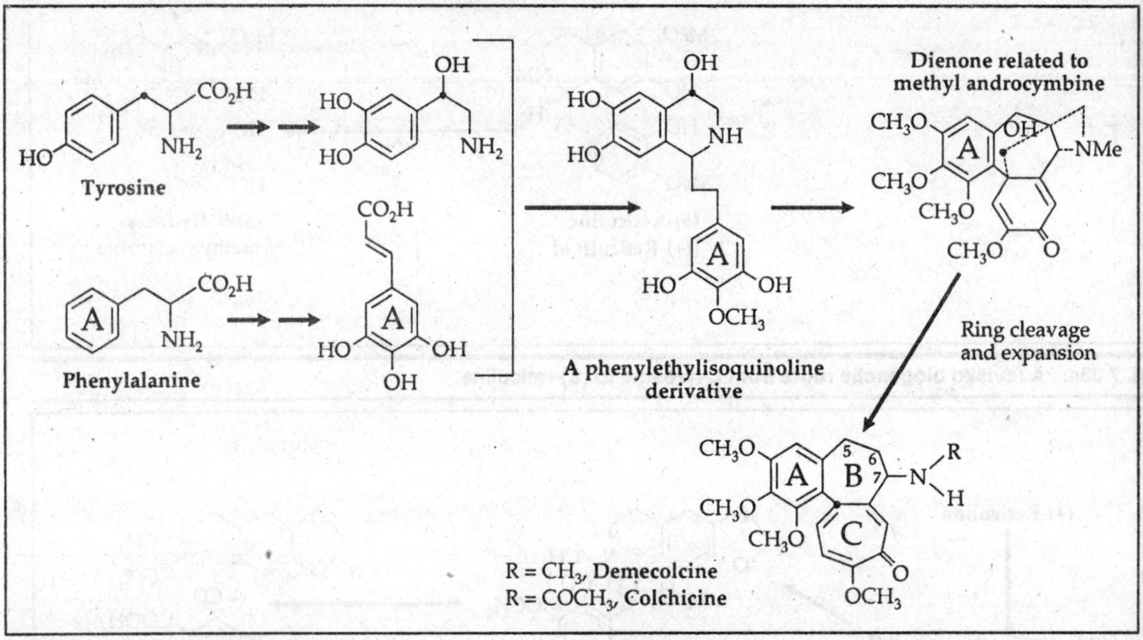

Fig. 7.35: Steps in the biogenesis of colchicine.

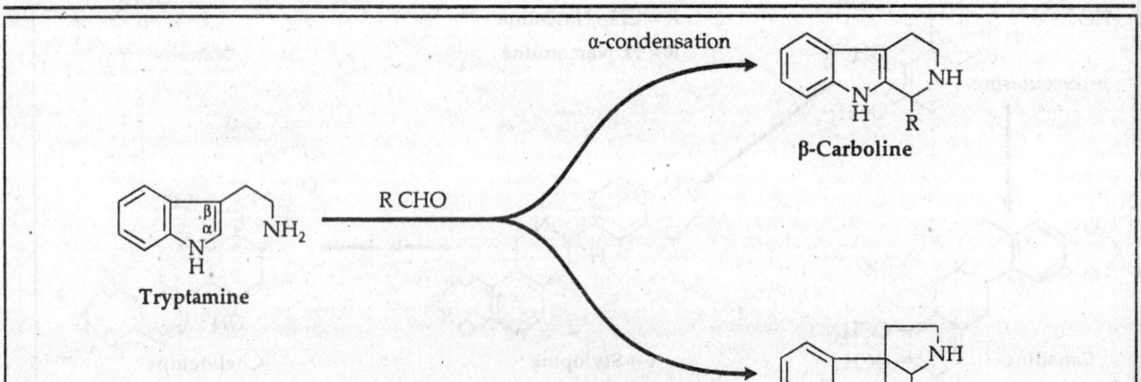

Fig. 7.36: Formation of the indole alkaloids.

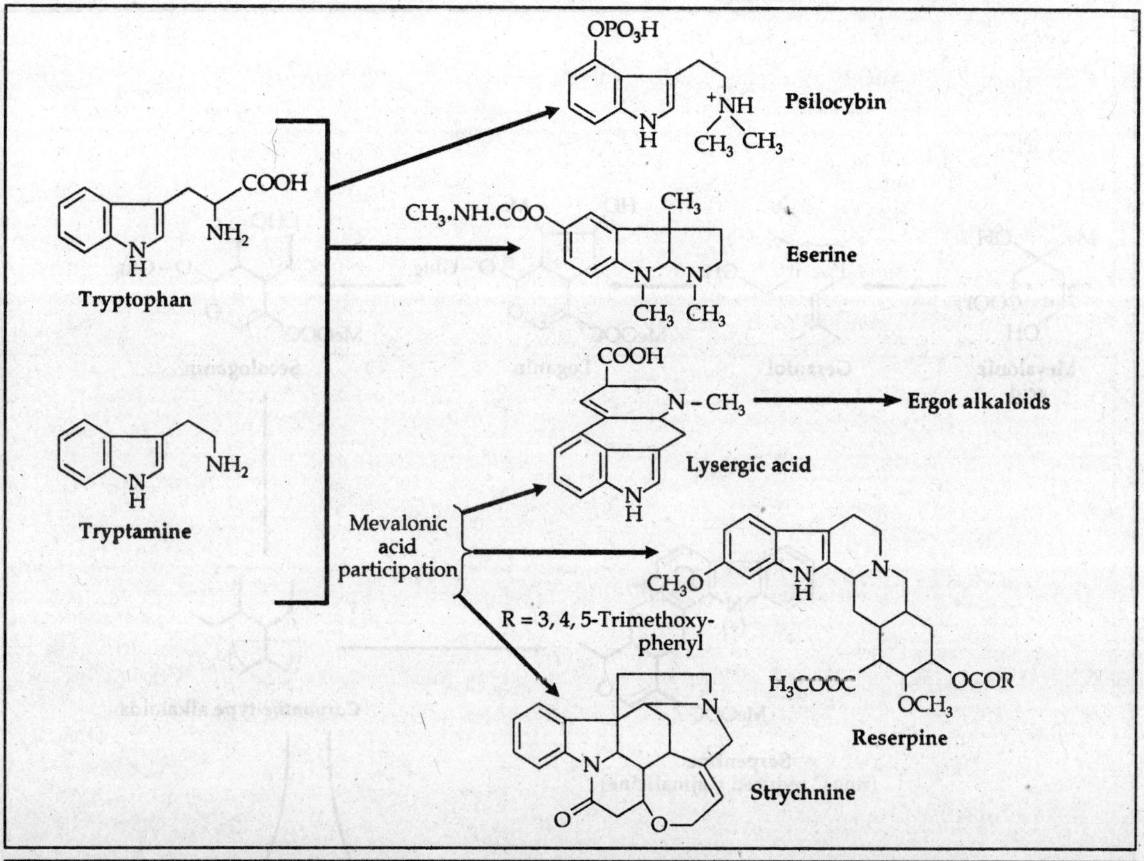

Fig. 7.37: Tryptophan and tryptamine as precursors of indole alkaloids.

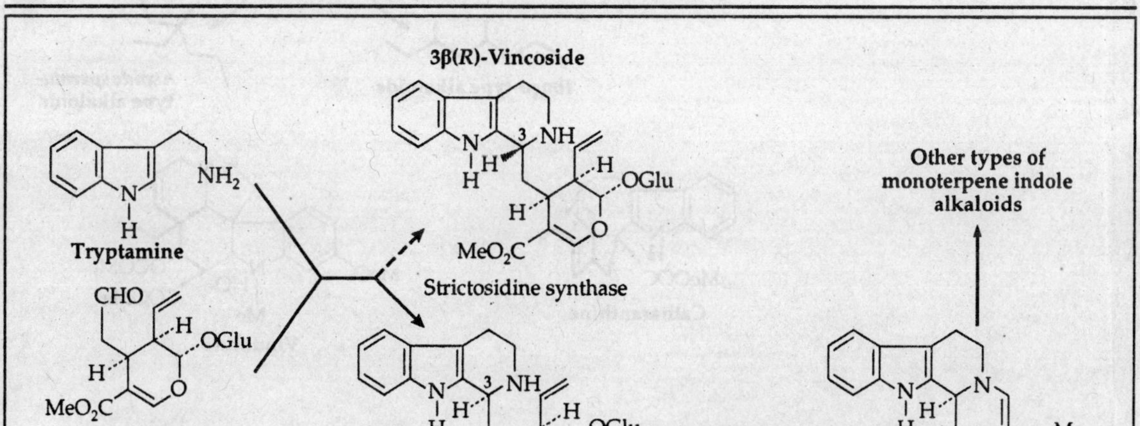

Fig. 7.38: Formation and metabolism of strictosidine.

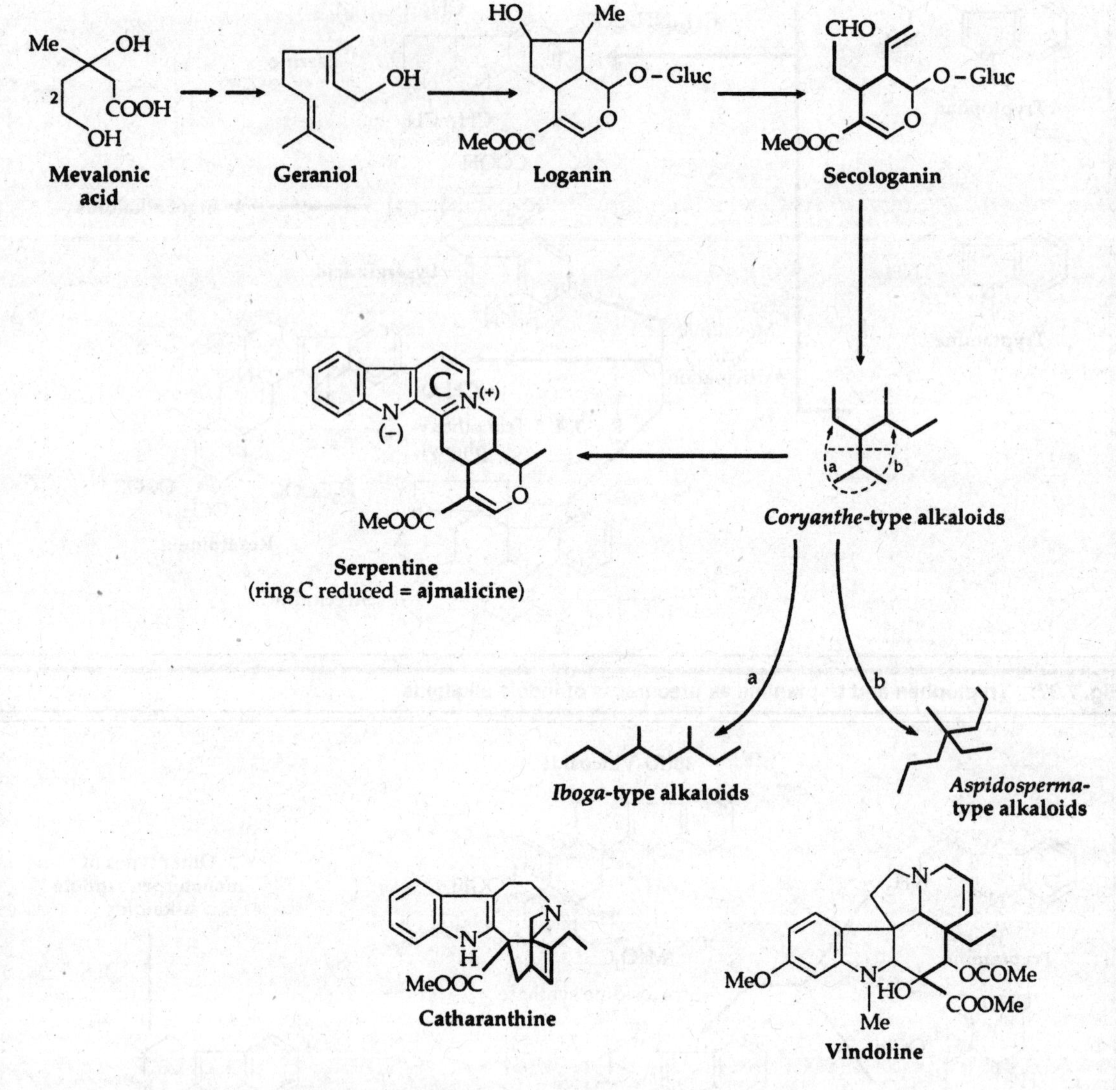

Fig. 7.39: Incorporation of mevalonic acid into some indole alkaloids.

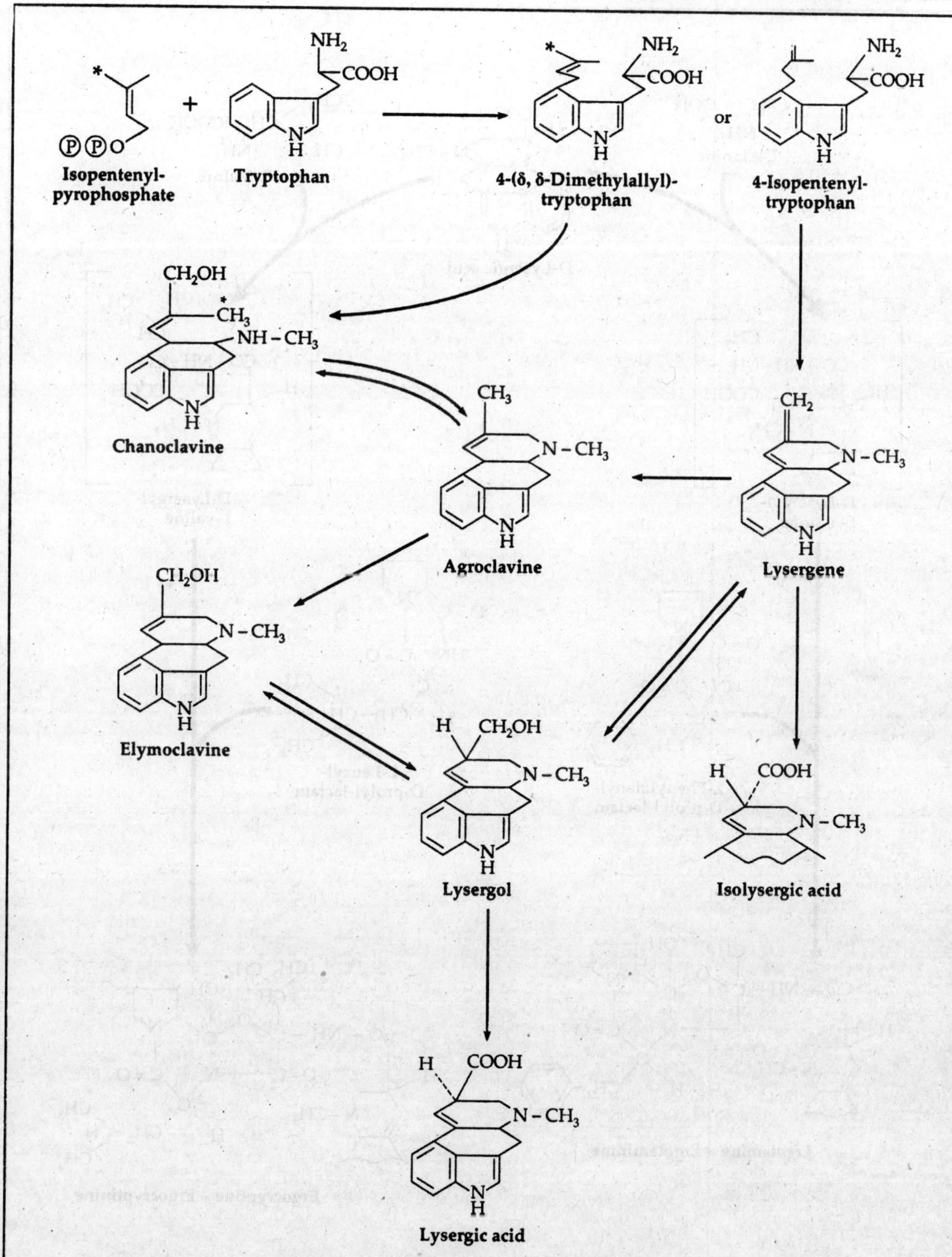

Fig. 7.40: Possible biogenesis of lysergic acid.

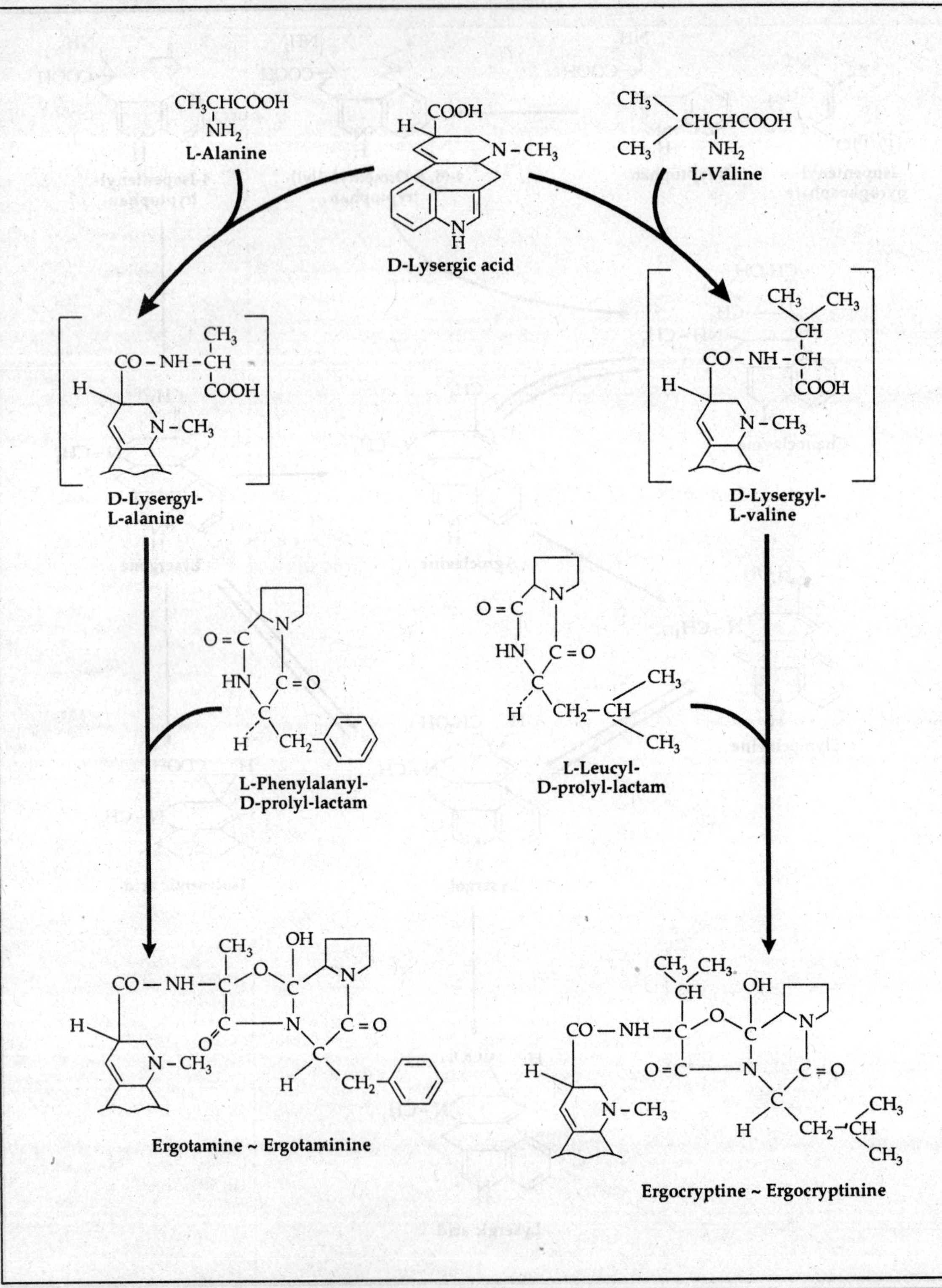

Fig. 7.41: Biogenesis of ergot peptide alkaloids.

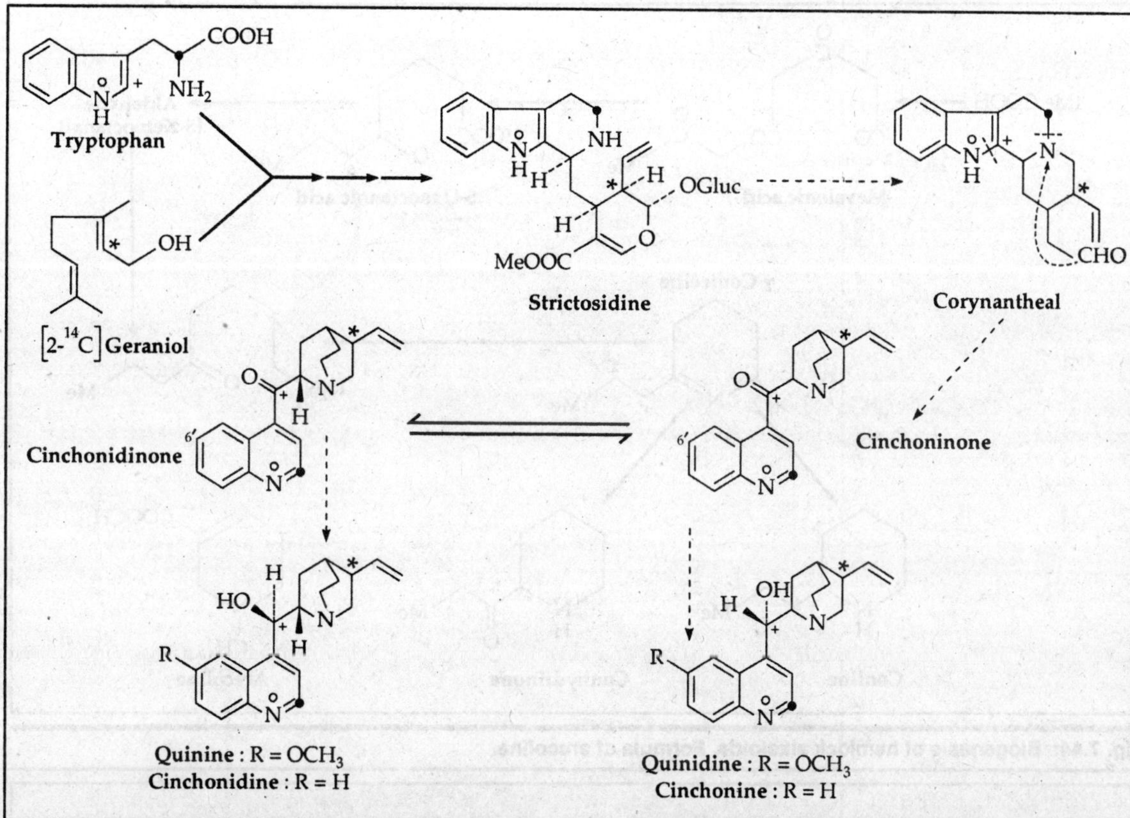

Fig. 7.42: Outline of possible biogenetic pathway for *Cinchona* alkaloids. Marked atoms illustrate the structural changes.

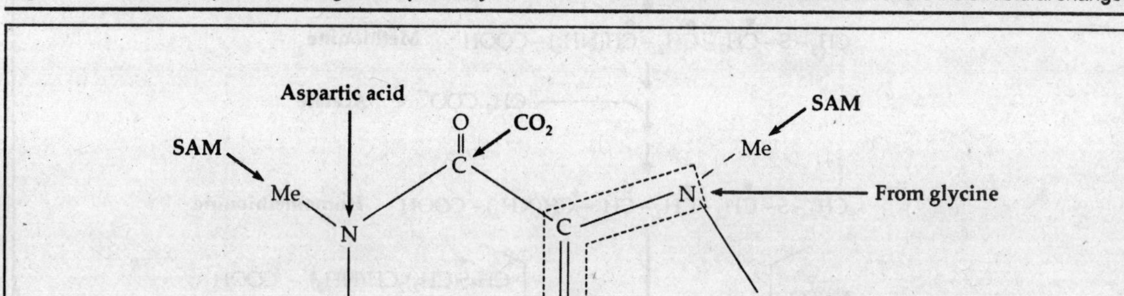

Fig. 7.43: Biogenetic origin of the caffeine molecule (SAM = S-adenosylmethionine).

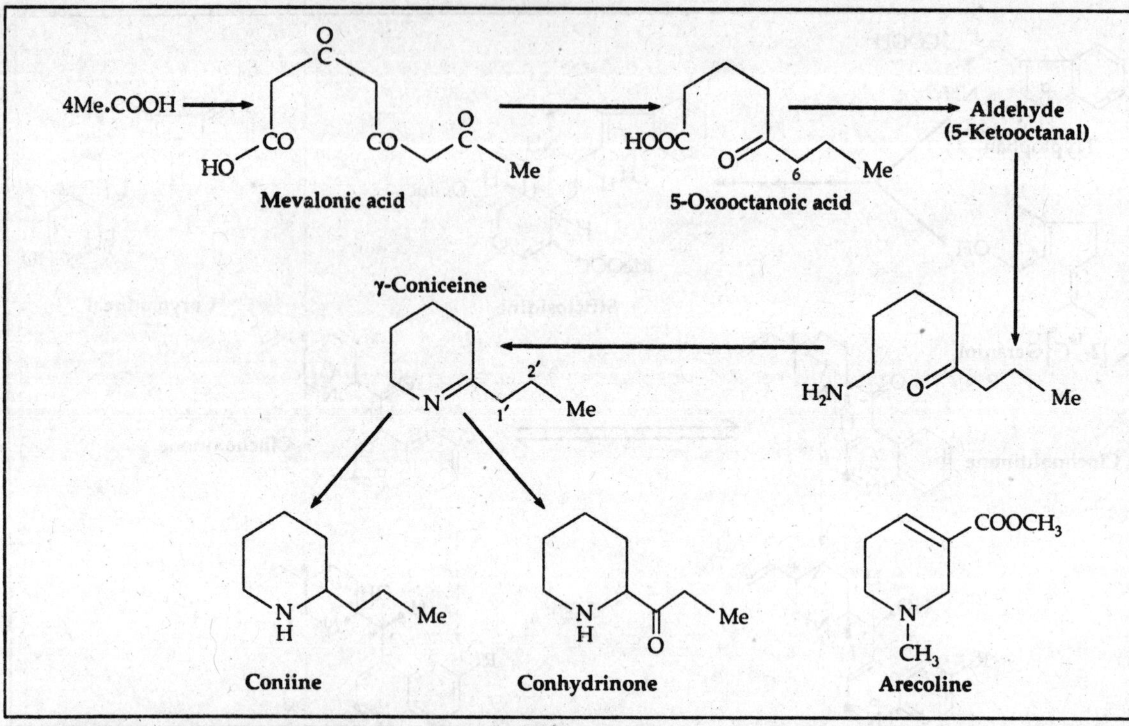

Fig. 7.44: Biogenesis of hemlock alkaloids. Formula of arecoline.

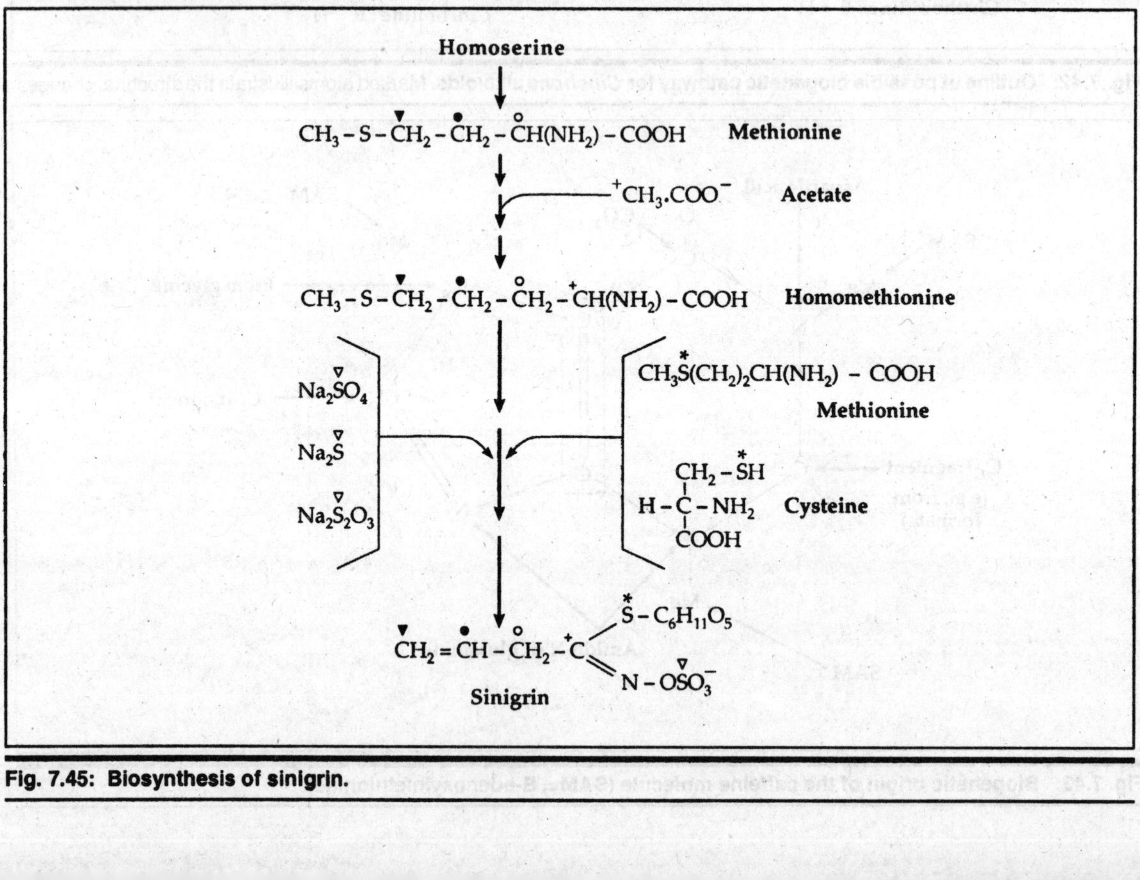

Fig. 7.45: Biosynthesis of sinigrin.

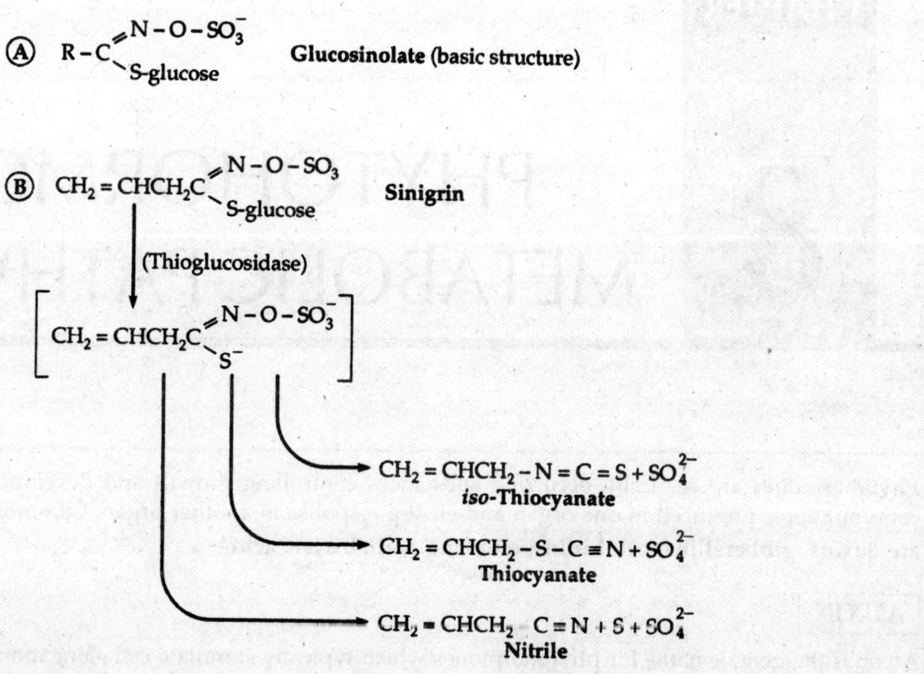

Fig. 7.46: Glucosinolates

(A) All glucosinolates have the same basic structure.

(B) Enzymatic hydrolysis of **sinigrin**, a glucosinolate from black mustard (*Brassica nigra*). Enzymatic removal of glucose creates an unstable product that spontaneously rearranges to form the pungent mustard oils **thiocyanate, isothiocyanate** or **nitrile**.

CHAPTER

8 PHYTOHORMONES: METABOLIC PATHWAYS

Phytohormones are naturally occurring substances controlling growth and development in very low concentrations, produced in one organ and elicit a response in another organ. Common phytohormones are **auxins, gibberellins, cytokinins, ethylene** and **abscisic acid**.

AUXIN

Auxin is the generic name for phytohormones which typically stimulate cell elongation along with wide range of growth responses. The principal native auxin in plant is indole acetic acid or **IAA** which is chemically with an indole ring having a acid side chain attached to carbon 3 of pyrrole.

IAA is biosynthesized from tryptophan (Trp) through several pathways. **Indole pyruvic acid pathway** involves deamination of Trp to IPA followed by decarboxylation and oxidation to IAA. In **tryptamine pathway**, decarboxylation of Trp to TAM is followed by deamination and oxidation to IAA. **Indole acetaldoxime pathway** involves decarboxylation of Trp followed by oxidation to IAO_X which is converted to IAN, nitrilase activity of IAN yields IAA. Through **tryptophol pathway,** deamination is followed by decarboxylation leading to synthesis of IAld which may be reduced to TOL or oxidized to IAA. TOL oxidase resynthesize IAld for conversion to IAA. **Tryptophan-independent pathway** of IAA synthesis exists in some plants. The suggested branch points from tryptophan biosynthetic pathway are at indole-3-glycerol phosphate or at indole with the possible intermediates of IAN and IPA.

Biodegradation of IAA occurs either through decarboxylation pathway by the activity of peroxidase or through non-decarboxylation routes into oxindole-3-acetic acid, the later is the most common metabolic pathway.

Various **bound auxins** like indole acetylaspartate, indole acetyl-β-D-glucose, indole-2-O-myoinositol, indoleacetyl-2-O-myoinositol arabinoside or galactoside are produced from IAA, the reactions are mostly reversible.

Figures: 8.1–8.5

GIBBERELLIN

Gibberellins are cyclic diterpenoid compounds derived from the condensation of basic **isoprene monomers**. All the carbon atoms of isoprene unit are derived from acetate. The condensing reaction followed by reduction and phosphorylation results in the formation of two C_5 isomers—IpPP and DMAPP. Geranyl pyrophosphate (C_{10}), produced by condensation of IpPP and DMAPP, undergoes subsequent elongation by further addition of IpPP units and produce diterpenoids. Geranyl geraniol pyrophosphate (C_{20}) produced, undergoes **cyclization reactions** in proplastids and converted to kaurene. **Oxidation reactions** of kaurene in endoplasmic reticulum lead to the successive formation of kaurenol, kaurenal, kaurenoic acid. The ring contraction of the later resulted in the formation of G_{12}-aldehyde. In the last stage G_{12}-aldehyde is converted to GA_{12}, the first formed gibberellin, from which other **gibberellins are synthesized** in cytosol. Hydroxylation pathway from GA_{12} to GA_{53} followed by a series of oxidations at C_{20} produce GA_{20} which on 3β-hydroxylation produce active gibberellin, GA_1.

Figures: 8.6–8.8

CYTOKININ

Cytokinins are N^6-substituted adenines which induce cytokinesis, e.g., zeatin, kinetin. The first step in biosynthesis involves the **transfer of an isopentenyl group** from isopentenyl pyrophosphate to AMP. Isopentenyl adenine ribotide produced will readily be converted to the **zeatin ribotide** and dihydrozeatin ribotide, which ultimately transform into free **base forms**—the naturally occurring cytokinins.

Figure: 8.9

ETHYLENE

Ethylene is a gaseous phytohormone that accelerates fruit ripening. The amino acid **methionine** is the precursor of ethylene. First methionine with the help of ATP is converted to *S*-adenosyl methionine which undergoes conversion to ACC and MTA, the **regulatory step** in ethylene biosynthetic pathway. ACC is finally oxidized to ethylene by molecular oxygen or may be converted to MACC to reduce the ethylene level. The CH_3-S group of methionine is recycled via the **Yang cycle** to methionine and thus conserved for continued synthesis of ethylene. Ethylene may undergo **oxidation** to ethylene oxide which on subsequent hydrolysis yields ethylene glycol, however, ethylene may also be completely oxidized to CO_2.

Figures: 8.10–8.11

ABSCISIC ACID

ABA, a sesquiterpenoid, is able to antagonize the promotive influence of growth-promoting hormones. ABA is biosynthesized via the **terpenoid pathway**. IPP (C_5) serves as the precursor for **zeaxanthin**, a C_{40} xanthophyll. Zeaxanthin is then converted to neoxanthin by a multistep process. The later is oxidatively cleaved to C_{15} xanthoxin which serves as immediate precursor of ABA-aldehyde. ABA-aldehyde is then oxidised to form ABA. **Catabolic pathway** of ABA inactivation includes its conjugation with monosaccharide to yield ABA-glucosyl ester or oxidation to phaseic acid and then dihydrophaseic acid.

Figure: 8.12

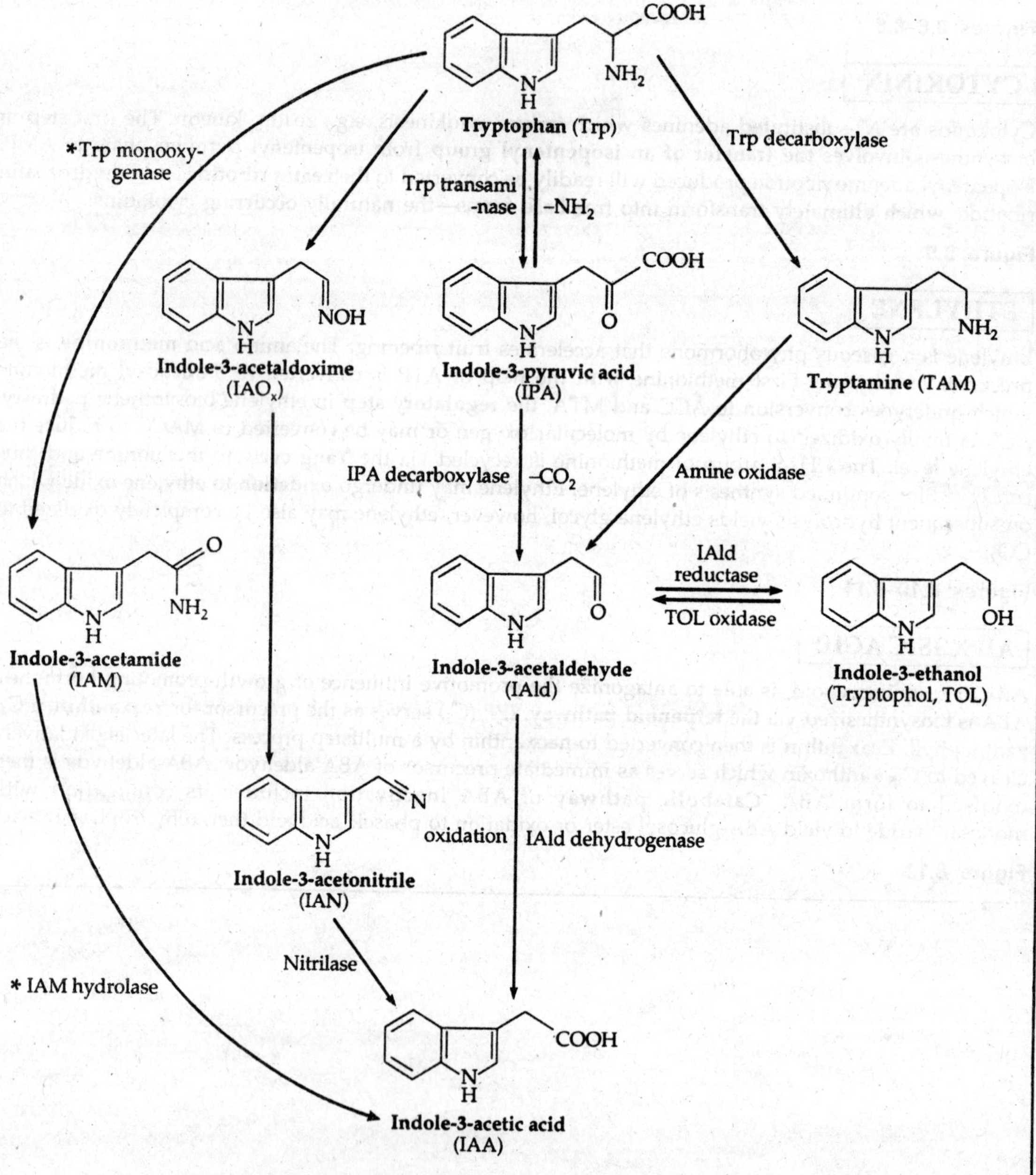

Fig. 8.1: Tryptophan-dependent pathways of IAA biosynthesis in plants and bacteria: The enzymes that are present only in bacteria are marked with an asterisk.

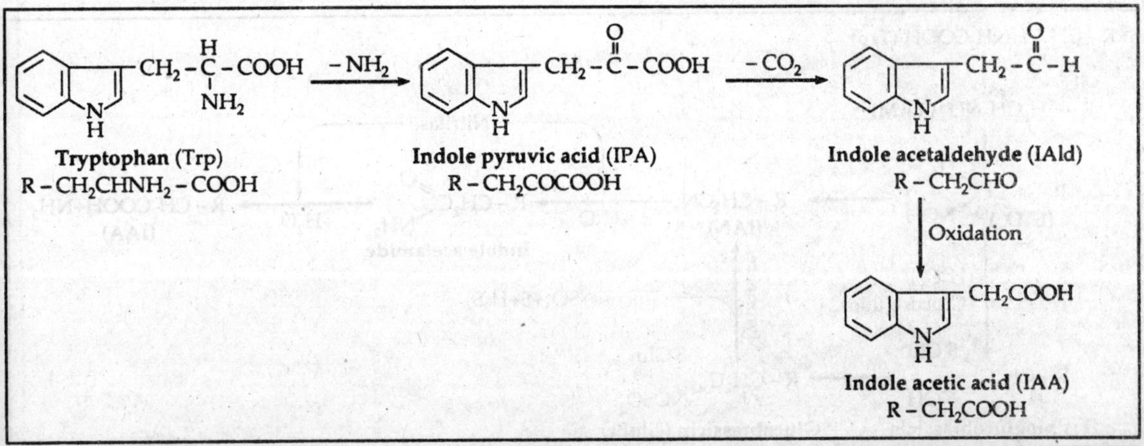

Fig. 8.2a: Indole pyruvic acid pathway of IAA synthesis.

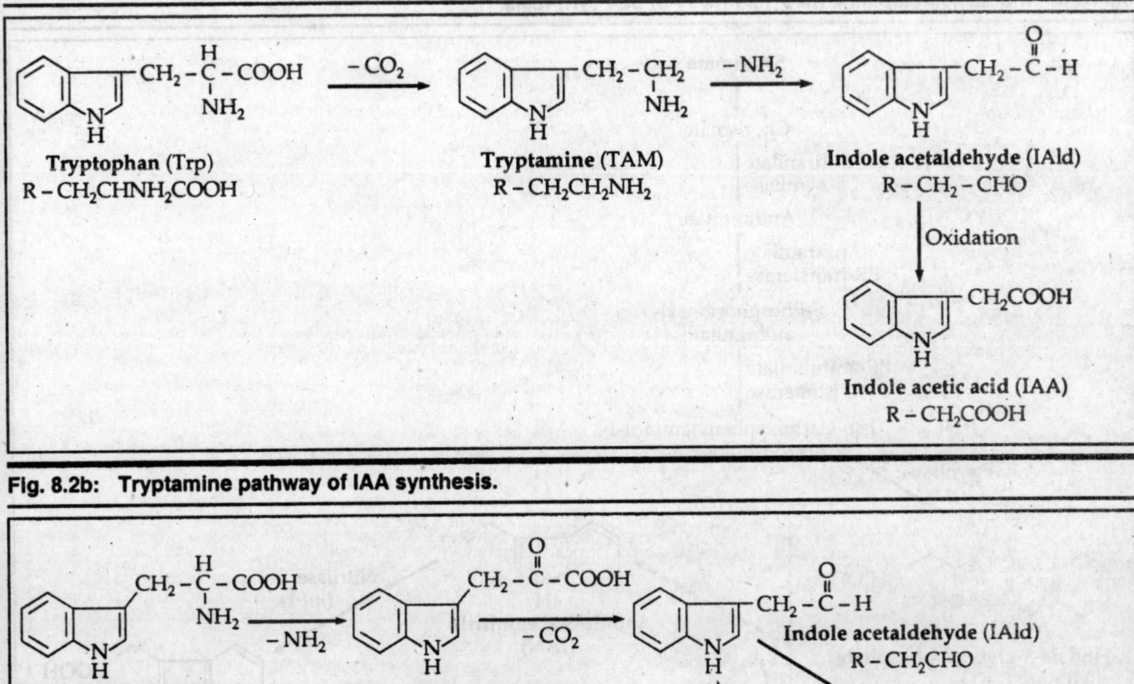

Fig. 8.2b: Tryptamine pathway of IAA synthesis.

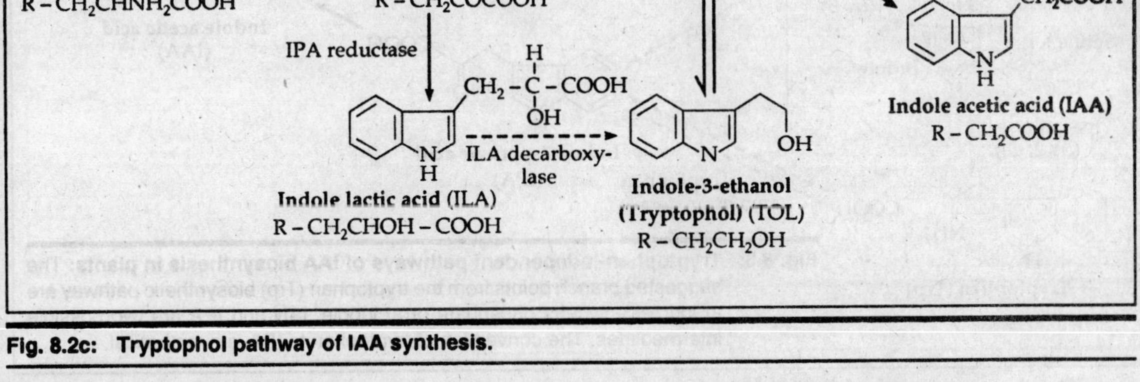

Fig. 8.2c: Tryptophol pathway of IAA synthesis.

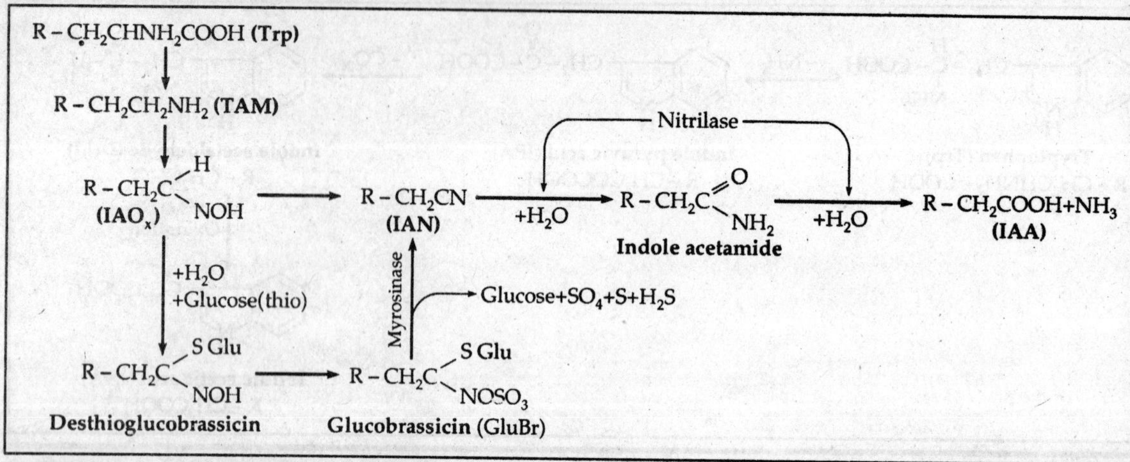

Fig. 8.2d: Indole acetaldoxime (IAO_x) pathway of IAA synthesis.

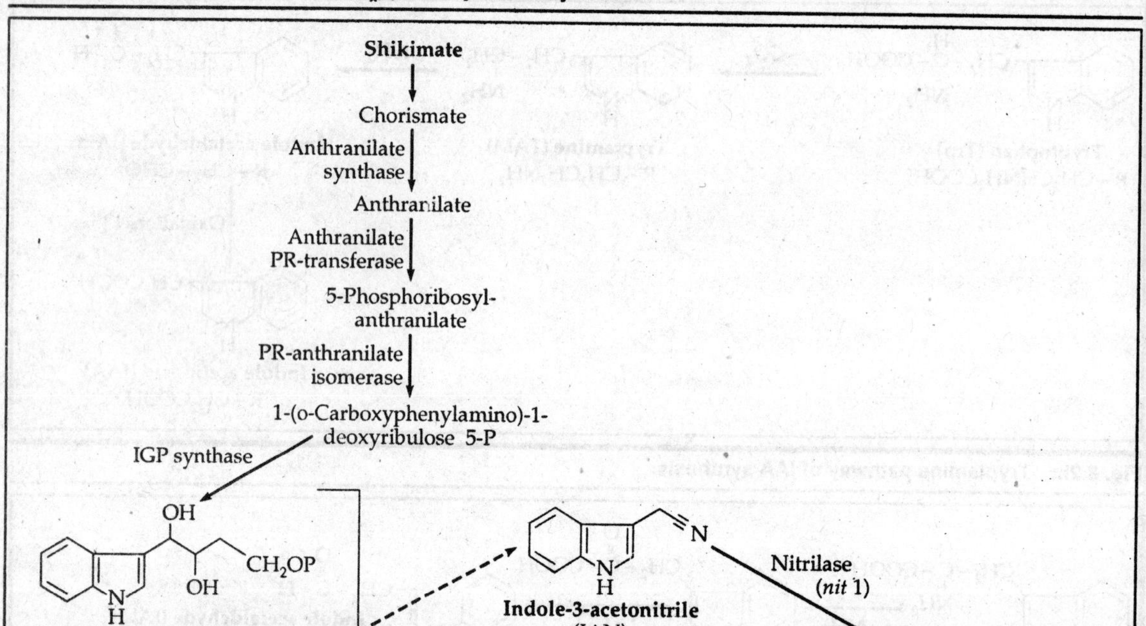

Fig. 8.3: Tryptophan-independent pathways of IAA biosynthesis in plants: The suggested branch points from the tryptophan (Trp) biosynthetic pathway are at indole-3-glycerol phosphate or at indole. IAN and IPA are two possible intermediates. The conversion of tryptophan to IPA is hypothetical.

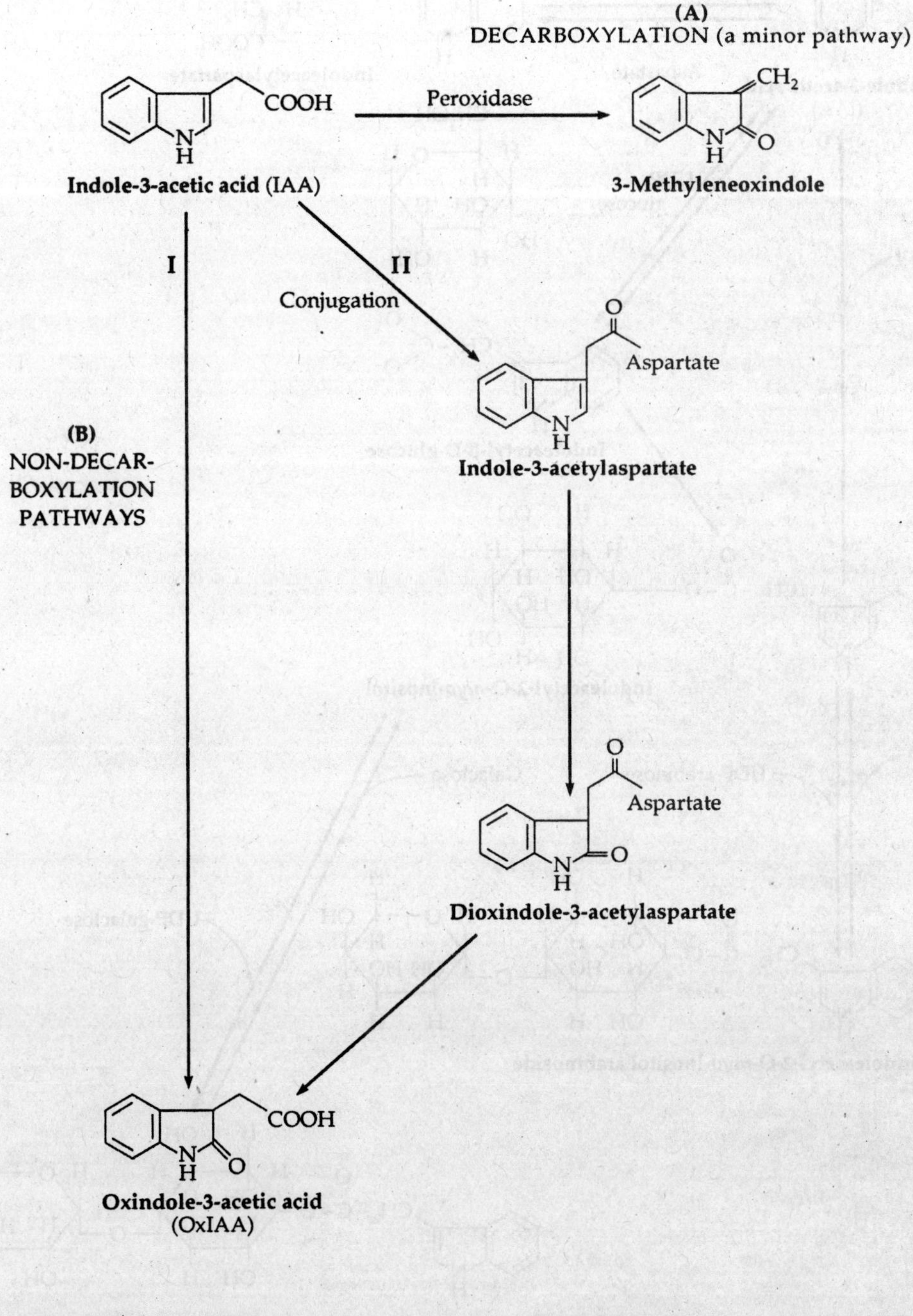

Fig. 8.4: Biodegradation of IAA:
(A) The peroxidase route (decarboxylation pathway) plays a relatively minor role.
(B) The two non-decarboxylation routes of IAA oxidative degradation, I and II, are the most common metabolic pathways.

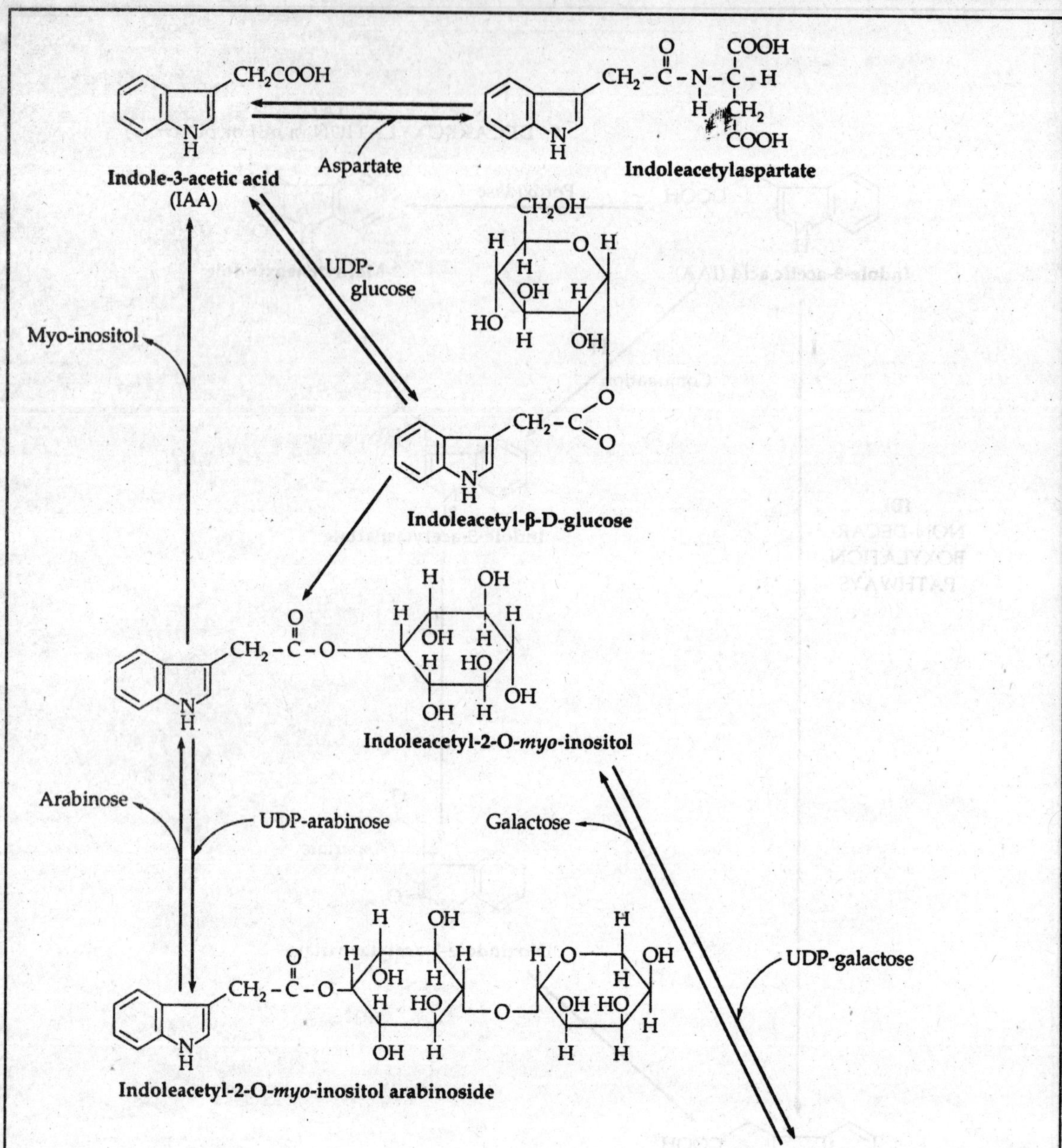

Fig. 8.5: Structure and proposed metabolic pathways of bound auxins: The diagram shows structures of various IAA conjugates and proposed metabolic pathways involved in their synthesis and breakdown. Single arrows indicate irreversible pathways, double arrows indicate reversible ones.

CH_3COOH + CoASH + ATP —(i)→ CH_3CO-S CoA + AMP + PPi
Acetic acid Coenzyme-A Acetyl-CoA

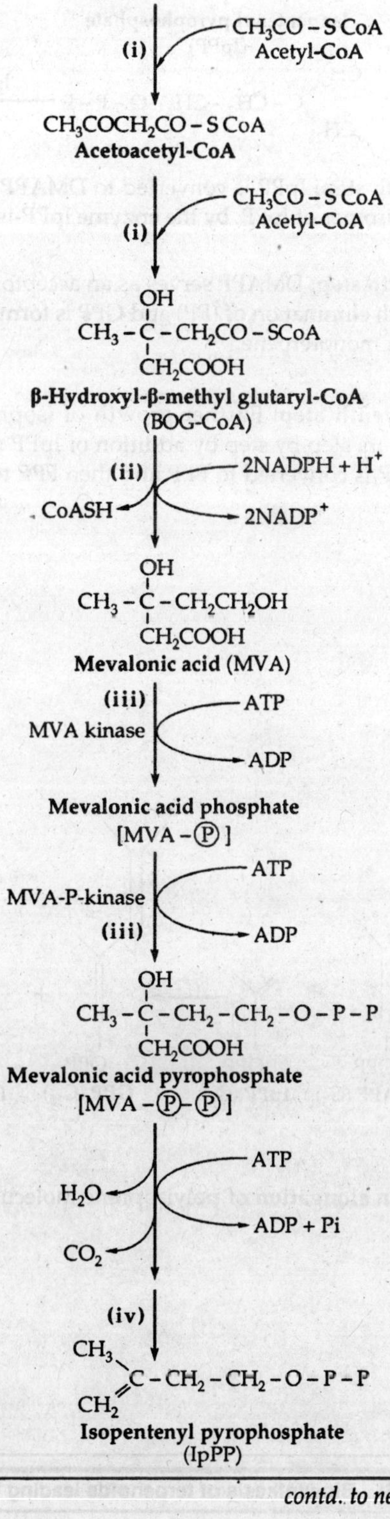

First step: The formation of Acetyl-CoA; 3 molecules of Acetyl-CoA in two successive condensation reactions form β-hydroxy-β-methyl glutaryl -CoA (BOG-CoA).

Second step: BOG-CoA is reduced to form mevalonic acid in two successive reactions with NADPH.

Third step: MVA is then phosphorylated in two steps, with the help of MVA-kinase and MVA–Ⓟ kinase and mevalonic acid pyrophosphate (MVA–Ⓟ–Ⓟ) is formed. These are ATP-requiring reactions.

Fourth step: MVA–Ⓟ–Ⓟ is converted to isopentenyl pyrophosphate with the elimination of carboxyl group as CO_2 and water. This is also an energy-requiring process.

contd. to next page

Contd. from previous page

Isopentenyl pyrophosphate (IpPP) → **Dimethylallyl pyrophosphate (isoprene) (DMAPP)**

$$CH_2{=}C(CH_3)-CH_2-CH_2-O-P-P \xrightarrow[\text{(v)}]{\text{IpPP isomerase}} (CH_3)_2C{=}CH-CH_2-O-P-P + P-P$$

Fifth step: IpPP is converted to DMAPP which is an isomer of IpPP, by the enzyme IpPP-isomerase.

Sixth step: DMAPP serves as an acceptor of IpPP, with elimination of (PP) and GPP is formed which is a monoterpene.

Seventh step: Further growth of isoprene chain occurs step by step by addition of IpPP molecules. GPP is converted to FPP and then FPP to GGPP.

DMAPP → (vi) + IpPP → **Geraniol pyrophosphate (GPP)** → (vii) + IpPP → **Farnesol pyrophosphate (FPP)** → (vii) + IpPP → **Geranyl geraniol pyrophosphate (GGPP)**

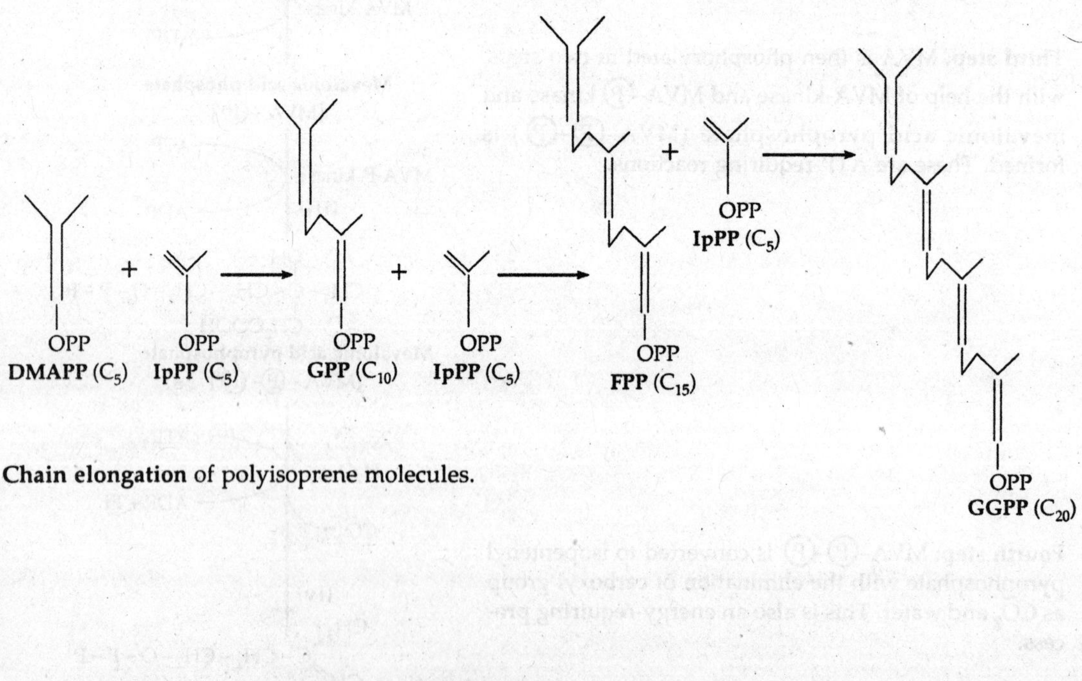

Chain elongation of polyisoprene molecules.

Fig. 8.6: Biosynthesis of terpenoids leading to gibberellins.

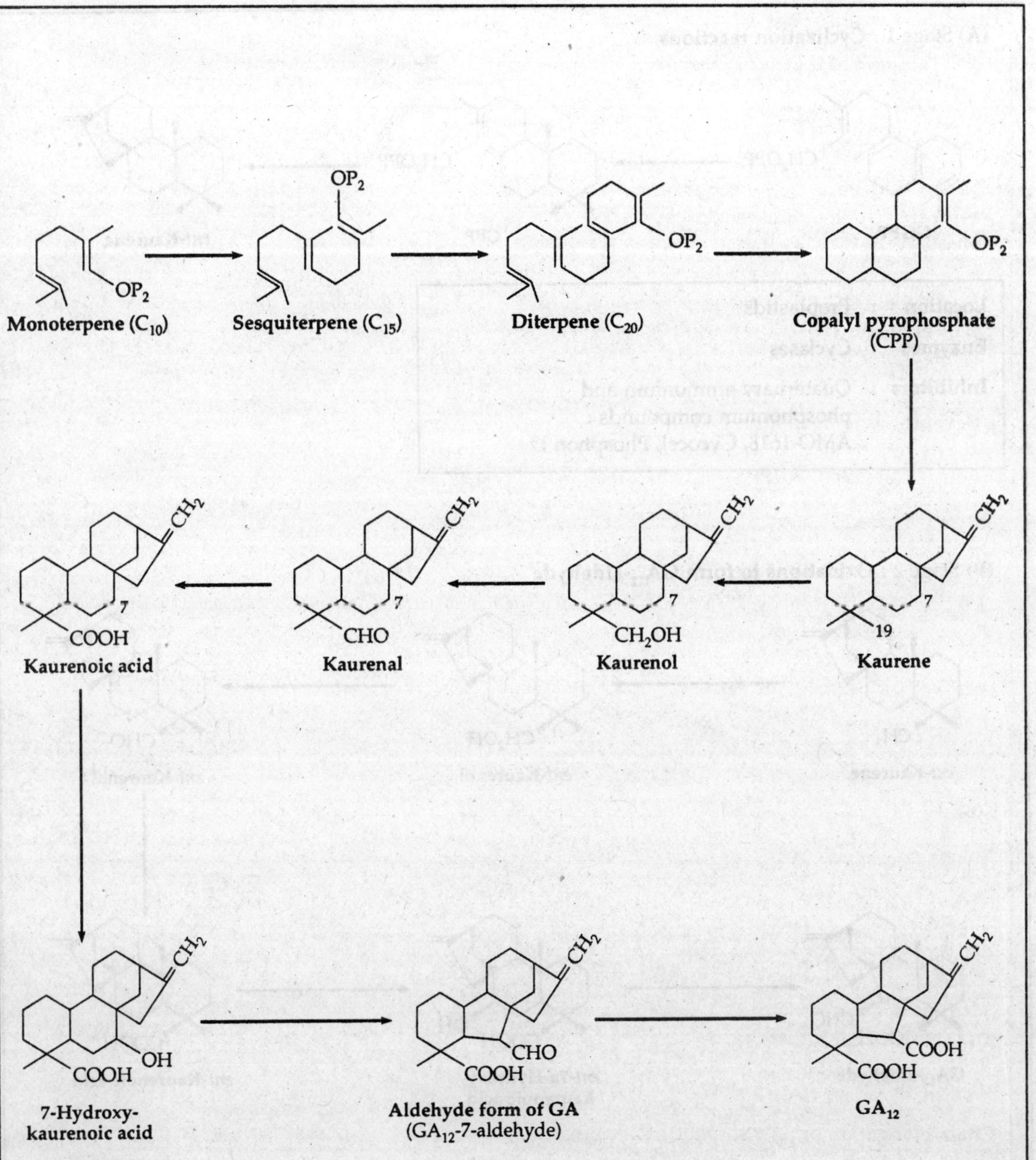

Fig. 8.7: Pathway of GA_{12} biosynthesis from monoterpene.

(A) Stage 1 : Cyclization reactions

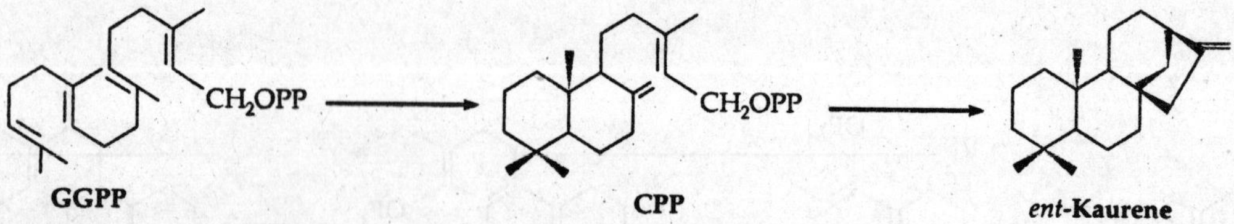

Location	:	Proplastids
Enzymes	:	Cyclases
Inhibitors	:	Quaternary ammonium and phosphonium compounds : AMO-1618, Cycocel, Phosphon D

(B) Stage 2 : Oxidations to form GA$_{12}$-aldehyde

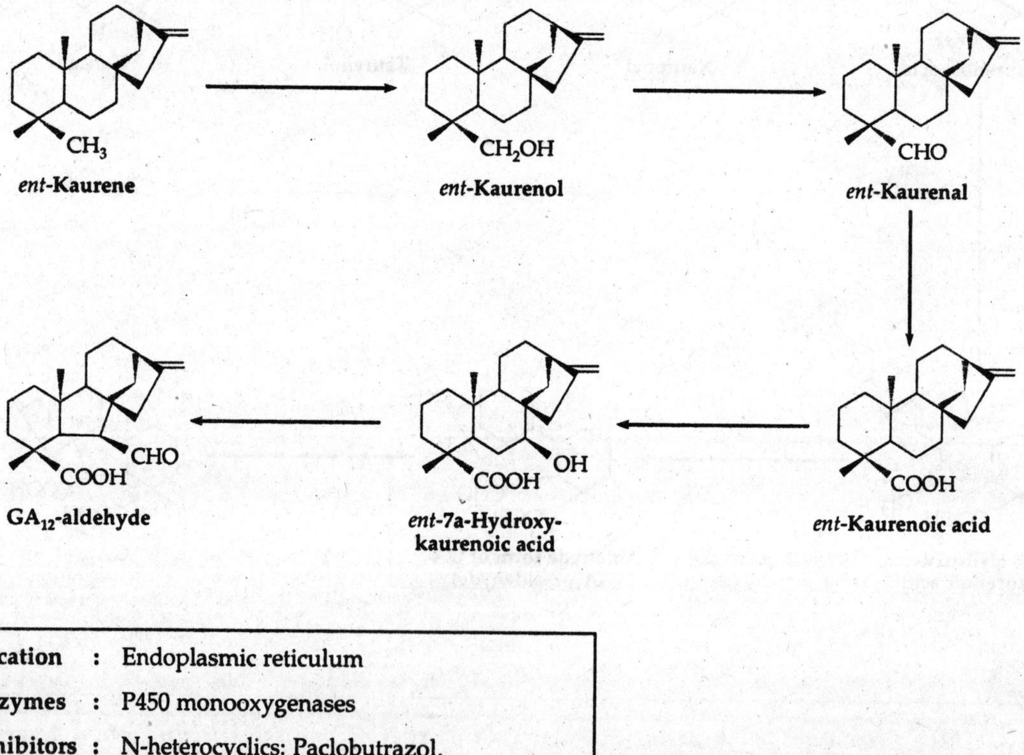

Location	:	Endoplasmic reticulum
Enzymes	:	P450 monooxygenases
Inhibitors	:	N-heterocyclics: Paclobutrazol, Uniconazole, Tetcyclacis

contd. to next page

Contd. from previous page

(C) Stage 3 : **Formation of all other GAs from GA_{12}-aldehyde**

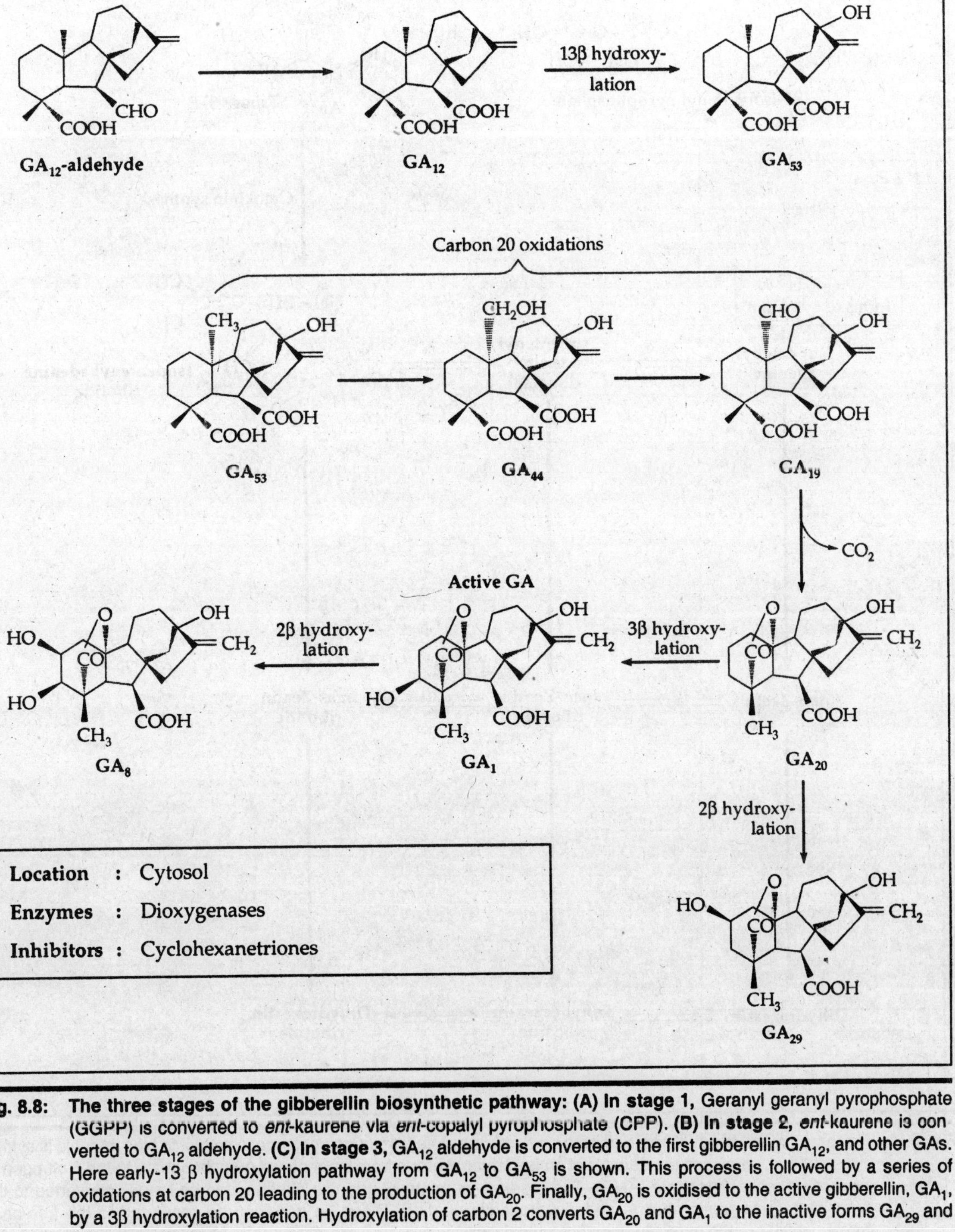

Location	:	Cytosol
Enzymes	:	Dioxygenases
Inhibitors	:	Cyclohexanetriones

Fig. 8.8: The three stages of the gibberellin biosynthetic pathway: (A) In stage 1, Geranyl geranyl pyrophosphate (GGPP) is converted to *ent*-kaurene via *ent*-copalyl pyrophosphate (CPP). **(B) In stage 2,** *ent*-kaurene is converted to GA_{12} aldehyde. **(C) In stage 3,** GA_{12} aldehyde is converted to the first gibberellin GA_{12}, and other GAs. Here early-13 β hydroxylation pathway from GA_{12} to GA_{53} is shown. This process is followed by a series of oxidations at carbon 20 leading to the production of GA_{20}. Finally, GA_{20} is oxidised to the active gibberellin, GA_1, by a 3β hydroxylation reaction. Hydroxylation of carbon 2 converts GA_{20} and GA_1 to the inactive forms GA_{29} and GA_8 respectively.

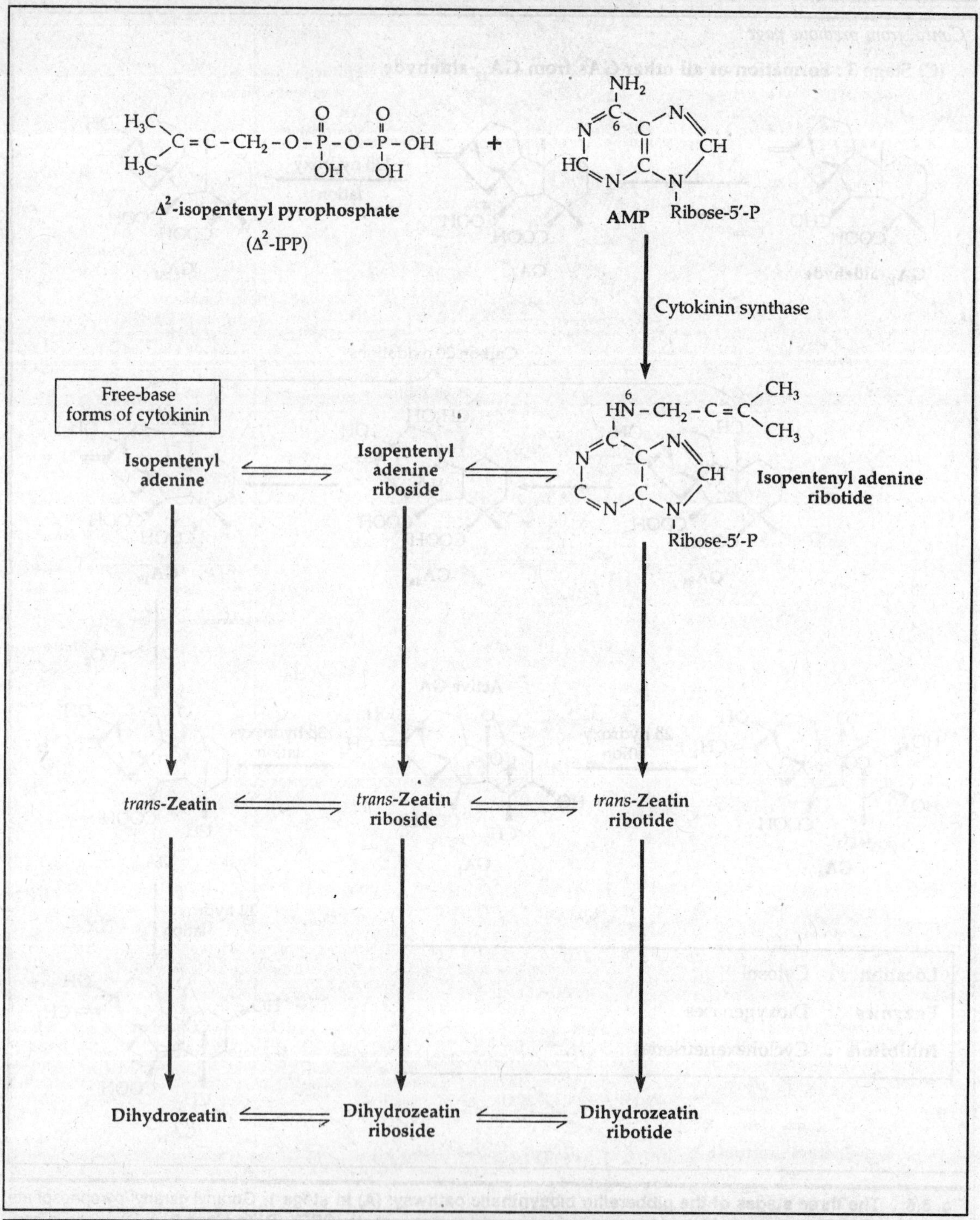

Fig. 8.9: Scheme for the biosynthesis of some cytokinins: The first unique enzyme in cytokinin biosynthesis is cytokinin synthase, catalyzes the transfer of an isopentenyl group from isopentenyl pyrophosphate to the 6 nitrogen of adenosine monophosphate (AMP). The product, isopentenyl adenine ribotide, can be readily converted to the cytokinins *trans*-zeatin ribotide and dihydrozeatin ribotide. These, in turn, can be converted to the free-base forms, which are the naturally occurring cytokinins.

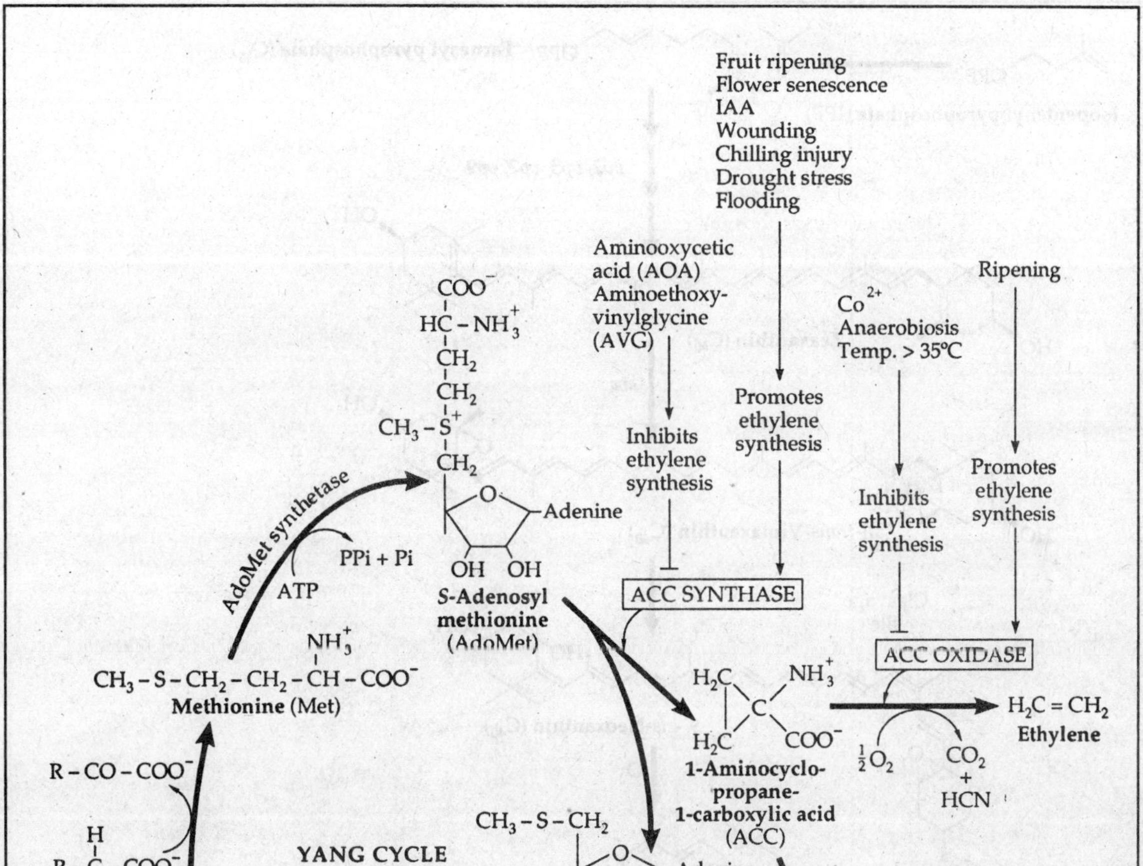

Fig. 8.10: Ethylene biosynthetic pathway and the Yang cycle: The amino acid methionine is the precursor of ethylene. The rate-limiting step in the pathway is the conversion of AdoMet to ACC, which is catalysed by the enzyme ACC synthase. The last step in the pathway, the conversion of ACC to ethylene, requires oxygen and is catalysed by enzyme ACC oxidase. The $CH_3 - S$ group of methionine is recycled via Yang cycle and thus conserved for continued synthesis. Besides being converted to ethylene, ACC can be conjugated to *N*-malonyl-ACC.

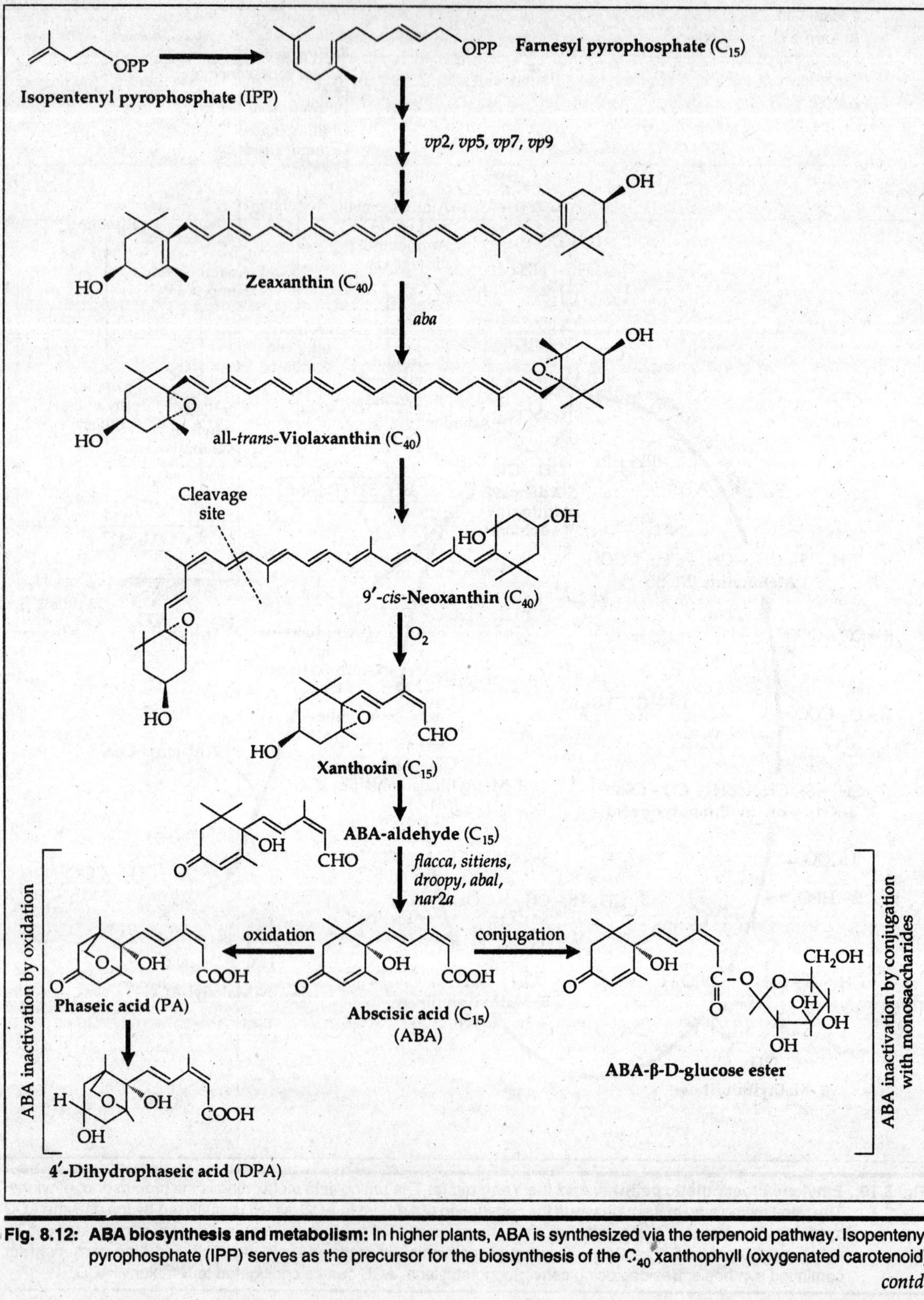

Fig. 8.12: ABA biosynthesis and metabolism: In higher plants, ABA is synthesized via the terpenoid pathway. Isopentenyl pyrophosphate (IPP) serves as the precursor for the biosynthesis of the C_{40} xanthophyll (oxygenated carotenoid)

contd.

zeaxanthin. Zeaxanthin is then converted by a multistep process to 9′-*cis*-Neoxanthin, which is oxidatively cleaved to form a C_{15} molecule **Xanthoxin**. Xanthoxin serves as the immediate precursor of ABA-aldehyde. ABA-aldehyde is then oxidised to form ABA. Some ABA-deficient mutants that have been helpful in elucidating the pathway are shown at the steps at which they are blocked. These include the *vp* mutants of corn, the *aba* mutants of *Arabidopsis*, the *flacca* and *sitiens* mutants of tomato, the *droopy* mutants of potato and the *nar* 2a mutant of barley. The pathways for ABA catabolism include conjugation to form ABA-β-D-glucosyl ester or oxidation to form phaseic acid and then dihydrophaseic acid.

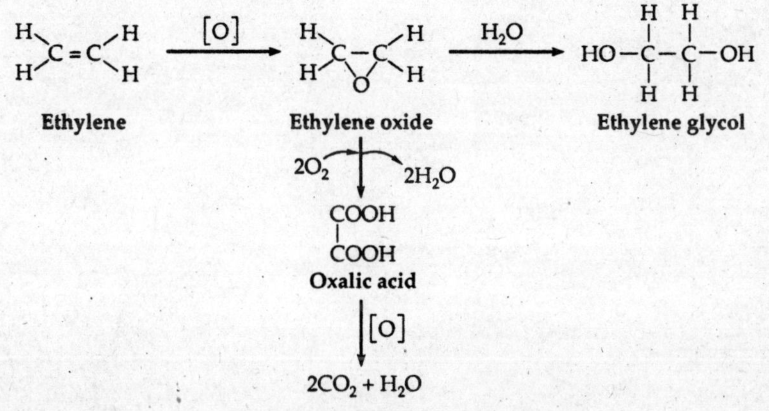

Fig. 8.11: Oxidation of ethylene.

BIOCHEMICAL ABBREVIATIONS

A	adenine, adenosine, or adenylate
ACP	acyl carrier protein
Acyl-CoA	acyl derivatives of coenzyme A (acyl-S-CoA)
ADH	alcohol dehydrogenase
adoHcy	S-adenosylhomocysteine
adoMet	S-adenosylmethionine (SAM)
Ala	alanine
AMP	adenosine 5¢-monophosphate
ADP	adenosine diphosphate
ATP	adenosine triphosphate
Arg	arginine
Asn	asparagine
Asp	aspartate
ATCase	aspartate transcarbamoylase
ATPase	adenosine triphosphatase
bp	base pair
1, 3-BPG	1,3-bisphosphoglycerate
C	cytosine, cytidine, or cytidylate
cAMP	3¢, 5¢-cyclic AMP
cGMP	3¢, 5¢-cyclic GMP
cDNA	complementary DNA
chl	chlorophyll
CL	cardiolipin
CMP	cytidine 5¢-monophosphate
CDP	cytidine diphosphate
CTP	cytidine triphosphate
CoA	coenzyme A (CoASH)
CoQ	coenzyme Q (ubiquinone, UQ)
CRP	cAMP receptor protein
Cys	cysteine
D	dihydrouridine
dADP	deoxyadenosine 5¢-diphosphate
dGDP	deoxyguanosine 5¢-diphosphate
dAMP	deoxyadenosine 5¢-monophosphate
dGMP	deoxyguanosine 5¢-monophosphate
dATP	deoxyadenosine 5¢-triphosphate
dGTP	deoxyguanosine 5¢-triphosphate
DEAE	diethylaminoethyl
DFP (DIFP)	diisopropylfluorophosphate
DHAP	dihydroxyacetone phosphate
DHF	dihydrofolate (H_2 folate)

DMS	dimethyl sulfate
DNA	deoxyribonucleic acid
DNase	deoxyribonuclease
DNP	2, 4-dinitrophenol
Dol	dolichol
DOPA	dihydroxyphenylalanine
EDTA	ethylenediaminetetraacetate
EF	elongation factor
ER	endoplasmic reticulum
FA	fatty acid
FAD, $FADH_2$	flavin adenine dinucleotide, and its reduced form
FBPase-1	fructose-1, 6-bisphosphatase
FBPase-2	fructose-2, 6-bisphosphatase
Fd	ferredoxin
FDNB	1-fluoro-2, 4-dinitrobenzene
f-Met	N-formylmethionine
FMN, $FMNH_2$	flavin mononucleotide, and its reduced form
F_0F_1	mitochondrial ATP synthase
FP	flavoprotein
FIP	fructose-1-phosphate
F6P	fructose-6-phosphate
Fru	D-fructose
DG	free energy change
DG¢^{0}	standard free energy change
G	guanine, guanosine, or guanylate
GABA	d-aminobutyric acid
Gal	D-galactose
GalN	D-galactosamine
GalNAc	N-acetyl-D-galactosamine
GAP	glyceraldehyde-3-phosphate
GDH	glutamate dehydrogenase
GLC	gas liquid chromatography
Glc	D-glucose
GlcA	D-gluconic acid
GlcN	D-glucosamine
GlcNAC	N-acetyl-D-glucosamine (NAC)
GlcUA	D-glucuronic acid
Gln	glutamine
Glu	glutamate
Gly	glycine

GMP, GDP, GTP	guanosine 5¢-mono-, di-, triphosphate
G1P	glucose-1-phosphate
GPI	glycosylated derivatives of phosphatidyl inositol
G6P	glucose-6-phosphate
GSH, GSSG	glutathione, and its oxidised form
DH	enthalpy change
H_2-folate	dihydrofolate (DHF)
H_4-folate	tetrahydrofolate (THF)
His	histidine
HMG-CoA	b-hydroxy-b-methylglutaryl-CoA
hnRNA	heterogenous nuclear RNA
HPLC	high-performance liquid chromatography
Hyp	hydroxyproline
I	inosine
IF	initiation factor
Ig	immunoglobulin
Ile	isoleucine
IMP, IDP, ITP	inosine 5¢-mono-, di-, triphosphate
IP_3	inositol 1,4,5-triphosphate
IS	insertion sequence
K	dissociation constant
Km	Michaelis constant
Kb	kilobase
Kbp	kilobase pair
a-KG	a-Ketoglutarate
lac	lactose
l	wavelength
LDH	lactate dehydrogenase
Leu	leucine
log	logarithm to the base 10
LTR	long terminal repeat
Lys	lysine
Man	D-mannose
Met	methionine
mRNA	messenger RNA
mDNA	mitochondrial DNA
Mur	muramic acid
Mur2Ac	N-acetyl muramic acid (NAM)
NAD^+, NADH	nicotinamide adenine dinucleotide, and its reduced form
$NADP^+$, NADPH	nicotinamide adenine dinucleotide phosphate, and its reduced form
NAG	N-acetylglucosamine (GlcNAc)
NAM	N-acetylmuramic acid (MurNAc)
Neu5Ac	N-acetylmuraminic acid (sialic acid)
NMN^+, NMNH	nicotinamide mononucleotide, and its reduced form
NMP, NDP, NTP	nucleoside mono-, di-, and triphosphate
NMR	nuclear magnetic resonance
NO	nitric oxide
OAA	oxaloacetate
Pi	inorganic orthophosphate
PAB or PABA	p-aminobenzoate
PAGE	polyacrylamide gel electrophoresis
PC	plastocyanin, phosphatidyl choline
PCR	polymerase chain reaction
PDE	cyclic nucleotide phosphodiesterase
PE	phosphatidylethanolamine
PEP	phosphoenolpyruvate
PFK	phosphofructokinase
2PGA	2-phosphoglycerate
3PGA	3-phosphoglycerate
pH	$\log 1/[H^+]$
Phe	phenylalanine
pI	isoelectric point
PI	phosphatidylinositol
PIP_2	phosphatidylinositol 4, 5-bisphosphate
PK	protein kinase, pyruvate kinase
PKA	cAMP-dependent protein kinase
PKC	Ca^{2+}/calmodulin-dependent protein kinase
PKG	cGMP-dependent protein kinase
PLC	phospholipase C
PLP	pyridoxal-5-phosphate
Pol	polymerase (DNA or RNA)
PPi	inorganic pyrophosphate
PQ	plastoquinone
Pro	proline
PRPP	5-phosphoribosyl-1-pyrophosphate
PS	phosphatidylserine
PUFA	polyunsaturated fatty acid
RC	photoreaction center
RER	rough endoplasmic reticulum
RF	release factor, replicative form
RFLP	restriction-fragment length polymorphism
RIA	radioimmunoassay
Rib	D-ribose
RNA	ribonucleic acid
RNase	ribonuclease
rRNA	ribosomal RNA
RTK	receptor tyrosine kinase

DS	entropy change
SAM	S-adenosyl methionine (adoMet)
SDS	sodium dodecyl sulfate
SER	smooth endoplasmic reticulum
Ser	serine
snRNA	small nuclear RNA
SRP	signal recognition particle
STAT	signal transducer and activator of transcription
STP	standard temperature and pressure
T	thymine, thymidine, or thymidylate
T	absolute temperature
THF	tetrahydrofolate (H_4-folate)
Thr	threonine
TLC	thin layer chromatography
TMP, TDP, TTP	thymidine 5¢-mono-, di-, triphosphate
TPI	triose phosphate isomerase
TPP	thiamine pyrophosphate
tRNA	transfer RNA
Trp	tryptophan
Tyr	tyrosine
U	uracil, uridine, or uridylate
UCP	uncoupling protein of the mitochondrial inner membrane
UDP-Gal	uridine diphosphate galactose
UDP-Glc	uridine diphosphate glucose
UMP, UDP, UTP	uridine 5¢-mono-, di-, triphosphate
UQ	coenzyme Q (ubiquinone, CoQ)
UV	ultraviolet
V_{max}	maximum velocity
V_0	initial velocity
Val	valine

REFERENCES

1. Buchanon, B.B., Gruissem, W and Jones, R.L.: *Biochemistry and Molecular Biology of Plants* (2000), American Society of Plant Physiologists. I.K. International.
2. Conn, E.E., Stumpf, P.K., Bruening, G. and Doi, R. H: *Outlines of Biochemistry* (1987), Wiley Eastern.
3. Evans, W.C.: *Trease and Evans Pharmacognosy* (2002), W.B. Saunders.
4. Goodwin, T.N. and Mercer, E.J.: *Introduction to Plant Biochemistry* (1982), Oxford, Pergamon.
5. Hatch, M.D. and Boardman, N. K. (Eds.): *The Biochemistry of Plants* (1981), Vol. 8, Academic Press.
6. Lea, P.J. and Leegwood, R.C.: *Plant Biochemistry and Molecular Biology*, John Wiley.
7. Mukherjee, S and Ghosh, A: *Plant Physiology* (2005), New Central Book Agency (P) Ltd.
8. Nelson, D. L. and Cox, M.M.: *Lehninger Principles of Biochemistry* (2003), Macmillan Press/Worth Publishers.
9. Stryer, L.: *Biochemistry* (1988), CBS Publisher and Distributors.
10. Stumpf, P.K. and Conn, E.E. (Eds.): *The Biochemistry of Plants* (1980), Vol. 2, Academic Press.
11. Taiz, L. and Zeiger, E.: *Plant Physiology* (2002), Saunderland: Sinauer Associates, Inc. Publishers.
12. Voet, D. and Voet, J. G., Judith, G. and Pratt, W.: *Biochemistry*, John Wiley.

Some exciting gifts are up for grabs with this book!!

A HANDBOOK OF PLANT BIOSYNTHETIC PATHWAYS

Dr Debabrata Chaudhuri, Dr Dipak Kr. Kar,
Dr Soma Halder

FREE Gifts
Rs. 1000/-

Dear Readers

By now you must have gone through this book in the entirety. So we invite your assessment on this title with absolute impartiality and candid observations. After all, we publishers blindly depend on your benchmarking. Friends, please fill up the open spaces below with CAPITAL LETTERS. You may add some extra sheets of paper if the space provided is not sufficient for you.

i. What are the important topics of your syllabus which are not adequately covered in this book?

ii. In which chapters and on what topics the treatment of the subject-matter by the author is not satisfactory and why?

iii. Have you come across the typos, factual inaccuracies, inconsistencies, and other forms of mistakes in the book? If yes, then please specify the mistakes and page numbers.

iv. Name top three books on the same subject, which, in your consideration, are better than this book. Could you provide some reasons?

Your 1st Preference: ....

Your 2nd Preference: ....

Your 3rd Preference: ....

v. Any suggestion to improve the quality of illustrations, diagrams etc.?

vi. How did you like the cover design? Any suggestions on colours?

vii. Suggestions for further improvement of production standard:

viii. Who recommended you this book:

☐ Teacher ☐ Friend ☐ Parents ☐ Bookseller ☐ Librarian

ix. Name of the person who has recommended:

Designation: ....

Address: ....

Phone No.: ....

e-mail: ....

x. Name and address of the bookseller from whom you have purchased this book:

....

....

....

xi. Name and address of the library where you would like to find the book:

....

....

xii. Name and address of your institution (Please mention the University or Board):

....

....

xiii. Your name, e-mail, and complete postal address:

....

....

....

xiv. A few other titles published by us that you would like to read:

....

....

The best assessment will be rewarded half-yearly. The award will be in the form of our publications, as decided by the Editorial and Marketing Operations Division, amounting to Rs. 1000/- (Rupees one thousand) only.

....

Please mail the filled up coupon at your earliest to :

Editorial & Marketing Operations Division

NEW CENTRAL BOOK AGENCY (P) Ltd

8/1 Chintamoni Das Lane, Kolkata 700 009

West Bengal, INDIA

Ph. No.: 033 + 22418411, 22418573, 22419094

Fax: 033 + 22419573, 22413852

e-mail: ncbapvtltd@eth.net, ncba@cal2.vsnl.net.in

Website: www.centralgroup.com